PHOTOVOLTAIC SOLAR ENERGY CONVERSION

PHOTOVOLTAIC SOLAR ENERGY CONVERSION

Technologies, Applications and Environmental Impacts

Edited by

SHIVA GORJIAN

Biosystems Engineering Department, Faculty of Agriculture, Tarbiat Modares University (TMU), Tehran, Iran

ASHISH SHUKLA

Centre for Built and Natural Environment, Faculty of Engineering, Environment and Computing, Coventry University, Coventry, United Kingdom

Academic Press is an imprint of Elsevier
125 London Wall, London EC2Y 5AS, United Kingdom
525 B Street, Suite 1650, San Diego, CA 92101, United States
50 Hampshire Street, 5th Floor, Cambridge, MA 02139, United States
The Boulevard, Langford Lane, Kidlington, Oxford OX5 1GB, United Kingdom

Notices
Knowledge and best practice in this field are constantly changing. As new research and experience broaden our understanding, changes in research methods, professional practices, or medical treatment may become necessary.

Practitioners and researchers must always rely on their own experience and knowledge in evaluating and using any information, methods, compounds, or experiments described herein. In using such information or methods they should be mindful of their own safety and the safety of others, including parties for whom they have a professional responsibility.

To the fullest extent of the law, neither the Publisher nor the authors, contributors, or editors, assume any liability for any injury and/or damage to persons or property as a matter of products liability, negligence or otherwise, or from any use or operation of any methods, products, instructions, or ideas contained in the material herein.

Library of Congress Cataloging-in-Publication Data
A catalog record for this book is available from the Library of Congress

British Library Cataloguing-in-Publication Data
A catalogue record for this book is available from the British Library

ISBN 978-0-12-819610-6

For information on all Academic Press publications
visit our website at https://www.elsevier.com/books-and-journals

Publisher: Brian Romer
Acquisitions Editor: Lisa Reading
Editorial Project Manager: Chris Hockaday
Production Project Manager: Prem Kumar Kaliamoorthi
Cover Designer: Victoria Pearson

Typeset by SPi Global, India

Contents

Contributors

Mohammadreza Aghaei Eindhoven University of Technology (TU/e), Design of Sustainable Energy Systems, Energy Technology, Department of Mechanical Engineering, Eindhoven, The Netherlands; Department of Microsystems Engineering (IMTEK), University of Freiburg, Solar Energy Engineering, Freiburg im Breisgau, Germany

Hamsa Ahmed Physics Department, Oldenburg University, Oldenburg, Germany

Francesco Calise Department of Industrial Engineering, University Federico II, Naples, Italy

Francesco Liberato Cappiello Department of Industrial Engineering, University Federico II, Naples, Italy

Juan M. Castañón Symtech Solar Group, Lewes, DE, United States; Department of Microsystems Engineering (IMTEK), University of Freiburg, Solar Energy Engineering, Freiburg im Breisgau, Germany

Shauhrat S. Chopra School of Energy and Environment, City University of Hong Kong, Kowloon, Hong Kong

Hossein Ebadi Biosystems Engineering Department, Faculty of Agriculture, Shiraz University, Shiraz, Iran

Aref Eskandari Electrical Engineering Department, Amirkabir University of Technology (Tehran Polytechnic), Tehran, Iran

Patrice Geoffron Paris-Dauphine University, PSL University, LEDa-CGEMP, UMR CNRS-IRD, Paris, France

Barat Ghobadian Biosystems Engineering Department, Faculty of Agriculture, Tarbiat Modares University (TMU), Tehran, Iran

Shiva Gorjian Biosystems Engineering Department, Faculty of Agriculture, Tarbiat Modares University (TMU), Tehran, Iran; Leibniz Institute for Agricultural Engineering and Bioeconomy (ATB), Potsdam, Germany

Faezeh Ketabchi Mechanical Engineering Department, Amirkabir University of Technology (Tehran Polytechnic), Tehran, Iran

Saber Khanmohammadi Mechanical Engineering Department, University of Kashan, Kashan, Iran

Anil Kumar Department of Mechanical Engineering; Centre for Energy and Environment, Delhi Technological University, Delhi, India

Nallapaneni Manoj Kumar School of Energy and Environment, City University of Hong Kong, Kowloon, Hong Kong

Uzoma Edward Madukanya Department of Microsystems Engineering (IMTEK), University of Freiburg, Solar Energy Engineering, Freiburg im Breisgau, Germany

Ladan MalehMirchegini Mechanical Engineering Department, Alzahra University, Tehran, Iran

Abdur Rehman Mazhar Centre for Built and Natural Environment, Faculty of Engineering, Environment and Computing, Coventry University, Coventry, United Kingdom

Saeid Minaei Biosystems Engineering Department, Faculty of Agriculture, Tarbiat Modares University (TMU), Tehran, Iran

Aline Kirsten Vidal de Oliveira Federal University of Santa Catarina, Florianopolis, Brazil

Pramod Rajput Department of Physics, Indian Institute of Technology Jodhpur, Jodhpur, Rajasthan, India

Geetam Richhariya Department of Energy, Maulana Azad National Institute of Technology, Bhopal, India

Samsher Department of Mechanical Engineering; Centre for Energy and Environment, Delhi Technological University, Delhi, India

Mahdi Shakouri School of Environment, College of Engineering, University of Tehran, Tehran, Iran

Redmond R. Shamshiri Leibniz Institute for Agricultural Engineering and Bioeconomy (ATB), Potsdam, Germany

Ashish Shukla Centre for Built and Natural Environment, Faculty of Engineering, Environment and Computing, Coventry University, Coventry, United Kingdom

Renu Singh Centre for Environment Science and Climate Resilient Agriculture, Indian Agricultural Research Institute, New Delhi, India

Max Trommsdorff Fraunhofer Institute for Solar Energy Systems ISE, Freiburg im Breisgau, Germany

Shima Vaezi Chemical Engineering Department, Islamic Azad University South Tehran Branch, Tehran, Iran

Maria Vicidomini Department of Industrial Engineering, University Federico II, Naples, Italy

Hyun Jin Julie Yu French Alternative Energies and Atomic Energy Commission (CEA Saclay), Institute for Techno-Economics of Energy Systems (I-tésé), Gif-sur-Yvette Cedex, France

CHAPTER

1

Introduction

Shiva Gorjian[a] *and Hossein Ebadi*[b]

[a]Biosystems Engineering Department, Faculty of Agriculture, Tarbiat Modares University (TMU), Tehran, Iran, [b]Biosystems Engineering Department, Faculty of Agriculture, Shiraz University, Shiraz, Iran

1.1 Introduction

The world is undergoing a global crisis because of increasing population, with an extra 1.7 billion people expected to be added by 2040. This will increase energy demand by more than a quarter, mostly driven by developing nations. Improvement in new energy sectors has decreased the relentless appetite for conventional energy sources, although robust data still suggest a continuous increase in coal, oil, and natural gas production. Based on the conclusion of the International Energy Agency (IEA) in its annual "World Energy Outlook" report (2018) [1], meeting future energy demands seems implausible to fulfill climate goals, unless governments take serious and forceful steps to employ carbon-free sources. The report warns that if the current policies scenario (CSP), the set of active policies in today's world, remains constant, energy security will be strained from almost all aspects. However, the gap between the outcome and the sustainable development scenario remains huge in addressing goals to halt climate change. Therefore, governments may be the only game-changer in making decisions for the future.

Data analyses estimate that there was a 1.6% rise in the amount of carbon dioxide (CO_2) released from the energy sector in 2017 globally, with this trend possibly causing millions of premature deaths per year as the fallout. Being aware of the tragedy behind the consequences of current policy measures, people are directing a global march toward clean energy at a higher but not sufficient pace. However, the coincidence of three main changes—*the drastic cost reduction in renewable energy technologies, the emergence of digital applications*, and *the growing role of electricity*—has triggered a notable prospect for meeting sustainable development goals.

Photovoltaic Solar Energy Conversion

https://doi.org/10.1016/B978-0-12-819610-6.00001-6

The deployment of renewable energy systems into the global energy mix is an essential and beneficiary tool in world progress strategy. Renewable energy sources are extensively distributed around the world, and every nation can harness a substantial share. The integration of these green sources with economic developments in remote and less-developed areas contributes to financial improvements, thanks to alleviating migration from rural regions and easing social problems. In this regard, considering current challenges and new advances in technology associated with economic issues and environmental impacts is crucial.

1.1.1 Basic energy concepts

Energy is a conventional concept that may be remembered by definitions such as "the potential for causing change," and "the capacity to do work." Energy comes in different forms such as thermal, chemical, mechanical, electrical, etc. There is also the term *power* that indicates "the rate at which energy is transformed, used or transferred." The source is the matter that energy comes from, and the mass of matter is proportional to the amount of energy transformed as work [2]. Some of the primary energy concepts are introduced in this section, as follows:

- **Energy conversion** is the transformation from one form to another, which is a common occurrence in nature. Sometimes there is a conversion process chain between a set of energy forms that means, in this process, that more than one form of energy has resulted from the device. In other cases, energy may be converted through several ways to yield the desired format [3].
- **Energy efficiency** comes as a measure to estimate the amount of output from a given input of energy. In a broader description from the European Union (EU) Energy Efficiency Directive (2015) [4]; "It means the ratio of the output of performance, service, goods, or energy to the input of energy."
- **Energy flow** refers to the flow of energy through different processes from the primary source to the final form of energy [5].
- **Energy balance** is a set of relations considering input and output energies in forms of supply, generation, consumption, and loss in a processor stage. This can be used as a technique providing overviews for analyzing energy positions such as management, optimization, and energy auditing tools.
- **Traditional energy** or traditional resource is another name for fossil fuels, which are nonrenewable resources and are formed by rotting

animals and plants buried under rock layers over millions of years. Three main types of traditional resources are coal, oil, and natural gas [5].

- **Commercial energy** includes sources of energy that have been monetarized, which are usually based on fossil fuels. However, there are some renewable forms of energy that are cost-competitive and have a considerable share of the world economy [5].
- **Cogeneration** is defined as the sequential use of energy to generate two different forms of energy originating from a primary source. In most cogeneration cases, electricity and heat are generated from the same process, and the overall efficiency of energy use can be as high as 85% [6].
- **Gross energy requirement (GER)** can be defined as the accumulation of resources coming from all nonrenewable energy forms that are used in yielding a product or service; it is denoted as the ratio between energy units and the physical unit of product or service delivered. GER is a metric tool to evaluate the consumption of nonrenewable resources. It exhibits the share of exhaustion of the Earth's inherited supply of nonrenewable energy in the manufacturing procedure of a product or service [7].
- **Greenhouse gases (GHGs)** are a set of gases that accumulate in the lower layer of the atmosphere, the troposphere, and absorb infrared radiation, which contributes to increasing the average temperature of the Earth's surface [8].
- **Renewable energy** comes from natural and renewable sources that are known as nonpollutant sources. These are continuously replenished and have a high capacity for being implemented in rural and remote areas with no access to local power grids [9].

1.1.2 Energy economics

The energy sector plays a crucial role in the entire economy, and what is known as stability or vibrancy is affected by energy under different states such as job creation, environmental considerations, and energy efficiency [7]. With the advent of concerns over climate change and its severe impacts on human welfare and the environment, policymakers and law enforcement are compelled to shift from fossil-based fuels to renewable energy systems. As the energy sector's contribution to gross domestic product (GDP) is 6% on average, governments must maximize the benefits of renewable energy deployment for their national economy [10]. Under world energy scenarios focusing on future energy, global energy consumption is predicted to undergo an annual 1.2% increase from 2010 to

2050. This energy consumption supports the economic growth of developing nations. Simultaneously, energy technology advancements will diminish energy consumption per unit of GDP, starting from 2.7 tons/US$10,000 to 1.4 tons/US$10,000, which reflects remarkable developments in the effectiveness of energy deployment [11].

As shown in Fig. 1.1, the rise of the average annual global economy is estimated to reach 3% from 2010 to 2050 while energy demand growth is estimated to increase to 1.2% and the average elasticity coefficient of energy consumption is expected to be 0.4.

This figure also demonstrates the rise of electrification to about 3.1% in the future, which surpasses the growth of global demand. This indicates the gradual strengthening of the electricity position in energy systems and the necessity of its prioritization in technological development [11].

According to recent reports published by IRENA [12], the transformation of the energy system influences both the energy importer and the exporter with different approaches. The increase in the share of renewable energy would be a boon for fossil fuel importers where the more significant source of energy reflects ripple effects on their economy by reducing reliance on foreign energy and investing in indigenous sources. However, new trade patterns do not seem favorable to fossil fuel exporters as their GDP has the vulnerability to export revenues, which has a significant impact on their economy.

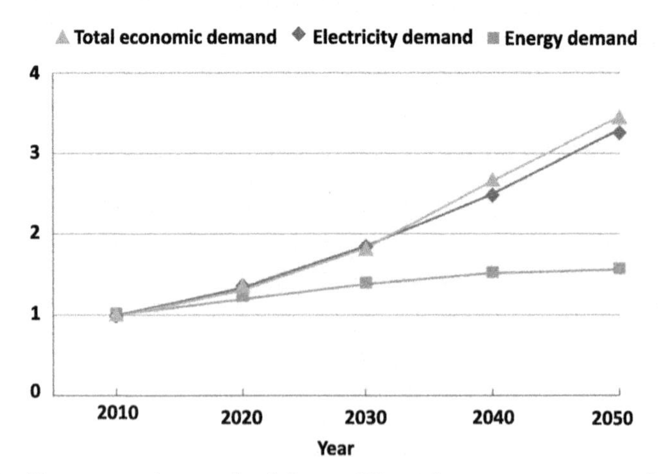

FIG. 1.1 The comparative trends of the world's total economy, energy, and electricity demand increase from 2010 to 2050 [11]. (Note: To make the values nondimensionalized, initial amounts are set at 1).

Renewable energy systems are affecting the economy in different ways and by various factors. In order to assess the renewable energy footprint on the economy, several indicators have been used by economists [13]:

- **Number of employees** is one of the most frequent indicators for evaluating the impacts on the economy, and it must be investigated in full-time equivalents or person-year.
- **Fiscal effects** represent the financial contribution share of the local or national government to the renewable energy industry.
- **Salaries of employees** reveal the quality of jobs created in the industry in which highly paid and low-paid employees are distinctive, and highly paid jobs render a more significant impact on the economy.
- **Landowner benefit** lies in land lease payments to the landowner from the plant operators, which chiefly are found in wind power plants and other space-intensive technologies such as solar photovoltaic (PV) farms.

Some other economic effects are also used to assess the impacts of renewable energy systems on the economy such as GDP, gross value added (GVA), the new democratic party (NDP), the rights of conserved water, the benefits of operators, etc., that have been assessed for the impacts of renewable energy systems on the economy.

1.2 Global trend of energy supply and demand

Human civilization, industrial evolution, and population growth are the leading causes behind the depletion of conventional energy sources through increasing the world's energy demand. In the case of energy reporting, fuels, and energy commodities, the energy flow refers to the following terminology [14]:

- **Primary energy** is dedicated to the means of energy commodity that is extracted from the original source and via a direct process with minimum unit operation.
- **Secondary energy** is defined as a means of fuel or energy commodities that are consumed as the energy source converted from a form (usually primary energy) to another form.
- **Final energy** is what the consumer buys at the marketplace.
- **Useful energy** is described as the input energy in an end-use application.

Reviewing the world's energy consumption reports [15] shows that by the end of 2018, the share of primary energy consumed by the world's population increased by 2.9% on an average annual scale. Oil and coal as primary fuels for energy consumption are decreasing, as coal has the lowest share by 27% in the last 15 years. However, natural gas has increased to 24% and is predicted to replace the coal consumption share in the near future. Hydro and nuclear energy shares remain at near 7% and 4%, respectively, while other renewable energies such as united sources have soared to 4%, just close to nuclear. The statistics prove that the total amount of energy consumed as fuel in 2018 is equal to 13,865 Mtoe, in which China had the most significant share with respect to the United States as the second-largest energy consumer. The development trend of world energy consumption for different fuels from 1965 to 2013 is depicted in Fig. 1.2.

In 2018, fossil fuel production increased as oil expanded by 2.2 Mb/d, natural gas improved by 5.2%, and coal rose by 4.3% [15]. However, according to the statistics reported by IRENA (2019), the total renewable energy capacity by the end of 2018 was measured as 2,356,346 MW and the amount of energy produced from renewable energy technologies until 2017 was 6,193,948 GWh. Of that, the hydropower share was 67%, the wind energy was 18%, and the total solar energy was 7%, with the remaining shares for marine, bioenergy, and geothermal. In 2017, 425,810 GWh of solar power was dedicated to PV and 11,476 GWh to concentrating solar power (CSP) plants [17]. Fig. 1.3 shows the global energy generation trends for each sector from 1990 to 2030.

According to Fig. 1.4, the primary energy demand is projected to increase consciously at a steady lower speed. In 2013, the global population was measured at 7.2 billion, where the total demand for primary energy reached about 19.5 billion tons of standard coal [11]. Thanks to the global energy agenda toward a low-carbon energy supply, there are

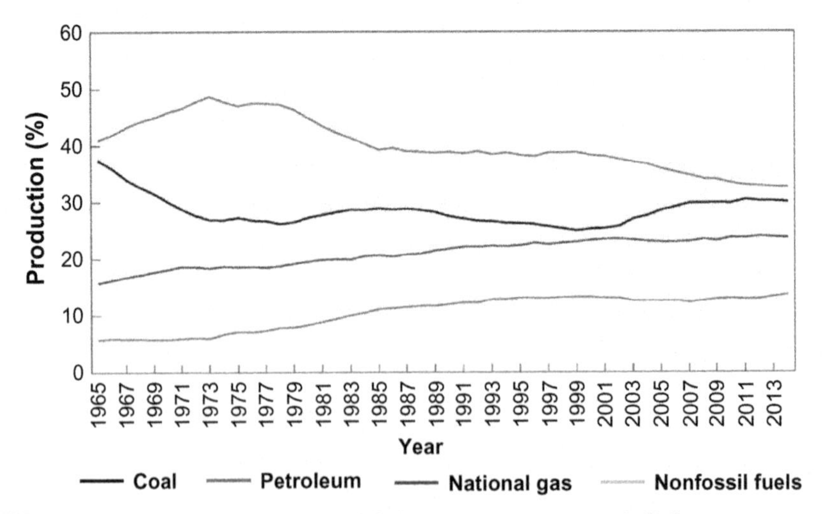

FIG. 1.2 Development trend of the global energy consumption mix [16].

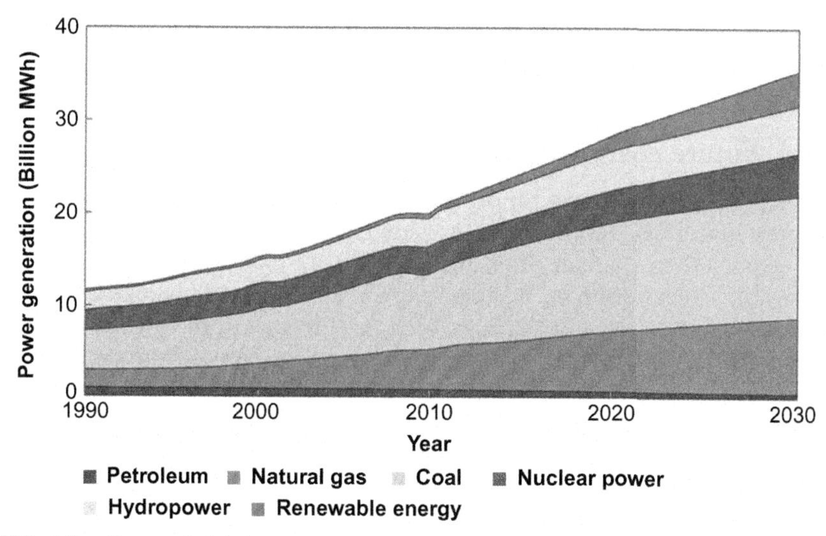

FIG. 1.3 Shares of global energy generation by fuels [16].

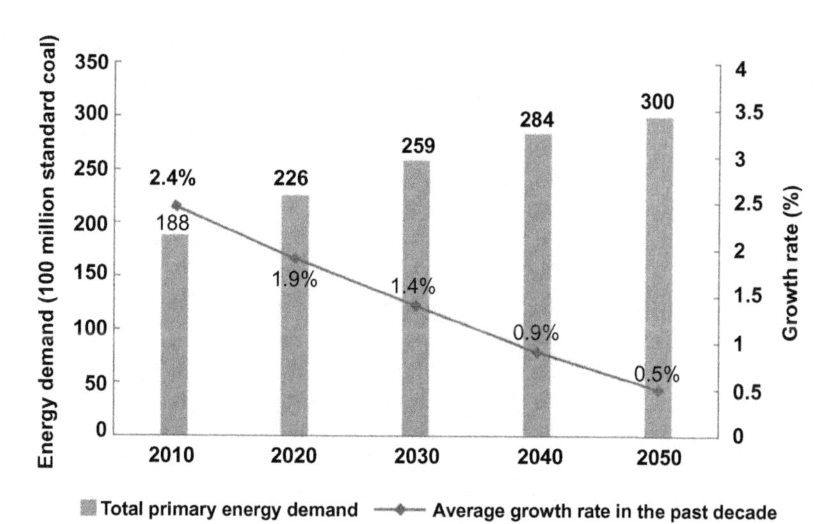

FIG. 1.4 Total primary energy demand and its growth rate [11].

mandates for renewable energy portfolios, and every nation is trying to formulate strategies, plans, and policies for avoiding an energy crisis and promoting sustainable development initiatives. Renewable energies have a more significant contribution to total energy supplies in developing countries than in developed countries. Statistical reports show that more than 75% of produced renewable energy is used by developing countries, where the most useable sources are hydropower and traditional biomass. Further analysis shows that in 2030, renewable energy is expected to have a considerable chunk, such as 7,775 TWh of the world's electricity generation and 539 Mtoe in energy generation for industries and buildings [18].

1.2.1 Future energy outlook

According to the latest report published by McKinsey [19], the global primary energy demand will have slowing growth in the future, when it is expected to plateau after 2035 due to the stronger penetration of renewable energy sources. It is also expected that electricity consumption will increase twofold until 2050, which rises from a shift toward electric transportation with 50% of total energy production comes from renewable energy. The report states that gas is the only fossil fuel that shows a growing trend in total energy demand, which is expected to plateau after 2035; however, in the long term, gas demand is expected to decline to 33% from 41% for the period from 2015 to 2050.

1.2.1.1 Challenges

Conventional sources of energy are emitting CO_2, which hinders the heat exchange between the Earth's surface and space; thus, absorbed energy remains on the earth, making it warmer. This excessive heat culminates in the climate-changing effect that is looming nowadays. So far, oil, coal, and natural gas have fulfilled human energy needs while they ultimately have elevated CO_2 emissions at a pace that natural photosynthesizing was not able to cover, and accumulation of CO_2 has been recognized in the atmosphere [20].

Although GHGs form 0.03% of the total atmospheric volume, they have a tremendous influence on the Earth's surface temperature. Usually, these gases include: (i) CO_2 emitted by combustion of fossil fuels, deforestation, and desertification, (ii) methane (CH_4) produced by rice cultivation, cattle rearing, coal mining, biomass burning, landfills, ventilation of natural gas, and using wood fuel, (iii) nitrous oxide (N_2O) released from agricultural practices as well as catalytic conversion through cars and fossil fuel combustion, and (iv) chlorofluorocarbons (CFC) discharged from air conditioners, freezers, solvents, and insulators [21].

Regarding a recent United Nations (UN) call for climate action, nations have been encouraged to propose plans not only to mitigate existing circumstances but also to show a way toward a full transformation of economics along with sustainable development goals, where the aim is to reduce GHG emissions by 45% in the next decade [22]. Some scientists believe that the best way to solve the energy crisis is to study interregional energy flows and connections by monitoring globalized villages. They put forth that it is becoming common that a product is produced using energy either exploited from domestic resources or traded from foreign countries and then is exported somewhere else for reprocessing or consumption [23].

1.2.1.2 Opportunities

GDP is a value for economic analysis and is a function of energy systems. Increasing living standards raises energy demands, which have resulted in GDP growth besides the rise in CO_2 emissions. However, the decoupling of industrial energy consumption and CO_2 emissions has shown its viability through a successful case in the energy-intensive industry in Scandinavia, which exemplified GDP growth that was decoupled from carbon emissions [24]. In the case of decoupling CO_2 emissions and GDP growth, two salient strategies of increasing the grid share of low-carbon generation sources and utilizing more efficient energy generation technologies are playing crucial roles [25].

There is growing recognition that renewable energies are becoming increasingly affordable, which leads us to leapfrog to more sustainable, resilient economies. Recent findings have proven that supplying energy systems by 100% renewable energy is achievable and the feasibility is well documented for around the world [26]. As was mentioned earlier, renewable energy utilization has accelerated in recent decades, and the plunging trend is expected to be perpetuated by technology maturity and significant price drops behind renewable energy implementation. Economic analysis indicates that in 2010 and 2017, the prices of PV and wind turbines have fallen by 80% and 50%, respectively [27].

In 2015, the world invested about $286 billion in wind, hydro, solar, and biofuel production. Patent data based on climate change mitigation technologies reveal that the solar PV fabrication industry, with a focus on cost-effective solar cells and hybrid solar technology, has the highest number at 15% of total patents [28]. It is also estimated that the renewable energy industry will provide 7.7 million jobs around the world, where PV cell manufacturers hire the largest employer [29].

In the context of future energy systems, the only way to strive for sustainable development goals as well as economic growth is orienting current infrastructure toward more sustainable systems via a transition pathway. Two main characteristics have been assigned for this transition pathway [30];

I. The impacts are long term due to the long lifetime of the used infrastructures and the effects that energy systems have on finite resources and climate change.
II. Regional characteristics are different, and transition pathways may have various starting points depending on the region.

In order to evaluate world energy systems, there are several models that must be assisted with life cycle assessment (LCA) for a comprehensive analysis of different scenarios. What has been added to this evaluation process is a large set of sustainable indicators that thoroughly deals with environmental and human life issues coupled with an investigation from raw material to the disposal of technology.

1.3 Overview of renewable energies

Usually found as parts of physical structures on our planet, various forms of renewable energies originating from sunlight, wind, water, biomass, and underground natural heat are inexhaustible and constantly renewed by natural phenomena. Based on recent advances, renewable energy sources are becoming cost-competitive and an alternative choice for conventional energy sources such as oil, gas, coal, etc. Despite the fact that these renewables are clean, they are not necessarily 100% eco-friendly and have some environmental impacts that vary with the location, technology, materials, and other factors. Their distinctive role appears when renewables reflect a very much lower carbon footprint compared to fossil fuels. They are a key element in hampering global climate change with minimizing air pollution. Different forms of renewable energies are depicted in Fig. 1.5 and concise descriptions of them are given in the following.

Solar energy is an unlimited energy flow from the sun, which runs most of the renewable energy sources. The energy is conveyed through sunlight, which is one of the clear examples of direct solar energy delivered to Earth. Some types are found in forms of stored solar energy, where rainfall or wind serves as short-term storage of solar energy, and some others such as biomass feature the long-term storage of solar energy. This source of renewable energy is described in Section 1.3.2 in detail.

Hydropower is a sustainable power production technique in which a dam is usually used to pass the stored water through a turbine, spinning

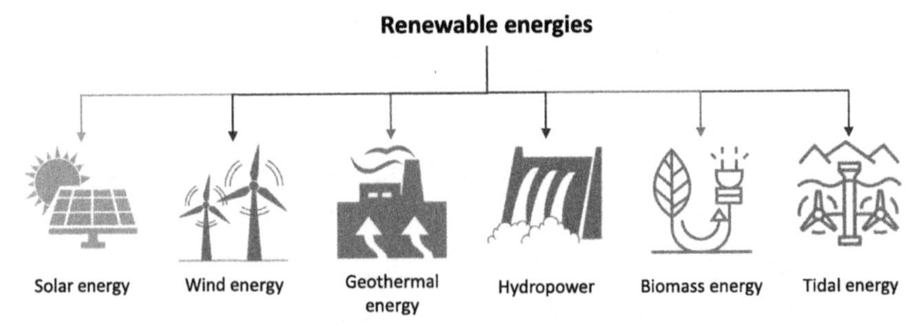

FIG. 1.5 Different types of renewable energies.

the blades and generating electricity. This technology is not limited to large dams, as there are many small stations that employ rivers, channels, or water canals to harness their kinetic energy. *Pumped storage plants* are the other form of hydroelectric unit, making it possible to store energy. In this type, electric power is first used from the utility grid to pump water from a low reservoir to the upper reservoir, a process that stores power. Finally, when water is released from the elevated reservoir, power is generated via the turbine placed along the flow line [31]. It is the world's largest source of renewable electricity. Hydropower generation capacity surged from 250GW in 1950 to nearly 1400GW in 2018 while the global pumped hydropower storage has reached 160.3GW [32].

Biomass energy; Biomass is an organic material that stores the sun's energy via photosynthesis as chemical energy; this will result in heat when burned. Moreover, biomass can be converted to a liquid form as biofuel or a gas as biogas to run engines or power plants [33]. There are numerous examples for bio-based products that are employed for different energy conversion systems such as woods, agriculture crops, food wastes, animal manure, municipal wastes, and human sewage, which undergo a digestion or degradation process inside the bioreactors to produce biogas or fuels. Heat energy derived from these biomaterials is usually used for residential water and space heating and cooking, or on a larger scale for industrial thermal processes. According to the latest reports from OECD/IEA,[a] bio-based renewable sources accounted for 12.4% of global energy consumption in 2017 while modern bioenergy accelerated its growth and increased its biopower capacity to nearly 130 GW in 2018 [34].

Wind energy is one of the renewable sources that can be harvested and used as a clean and favorable form of energy to generate electricity. This mature technology is a zero-pollutant and completely affordable in large scale. Inexhaustible wind power is harnessed via wind turbines to drive an electric generator for electrification purposes. One of the constraints in the wind industry is land use issues that have yielded offshore wind farms, a plausible solution targeted at bolstering the capacity of renewable energy sources in the world's electricity supply. In 2018, 46.8GW onshore wind power out of a total of 51.3GW was added to global installations reaching 591GW [35].

Geothermal is a thermal source of energy that comes from the earth's core, producing a high-temperature condition that is ideal for steam generation in power plants. This technology is put to work around the world where geographical factors are favorable. It can be found in two main types: dry steam or hot water resources. In dry steam, natural steam is

[a]Organization for Economic Cooperation and Development/International Energy Agency.

extracted from deep wells and used to drive turbine generators while in hot water reservoirs, the saturated wells keep hot water under pressure and in liquid form so that piping the hot water to low-pressure separators will change the water to vapor. Powering the turbine generators by the resulting steam makes electricity, where cooled and condensed water can be returned back to the reservoirs. In 2018, a new 0.5 GW geothermal power capacity was added to the global energy market, making the total accumulated capacity 13.3 GW [34].

The **ocean** enjoys two prime sources of energy in the forms of thermal and mechanical. Absorbing solar thermal power and storing the heat within its layers, the ocean has marine thermal energy that can be harvested from the temperature gradient between the surface and deep-water layers to produce electricity or desalinate ocean water. Tidal power is the other form of energy inside oceans that generates electricity out of the repeated rise and fall of water levels. Tidal barrages and fences are utilized to power electric generators and turbines by converting the mechanical energy of tides. Wave power harvesting has also loomed in recent years and new methods are under development to enhance the efficiency of energy conversion from wave energy to electricity. Although ocean thermal power plants are still at their infancy, wave and tidal energy systems are exhibiting an increasing rise where the global installed marine energy capacity reached 532 MW in 2018 [36].

1.3.1 Global market for renewable energies

To achieve social and economic development, the global demand for energy and related services has been increased. As a consequence of energy service provisions, GHG emissions have dramatically increased. It has been confirmed that burning fossil fuels accounts for the largest part of the GHG emissions originating from human activities. In the majority of cases, an increase in the share of renewable energies in the energy mix can significantly mitigate destructive impacts and provide vast benefits. The use of different renewable energy sources and technologies is envisioned to be further expanded and employed at an increasing rate in the near future [37].

Fig. 1.6 provides the carbon emissions from the energy providers in the power sector, which reflects an ascending trend with a 2.7% increase in 2018. It shows that despite the rapid growth in renewable energy, the significant power demand has resulted in substantial carbon emissions. This crucial vector has resulted from the businesses and factories that have been electrified without decarbonization. In this way, it is essential to improve the share of the renewable energy technologies in the power sector; however, to win the Paris agreement, the world needs to incorporate a range of fuels and technologies [15].

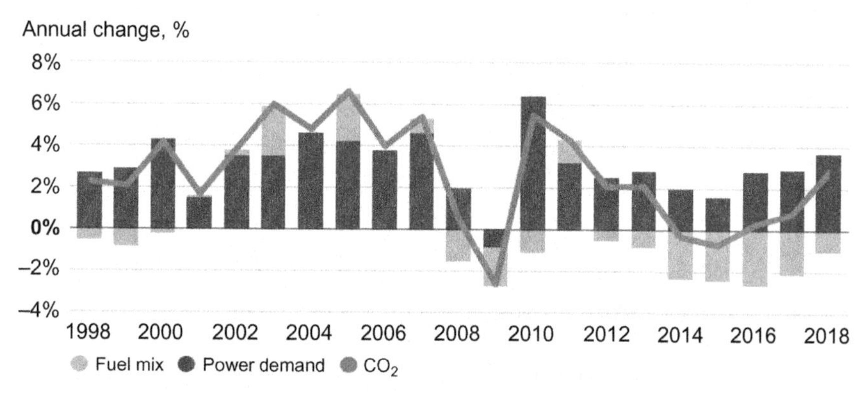

FIG. 1.6 Carbon emissions from the power sector [15].

After a long period of global slowdown, the growth of renewable energy sources is being spurred in the energy transition from fossil fuels to low-carbon strategies. According to the latest report published by REN21 [34] in 2018, 181 GW of renewable power has been added to the world's renewable energy systems, reaching a total power capacity of 2378 GW. Although the annual investment in renewable power and fuel decreased to $37 billion in 2018, energy efficiency augmentation, energy access expansion, and the progress in renewable energy utilization have driven this growth. Power generation has been the most dominant market in the renewable energy industry in recent decades. But the heating and cooling sector is also a critical segment where renewable energy has the least penetration due to the lack of implementing policies [34].

In particular, by 2016, modern renewable energy made up just 10% of the total heating and cooling demand. Therefore, more incentives and policies are necessary, especially in building energy codes to meet the rising trend of energy demand by the integration of modern renewable energy technologies in this sector [34]. Transportation is another sector in which renewable energy integration has been less than expected, although biofuels and a new generation of electric vehicles (EVs) are booming in this respect. Major uncertainties about the utilization of biofuels still confine the vast use of this liquid renewable energy source so that in 2018, only 3.3% of the total transportation energy market was provided by renewables [34].

Fig. 1.7 depicts the share of each source in the total power generated by renewables in 2018, in which hydropower capacity is still the leading source of energy, followed by wind power and solar PV. However, data analysis revealed that solar PV capacity has had the highest pace among these renewable sources by making 55% of the total 181 GW increase in the power sector [34]. Unlike the deployment of some renewable energy

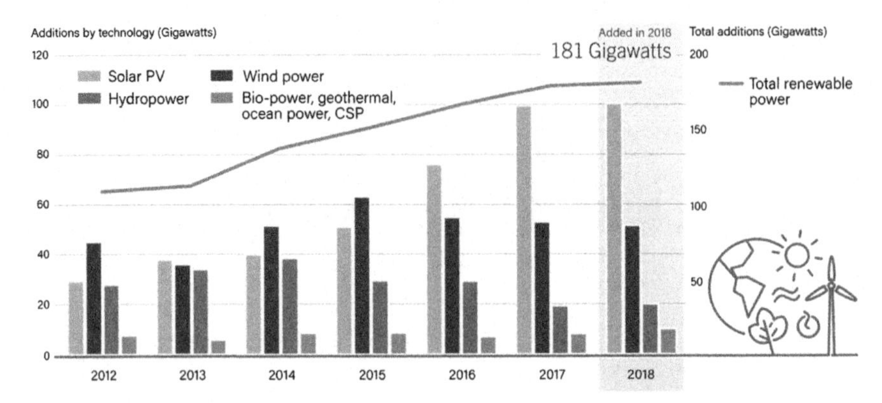

FIG. 1.7 Yearly renewable power addition by technology from 2012 to 2018 [34]. Note: Solar PV capacity data are provided in direct current (DC).

sources that are on track, such as solar PV and wind power technologies, other systems need a spike in technology penetration to considerable shares such as biofuels for transportation and solar heat for industrial processes [38].

1.3.2 Solar energy

Both renewable and fossil fuels originate from the abundant energy source of the sun. The sun, as a blackbody with a 5777 K surface temperature, encompasses several fusion reactions that trigger its abundant contained energy. This continuous fusion reactor is based on its constituent gases, referred to as a "containing vessel" retained by gravitational forces [39]. The released energy is vast in quantity and can meet human needs by several billion times where the solar power output is estimated as 3.86×10^{12} MW. Traveling near 1.495×10^{11} m, this high potential energy reaches the Earth at a constant rate of 1.37 kW per m^2 [40]. After the direct sunlight bathes the Earth's atmosphere, a fraction of it is blocked from penetration into the atmosphere layers. Solar energy can be harnessed by utilizing various technologies including solar thermal collectors, concentrators, and solar cells to produce heat and electricity.

1.3.2.1 Properties of sunlight

In space, the solar spectrum is more like the radiation of a black body and covers different wavelengths. However, the Earth's surface absorbs the selected sunlight, regulated by the atmosphere at a certain wavelength. Sunlight comprises photons, minuscule particles that carry the electromagnetic waves originating from the sun and traveling through space [4]. All the photons reaching a solar cell can be converted into electricity, except those with lower energy levels. For successful energy conversion, those with higher energy levels have to decrease their energy content to the gap energy by utilizing the thermalization of the photogenerated carriers [5].

According to Fig. 1.8, the solar spectral irradiance curve covers a broad range of wavelengths and forms the distribution of extraterrestrial radiation. However, the atmosphere attenuates several parts of the spectrum and regulates the solar radiation delivered to the Earth's surface by the elimination of x-rays, for instance.

Air mass (AM) is defined as the atmospheric parameter that usually has a strong influence on the solar spectrum. It depends directly on the shortest distance that sunrays travel through the atmosphere while reaching the Earth's surface. Therefore, as the sun angles approach the horizon, the solar intensity reaching the ground becomes weakened due to the greater air mass in the way of the rays. The apex for solar intensity at the Earth's surface is usually assumed as $1\,kW/m^2$, which varies by a set of factors such as cloudiness and climatic conditions as well as seasonal changes [42].

The solar radiation that reaches space is highly dissimilar to the portion that reaches Earth. In space, the solar spectrum is more like the radiation of a black body and covers different wavelengths; however, the Earth's surface absorbs the selected sunlight, regulated by the atmosphere at a certain wavelength [43].

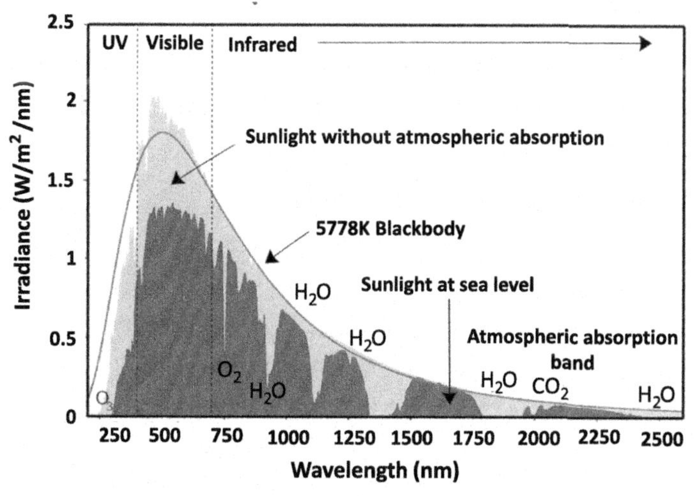

FIG. 1.8 The spectrum of solar radiation on Earth [41].

1.3.2.2 *Solar photovoltaic energy conversion technology*

Photovoltaic effect

The photovoltaic effect is a process in which light is converted into electricity. It was first experienced in 1839 by Henri Becquerel when he immersed a sheet of platinum (Pt) coated with a thin layer of silver chloride in an electrolytic solution, and then illuminated the sheet while it was connected to a counterelectrode [44]. In PV technology, direct current (DC) electrical power, indicated by watts (W) or kilowatts (kW), is generated from semiconductor materials as they receive photons in an illumination process [45]. Functionally, individual PV elements mainly known as solar cells include a p-n junction in a semiconductor material where light absorption has occurred. Solar cells never need recharging to produce electricity again, like what is happening in a battery. Therefore, electric power generation continues as long as the light is cast on a solar cell. Once the illumination is interrupted, the electricity generation is also stopped [44, 45].

A solar cell is as simple as a semiconductor diode in which careful design and fabrication have made it possible to obtain and use photonic energy conveyed by radiant light from the sun to generate electrical energy in an efficient way. The key physics of a simple conventional solar cell are demonstrated in Fig. 1.9. First, the solar incident is delivered on the top surface of a solar cell. One of the electrical contacts of the diode that is formed by a metallic grid allows light to reach the semiconductor and lies between the grid lines to be absorbed and consequently produce an electric current. In order to improve the amount of transmitted light in the process, an antireflective layer can be used between the grid lines. As an n-type semiconductor becomes adjunct to a p-type semiconductor, shaping the metallurgical junction, the semiconductor diode is formed.

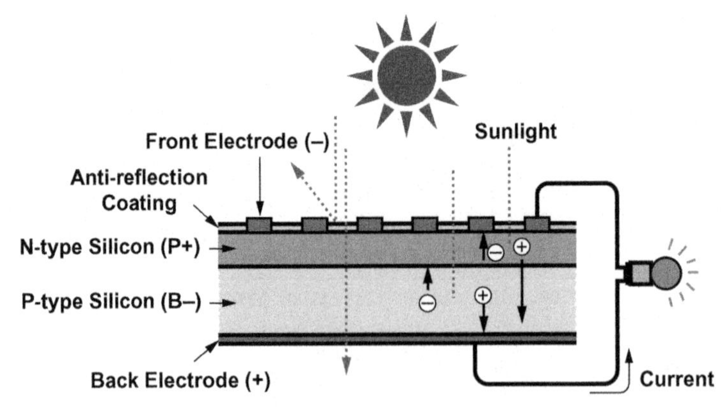

FIG. 1.9 Photovoltaic effect in a solar cell [46].

Another metallic layer attached to the back surface of solar cells forms the diode's other electrical contact [45].

1.3.2.3 *Development history of solar cells*

In essence, the study of PV technology began in the 1800s, and research continued until 1904 to complete the discovery phase of PV. In 1905, a scientific foundation was formed for technology development until the 1950s [47], where Bell labs introduced the first silicon solar cell in 1954 [48]. After this breakthrough, the first practical PV devices emerged drastically for a decade. From 1960 to 1980, PV technology obtained a global scale, and its utilization was extended to the power scale while the number of devices integrated with this technology rose. PV developments slowed during 1980–2000, which was rooted in the political and energy independence strategies [47]. However, after 2000, the technology accelerated, fueled by the considerable cost reduction and great improvements in the efficiency of commercial cells. This led to the advent of new cell generation and advanced deployment of the technology. Fig. 1.10 summarizes the historical timeline of the development of PV technology.

First generation of PV cells

The first-generation PV cells are based on crystalline silicon technology in which silicon, one of the most abundant materials on the Earth, with a 1.1 eV bandgap is used to exploit photon energy coming from solar irradiance. After the purification, the formation of silicon ingots and wafers are the steps that account for the production costs and have undergone numerous improvements [49]. This mature technology has provided two different types of silicon solar cells: single (mono)-crystalline silicon, multi (poly)-crystalline silicon. This technology was profoundly improved in 1990, where the efficiency reached 35% under lab tests. Due to easier manufacturing process, the poly-silicon crystalline solar cells were expanded rapidly by a commercial efficiency as 21.5% [50].

Second generation of PV cells

New explorations for a reduction in material consumption and a feasibility study for new materials led to the advent of second-generation PV cells. Thin-film technology was developed for easier and cheaper manufacturing of PV modules, opening new opportunities to industry. Several key technologies such as amorphous (a-Si), copper indium gallium selenide ($CuIn_xGa_{1-x}$ Se_2, CIGS), and cadmium telluride (CdTe) emerged in this respect. Although a-Si is a widespread technology due to the ability of silicon-based manufacturing, low absorption and

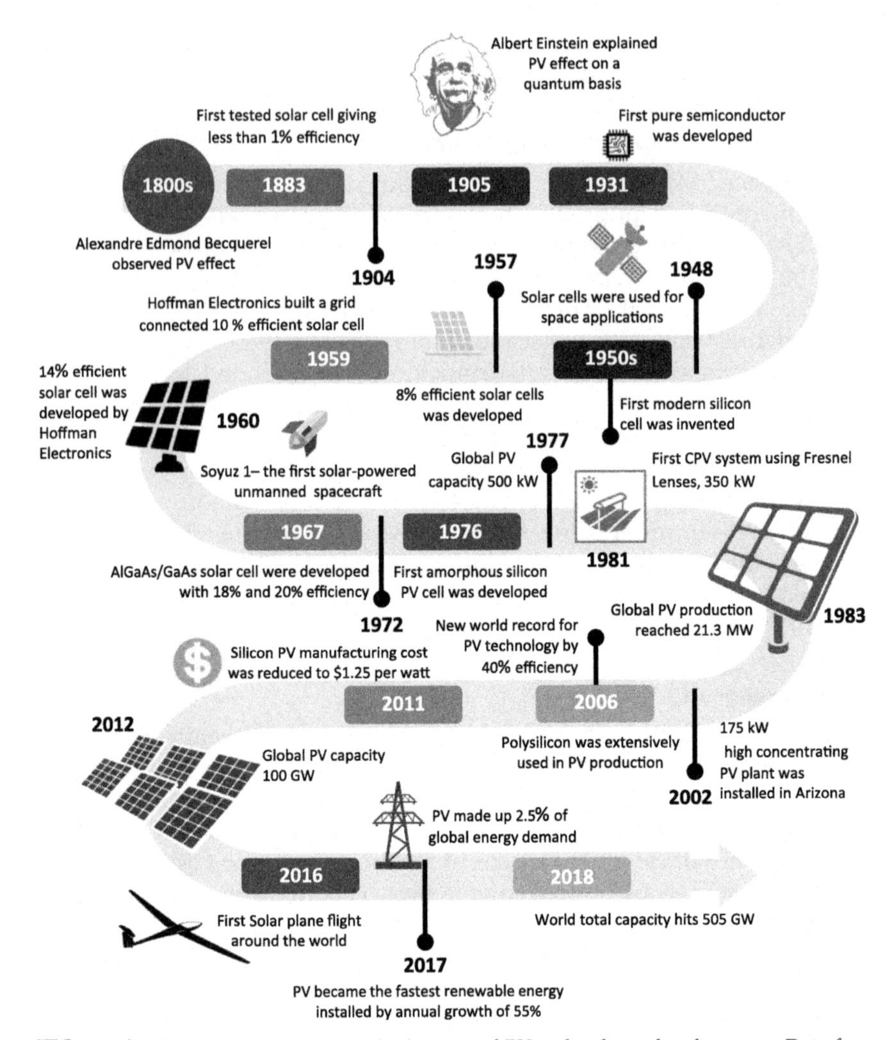

FIG. 1.10 The key milestones in the history of PV technology development. *Data from Fraas LM. History of solar cell development. Low-cost sol. electr. power. Cham: Springer International Publishing; 2014. p. 1–12. https://doi.org/10.1007/978-3-319-07530-3_1; International Renewable Energy Agency (IRENA). Global energy transformation: a roadmap to 2050 (2019 edition); 2019; Masson G, Kaizuka I. The 23rd international survey report on trends in photovoltaic (PV) applications; 2018.*

high vulnerability are its alarming disadvantages [49]. One of the merits of thin-film technology is the flexibility in the substrate materials ranging from metal to fabrics, which broadens its viability with new applications. On the other hand, using much fewer materials, the payback period of this technology is merely 6 months that increases its investing interests [51].

Third generation of PV cells

Environmental and economic concerns shifted the PV conversion technology toward cheaper and more ecofriendly PV cells. Analogous to the second generation, this technology is based on the thin-film deposition technique in the manufacturing process. One of the perspectives of this concept is to use nontoxic and abundant materials to be plausible in large-scale production [52]. Third-generation technology is based on novel thin-film devices, including dye-sensitized, organic, and quantum dot solar cells as well as other "exotic" concepts consisting of spectral-splitting devices, hot-carrier collection, carrier multiplication, and thermophotovoltaics. The prime aim behind the third-generation technology is to achieve low-cost production such as $0.2 per watt by increasing the efficiency beyond the Shockley-Queisser limit for single bandgap devices [53–55].

1.3.2.4 PV technology landscape

Fig. 1.11 represents the technology status of three PV generations in a plot of efficiency versus area cost. Single bandgap technologies suffer from a power-loss mechanism, which limits their efficiency below the shaded area. In these technologies, the photons exceeding the bandgap account for half the solar incident, resulting in thermalization and a major loss in light-to-electricity conversion. The thermodynamic limit is the other constraint applied to the third-generation technologies (multiple threshold approaches) and hinders their efficiency improvement by 67% or 86.8%, depending on the concentration [52]. This figure also indicates that all technologies shift toward the upper-left corner, as the efficiency enhances and the cost reduces progressively [53].

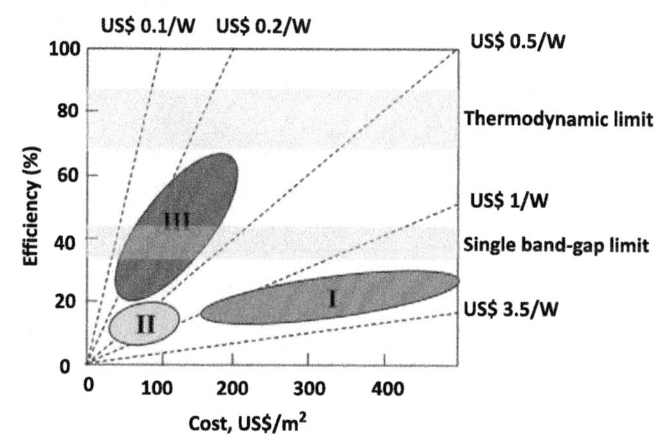

FIG. 1.11 Projection of efficiency and costs for three solar cell generations [52].

1.3.2.5 *Global PV market and industry*

As Fig. 1.12 illustrates, the annual PV industry has shown stable growth in recent years, surpassing 100 GW (including on/off-grid capacity), and reaching a total capacity of 505 GW compared to the global total of 15 GW only a decade ago. Moreover, in 2018, the rising tide of renewable energy demand in emerging markets such as Europe and some developing countries offset the major declines in some of the older markets such as China, where a number of countries have turned to gigawatt-scale markets [56].

In the PV technology market, there are seven categories of applications for PV power systems, ranging from small pico systems with a few watts to very large-scale PV plants with hundreds of MW [57]:

- **Pico PV** systems emerged as small PV systems with a power output of a few watts (1–10 W) integrated with small LEDs, charge controllers, and efficient batteries. They were introduced into the market as small kits to enable services for lighting, phone charging, and powering radios or small computers. These power systems are in extensive use for off-grid electrification, especially in developing nations.
- **Off-grid domestic** systems supply electricity to residential households and villages that lack a permanent connection to utility power networks. The given energy comes as lighting, refrigeration, and other loads to meet the energy demand from off-grid users. Most off-grid domestic applications are mostly scaled to 5 kW.
- **Off-grid nondomestic** systems emerged as the initial widespread utilization of terrestrial PV systems, empowering vast applications in remote regions where such a small system was able to bring significant developments such as telecommunications, water pumping, vaccine refrigeration, and navigational aids.

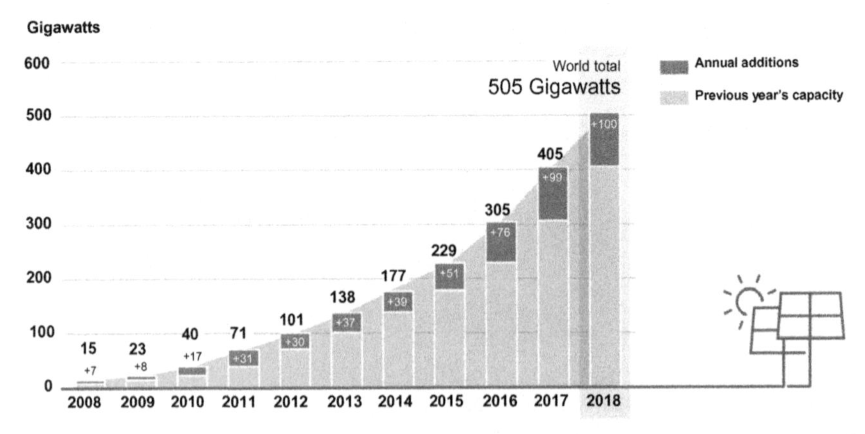

FIG. 1.12 Solar PV global capacity and annual additions [34]. *Source: Becquerel institute and IEA PVPS.*

- **Hybrid** systems include minigrids that benefit from both PV systems and diesel generators. The synergy comes out of the combination, which leads to a major fuel price reduction and delivers higher-quality service. With the advent of microhybrid systems, new possibilities emerged in the development of telecom base stations and rural electrification. However, large hybrid systems are able to power large cities far from the grid, using utility-scale battery storage.
- **Grid-connected distributed PV** systems are a set of PV systems on or integrated on residential, industrial, or commercial buildings, or merely in the built environment, to produce electricity for a wide range of demands. These systems supply power to end users through the utility network or directly to the electricity network. For integration into buildings, the PV system may be building applied photovoltaic (BAPV) or building-integrated photovoltaic (BIPV), where BAPV comes from installation on existing buildings while BIPV refers to the replacement of building materials with PV.
- **Grid-connected centralized PV** systems work like the centralized power stations where the supplied power is not directed to specific functions on the network and is usually in the form of bulk power. Most systems are ground-mounted; however, recent advances have introduced floating PV systems, enabling harnessing of the abundant solar energy in the dam vicinity, and agricultural PV systems that provide electricity generation alongside crop production.
- **ViPV** (vehicle integrated PV) is the latest segment of PV application, which has displayed significant potential on cars, trucks, ships, etc. The embedded PV may result in a significant reduction in carbon emissions that comes from the transportation sector.

Thanks to enormous cost reduction and widespread production technology, the PV industry has evolved drastically in recent years. In Fig. 1.13, the yearly PV production, production capacity, and installation are shown. The trend indicates that in just 2017, 28GW was added to global production capacity, where the most significant increase was reported in China's photovoltaic industry. It also suggests that the increase in manufacturing capacity is a product of the incorporation of new factories, besides the procurement of closed factories or inaugurating joint ventures with other companies [56].

The PV manufacturing industry covers the value chain from raw material to maintenance and installation. Although the PV industry is mostly based on solar cells and module manufacturing, there are some upstream industries, including materials and equipment manufacturing, as well as some downstream industries consisting of battery manufacturing and other maintenance and installation services. In the history of the PV

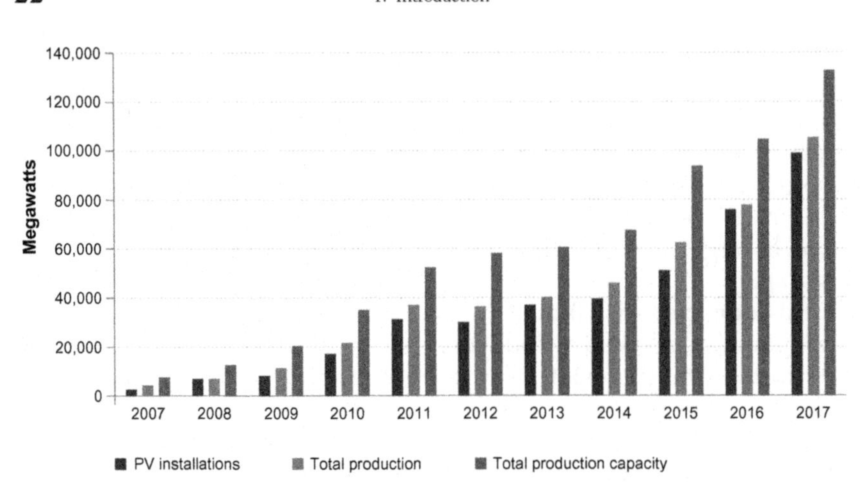

FIG. 1.13 Annual PV installation, production, and production capacity from 2007 to 2017 (MW) [56].

industry and technology development, there are some key parameters that have influenced the market rigorously.

After 2008, the price of silicon raw materials underwent a drastic fall, resulting in much lower wafer-based silicon solar cells. Achieving up to 95% of the total PV market in 2017 led this technology to become the main PV industry for years. With recent advances, the efficiency of commercial modules has reached 12–22%, whereas the efficiency of monocrystalline modules varies from 16% to 22% while the efficiency of polycrystalline modules performs at 12–18%. However, the fastest-growing segment in the PV power plant sector is PV technology with tracking systems, and it has been expected that this technology will cover 40% of the PV market by 2020, continuing its share of 20% in 2016 [58]. Thin-film solar module manufacturing is another significant industry, which started in 2005 and reached 20% of PV markets in 2009. However, in 2017, this technology declined to less than 5% due to the lack of progress in efficiency. Concentrating PV (CPV) is the other segment suffering from a sharp fall in the number of manufacturing companies due to the lack of considerable cost reductions in processing. In this technology, optical elements such as lens and tracking systems play major roles. Moreover, new nonconcentrating high-efficiency concepts are emerging to increase the cell and module efficiencies [58].

1.4 Conclusions and goals of the book

"Energy transition" is a pathway toward transferring the global energy sector from fossil fuels to renewables. In order to confine climate change and its consequences, energy sector decarbonization needs urgent action on a global scale. In this way, two known measures are renewable energy and energy efficiency, which can potentially achieve 90% of the essential carbon decrements.

Approved scenarios of "100% renewable electricity" have recently focused on the significance of solar PV technology as a necessity to achieve this goal and to decarbonize the power sector in a cost-competitive way.

PV systems are becoming more widespread as their advantages become apparent and costs fall. The development of new technologies and the establishment of new policies are required by the PV industry, grid operators, and utilities to facilitate the deployment of large-scale PV into flexible, efficient, and smart grids. Enjoying numerous advantages, PV is one of the most sustainable, elegant, and benign technologies in power generation and has strong superiority over conventional methods. PV power has the flexibility to be generated in any place with a high potential of solar radiation, regardless of urban or rural limitations, and at any operating mode ranging from distributed to grid-feeding. Having no dependency on conventional fuels and as a result of a sustainable production system, PV would play a major role in national energy security scenarios and ameliorate the CO_2 impacts in the future.

Empowering electrical equipment from an inexhaustible source like the sun is always a laudable idea. This clean and abundant energy with the least harm to our planet comes free and is needless of any power socket when it appears as PV. There is a significant need for accelerating R&D studies by governments and industry to cut production costs and support long-term technology innovation to achieve deep PV penetration. International collaboration is a vital step in conducting PV research, development, capacity scaling, and funding for better learning and averting parallel efforts. To the current technology, solar electrification is demonstrated as the most economical method to generate electricity.

This book provides up-to-date knowledge of the photovoltaic solar energy conversion technology along with the most recent applications in different fields, the environmental impacts, and the global market and policies. This reference offers engineers, scientists, specialists, etc., a comprehensive knowledge of different aspects of PV technology in the context of the most recent advances in science and technology, providing insights into future developments in the field of photovoltaics. In this regard, this book covers four main areas: *PV Cells and Modules*, *Applications of PV Systems*, *Life Cycle and Environmental Impacts*, and *PV Market and Policies*.

The contents of this book encompass all the mentioned topics through 13 chapters, briefly introduced here. This chapter represents a general view of the global energy status and outlook, the basics of renewable energies, the operating principles of PV cells, and the historical background and global market for PV energy conversion technology. Chapter 2 introduces all generations of solar cells along with the most recent advances in this field. Chapter 3 studies the conventional and emerging PV module technologies as well as their possible failures and defects, and discusses the monitoring and inspection methods of PV modules. Chapter 4 covers the most recent advances in photovoltaic thermal (PVT) technology along with their various applications. Chapter 5 studies different configurations and design principles of the PV systems, including on-grid, off-grid, hybrid, and their main components. Chapter 6 investigates the employment of PV systems in farm applications, including pumping and irrigation, drying, greenhouse cultivation, dairy farming practices, etc. Chapter 7 reviews novel applications of PV systems in precision agriculture (PA) by introducing solar farm robots, solar tractors, solar-powered wireless sensor networks (WSNs), etc. In Chapter 8, applications of solar PV systems in desalination technologies are presented and discussed, including commercial desalination techniques as well as solar distillation systems. Chapter 9 studies the applications of solar PV systems in hydrogen production with the most focus on the electrolysis water-splitting method. Chapter 10 represents the developments of PV power plants, distributed PV generation stations, and the autonomous monitoring concept. Chapter 11 introduces new concepts and applications of PV systems and their use in buildings, cars, spacecraft, gadgets, etc. The life cycle assessment (LCA) methods and environmental impacts of PV systems are discussed in Chapter 12. Finally, the global market and policy mechanisms of PV technology are presented in Chapter 13.

References

[1] IEA. World Energy Outlook—Executive Summary. Oecd/Iea 2018; 2018. p. 11.
[2] energy | Definition, Types, & Examples | Britannica.com; n.d.
[3] Sørensen B. The energy conversion processes. Renew Energy 2017;357–567. http://dx.doi.org/10.1016/B978-0-12-804567-1.00004-9.
[4] Energy efficiency—targets, directive and rules | Energy; n.d.
[5] Chapter 1—Basic energy concepts*; n.d.
[6] Bhatia SC. Cogeneration. Adv Renew Energy Syst 2014;490–508. http://dx.doi.org/10.1016/B978-1-78242-269-3.50019-X.
[7] Energy Agency I. World Outlook Energy 2015; 2015.
[8] Greenhouse Gas Definition; n.d.
[9] Renewable Energy Definition and Types of Renewable Energy Sources | NRDC; n.d.
[10] Energy industry's share of GDP by select country 2015 | Statista; n.d.
[11] Liu Z. Supply and demand of global energy and electricity. Glob Energy Interconnect 2015;101–82. http://dx.doi.org/10.1016/b978-0-12-804405-6.00004-x.
[12] Ferroukhi R, Lopez-Peña A, Kieffer G, Nagpal D, Hawila D, Khalid A, et al. Renewable Energy Benefits: Measuring the Economics. IRENA International Renewable Energy Agency; 2016. p. 92.
[13] Jenniches S. Assessing the regional economic impacts of renewable energy sources—a literature review. Renew Sustain Energy Rev 2018;93:35–51. http://dx.doi.org/10.1016/j.rser.2018.05.008.

[14] Reporting energy production and consumption guideline; 2017. p. 0–20.
[15] BP Statistical Review of World Energy Statistical Review of World; 2019.
[16] Review and outlook of world energy development. In: Non-Fossil Energy Development in China. Elsevier; 2019. p. 1–36. http://dx.doi.org/10.1016/b978-0-12-813106-0.00001-5.
[17] IRENA. Renewable Energy Statistics 2019. 1; 2019.
[18] Co-operation E. 2. Energy supply and demand; 2010. p. 5–20.
[19] McKinsey Global Institute. Global Energy Perspective 2019: Reference Case. Energy Insights; 2019.
[20] Introduction. Solar Electric Power Generation, Berlin, Heidelberg: Springer Berlin Heidelberg; 2006, p. 1–18. https://doi.org/10.1007/978-3-540-31346-5_1.
[21] Mukhopadhyay R, Karisiddaiah SM, Mukhopadhyay J. Introduction. In: Clim. Chang. Elsevier; 2018. p. 1–13. http://dx.doi.org/10.1016/b978-0-12-812164-1.00001-3.
[22] UNITED NATIONS Climate Change Summit n.d.
[23] Chen GQ, Wu XF. Energy overview for globalized world economy: source, supply chain and sink. Renew Sustain Energy Rev 2017;69:735–49. http://dx.doi.org/10.1016/j.rser.2016.11.151.
[24] Enevoldsen MK, Ryelund AV, Andersen MS. Decoupling of industrial energy consumption and CO2-emissions in energy-intensive industries in Scandinavia. Energy Econ 2007;29:665–92. http://dx.doi.org/10.1016/j.eneco.2007.01.016.
[25] Mikulčić H, Ridjan Skov I, Dominković DF, Wan Alwi SR, Manan ZA, Tan R, et al. Flexible carbon capture and utilization technologies in future energy systems and the utilization pathways of captured CO2. Renew Sustain Energy Rev 2019;114:109338. http://dx.doi.org/10.1016/j.rser.2019.109338.
[26] Connolly D, Mathiesen BV, Østergaard PA, Möller B, Nielsen S, Lund H, et al. Heat Roadmap Europe 2050 : Second Pre-study for the EU27. Aalborg University; 2013.
[27] Dominkovic DF. Modelling energy supply of future smart cities; 2018. http://dx.doi.org/10.11581/DTU:00000038.
[28] Mukhopadhyay R, Karisiddaiah SM, Mukhopadhyay J. Threat to opportunity. In: Altern Gov Policy South Asia; 2018. p. 99–117. http://dx.doi.org/10.1016/B978-0-12-812164-1/00005-0.
[29] Shukla AK, Sudhakar K, Baredar P. Renewable energy resources in South Asian countries: challenges, policy and recommendations. Resour Technol 2017;3:342–6. http://dx.doi.org/10.1016/j.reffit.2016.12.003.
[30] Volkart K, Mutel CL, Panos E. Integrating life cycle assessment and energy system modelling: methodology and application to the world energy scenarios. Sustain Prod Consum 2018;16:121–33. http://dx.doi.org/10.1016/j.spc.2018.07.001.
[31] Hydropower Tech—Renewable Energy World; n.d.
[32] IHA. Hydropower status report 2019: Sector trends and insights; 2019. p. 1–83. http://dx.doi.org/10.1103/PhysRevLett.111.027403.
[33] Energy Information Administration (EIA). Unpublished; 2014. http://www.eia.gov/countries/cab.cfm?fips=ir [Accessed 19 August 2014].
[34] Members REN. Renewables 2019 global status report; 2019.
[35] GWEC. Global wind energy council report 2018. Global Wind Energy Council; 2019. p. 1–61.
[36] Marine energy capacity worldwide 2018 | Statista; n.d.
[37] Edenhofer O, Pichs-Madruga R, Sokona Y, Seyboth K, Eickemeier P, Matschoss P, et al. IPCC, 2011: summary for policymakers. In: IPCC special report on renewable energy sources and climate change mitigation; 2011. http://dx.doi.org/10.5860/CHOICE.49-6309.
[38] Gielen D, Boshell F, Saygin D, Bazilian MD, Wagner N, Gorini R. The role of renewable energy in the global energy transformation. Energy Strateg Rev 2019;24:38–50. http://dx.doi.org/10.1016/j.esr.2019.01.006.
[39] Duffie JA, Beckman WA. Solar engineering of thermal processes. 4th ed. John Wiley and Sons; 2013. http://dx.doi.org/10.1002/9781118671603.

[40] Hersch P, Zweibel K. Introduction. In: Basic photovolt. princ. methods. Technical Information Office; 1982. p. 5–8 [chapter 1], https://doi.org/10.2172/5191389.
[41] Solar Spectrum-Wikimedia Commons n.d. https://commons.wikimedia.org/wiki/File:Solar_Spectrum.png [Accessed 30 October 2019].
[42] Riverola A, Vossier A, Chemisana D. Fundamentals of solar cells. In: Nanomaterials for solar cell applications. Elsevier; 2019. p. 3–33. http://dx.doi.org/10.1016/B978-0-12-813337-8.00001-1.
[43] Bayod-Rújula AA. Solar photovoltaics (PV). In: Sol. Hydrog. Prod. Elsevier; 2019. p. 237–95. http://dx.doi.org/10.1016/b978-0-12-814853-2.00008-4.
[44] Pietruszko SM. 1 What is photovoltaics ? In: Springer Ser Opt Sci., 112; 2005. p. 1–10. http://dx.doi.org/10.1007/3-540-26628-3_1.
[45] Luque A, Hegedus S. Handbook of photovoltaic science and engineering. Chichester, UK: John Wiley & Sons; 2003. http://dx.doi.org/10.1002/9780470974704.
[46] Electrochemistry—Can static electricity damage solar cells?—Chemistry stack exchange; 2019. https://chemistry.stackexchange.com/questions/113020/can-static-electricity-damage-solar-cells [Accessed 28 October 2019].
[47] Fraas LM. History of solar cell development. In: Low-cost sol. electr. power. Cham: Springer International Publishing; 2014. p. 1–12. http://dx.doi.org/10.1007/978-3-319-07530-3_1.
[48] Chapin DM, Fuller CS, Pearson GL. A new silicon p-n junction photocell for converting solar radiation into electrical power. J Appl Phys 1954;676–7.
[49] Sundaram S, Benson D, Mallick TK. Overview of the PV industry and different technologies. Sol Photovolt Technol Prod 2016;7–22. http://dx.doi.org/10.1016/b978-0-12-802953-4.00002-0.
[50] Green MA, Emery K, Hishikawa Y, Warta W, Dunlop ED. Solar cell efficiency tables (Version 45). Prog Photovoltaics Res Appl 2015;23:1–9. http://dx.doi.org/10.1002/pip.2573.
[51] Upadhyaya HM, Sundaram S, Ivaturi A, Buecheler S, Tiwari AN. Thin-film PV technology. In: Energy effic. renew. energy handb. 2nd ed. ROUTLEDGE in association with GSE Research; 2016. http://dx.doi.org/10.9774/GLEAF.9781466585096_46.
[52] Conibeer G. Third-generation photovoltaics. Mater Today 2007;10:42–50. http://dx.doi.org/10.1016/S1369-7021(07)70278-X.
[53] Jean J, Brown PR, Jaffe RL, Buonassisi T, Bulović V. Pathways for solar photovoltaics. Energy Environ Sci 2015;8:1200–19. http://dx.doi.org/10.1039/c4ee04073b.
[54] Green MA. Third generation photovoltaics: solar cells for 2020 and beyond. Phys E Low-Dimens Syst Nanostruct 2002;14:65–70. http://dx.doi.org/10.1016/S1386-9477(02)00361-2.
[55] Green MA. Third generation photovoltaics: ultra-high conversion efficiency at low cost. Prog Photovolt Res Appl 2001;9:123–35. http://dx.doi.org/10.1002/pip.360.
[56] Masson G, Kaizuka I, Cambiè C, Brunisholz M. The International Energy Agency (IEA)—Photovoltaic Power Systems Programme—2018 Snapshot of Global Photovoltaic Markets; 2018. https://dx.doi.org/978-3-906042-58-9.
[57] Masson G, Kaizuka I. The 23rd nternational survey report on trends in photovoltaic (PV) applications; 2018.
[58] Joint Research Centre. European Comission. PV status report 2018 2018. http://dx.doi.org/10.2760/826496.

Further reading

International Renewable Energy Agency (IRENA). Global energy transformation: a roadmap to 2050. 2019 edition; 2019.

CHAPTER

2

Solar cell technologies

Geetam Richhariya[a], *Anil Kumar*[b,c], *and Samsher*[b,c]

[a]Department of Energy, Maulana Azad National Institute of Technology, Bhopal, India, [b]Department of Mechanical Engineering, Delhi Technological University, Delhi, India, [c]Centre for Energy and Environment, Delhi Technological University, Delhi, India

2.1 Introduction

Globally, increasing growth, improved industrialization and an increase in people's living standard have raised energy demands over the past few decades. The International Energy Agency (IEA) projected that developing nations are increasing their energy consumption at a quicker rate than advanced ones and will need nearly twice their current installed generation ability by 2020 to meet that demand [1]. Solar cells, commonly referred to as photovoltaic (PV) cells, are in fact electrical devices that convert solar energy into direct current (DC). When these cells are exposed to sunlight, photons are absorbed, and the electrical current begins to flow after completing the gap between two poles, as shown in Fig. 2.1 [2].

These cells are useful in controlling the electricity voltage generated by positive and negative cell reactions. A few decades ago, when researchers tried to find an affordable and efficient way to produce energy through the use of renewable resources, this field of technology came into practical use. Solar power, however, remains the most productive source of renewable energy. Small business owners can generate their own electricity for personal use at lower rates than the local service provider due to the use of solar technology. To date, solar energy has faced strong competition from hydroelectric and wind power, accounting for only a small proportion of total energy production. Over time, however, solar power has emerged as perhaps the greenest form of renewable energy and has seen growing worldwide demand. The solar industry has made significant progress in reducing solar electricity costs, and in many parts of the world, it is

https://doi.org/10.1016/B978-0-12-819610-6.00002-8

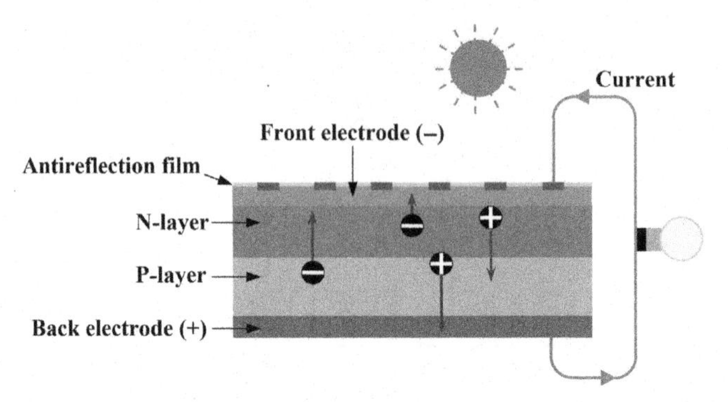

FIG. 2.1 Photovoltaic effect of a solar cell [2].

now competing with grid electricity in terms of costs. Because it requires less infrastructure, solar power can also be used in areas where conventional electricity is not an easy option [3]. Recent innovations in alternative PV technologies, however, have opened up the possibility of solar panels with features such as flexibility, a custom shape, and transparency [4]. Because of the lack of heavy glass sheets and metal frames, flexible solar cells are lightweight, significantly reducing the transport and deployment costs. Specifications, limitations, and advances for transforming solar energy into electricity using solar cells have been processed in order to provide four generations with specific improvements in their characteristics, such as solar range, cost, protection, durability, and efficiency. In the near future, the next generation of more exotic solar cells will be in the pipeline. There are four generations of solar cells: crystalline solar cells, thin-film solar cells, dye solar cells, and perovskite solar cells. This means that different types of solar cells can be used according to needs and preferences. There is progress in the research and development linked to distinct kinds of solar cell materials. The recent developments in solar cell efficiency are shown in Fig. 2.2 [5]. In this chapter, some of the important ones are described, as shown in Fig. 2.3.

2.2 Solar cell technologies

2.2.1 Crystalline silicon solar cells

Single junction solar cells based on silicon wafers along with single crystal and multicrystalline silicon oversee recent solar photovoltaic development. This sort of single-junction, silicon wafer system is now frequently referred to as the first generation of solar PV technology, much of which was based on screen-based systems comparable to those shown in Fig. 2.4 [6].

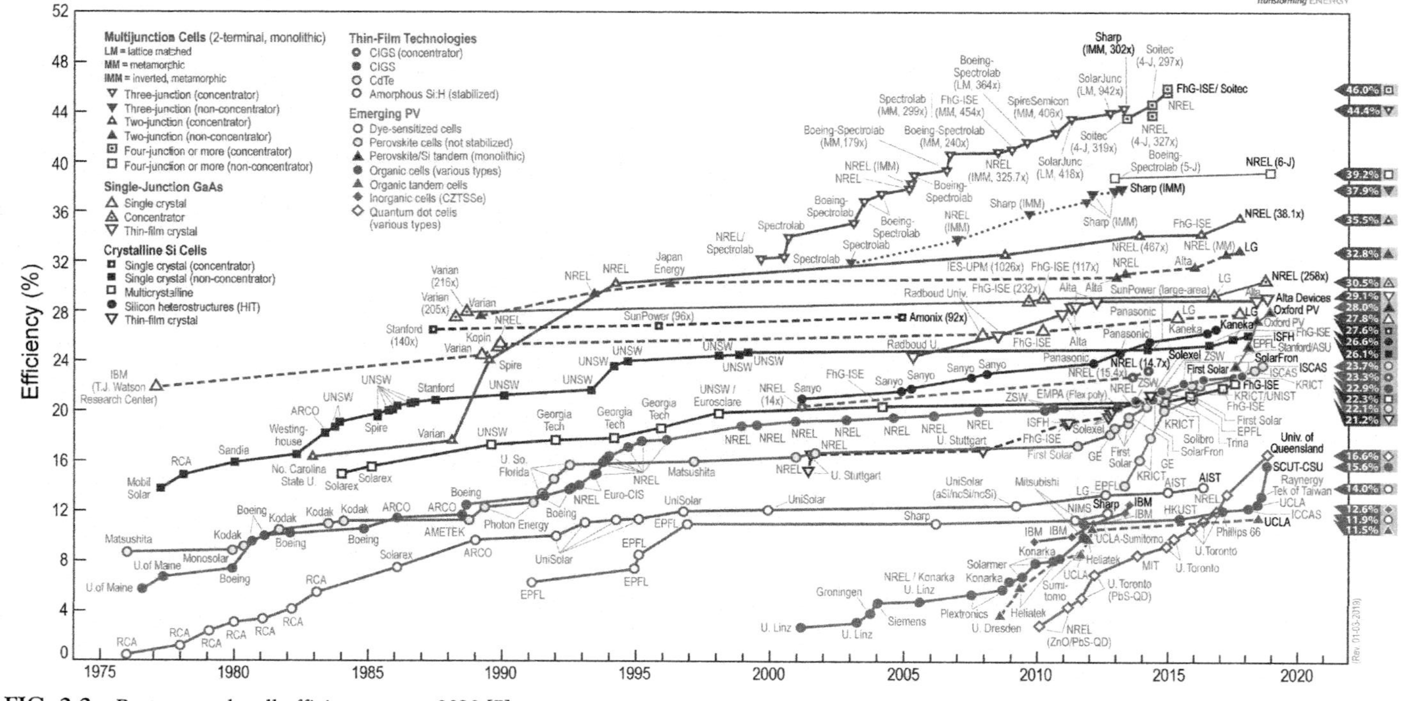

FIG. 2.2 Best research cell efficiency up to 2020 [5].

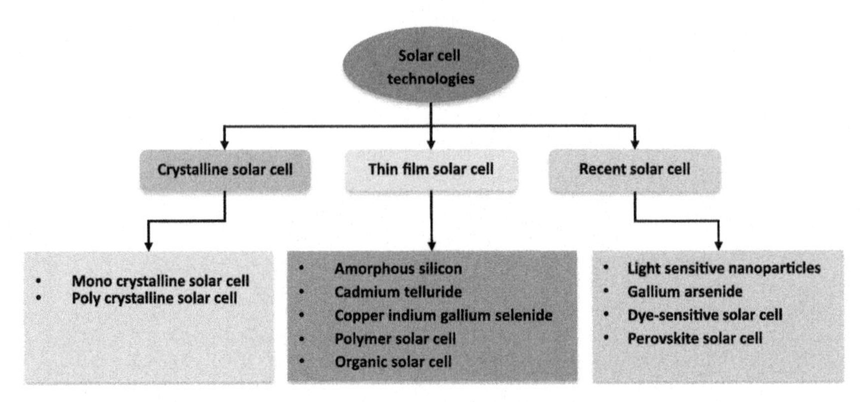

FIG. 2.3 Solar cell technologies.

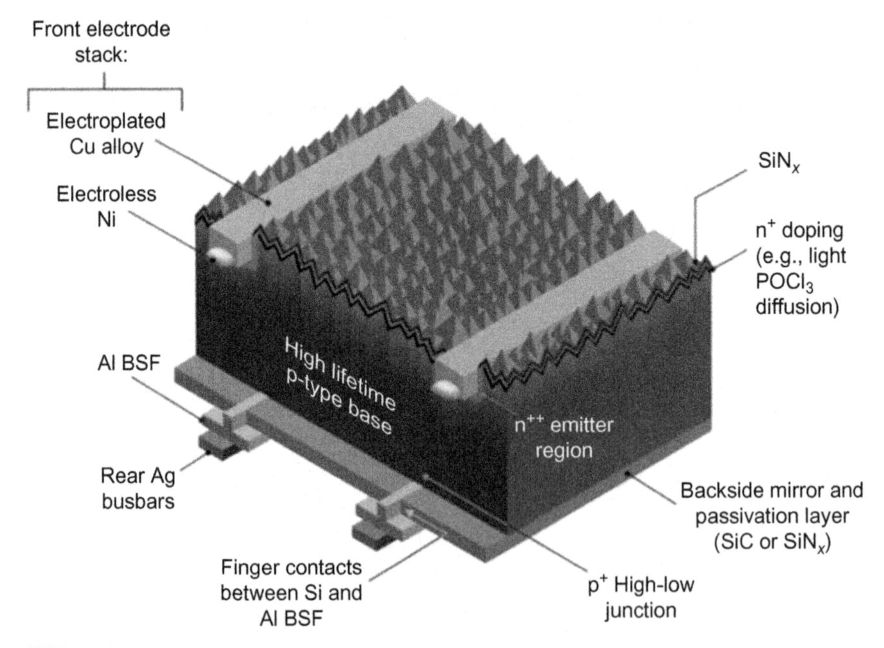

FIG. 2.4 Schematic of a first-generation mono-Si PV cell [6].

Crystalline silicon solar cells have numerous benefits, such as better efficiency than other sorts of solar cells and simple accessibility; these benefits have compelled producers to use them as a prospective material for solar cells [7]. In many cases, the mono-Si solar cells are employed due to high efficiency; however, the high material prices remain a point of concern for producers and end users. As a result, the production unit could use a substitute, and thus poly-Si solar cells may be the next alternative that is more inexpensive than the monocrystalline solar cells.

2.2.1.1 Mono-Si solar cells

Mono-Si solar PV cells are a photovoltaic technology of the first generation. They have been around for a long time, proving their reliability, durability, and longevity. The technology, installation, and performance issues are all understood. Many of the early PV cells that were installed in the 1970s still produce electricity today. These are manufactured from the method of Czochralski (CZ) [8]. A molten vat is made of a silicon ingot. It is then sliced into a number of ingots that form the solar cell substrate. The main demerit of mono-Si solar cells is their high cost due to the costly manufacturing process. A second demerit is its reduction in efficiency as the temperature drops to approximately 25°C. For panel installation, proper maintenance is required to provide air circulation over the panel.

2.2.1.2 Poly-Si solar cells

Polycrystalline Si (Poly-Si) solar cells are less expensive than mono-Si solar cells, but they are also less efficient because of their nonuniform lattices. They have a number of ingots drawn from a molten vat rather than a single large ingot [9].

Polycrystalline silicon is a high purity, poly-si form (Fig. 2.5) used by the PV industry as a raw material. A chemical purification process known as the Siemens process produces poly-Si from metallurgical-grade silicon. This process involves distilling volatile silicon compounds at high temperatures and decomposing them into silicon [10]. In addition, the solar PV industry generates upgraded metallurgical-grade silicon using metallurgical processes in place of chemical purification. Poly-Si contains adulteration levels of less than one fraction per billion when produced for the

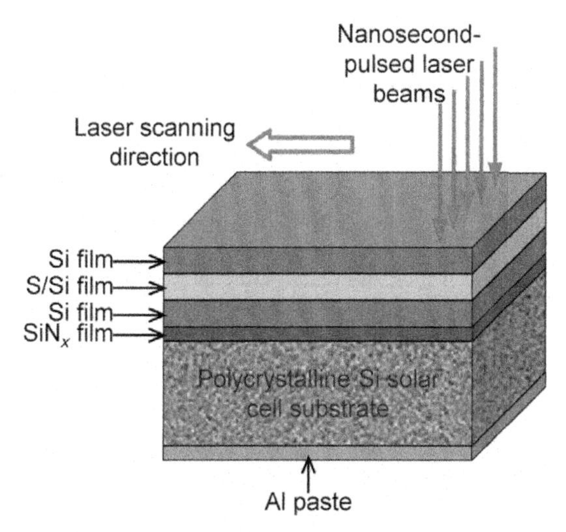

FIG. 2.5 Structure of a typical poly-Si solar cell [9].

solar PV industry, whereas poly-Si solar grade silicon is usually less pure. Poly-Si consists of minute crystals, also known as crystallites, giving a typical metal flake effect to the material. While poly-Si or multi-Si are frequently used as synonyms, multi-Si usually refers to crystals larger than 1 mm. Multi-Si solar PV cells are the most common type of solar cells in the fast-growing PV market and consume most of the worldwide produced poly-Si. Poly-Si is distinct from mono-Si silicon and amorphous silicon. Initially built using single-crystal wafer silicon and processing technology from the integrated circuit industry, it is apparent that the first generation of solar cell technology greatly benefited from its symbiosis with the integrated circuit industry, which provided the essential materials, processing know-how, and manufacturing tools to enable a rapid transition to large-scale production.

2.2.2 Thin-film solar cells

Thin-film solar cells are considered to be the second generation of solar cells. Fig. 2.6 shows a structural perspective of the thin-film solar cell. These are more inexpensive than silicon crystalline solar cells, but they have revealed lower efficiencies.

In such cell types, a thin layer of semiconductor PV material is deposited onto the metal, glass, or plastic foil. Thin films experience low efficiency and also need a wider range of fields and thus also increase associated costs such as mounting due to nonsingle-crystal structure. Examples of thin-film solar cells for outdoor uses are amorphous silicon (a-Si), cadmium telluride (CdTe), copper indium gallium selenide (CIGS), polymer, and organic solar cells.

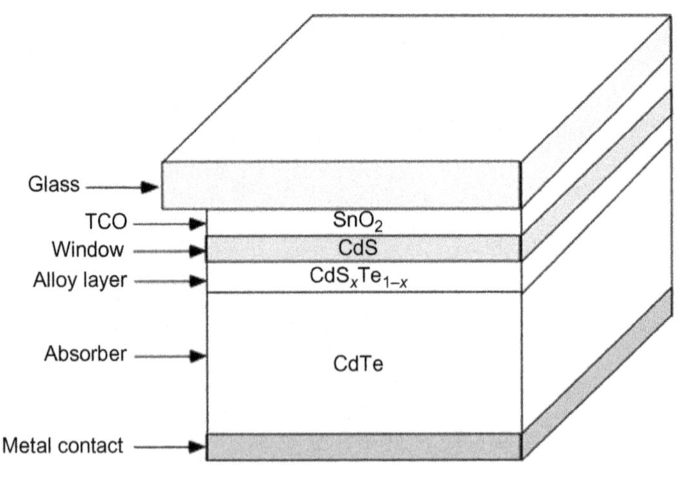

FIG. 2.6 Structure of a thin-film solar cell [11].

2.2.2.1 *Amorphous silicon solar cells*

Because amorphous silicon is a noncrystalline and disordered silicon structure, the absorption rate of light is 40 times higher compared to the mono-Si solar cells [12]. Therefore, amorphous silicon solar cells are more eminent as compared to CIS, CIGS, and CdTe solar cells because of higher efficiency. Such types of solar cells are categorized as thin-film Si solar cells, where one or numerous layers of solar cell materials are deposited onto the substrate. To form a module based on the a-Si solar cell, a thin layer of silicon (approximately 1 μm) is vapor deposited onto the substrate such as metal or glass. For a plastic substrate, a-Si thin film is deposited at a very low temperature [13]. The structure of an a-Si thin film consists of layers of p-i-n in a single sequence. However, such solar cells suffer from large degradation when exposed to sunlight. The process of degradation is called the Staebler-Wronski effect [14]. The application of thinner layers improves the stability, but decreases the absorption of light and hence affects the PV cell efficiency [15–17]. This has led industries to produce tandem and triple-layer devices that consist of p-i-n cells assembled one on the apex of others. Thin-film solar cells using p-i-n layer hydrogenated amorphous silicon oxide (a-SiO_x:H) were fabricated to enhance transmittance in visible ranges of 500–800 nm. At an R (CO_2/SiH_4) ratio of 0.2, the highest figure of merit that was achieved was greater than the conventional a-Si PV solar cell, as shown in Table 2.1 [18].

The parasitic absorption in the doped layers is reduced by using nanocrystalline silicon oxide-doped layers with a wide bandgap in case of an ultrathin a-Si:H p-i-n solar cell. A conversion efficiency of 8.79% was achieved for a 70 nm thickness of i layer, as discussed in Table 2.2 [19].

Doping of Ge in traditional a-Si:H thin-film solar cells improves the power conversion efficiency of a solar cell. An open circuit voltage of 0.58 V, a short current density of 20.14 mA/cm^2 and a fill factor of 0.53, which offer a high power conversion efficiency of up to 6.26%, as shown in Table 2.3 [20].

TABLE 2.1 Efficiency, average transmittance, and figures of merit for semitransparent solar cell performance [18].

Sample (CO_2/SiH_4)	η (%)	Average transmittance	Figure of merit
0	5.52	17.49	96.5
0.2	5.71	18.94	108.1
0.4	5.27	19.45	102.5
0.6	4.91	20.99	103.1

TABLE 2.2 Performance parameter characteristics of ultrathin a-Si:HSC with a distinct i layer thickness [19].

i-Thickness (nm)	V_{oc} (V)	J_{sc} (mA/cm^2)	Fill factor (FF)	η (%)
70	0.91	13.83	0.69	8.79
50	0.89	12.82	0.66	7.65
20	0.86	10.02	0.61	5.32

TABLE 2.3 Photovoltaic properties of a-Si:Ge alloy thin-film solar cell [20].

Device	V_{oc} (V)	J_{sc} (mA/cm^2)	Fill factor (FF)	η (%)
a-Si:Ge alloy	0.58	20.14	0.53	6.26

2.2.2.2 *Cadmium telluride (CdTe) solar cells*

Cadmium telluride (CdTe) solar cells contain thin-film layers of cadmium telluride materials as a semiconductor to convert absorbed sunlight and hence generate electricity. In these types of solar cells, the one electrode is prepared from copper-doped carbon paste while the other electrode is made up of tin oxide or cadmium-based stannous oxide. And between the two electrodes, cadmium sulfide is placed. The cadmium telluride photovoltaic solar cells are the next most ample solar cell photovoltaic technology after crystalline silicon-based solar cells in the world market. CdTe thin-film PV solar cells can be assembled rapidly and as long as an economical substitute for conventional silicon-based PV technologies. CdTe thin-film photovoltaic solar cells can be assembled easily and as long as they are an economical replacement for traditional silicon-based photovoltaic technologies. However, developing a stable low-resistivity back contact to the CdTe solar cells is still an issue. The back contact was activated by rapid thermal processing (RTP), resulting in spectacular improvement in key device performance indicators. A power conversion efficiency of 11.25% was obtained with an open-circuit voltage of 0.72 V, as shown in Table 2.4 [21].

TABLE 2.4 Photovoltaic parameters of a CdTe solar cell using the RTP process [21].

RTP process	V_{oc} (V)	J_{sc} (mA/cm^2)	Fill factor (FF)	η (%)
360°C	0.63	21.95	0.66	9.33
380°C	0.72	21.99	0.70	11.25
400°C	0.76	21.40	0.66	10.88

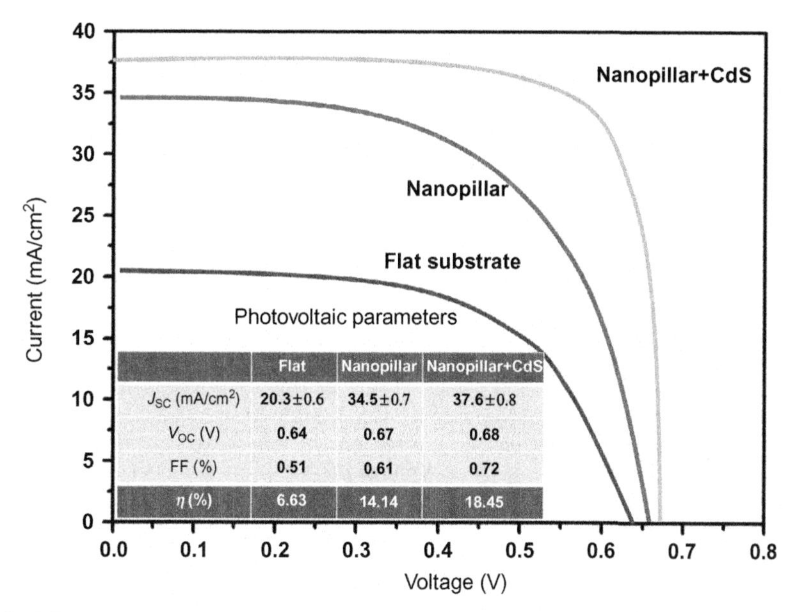

	Flat	Nanopillar	Nanopillar+CdS
J_{SC} (mA/cm²)	20.3±0.6	34.5±0.7	37.6±0.8
V_{OC} (V)	0.64	0.67	0.68
FF (%)	0.51	0.61	0.72
η (%)	6.63	14.14	18.45

FIG. 2.7 *I-V* characteristic of a nanopillar with CdS solar cells [22].

The introduction of a nanopatterned structure and CdS quantum dots in the n-i-p structure amorphous silicon solar cells enhances the photovoltaic properties. The short circuit current was reached up to the value of $37.6 \pm 0.8\,mA/cm^2$ from $20.3 \pm 0.6\,mA/cm^2$, as shown in Fig. 2.7 [22].

2.2.2.3 *Copper indium gallium selenide (CIGS) solar cells*

The copper indium gallium selenide-based solar cells have the largest energy generation compared to any thin-film-based solar cell technology. On the glass substrate, the power conversion efficiency reached nearly 21%. Current developments in the area of CIGS technology have been perceived as a shift toward flexible photovoltaic cells, with metal foil or polyamide substrates. The flexible structure, high defined power, and resistance to the intensity of solar radiation have led CIGS solar cells to become progressively more used for space applications. CIGS PV solar cells have an extremely high absorption coefficient at the forbidden energy band gap of 1.5eV, resulting in an incredibly strong absorption of the sunlight spectrum. A new solar cell configuration was assembled by taking into consideration nontoxic indium selenide (In_2Se_3) as the photosensitizer while a copper zinc tin sulfide ($Cu_2ZnSnS_2Se_2$)-coated carbon fabric was used as the counter electrode. Photovoltaic solar cell with TiO_2/In_2Se_3-S_2-/poly(hydroxyethyl methacrylate) gel-$Cu_2ZnSnS_2Se_2$/carbon fabric structure offers a conversion efficiency of 4.15%, which means that quasisolid-state quantum dot solar cells with nontoxic components can be easily manufactured and still

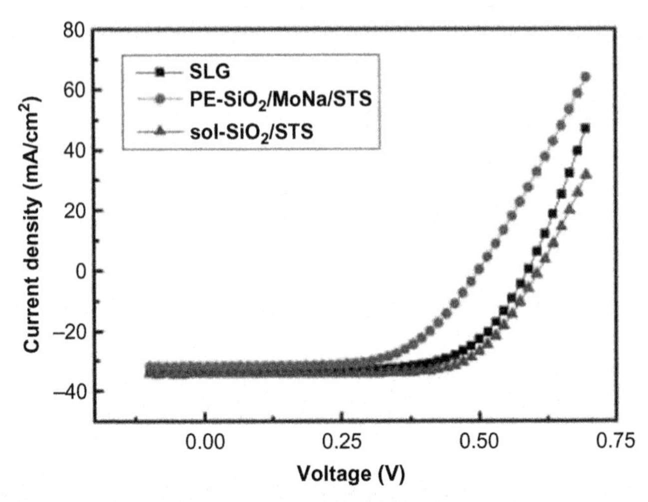

FIG. 2.8 Current-voltage characteristics of CIGS SC [24].

scaled up for current applications [23]. Solar cells using flexible CIGS achieved a conversion efficiency of 14% using a stainless steel substrate enclosed with a layer of insulating material (Fig. 2.8). Moreover, the open-circuit voltage was increased due to a grown double-graded gallium composition in the film of CIGS on stainless steel [24].

The x-ray diffraction and energy dispersive spectroscopy techniques were used to observe the crystal structure and the ratio of the atomic composition of elements in $Cu_3In_5Se_9$, $Cu_3Ga_5Se_9$, $Cu_3In_5Se_9$, and $Cu_3In_5Te_9$ crystals [25]. The chemical bath deposition process was employed to fabricate zinc oxysulfide (Zn(S,O)) thin films onto glass substrates as well as the surface of CIGS for the fabrication of solar cells. The power conversion efficiency of an annealed ZnO (S,O)-based copper indium gallium selenide solar cell was recorded as 6.15%, which was much higher than nonannealed ZnO (S,O) (4.56%), as shown in Fig. 2.9 [26].

Microelectronic and photonic structure (AMPS-1D) simulation software was used to analyze the photovoltaic properties of CdTe/Si tandem solar cells with the configuration of n-SnO_2/n-CdS/p-CdTe/p+CdTe/n+-Si/n-Si/p-Si/p+-Si/Al. A power conversion efficiency of 28.457% was obtained from the CdTe/Si tandem structure that was higher than the baseline CdTe solar cell efficiency of 19.701%, as mentioned in Fig. 2.10 [27].

2.2.3 Polymer solar cells

A polymer solar cell is a sort of flexible PV solar cell prepared with polymers, large molecules of persistent structural units, that generates an electric current from sunlight by the photovoltaic effect. A schematic diagram of a polymer solar cell is shown in Fig. 2.11.

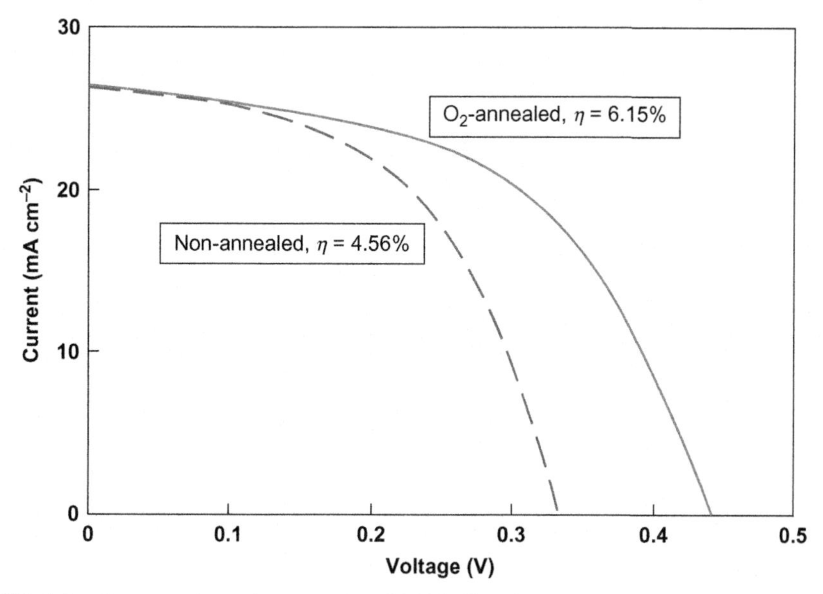

FIG. 2.9 Current-voltage characteristics of ZnO(S,O) and oxygen annealed ZnO (S,O)-based CIGS solar cells [26].

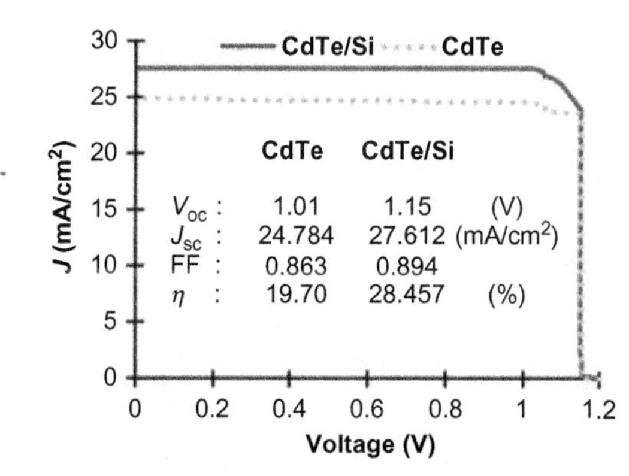

FIG. 2.10 *I-V* characteristics of CdTe and CdTe/Si solar cells [27].

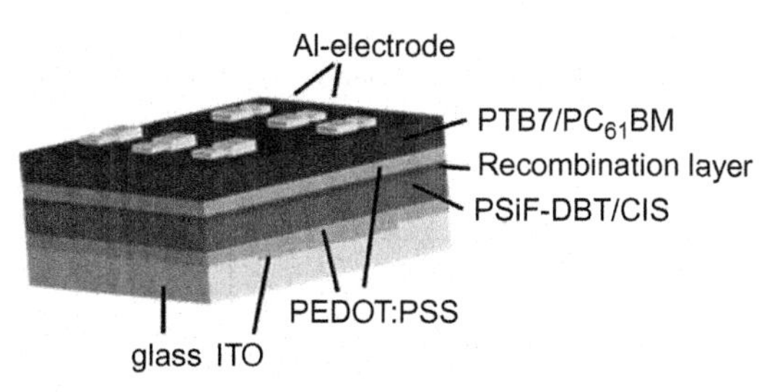

FIG. 2.11 Schematic diagram of a polymer solar cell [28].

The commercially available solar cells are prepared from a refined, purified silicon crystal such as the materials used in fabricating computer chips and integrated circuits. The uneconomical nature as well as the complicated production process have led to interest in alternative technologies. The light weight of such types of solar cells reveals applications in autonomous sensors. As compared to other solar cell devices, these solar cells are economical, disposable, flexible in nature, and have less of an effect on the environment. The transparent nature of polymer solar cells results in applications in walls, windows, and flexible items. Four novel random polymers were synthesized using accepting units such as benzothiadiazole and benzotriazole while the donor unit was benzodithiophene. The polymer-based solar cells P1, P2, P3, and P4 revealed 4.10%, 3.84%, 1.60%, and 3.83% power conversion efficiency, respectively, as shown in Table 2.5 [29].

Two-dimensional conjugated polymers named PBDTT-4S-TT and PBDTT-4S-BDD were fabricated and synthesized using a benzo[1,2-b:4,5-b′] dithiophene unit with 4-methylthio substituted thiophene side chains. PBDTT-4S-TT and PBDTT-4S-BDD showed power conversion efficiencies of 6.6% and 7.8%, respectively (Table 2.6). The open-circuit voltage (V_{oc}) was increased due to the substitution of 4-methylthio on the thiophene side chain of the two-dimensional polymers [30].

With the addition of benzothienoisoindigo as the electron, two conjugated polymers have been fabricated and synthesized. A high open-circuit voltage of 0.96 V was shown by PBDTT-TBID based on an alkylthienyl-substituted benzo[1,2-b:4,5-b′]dithiophene (BDT) derivative, as shown in Table 2.7 [31].

TABLE 2.5 Photovoltaic properties of polymer-based solar cells [29].

Polymer	V_{oc} (V)	J_{sc} (mA cm^{-2})	Fill factor (FF)	η (%)
P1	0.77	9.51	0.56	4.10
P2	0.76	9.00	0.56	3.84
P3	0.64	3.80	0.65	1.60
P4	0.66	8.84	0.65	3.83

TABLE 2.6 Performance parameters of 4-methylthio-based polymer solar cells [30].

Polymer	V_{oc} (V)	J_{sc} (mAcm^{-2})	Fill factor (FF)	η (%)
PBDTT-4S-TT	0.94	12.3	0.57	6.6
PBDTT-4S-BDD	0.98	11.9	0.67	7.8

TABLE 2.7 Photovoltaic properties of polymer-based solar cells [31].

Polymer	V_{oc} (V)	J_{sc} (mAcm^{-2})	Fill factor (FF)	η (%)
PBDT-TBID	0.85	11.81	0.58	5.87
PBDTT-TBID	0.96	09.77	0.51	4.83

The effect of three different additives (1-chloronaphthalene, 1,8-diiodooctane, diphenylether) was carried out on the performance of polymer-polymer solar cells based on a BHJ blend comprising a donor such as poly[4,8-bis(5-(2-ethylhexyl) thiophen-2-yl) benzo[1,2-b,4,5-b′] dithiophene-alt-3-fluorothieno [3,4b] thiophene-2-carboxylate] (PTB7-Th) while the acceptor was poly [[N,N′-bis (2-octyldodecyl)-naphthalene1,4,5,8-bis (dicarboximide)-2,6-diyl]-alt-5, 5′-(2,2′-bithiophene)] (P(NDI2OD-T2)), as shown in Fig. 2.12. A slightly improvement in the performance of the PV solar cell was observed with additives as compared to without additives, as shown in Table 2.8 [32].

2.2.4 Organic solar cells

An organic solar cell (OSC) is a variety of the PV solar cell that employs organic electronics. The flexibility of organic molecules and the cost

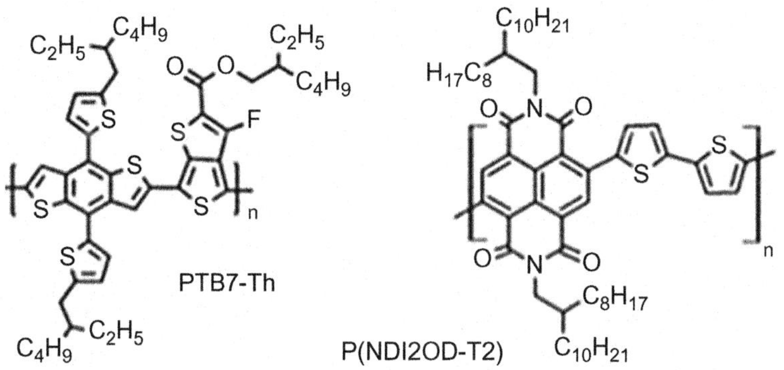

FIG. 2.12 Chemical structure of photovoltaic materials [32].

TABLE 2.8 Performance parameters of polymer solar cells without and with additives [32].

Condition	V_{oc} (V)	J_{sc} (mAcm^{-2})	Fill factor (FF)	η (%)
W/o additive	0.80	11.92	0.47	4.51
With additive	0.80	12.64	0.55	5.41

effectiveness are the main advantages of such solar cells. Also, they have the greatest optical absorption coefficient, thus maximum light can be trapped. However, they suffer from the severe drawbacks of less efficiency and less stability as compared to a silicon-based solar cell with an inorganic structure. The inverted organic solar cells using hafnium-indium-zinc-oxide (HIZO) act as an electron transport layer (ETL) with an active layer of a bulk heterojunction of Poly({4,8-bis[(2-ethylhexyl) oxy] benzo [1,2b:4,5-b′] dithiophene-2,6-diyl} {3-fluoro-2-[(2-ethylhexyl) carbonyl]thieno[3,4-b]thiophenediyl}) (PTB7) and [6,6]-phenyl C71 butyric acid methyl ester (PC70BM), as shown in Fig. 2.13. The degradation of encapsulated HIZO-iOSC was seen at a slow rate compared to an inverted organic solar cell using poly [(9,9-bis(30(N,N-dimethylamino)-propyl)-2,7-fluorene)-alt-2,7-(9,9-dioctylfluorene)] (PFN) as the electron transport layer [33].

The organic cells based on PTB7Th:PC71BM utilized thiophene (TH) and diphenyl ether (DPE) as halogen-free processing solvents. The performance of the cell is compared with chlorobenzene (CB) and 1,8-diiodineoctane (DIO) combination. A power conversion efficiency of 9.02% was achieved by TH while only 7.41% was obtained from CB, as mentioned in Table 2.9 [34].

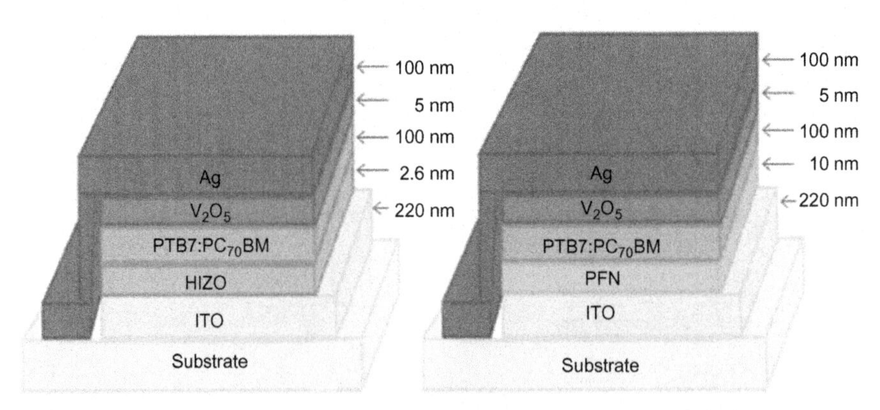

FIG. 2.13 Schematic structure of iOSCs HIZO as ETL and PFN as ETL [33].

TABLE 2.9 Photovoltaic parameters of OSCs using CB and TH [34].

Solvent	V_{oc} (V)	J_{sc} ($mAcm^{-2}$)	Fill factor (FF)	η (%)
CB	0.76	15.40	0.62	7.41
TH	0.81	16.60	0.66	9.02

The donor moiety was connected via a nonconjugated σ-linker to a PC61BM-derived block based on push-pull π-conjugated systems. Bithiophene (BT), thienothiophene (TT), and cyclopentadithiophene (CPDT) (Fig. 2.14) units were used as π-connectors, constituting a push-pull unit to evaluate the effect of the chemical bridging (CPDT) or fusing (TT) on the electrochemical and electronic *I-V* characteristics of the cell [35].

2.3 Recent advances in solar cells

There has been continuous progress in the use of renewable energy in a wide spectrum and solar energy in a diligent manner for useful applications. Recently, various novel solar cells have been developed such as gallium arsenide as well as perovskite and dye-sensitized solar cells, which are discussed in detail in the following.

2.3.1 Gallium arsenide

Gallium arsenide is considered the second material after silicon in terms of development and properties. Nevertheless, there are demerits such as recombination process, deficiency, and constant content. The conversion efficiency of 5.3% with an open-circuit voltage (V_{oc}) and a short-circuit current density (J_{sc}) of 0.35 V and 2.14 mA/cm^2 was achieved using graphene/gallium arsenide Schottky solar cell junction as shown in Table 2.10 [36].

Polytype gallium arsenide (GaAs) nanowires (NWs) were investigated for solar photovoltaic applications. The different configurations of polytype defects along the wire were investigated using a rate equation model together with surface and temperature effects. The peak positions of bandgap and below bandgap transitions of GaAs NW are shown in Fig. 2.15 [37].

A novel configuration of graphene-based gallium arsenide (Gr-GaAs) solar cells using poly(3hexylthiophene) (P3HT) as the hole transport layer was reported. It is observed that the open-circuit voltage and short-circuit current of the devices notably improves due to the P3HT layer, as shown in Table 2.11 [38].

2.3.2 Dye-sensitized solar cells

A dye-sensitized solar cell (DSSC) is an economical solar cell that comes in the category of thin-film solar cells. The DSSC has various striking merits: easy fabrication using the roll-to-roll technique as well as an economical and environmentally friendly nature, in the case of natural dyes. DSSC cells are designed and fabricated using four natural dyes: Celosia

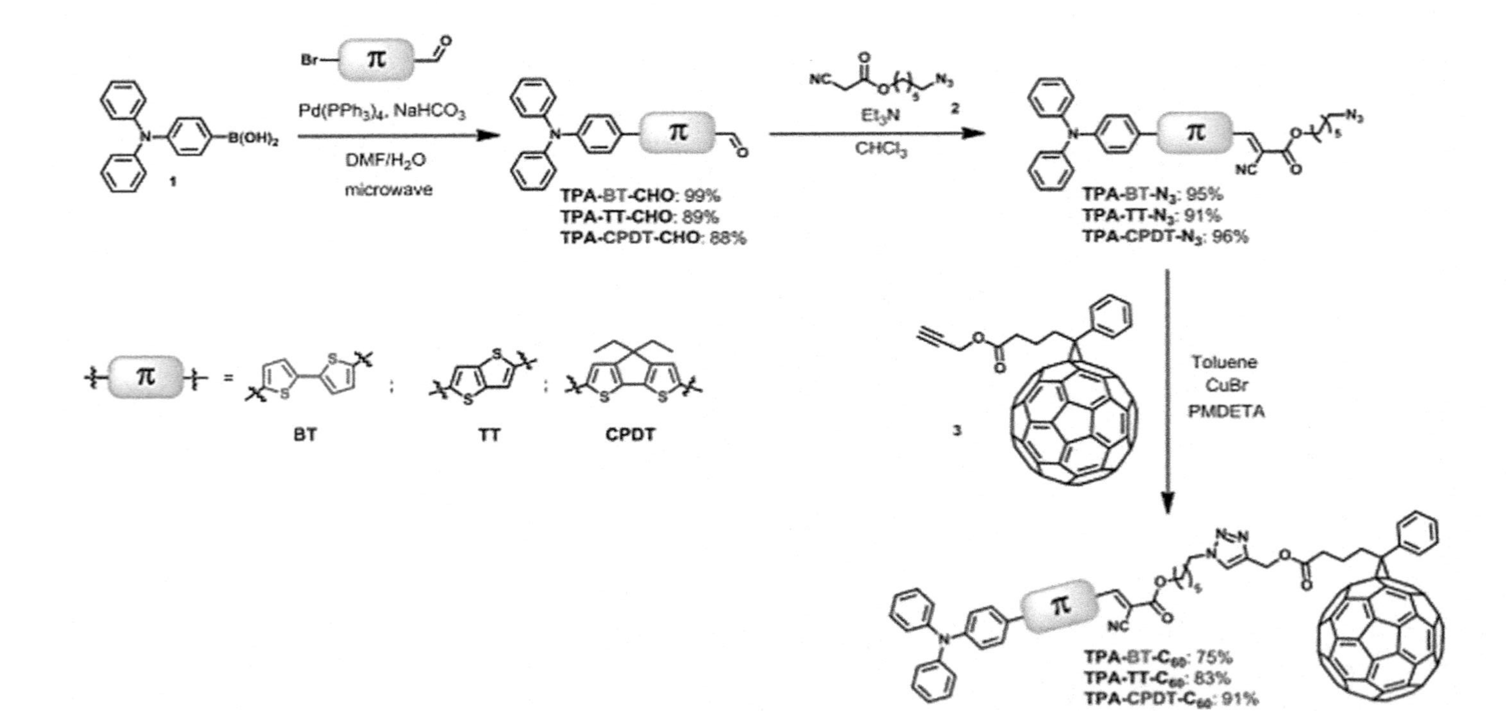

FIG. 2.14 Synthetic route to TPA-BT-C60, TPA-TT-C60, and TPA-CPDT-C60 [35].

TABLE 2.10 Photovoltaic parameters of a gallium arsenide solar cell [36].

Structure of device	V_{oc} (V)	J_{sc} (mA/cm^{-2})	Fill factor (FF)	η (%)
Graphene/SiO_2/ n-GaAs/metal	0.35	2.14	0.69	5.3

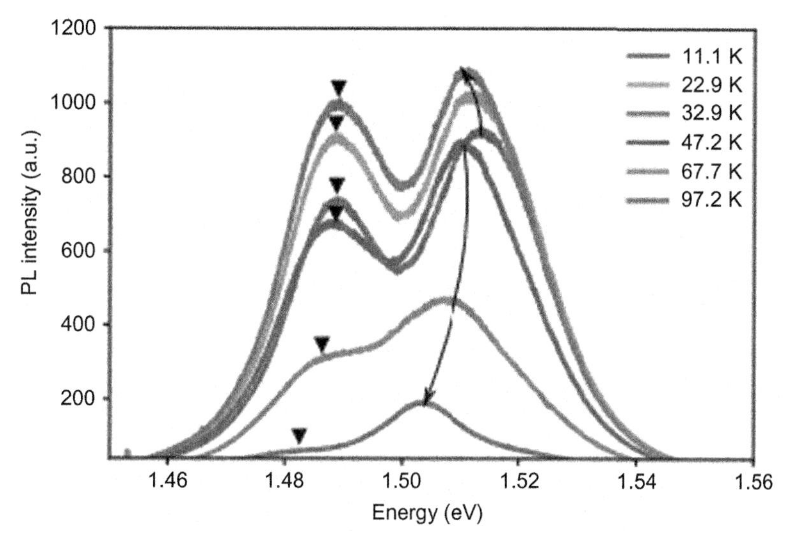

FIG. 2.15 Peak positions of bandgap and below bandgap transitions of GaAs NW [37].

TABLE 2.11 Photovoltaic parameters of devices [38].

Devices	V_{oc} (V)	J_{sc} (mAcm^{-2})	Fill factor (FF)	η (%)
P3HT-GaAs	0.51	0.10	0.54	0.03
Gr-GaAs (without P3HT layer)	0.60	12.44	0.61	4.63
Gr-GaAs (with P3HT layer)	0.72	16.59	0.56	6.84

Cristata, Saffron, Cynoglossum, and eggplant peel. A maximum energy conversion efficiency of 1.38% was achieved by DSSC using Celosia Cristata as dye with a maximum absorbance of 510 nm as shown in Fig. 2.16 [39].

A dye-sensitized solar cell was sensitized using single-step hydrothermally synthesized CDOTs. A power conversion efficiency of 0.10% was

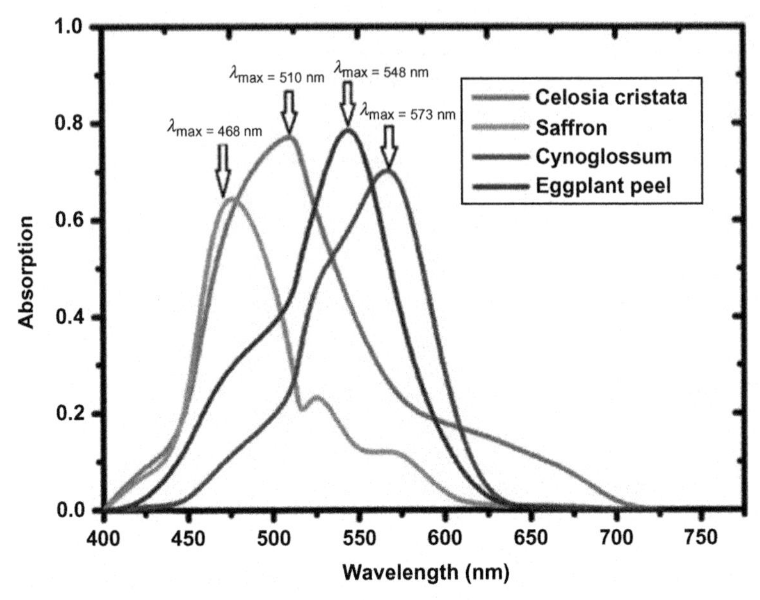

FIG. 2.16 UV-Vis spectrum of natural dyes [39].

TABLE 2.12 *I-V* characteristics of CDOT DSSCs [40].

Dye	V_{oc} (V)	J_{sc} ($mAcm^{-2}$)	Fill factor (FF)	η (%)
CDOT	0.49	0.89	0.22	0.10
N719	0.57	0.80	0.30	1.35
CDOT/ N719	0.89	0.89	0.24	0.19

observed by a carbon dot-sensitized solar cell while 0.19% efficiency was observed in the case of a CDOT- and N719-based dye solar cell, as shown in Table 2.12 [40].

ZnO- and TiO_2-based dye solar cells were fabricated using the peels of Malus Domestica, Punicagranatum, Phoenix dactylifera, Raphanus sativus, and Solanum melongena as natural sensitizers. A power conversion efficiency of 1.10% was observed by Solanum melongena dye using a TiO_2 photoelectrode DSSC, as shown in Table 2.13 [41].

A flexible quasisolid DSSC was reported using eco-friendly, economical stainless steel mesh with a photoelectrode of electrospun ZnO

TABLE 2.13 Photovoltaic parameters by various natural dye-sensitized TiO_2-DSSCs [41].

Name of peel	V_{oc} (V)	J_{sc} (mAcm^{-2})	Fill factor (FF)	η (%)
Malus domestica	0.32	4.48	0.59	0.85
Punicagranatum	0.32	4.50	0.70	1.008
Phoenix dactylifera	0.33	4.21	0.58	0.81
Raphanus sativus	0.33	3.12	0.71	0.73
Solanum melongena	0.33	4.51	0.74	1.10

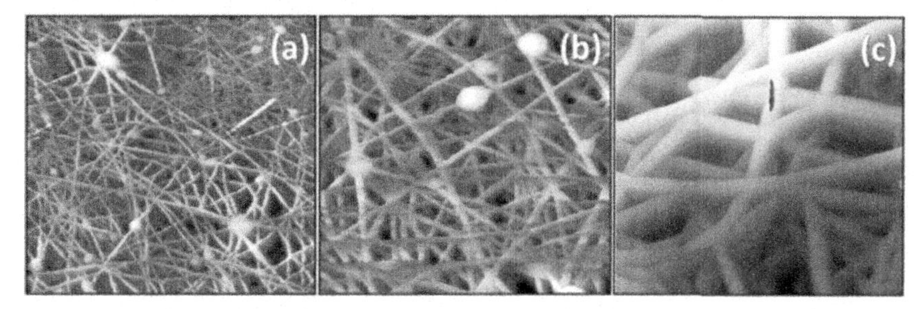

FIG. 2.17 (A)–(C) AFM images of the ZnO nanofibers [42].

nanofibers. The surface-to-volume ratio was improved because of the higher porosity of electrospun ZnO nanofibers having a 12–20 nm grain size. AFM images of the ZnO nanofibers are shown in Fig. 2.17. A power conversion efficiency of 0.13% was achieved [42].

2.3.3 Perovskite solar cells

A perovskite solar cell includes a perovskite structured compound as a light-harvesting active layer. Methylammonium lead halide materials are used as perovskite materials that are less costly to obtain and easy to fabricate. The power conversion efficiency with such materials was considered as 3.8% in 2009 [43]. The use of NiO nanoparticles as a hole transport layer in carbon-based solar polymer cells enhances stability and also increases the performance of the solar cell as compared to the simple carbon electrode that causes recombination problems. A scanning electron microscopy image of perovskite solar cells with a carbon/NiO deposition is shown in Fig. 2.18 [44].

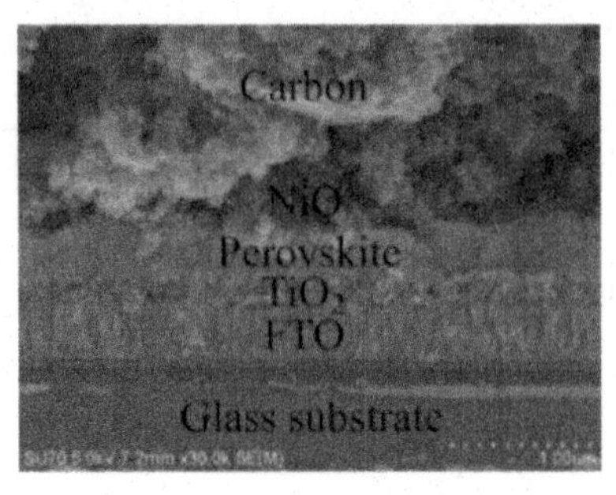

FIG. 2.18 Scanning electron microscopy image of a perovskite solar cell [44].

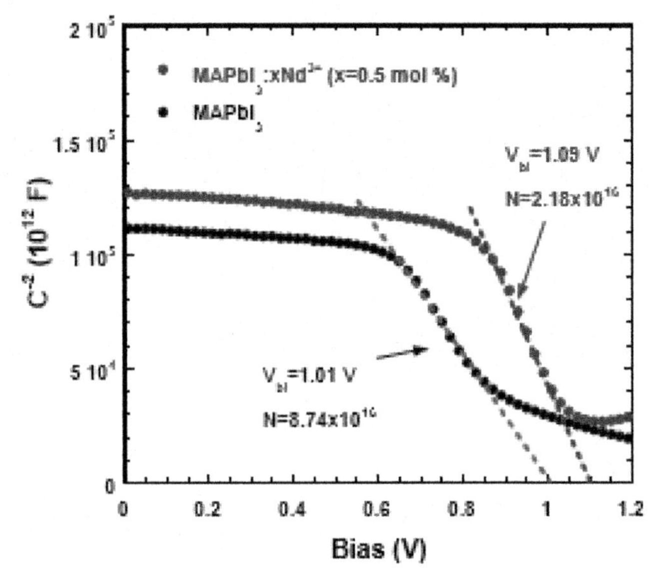

FIG. 2.19 Mott-Schottky analysis of the PSCs derived by either pristine MAPbI3 thin film or the MAPbI3:xNd3+ (x=0.5mol%) thin film [45].

Significantly, there was enhanced and balanced in charge carrier mobilities due to Nd3$^+$-doped hybrid perovskite materials (Fig. 2.19) that have better film quality with extremely reduced trap states [45].

The PV performance has been improved by the utilization of perovskite compound-based solar cells with metal-free phthalocyanine, silicon phthalocyanine, or germanium naphthalocyanine complexes. The best photovoltaic performance has been accounted for due to the small addition of silicon phthalocyanine into the perovskite layer, which suppresses defects and pinholes in the surface layer. Fig. 2.20 shows the fabrication process of the perovskite solar cell [46].

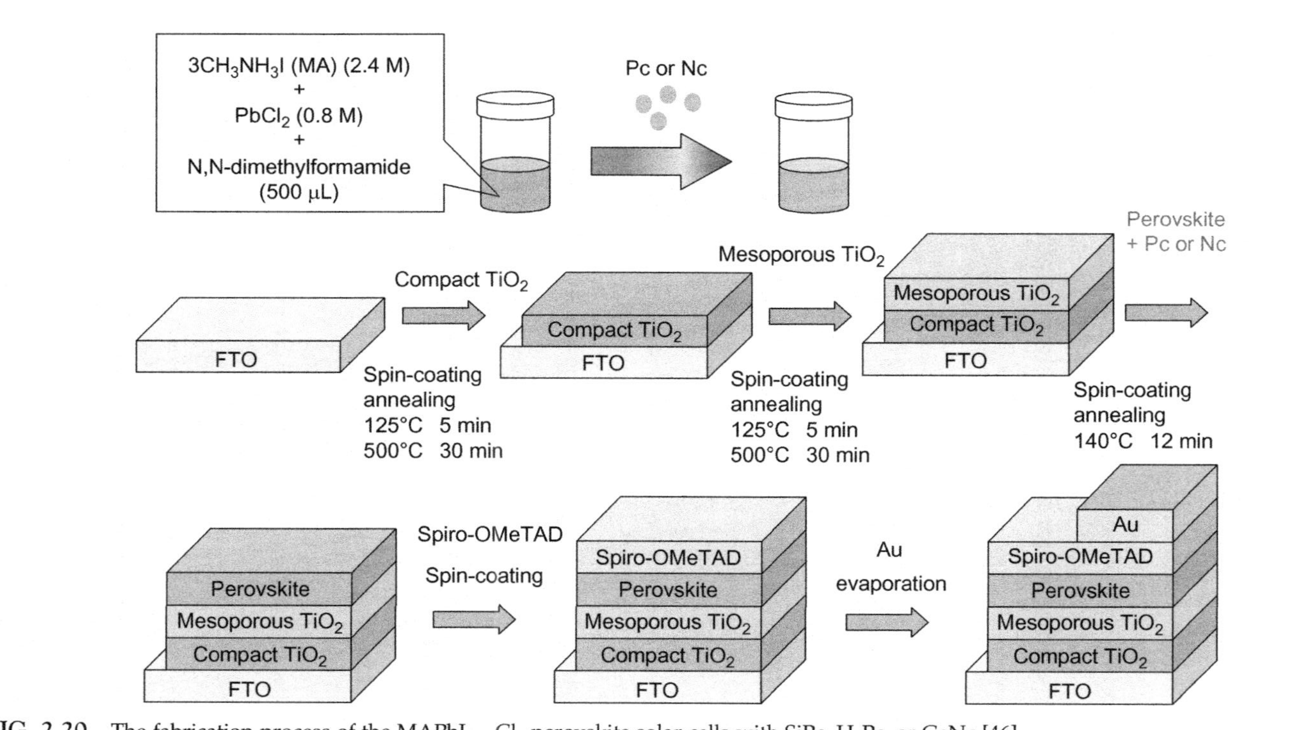

FIG. 2.20 The fabrication process of the $MAPbI_{3-x}Cl_x$ perovskite solar cells with SiPc, H_2Pc, or GeNc [46].

2.4 Conclusion

The creation of various solar cell technologies has been explored in this chapter and will continue to play a key role in the advancement of different technologies. The rising awareness of the need for sustainable and clean energy sources has positively helped in supporting the continuing research and development of solar cell technologies. Development of PV cells has been embraced by viable companies rather than by educational laboratories in case of traditional solar cell technologies. The same can be expected to be done with solar cell technologies that are similar to commercialization but for which current cell production is dominated by academia.

References

[1] Pandey AK, Tyagi VV, Selvaraj J, Rahim NA, Tyagi SK. Recent advances in solar photovoltaic systems for emerging trends and advanced applications. Renew Sust Energy Rev 2016;53:859–84.

[2] Li Y, He C, Lyu Y, Song G, Wu B. Crack detection in monocrystalline silicon solar cells using air-coupled ultrasonic lamb waves. NDT& E Int 2019;102:129–36.

[3] Kumar L, Hasanuzzaman M, Rahim NA. Global advancement of solar thermal energy technologies for industrial process heat and its future prospects: a review. Energy Convers Manage 2019;195:885–908.

[4] Al-Saqlawi J, Madani K, Dowell NM. Techno-economic feasibility of grid-independent residential roof-top solar PV systems in Muscat, Oman. Energy Convers Manage 2018;178:322–34.

[5] Laboratory NRE. Best research cell efficiencies, http://www.nrel.gov/ncpv/images/efficiency_chart.jpg; 2020.

[6] Tobnaghi DM, Madatov R, Naderi D. The effect of temperature on electrical parameters of solar cells. Int J Adv Res Elect Electro Inst Eng 2013;6404–7.

[7] Pavlovic T, Milosavljevic D, Radonjic I, Pantic L. Application of solar cells made of different Materials in 1 MW PV solar plants in Banja Luka. Contemp Mater (Renew Energy Sources) 2011;2:155–63.

[8] Qu Y, Huang X, Li Y, Lin G, Guo B, Song D, Cheng Q. Chemical bath deposition produced ZnO nanorod arrays as an antireflective layer in the polycrystalline Si solar cells. J Alloys Compd 2017;698:719–24.

[9] Wen C, Yang YJ, Ma YJ, Shi ZQ, Wang ZJ, Mo J, Li TC, Li XH, Hu SF, Yang WB. Sulfur-hyperdoped silicon nanocrystalline layer prepared on polycrystalline silicon solar cell substrate by thin film deposition and nanosecond-pulsed laser irradiation. Appl Surf Sci 2019;476:49–60.

[10] Ramos A, Filtvedt WO, Lindholm D, Ramachandran PA, Rodriguez A, Canizo CD. Deposition reactors for solar grade silicon: a comparative thermal analysis of a Siemens reactor and a fluidized bed reactor. J Cryst Growth 2015;431:1–9.

[11] Richhariya G, Kumar A, Tekasakul P, Gupta B. Natural dyes for dye sensitized solar cell: a review. Renew Sustain Energy Rev 2017;69:705–18.

[12] Lee TD, Ebong AU. A review of thin film solar cell technologies and challenges. Renew Sustain Energy Rev 2017;70:1286–97.

[13] Green MA. Thin-film solar cells: review of materials, technologies and commercial status. J Mater Sci 2007;18:15–9.
[14] Liu P, Yang S, Ma Y, Lu X, Jia Y, Ding D, Chen Y. Design of Ag nanograting for broadband absorption enhancement in amorphous silicon thin film solar cells. Mater Sci Semicond Process 2015;39:760–3.
[15] Lin Q, Lu L, Tavakoli MM, Zhang C, Lui GC, Chen Z, Chen X, Tang L, Zhang D, Lin Y, Chang P, Li D, Fan Z. High performance thin film solar cells on plastic substrates with nanostructure-enhanced flexibility. Nano Energy 2016;22:539–47.
[16] Schropp REI, Von der Linden MB, Ouwens JD, Gooijer HD. Apparent "gettering" of the Staebler-Wronski effect in amorphous silicon solar cells. Sol Energy Mater Sol Cells 1994;34:455–63.
[17] Chen S, Gong XG, Walsh A, Wei SH. Crystal and electronic band structure of Cu_2ZnSnX_4 (X=S and Se) photovoltaic absorbers: first-principles insights. Appl Phys Lett 2009;94.
[18] Yang J, Jo H, Choi SW, Kang DW, Kwon JD. All p-i-n hydrogenated amorphous silicon oxide thin film solar cells for semi-transparent solar cells. Thin Solid Films 2018;662:97–102.
[19] Fang J, Yan B, Li T, Wei C, Huang Q, Chen X, Wang G, Hou G, Zhao Y, Zhang X. Improvement in ultra-thin hydrogenated amorphous silicon solar cells with nanocrystalline silicon oxide. Sol Energy Mater Sol Cells 2018;176:167–73.
[20] Yu Z, Zhang X, Zhang H, Huang Y, Li Y, Zhang X, Gan Z. Improved power conversion efficiency in radial junction thin film solar cells based on amorphous silicon germanium alloys. J Alloys Compd 2019;803:260–4.
[21] Ulicna S, Isherwood JM, Kaminski PM, Walls JM, Li J, Wolden CA. Development of ZnTe as a back contact material for thin film cadmium telluride solar cells. Vacuum 2017;139:159–63.
[22] Li W, Cai Y, Wang L, Pan P, Li J, Bai G, Ren Q. Fabrication and characteristics of N-I-P structure amorphous silicon solar cells with CdS quantum dots on nanopillar array. Phys E Low Dimens Syst Nanostruct 2019;109:152–5.
[23] Kokal RK, Saha S, Deepa M, Ghosal P. Non-toxic configuration of indium selenide nanoparticles—$Cu_2ZnSnS_2Se_2$/carbon fabric in a quasi solid-state solar cell. Electrochim Acta 2018;266:373–83.
[24] Kim KB, Kim M, Lee HC, Park SW, Jeon CW. Copper indium gallium selenide (CIGS) solar cell devices on steel substrates coated with thick SiO_2-based insulating material. Mater Res Bull 2017;85:168–75.
[25] Gasanly NM. Far-infrared lattice vibration spectra of copper gallium (indium) ternary selenides (tellurides): a consequence of trivalent cation and anion interrelation. Optik 2017;143:19–25.
[26] Hsieh TM, Lue SJ, Jianping A, Sun Y, Feng WS, Chang LB. Characterizations of chemical bath-deposited zinc oxysulfide films and the effects of their annealing on copper indium gallium selenide solar cell efficiency. J Power Sources 2014;443–8.
[27] Enam FMT, Rahman KS, Kamaruzzaman MI, Sobayel K, Chelvanathan P, Bais B, Akhtaruzzaman M, Alamoud ARM, Amin N. Design prospects of cadmium telluride/silicon (CdTe/Si) tandem solar cells from numerical simulation. Optik 2017;139:397–406.
[28] Kaltenhauser V, Rath T, Edler M, Reichmann A, Trimm G. Exploring polymer/nanoparticle hybrid solar cells in tandem architecture. RSC Adv 2013;3:18643–50.
[29] Bagher AM, Vahid MMA, Mohsen M. Types of solar cells and application. AJOPR 2015;3:94–113.
[30] Zhang L, Liu X, Sun X, Duan C, Wang Z, Liu X, Dong S, Huang F, Cao Y. 4-Methylthio substitution on benzodithiophene based conjugated polymers for high open-circuit voltage polymer solar cells. Synth Met 2019;254:122–7.

[31] Lu J, Zhou H, Yu Y, Liu H, Peng L, He P, Zhao B. Benzothienoisoindigo-based polymers for efficient polymer solar cells with an open-circuit voltage of 0.96V. Polymer 2019;175:339–46.
[32] Tran HN, Kim DH, Park S, Cho S. The effect of various solvent additives on the power conversion efficiency of polymer-polymer solar cells. Curr Appl Phys 2018;18:534–40.
[33] Como MR, Balderrama VS, Sacramento A, Marsal LF, Lastra G, Estrada M. Fabrication and characterization of inverted organic PTB7: PC70BM solar cells using Hf-In-ZnO as electron transport layer. Sol Energy 2019;181:386–95.
[34] Xu Y, Sun L, Wu J, Ye W, Chen Y, Zhang S, Miao C, Huang H. Thiophene: an eco-friendly solvent for organic solar cells. Dyes Pigments 2019;168:36–41.
[35] Labrunie A, Habibi AH, Seignon SD, Blanchard P, Cabanetos C. Exploration of the structure-property relationship of push-pull based dyads for single-molecule organic solar cells. Dyes Pigments 2019;170.
[36] Ansari ZA, Singh TJ, Islam SM, Singh S, Mahala P, Khan A, Singh KJ. Photovoltaic solar cells based on graphene/gallium arsenide Schottky junction. Optik 2019;182:500–6.
[37] Vulic N, Goodnick SM. Analysis of recombination processes in polytype gallium arsenide nanowires. Nano Energy 2019;56:196–206.
[38] He H, Yu X, Wu Y, Mu X, Zhu H, Yuan S, Yang D. 13.7% efficiency graphene–gallium arsenide Schottky junction solar cells with a P3HT hole transport layer. Nano Energy 2015;16:91–8.
[39] Hosseinnezhad M, Rouhani S, Gharanjig K. Extraction and application of natural pigments for fabrication of green dye-sensitized solar cells. Opto-Electron Rev 2018;26:165–71.
[40] Ghann W, Sharma V, Kang H, Karim F, Richards B, Mobin SM, Uddin J, Rahman MM, Hossain F, Kabir H, Uddin N. Int J Hydrogen Energy 2019;44:14580–7.
[41] Pervez A, Javed K, Iqbal Z, Shahzad M, Khan U, Latif H, Shah SA, Ahmad N. Fabrication and comparison of dye-sensitized solar cells by using TiO_2 and ZnO as photo electrode. Optik 2019;182:175–80.
[42] Dinesh VP, Sriramkumar R, Sukhananazerin A, Sneha JM, Kumar PM, Biji P. Novel stainless steel based, eco-friendly dye sensitized solar cells using electrospun porous ZnO nanofibers. Nano-Struct Nano Objects 2019;19:100311.
[43] Mariska DWS. Energy payback time and carbon footprint of commercial photovoltaic systems. Sol Energy Mater Sol Cells 2013;119:296–305.
[44] Cai C, Zhou K, Guo H, Pei Y, Hu Z, Zhang J, Zhu Y. Enhanced hole extraction by NiO nanoparticles in carbon-based perovskite solar cells. Electrochim Acta 2019;312:100–8.
[45] Wang K, Zheng L, Zhu T, Yao X, Yi C, Zhang X, Cao Y, Liu L, Hu W, Gong X. Efficient perovskite solar cells by hybrid perovskites incorporated with heterovalent neodymium cations. Nano Energy 2019;61:352–60.
[46] Suzuki A, Okumura H, Yamasaki Y, Oku T. Fabrication and characterization of perovskite type solar cells using phthalocyanine complexes. Appl Surf Sci 2019;488:586–92.

CHAPTER

3

Solar PV module technologies

Nallapaneni Manoj Kumar[a], Shauhrat S. Chopra[a], Aline Kirsten Vidal de Oliveira[b], Hamsa Ahmed[c], Shima Vaezi[d], Uzoma Edward Madukanya[e], and Juan M. Castañón[e,f]

[a]School of Energy and Environment, City University of Hong Kong, Kowloon, Hong Kong, [b]Federal University of Santa Catarina, Florianopolis, Brazil, [c]Physics Department, Oldenburg University, Oldenburg, Germany, [d]Chemical Engineering Department, Islamic Azad University South Tehran Branch, Tehran, Iran, [e]Department of Microsystems Engineering (IMTEK), University of Freiburg, Solar Energy Engineering, Freiburg im Breisgau, Germany, [f]Symtech Solar Group, Lewes, DE, United States

3.1 Introduction

Photovoltaics (PV) are one of the fastest-growing energy generation options that we have in the modern energy sector. Because of favorable energy policies, PV is one of the main renewable energy sources to be adapted around the globe in trying to decrease greenhouse gas emissions in the energy sector. Its unique selling point of scalability and adaptability to any regional condition is added to the fast dissemination and successful implementation in different global regions. PV systems range from small isolated home systems to multimegawatt centralized power plants. These plants can be installed in hot, sandy climate areas as well as in cold regions with high altitudes. The adaption of PV to different needs and conditions has spurred the development of diverse PV technologies. Additionally, significant economies of scale have been achieved that have led to a rapid fall in PV system costs. In the last 5 years, module prices decreased by 8%

Photovoltaic Solar Energy Conversion

https://doi.org/10.1016/B978-0-12-819610-6.00003-X

from $0.53/W to an average of $0.49/W in 2019 [1]. A vivid scientific community provides technological innovations and a steep learning curve. These innovations are rapidly adopted and converted into new products. For example, the traditional aluminum-covered back surface solar cell (Al-BSF) is being replaced by passivated emitter rear contact (PERC) cells, achieving cell efficiencies of up to 25%. A decade ago, the average cell efficiency of a solar cell was at 16% [2]. The gain in efficiency is accompanied by falling PV module prices. This has led to gains in the efficiency of PV modules and the development of new materials and technologies. Apart from traditional crystalline silicon PV modules, much effort has been put into improving other PV technologies, such as thin-film or concentrating PV (CPV) modules. Nevertheless, crystalline silicon PV modules remain the main product for PV installations. In addition, a few novel solar PV modules have emerged recently, including flexible modules, bifacial modules, double glass modules, antireflection coated glass, light-capturing ribbons, light-reflective films, smart wire, multibus bars, and smart PV modules.

The chapter briefly explains the following concepts in subsequent sections:

- Performance characteristics of the PV module parameters
- PV module technologies and their classification
- Service life and reliability of PV modules with a focus on faults and their identification methods
- PV module testing methods
- Novel PV modules

3.2 PV module performance parameters

PV module performance parameters are evaluated based on *I-V* and *P-V* curves, where "*I*" represents the current, "*V*" represents the voltage, and "*P*" represents the power, as shown in Fig. 3.1. *I-V* and *P-V* curve tracing are the main methods for PV performance analysis. They determine the short-circuit current, the open-circuit voltage, the fill factor, and the maximum power point.

Various performance characteristic parameters are briefly defined below. The mathematical formulae used for evaluating PV module performance are highlighted in Table 3.1.

Open-circuit voltage (V_{oc}): This is the value of voltage measured under the STC or real-time operating conditions by putting the PV cell terminals in open-circuit condition. The value of the voltage recorded at this condition will be generally higher than the maximum voltage attained at the maximum point [3].

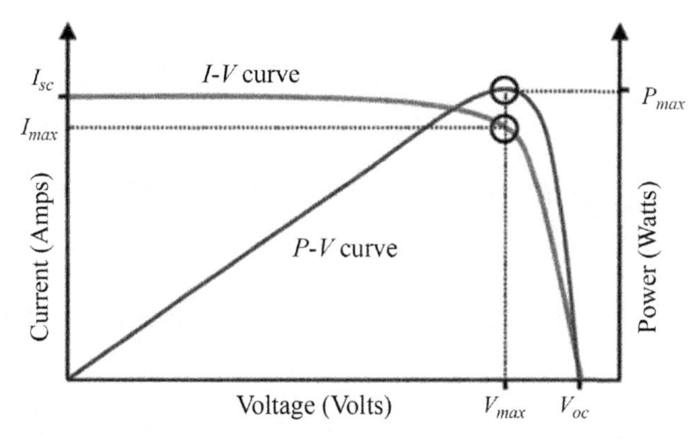

FIG. 3.1 *I-V* and *P-V* curves indicating the maximum power point as well as the open-circuit voltage and short circuit current [3].

TABLE 3.1 Description and the mathematical relation of major performance parameters [3–5].

Parameter	Mathematical relation	Description
Power input	$P_{in} = \text{Area} \times I$	It is the amount of solar energy potential available and is the product of the area of the PV module to the incident solar radiation. Units: kW
Power output	$P_{out} = V_{max} \times I_{max}$	It is the product of maximum current and maximum voltage at any given time. Simply referred to as the power output at the terminals. Units: kW
Power conversion efficiency	$PEC = \left[\frac{P_{out}}{(\text{Area} \times I)}\right] \times 100$	It is the ratio of power output to the power input. It generally represents the performance of PV cells. Units: %
Fill factor	$FF = \frac{P_{out}}{(V_{OC} \times I_{SC})}$	It is one factor that represents the quality of the PV cell. It is simply given as the ratio of power output to the product of open-circuit voltage and current Units: no

Short-circuit current (I_{sc}): The maximum current that is generated when a solar PV cell is made to operate in STC, or real-time conditions, is referred to as I_{sc}. While measuring this current, the PV cell terminals are short-circuited by means of some load. The current value at this point will be higher to the current that is achieved at maximum power point [3].

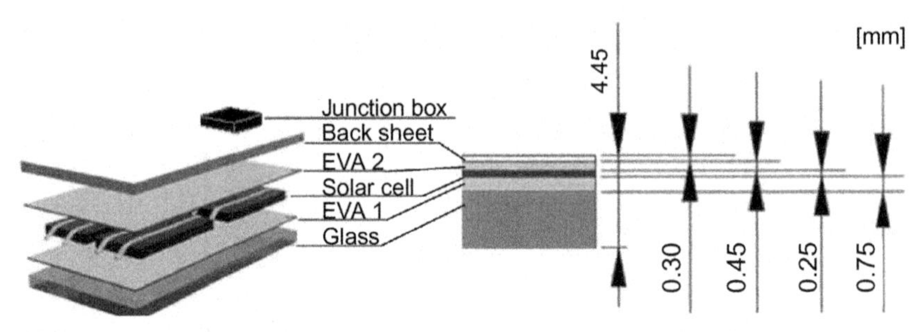

FIG. 3.2 Structure of a c-Si based PV module [6].

Maximum power point (M_{pp}): A PV cell is capable of producing maximum power at one particular point, and such a point is called the maximum power point. The voltage and current recorded at this point are generally termed the maximum voltage and maximum current. These values could vary based on solar radiation intensity as well as with other local weather parameters [3].

3.2.1 Crystalline silicon PV module

In this section, an overview of the crystalline silicon (c-Si) PV modules is provided. These PV modules are classified as the first generation of solar modules. At present, the PV market share is dominated by c-Si modules. Currently in the market, two different types of c-Si modules are available: monocrystalline (mono c-Si) and polycrystalline (poly c-Si). These modules are fabricated by joining the c-Si PV cells in series and parallel configurations [3]. Considering the cell level structure, c-Si PV cells consist of silicon slices that are generally visible as wafers. These wafers are used to fabricate the c-Si cells, which are then mechanically assembled by providing the electrical contacts. A simple stricture of the crystalline silicon PV module is shown in Fig. 3.2 [6].

3.2.1.1 *Types of crystalline PV modules*

There are two types of thin-film modules:

Monocrystalline silicon (mono c-Si): This type of c-Si module is widely used and will continue to be the leader of the PV market. At present, these modules seem to be readily available and the existing benefits are numerous. The only major driving factor is the low cost. The cells used in manufacturing c-Si modules consist of p-doped wafers with p–n junctions. At first, in the fabrication process, the c-Si ingot is manufactured. This c-Si ingot is then sized into wafers with a size less than 0.3mm. This forms the overall solar cell structure, which is then capable of generating 35mA current, 0.55V under the full illumination conditions. These modules generally have a special textured surface that most commonly looks like a pyramid structure. Depending upon the required voltage and currents, such solar cells are grouped together to form a PV module. While fabricating the module, these cells are arranged in series and parallel configurations with conductive contacts [7].

Polycrystalline silicon (poly c-Si): The next type of c-Si PV module is the poly c-Si, whose market is a bit lower when compared to mono c-Si. The problem of metal contamination exists in mono c-Si and to limit such problems, the industry has come up with poly c-Si cells. Like the mono c-Si, here also modules are fabricated by arranging cells in series and parallel configurations. The cells are manufactured such that they have different crystal structures. Melting and solidifying silicon takes place. In this process, crystals that orient only in one direction are produced, and are then made into thin blocks and then wafers. These solar cells have random structures on their surfaces, and they come with an additional layer that minimizes the light reflection [7].

3.2.2 Thin-film PV modules

In this section, an overview of the thin-film solar modules is provided. Thin-film PVs are classified as the second generation of solar systems. A thin-film module consists of several solar cells that are wired together. Fig. 3.3 shows a generic anatomy configuration of a thin layer solar cell. A typical thin-film solar panel is composed of the semiconductor sandwiched between two sheets of glass and sealed with an industrial laminate. Usually, an antireflecting coating is used to reduce the reflection

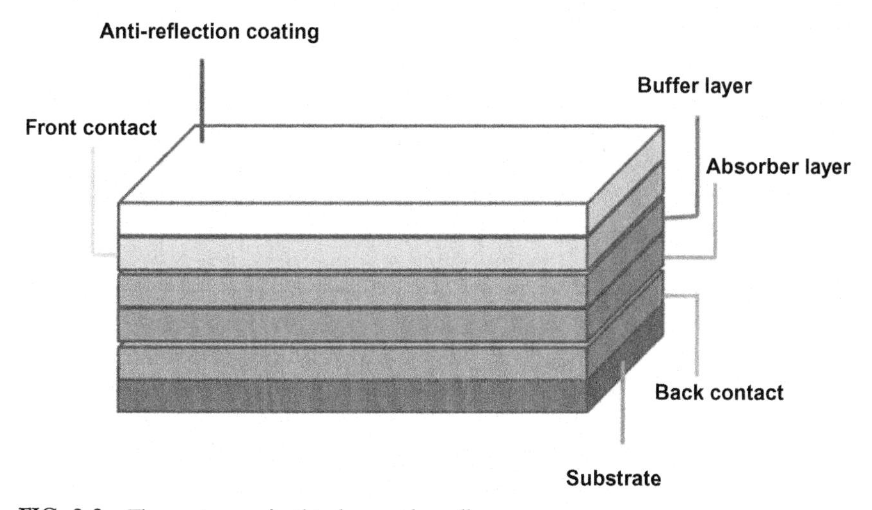

FIG. 3.3 The anatomy of a thin layer solar cell.

of light from the surface of the panels. Thin-film modules are defined as modules that use several thin absorbing layers, which are more than 300 times less than conventional crystalline silicon (c-Si). The thickness of the PV active layers varies from a few nanometers to tens of micrometers. One of the most important goals to be achieved by manufacturing the thin-film modules is reducing the high cost needed to produce the monocrystalline and polycrystalline silicon modules. However, thin-film modules have lower power and durability than the first-generation c-Si solar panels.

3.2.2.1 *Types of thin-film PV modules*

There are several types of thin-film modules:

Cadmium telluride (CdTe): This PV cell has a 1.45eV direct bandgap, allowing high absorption of light. This makes CdTe a leading thin-film technology, as it occupies half of the thin film available market. The most common techniques used for deposition of the CdTe thin layer are the closed-space sublimation (CSS), simple evaporation, electrodeposition, chemical-vapor deposition, metal-organic spray deposition, and screen-print deposition [8]. The most efficient modules are produced by the CSS technique. The highest recorded efficiency for the CdTe modules is $18 \pm 0.5\%$ in 2015 [9]. One of the main drawbacks of these modules is the toxicity of the Cd material.

Copper indium gallium selenide (CIGS): It is a direct bandgap semiconductor material that has a chalcopyrite crystal structure and polycrystalline interchangeable material. It has the stoichiometry of $CuIn_{(1-x)}Ga_{(x)}Se_2$, where x could vary from 0 to 1, depending on the amount of gallium and indium; hence, the bandgap can be tuned from 1.0 to 1.7eV. The record efficiency for this module is $19 \pm 0.5\%$ [9]. Different processes are involved in growing CIGS such as the coevaporation process and the sequential process. In order to get high-quality film and fewer shunts, selenium-rich content and copper-poor amount less than 25at.% is used. One of the most efficient approaches to boost the efficiency of this technology is grading the bandgap by varying the amount of Ga and Se. Another approach to improve the efficiency of CIGS is incorporating alkali material such as sodium in the CIGS layer, as Na passivates the defects in the grain boundary; this can be considered one of the main issues that limits CIGS efficiency [10].

Amorphous Si (a-Si:H): It is the noncrystalline form of Si. The bandgap can take values from 1.7 to 1.9eV [11]. The deposition techniques that are used to fabricate a-Si are magnetron evaporation, plasma-enhanced chemical vapor deposition (PECVD), and very high-frequency CVD. a-Si:H has several advantages including the ease of gas doping, stacking ability without material restrictions, and the ability of deposition on flexible substrates. Its drawbacks include sensitivity to the high solar irradiation in addition to the short diffusion length and the deficient charge mobility. The a-Si solar module efficiency is 9.8% [11].

Gallium arsenide (GaAs): This is a direct bandgap system 1.43 ev technology. The well-known deposition techniques in this technology are epitaxial growth, close-spaced vapor transport (CSVT), metal-organic vapor phase epitaxy (MOVPE), and the recently used cheap technique, hydride vapor phase epitaxy (HVPE) [12]. The module efficiency could reach up to $25.1 \pm 0.8\%$. Recently, a few other promising thin-film technologies have been developed on the lab scale or in the early stages of commercialization such as organic, perovskite, copper zinc tin sulfide, and quantum dot solar cells.

3.2.2.2 *Advantages of thin-film modules*

Generally, the advantages of thin film over other module types are:

- Less material usage due to the fact that they consist of a thin film rather than bulk material; as a consequence it is less expensive. However, thin-film modules require a larger space.
- High temperature and low light tolerance: Thin-film solar modules have a low-temperature coefficient compared to a-Si panels. Thin-film modules are less influenced by environmental effects such as low light radiation and shading.
- Flexibility: one of the great advantages of thin-film modules is that their shape can be flexible as they do not need a rigid substrate, which may widen their applications.
- Each thin-film module type has its own pros and cons. In solar PV module technologies, the trade-off between the efficiency and cost exists, such as in the comparison between CIGS and GaAs. For environmental concerns, the toxicity of Cd in CdTe modules has to be taken into account.
- Thin-film market and increased production: Although solar PV power represents a small fraction of the total renewable energy, many countries are focusing on its economic potential. Fig. 3.4 shows the growing installation of global solar PV. In 2013, almost 140 GW was installed worldwide [13]. This is related to the reduced costs of the modules and batteries. China achieved the largest amount of solar PV installations (13 GW in 2013) [13]. However, during the same period, PV installations in Europe decreased. The market share of all thin-film modules is about 5% (4.5 GWp) of the global total annual production in 2017 [14]. In 2017, contributions to total PV production from CdTe, CIGS, and a-Si were 2.3 GW, 1.9 GWp, and 0.3 GWp, respectively. The total annual production of CdTe and CIGS modules has increased since 2011 while the production of a-Si has decreased [14].

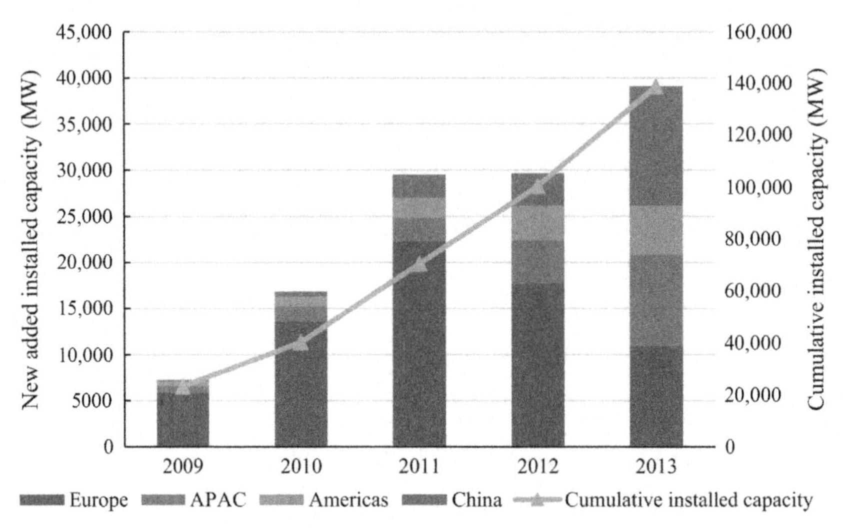

FIG. 3.4 World cumulative PV installed capacity from 2009 to 2013. *Data Resources: Global Market Outlook for Photovoltaic 2014–18, European Photovoltaic Industry Association (EPIA) Zou H, Du H, Ren J, Sovacool BK, Zhang Y, Mao G. Market dynamics, innovation, and transition in China's solar photovoltaic (PV) industry: a critical review. Renew Sustain Energy Rev 2017;69 (November 2016): 197–206.*

3.2.3 Concentrating PV modules

Concentrating PV (CPV) makes use of mirrors and optics to amass high levels of sunlight, contrary to traditional flat crystalline silicon modules that exclude any concentrator in their design. At present, the technology has matured and achieves the highest level of efficiency. At the same time, the cost of these modules is very high, which makes them unaffordable for conventional PV applications. However, in order to justify their cost, these modules are mostly used for space applications. The solar cells that are typically used in high concentrator photovoltaics (HCPV) are multijunction cells belonging to the III-V semiconductor group (alloy, containing elements from groups III and V in the periodic table), making this technology a more complex product to mass produce. This chapter will briefly describe the low- and high-concentration PVs (LCPV and HCPV) and compound parabolic concentrators (CPCs). Since the inclusion of PV into the electricity market, various types of systems and technologies have emerged. CPV modules are typically made with high-efficiency multijunction cells, which are optically addressed in order to concentrate the foremost amount of sunlight [15]. Although CPV modules have a reduced cell area, these modules work at preferred optimal conditions in geographic areas that are exposed to high levels of direct normal irradiance (DNI), typically at values over 2000 kWh/(m^2a) [16]. CPV modules concentrate the incoming irradiance by the use of collectors and mirrors.

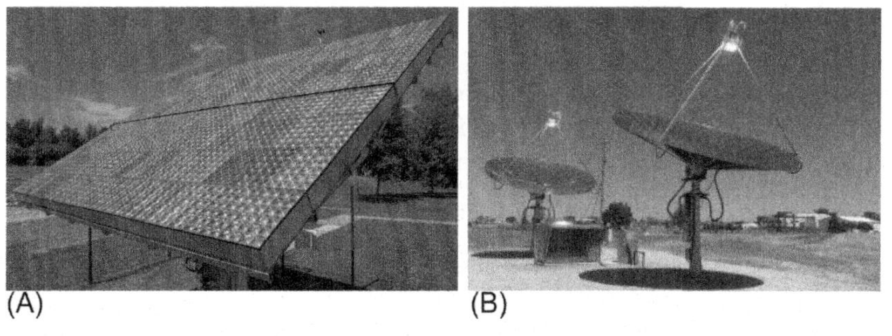

FIG. 3.5 (A) Solar concentrator including point-focus Fresnel lenses; (B) parabolic dish concentrator.

Due to the nature of their optics, these can include Fresnel lens, Cassegrain primary reflectors, and waveguiding designs [17]. These will depend on the design strategy of the manufacturer. A limitation of CPV systems would be the addition of active components such as a sun tracker, which is required by these systems to follow the position of the sun throughout the day. CPVs can be classified into two main levels of concentrators: HCPVs and LCPVs. HCPVs include solar cells from the III-V semiconductor group and are built with double-axis sun trackers, concentrating the light with an intensity magnitude between 300 and 1000 suns. LCPV systems are composed of c-Si or other semiconductor materials and may include double or single-axis trackers concentrating the light with an intensity in the magnitude below of 100 [18, 19]. Fig. 3.5 shows the typical examples of solar concentrators as well as the parabolic dish concentrator. In the below section, three different types of CPVs are briefly discussed.

3.2.3.1 *High concentrator PV (HCPV) modules*

HCPV modules have achieved high levels of module efficiency due to their concentrator design and type of cell (multijunction). Some efficiencies recorded have reached the 40% mark. In recent years, these systems have demonstrated the potential of being a financially viable option due to the reduction of their footprint when installed while utilizing fewer semiconductor materials for their cells, hence offering a balance to the overall cost [20].

3.2.3.2 *Low concentrator PV (LCPV) modules*

LCPV modules concentrate the light intensity below the factor of 100 suns. However, these modules are more cost-effective than HCPV and can be very reliable if their design allows good uniform irradiance collection and temperature control. These modules are made in some occasions with c-Si cells, which make them a very attractive option for the market [20].

3.2.3.3 *Compound parabolic concentrators (CPC)*

CPCs are the nonimaging category of CPVs. They are designed in a way that the reflecting ratio of the incident light onto the absorber is very high. They are capable of having higher efficiencies as they try to focus the light within wide limits. Because their design is concave, these systems do not require tracking. By concentrating a higher level of irradiance, these collectors improve system efficiency and at the same time reduce costs for PV and thermal applications. The current installed capacity of this type of CPV plant is circa X MWs [20].

3.3 Service life and reliability of PV modules

This section discusses the various common defects and faults seen in PV modules and methods to identify them. In addition, PV module durability, quality control, and service life prediction concepts are discussed.

3.3.1 Common defects and faults in PV modules

A fault is defined as anything that might decrease the PV module performance or change its characteristics. On the other hand, a defect is an unexpected thing that was not observed before on the module and does not necessarily imply power loss. Table 3.2 summarizes defects that are commonly found in PV modules.

3.3.1.1 *Methods for defect and fault detection on PV systems*

With the more widespread adoption of PV technology, the importance of O&M of power plants increases. An investigation of novel monitoring methods is a critical issue for solar energy market growth. With the advance of data analytics, sensors, and artificial intelligence, one applied solution is a system supervisory platform. However, these systems normally measure power, current, and voltage of each string, making it hard to detect faults that affect only one module. They also cannot detect nonenergy or latent risk factors such as hot spots or substring failures that do not reduce power production in the short term but can cause damage to the system in the long run. Therefore, onsite inspections are still necessary. The most traditional diagnostic tools to detect defects and failures on PV systems are visual inspection, *I-V* curve measurements, and infrared thermography analyses. Recently, new methods have gained attention because of their lower costs, such as electroluminescence (EL). The following subsections describe some of these methods.

Visual assessment: It is the first step and also the quickest and most effective evaluation tool/technique. Visual inspection can detect yellowing/browning,

TABLE 3.2 Common defects and degradation of PV modules [21, 21a, 21b].

Defect	Possible reason	Effect on PV module
Faults found in all PV modules		
Delamination	Problem of adhesion between the glass, encapsulant, active layers, and back layers, caused by contamination or environmental factors	Humidity ingress, corrosion, reduced module performance, isolation fault
Back sheet adhesion loss	Mechanical stress, UV radiation, heat, humidity	Corrosion, mechanical stress, risk of electric arcs
Junction box failure	Poor fixing, poor manufacturing, bad wiring	Higher resistance, risk of thermal damage
Frame breakage	Snow loads or other mechanical loads and extreme weather conditions	Humidity ingress, reduced module performance, isolation faults
Encapsulation discoloration	UV radiation, humidity, temperature	Higher light absorption in polymer, decreasing module performance
Encapsulation embrittlement	UV radiation, temperature	Higher stress in solar cells, cell breakage, module delamination
Cell cracks	Mechanical stress induced by mechanical or thermal loads	Insulated cell areas, reduced module performance
Broken interconnection	Mechanical stress induced by mechanical or thermal loads	Higher resistance, insulated cells, and strings, risks of hot spots
Light-induced power degradation (LID)	Light exposure	Reduced module performance
Series and shunt resistance deviations	Series resistances in the module, mismatch of the individual cell characteristics, shunts, interconnections resistance, corrosion, glass problems, delamination, broken cells	Hot spots, reduced module performance, fire hazards

Continued

TABLE 3.2 Common defects and degradation of PV modules [21, 21a, 21b]—cont'd

Defect	Possible reason	Effect on PV module
Faults found in silicon wafer-based PV modules		
Snail trails	Gray/black discoloration of the silver paste of the front metallization of screen-printed solar cells	Higher resistance, insulated cell areas, reduced module performance
Burn marks	Solder bond failure, ribbon breakage, thermal fatigue	Higher resistance, insulated cell areas, reduced module performance
Potential induced degradation (PID)	Migration of ions from the front glass through the encapsulant to the antireflective coating at the cell surface driven by the leakage current in the cell to the ground circuit	Hot spots, low shunt resistance, module delamination, reduced module performance
Disconnected cell or string interconnect ribbons	Problems in the manufacturing process, mechanical stress, thermal stress	Bypass diode activation, reduced module performance, hot spots, glass breakage
Defective bypass diode	High voltage discharges and mechanical stress	Hot spots, back sheet burns
Broken glass	Previous glass damage, mechanical load, thermal stress	Corrosion, loss of mechanical stability, cell breakage
Faults found in thin-film PV modules		
Micro arcs at glued connectors	Problems in the manufacturing process (lack of pressure on the connection area in the conductive gluing)	Reduced module performance (affects mainly the FF of the *I-V* curve)
Shunt hot spots	Problems in the manufacturing process, reverse bias operating of cells (caused by shading or soiling)	Hot spots, reduced module performance, glass breakage, loss of mechanical stability
Back contact degradation	Poor manufacturing, high temperatures, prolonged open-circuit conditions	Reduced module performance

Adapted from Köntges M, et al., Review of failures of photovoltaic modules; 2014, Aghaei M, Novel methods in control and monitoring of photovoltaic systems. Politecnico di Milano; 2016, Renganathan RMV, Natarajan V, Jahaagirdar T, Failure mechanism in PV systems. Grad. course MAE 598 Sol. Commer; 2016. pp. 1–17, Rajput P, Malvoni M, Kumar NM, Sastry OS, Tiwari GN. Risk priority number for understanding the severity of photovoltaic failure modes and their impacts on performance degradation. Case Stud Therm Eng 2019; 100563.

delamination, bubbles, cell cracks, misalignments, bad connections, and burnt cells. The inspection is performed from the front and the back of the module and can detect faults such as bubbles, delamination, yellowing, browning, broken or cracked cells, and burned or oxidized cell metallization. Other faults can be detected on the frame that can be bent, broken, scratched, or misaligned and on a junction box that can be loose or oxidized. The wires and connectors are also observed to detect detachments, brittleness, and insulation ruptures. For the assessment, the standards require more than 1000 lx of natural illumination [21, 22].

I-V curve: Historically, *I-V* curve tracing or electrical inspections are the main methods for PV performance analysis. They determine the short-circuit current, o the pen-circuit voltage, the fill factor, the maximum power point, and other parameters. As the reduction of power production is the main evidence of a problem, an *I-V* measurement curve gives sufficient information about the PV module condition. Fig. 3.6 presents the *I-V* curve of a module with a bypassed substring. The performance of the faulty module is reduced by 35.8% in comparison to a neighboring healthy module (reference module).

The *I-V* measurement curve gives enough information about the PV module condition. Fig. 3.2 presents the *I-V* curve of a module with a bypassed substring. The performance of a PV module would differ based on the condition of the module. Here, the condition refers to whether it is faulty or experiencing some adverse situation that would have an impact on the overall power outputs. The performance of the faulty module is reduced by 35.8% in comparison to a neighboring healthy module (reference module).

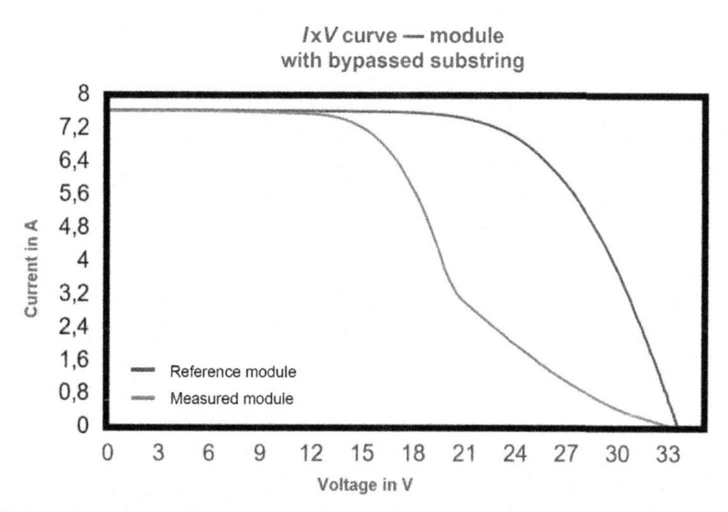

FIG. 3.6 *IV* curve of a PV module with a bypassed substring in comparison with the *IV* curve of a healthy module.

Normally, *I-V* curves are measured with solar simulators under standard test conditions (STC—cell junction temperature—25°C, irradiance—1000 W/m^2, the spectral distribution of irradiance air mass—1.5). They can also be performed under real sunlight exposure, where the PV modules do not meet the STC during the test measurement, with measurements corrected for STC. Usually, the radiation is less than 1000 W/m^2, the cell junction temperature is higher than 25°C, and the air mass is hardly ever 1.5. The measured curves can be corrected to STC and can be used as an ideal reference for comparison of different PV modules [21–23].

In power plants, these inspections can be performed by an *I-V* curve tracer on all or a subset of strings on a site. As the method is expensive and labor-intensive to be applied to an entire site, a typical power plant has 25% or less of its strings verified per measurement campaign. This means that many defects are ignored. The inspection also requires the opening of the fuse circuit, leading to a higher likelihood of a fuse failure [24]. For *I-V* characteristic measurements, there are many aspects affecting measurement accuracy, such as calibration of sensors, varying meteorological conditions during the inspection, variable soiling of modules, module mismatch, and interstring temperature variations.

Infrared thermography (IRT): This measures the radiation emitted by the surface of anything in the infrared wavelength spectrum between 1.4 and 15 μm. The infrared thermography employed for PV applications usually detects wavelengths in the mid-infrared wavelength range of 7–14 μm, which is an equilibrium between costs, availability, and measuring conditions of IRT sensors [25, 26]. There are basically three different types of thermography: the steady-state thermography, often called "passive thermography," and two "active thermography" techniques, pulse and lock-in thermography, that evaluate the time dependence of temperature distributions [27]. The three techniques are described as follows:

Thermography under steady-state conditions: Buerhop et al. [28, 29] showed that IRT is a reliable fault diagnosis method that requires minimum instrumentation and can be applied without interrupting the operation of the PV plant. IRT is a nondestructive and contactless method that can provide information about the exact physical location of an occurring fault, which allows for posterior electrical diagnosis of the problem. As in a PV plant, all PV modules receive in principle the same amount of irradiance, and the modules or cells that are not converting photons into electricity will turn into heat. Therefore, energy losses will appear in IRT images as temperature differences. IRT can be used for the detection of a great number of defects in PV cells, modules, and strings because the majority of the defects have an impact on the cell thermal behavior. Such thermal patterns have been identified and classified in previous studies [21, 23, 28, 29] and are now standardized in the international standard IEC TS 62446-3 Edition 1.0 2017-06, as shown in Table 3.3. IRT under

TABLE 3.3 Examples of thermal abnormalities in PV modules.

Thermogram	Description	$\triangle T$ (K)	Recommendation
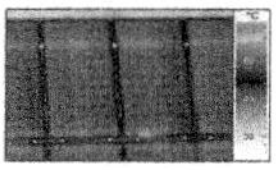	Suspicious conductor strip	>3K	• In-depth visible check. • Evaluation by a level 2 certified thermographer based on a detailed high-resolution backside thermogram/photo.
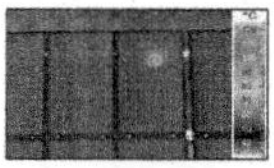	Overheated cell	>20K	• Check for shading or soiling effect. • Replace the module.
	Module with broken front glass	0–7K	• Replace the module.
	Heated module junction box	>3K higher temperature compared to nearby junction box	• Personal review by a PV expert or thermographer level 2.
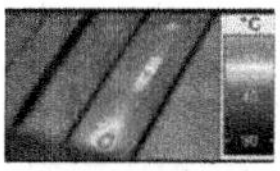	Substring in short circuit	Averaged 2–7K higher than substring	• Check module and bypass diodes for proper function under reverse biasing.
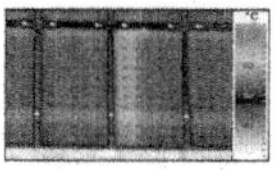	Bypassed substring (open-circuited)	2–7K	• Replace the module.
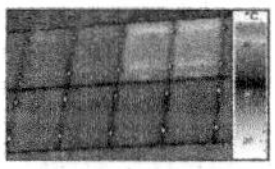	Several substrings bypassed	2–7K	• Replace the module.
	Modules in open circuit	2–7K	• Check module. • Check the state of operation of the inverter. • Check the condition of cabling, connectors, and fuses.
	Modules in short circuit	2–7K	• Check module. • Check cabling.
	Polarity of connector reversed	3–12K	• Check polarity of module connectors. • Check polarity of string connectors.
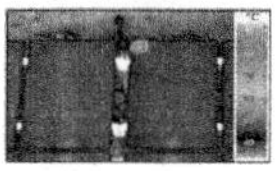	Single-cell is partly shaded (not a defect)	20–70K	• Remove the object that is causing the shadow.

steady-state conditions is normally performed with a portable uncooled IR camera [30, 31]. The measurements are performed under clear sky conditions with minimal irradiance of 700 W/m^2 on the plane of the PV array under inspection. The angle of the IRT sensor should be oriented perpendicular to the PV modules, and shading and reflections should be avoided. In order to increase the cost-effectiveness and employ IRT for large-scale PV plants or roof-mounted PV systems with limited access, IRT can be combined with aerial technologies such as unmanned aerial vehicles (UAVs) [32].

Pulsed thermography is a technique that can detect faults on interconnections as well as inhomogeneities on material properties and on the underlying layers and structures through an opaque back sheet [21]. It basically consists of heating the PV module from 1 to 5 K and then registering its temperature decay curve. It is performed using an external pulsed heat source, for example, a flashlight, to create dynamic heat flux on the module surface. The IR images are taken from the rear side of the module with a high acquisition frequency camera (more than 100 Hz). The faults will affect the dynamic local heat flux and are detected with the time evaluation of the surface temperature [27, 33]. The lock-in thermography is similar to the pulsed thermography, but in this technique, the heat is provided periodically in a lock-in frequency. The local surface temperature modulation is then evaluated and averaged over a number of periods. For the detection of different faults, several measurements applying different lock-in frequencies are performed. Therefore, the measurement time is considerably longer [27]. The lock-in thermography can detect shunts, interconnect problems, bubbles, and cracks [21].

Electroluminescence (EL) and photoluminescence (PL) imaging techniques: EL and PL characterizations are noninvasive methods for the detection of cell and module defects at any stage in the lifetime of a PV module. These techniques provide local information about shunts, contact and series resistances, microcracks and snail trails, inactive cell areas, inhomogeneity of materials, and potential induced degradation [34]. Electroluminescence is obtained from photons emitted by the radiative recombination of charge carriers excited under forwarding bias. Typically, an external power supply electrically stimulates the module under investigation to emit the electroluminescent radiation. For c-Si, the radiation ranges in the near-infrared at wavelengths around 1150 nm. Cell defects locally reduce or prevent the emission of EL radiation and are thus revealed and visualized as black areas by cameras that are sensitive to these wavelengths. Usually, such measurements are carried out in the dark [22, 34–36]. Fig. 3.7 shows an EL image of a PV cell presenting some cracks that are not affecting the current flow over the cell at the present stage, but which might evolve into active defects. The excitation can also be obtained by means of radiation

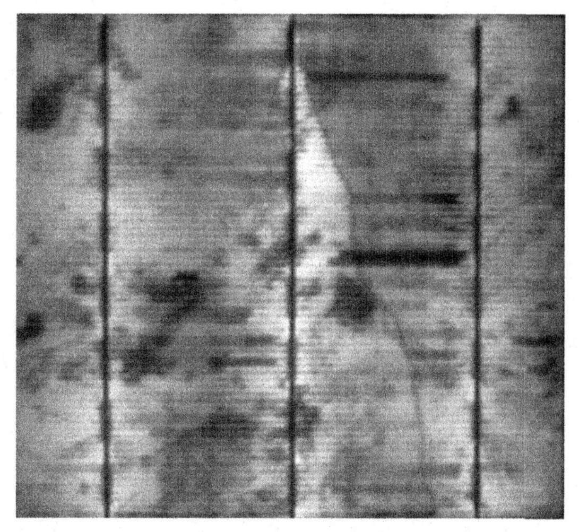

FIG. 3.7 EL image of a PV cell with cracks and dead fingers [22].

incident over the solar cell, in which case the light obtained is due to the PL effect [22].

UV fluorescence: UV fluorescence is a method capable of detecting degradation effects of the polymeric encapsulant of PV modules on the field. The measurements are performed in a dark environment under UV light, which reacts with the encapsulant and generates a fluorescence that can be detected by a photographic camera. The fluorescence does not occur along cracks or edges of a PV cell, due to the reaction to the oxygen that infiltrates the semiconductor through the encapsulant in these areas. The difference between fluorescent and nonfluorescent areas of a PV cell is used to detect cell cracks and to differentiate old cracks from new ones [37, 38].

3.3.1.2 PV module durability, quality control, and service life prediction

The long durability of PV modules guaranteed by the manufacturers is claimed to be their main advantage. Currently, some manufacturers offer warranty for up to 30 years, which means that during this period the module will keep safely producing energy with a small degradation within the range previously specified. The lifetime of a PV module depends on its reliability, which refers to the probability that an item will perform a required function and does not fail prematurely; its durability, which refers to continuous degradation that results in unacceptable levels of production reduction; soiling; incorrect installation; and many other issues. These characteristics have a direct effect on cost-effectiveness, energy payback balance, and public acceptance of PV energy. Therefore, the module durability, the quality assurance, and the correct lifetime prediction of the PV module are very important fields of study [39, 40].

For quality assurance in PV modules, several organisms have design methods and protocols. The lifetime of a module is normally predicted by a set of accelerated stress tests that mimic the real operation conditions. After installation, PV modules are exposed to mechanical loads, thermal cycling, UV exposure, humidity, soiling, and shading, which intensify degradation mechanisms such as encapsulation degradation, delamination, corrosion, etc. [39]. The tests can have a destructive or a nondestructive character, and simulate these stress factors using damp heat, thermal cycling, load tests, vibration tests, and others [41].

The International Electrotechnical Commission (IEC) is the main organization that established qualification, performance, and safety standards for PV modules (IEC 61215, 61646, 62108, 61730, 60904, 60891, 61724, 61829, 61853, 62670 and 62892) [31]. The IEC 61215, for example, specifies the procedures for qualifying commercial PV modules and includes UV exposure, thermal cycling in the climatization chamber, humidity freeze cycling, damp heat, pressure, hail impact test, outdoor exposure, hot spot tests, and others. The tests are performed to a sample of PV modules and they are considered qualified if no major failures appear after the tests [39].

3.4 PV module testing and standards

As the penetration of PV power in the electric grid continues on a global scale, standardization and testing become of utmost importance. For investors, reliable performance data are important because minor deviations from the expected energy yield can have a great impact on the financial revenue of PV projects. Because the upfront costs of these projects are considerably high, one seeks to have an optimum cost-benefit relationship. In the case of overestimation, the capital expenditure is higher than needed, as inverters, cabling, and other balance-of-system components are oversized. Consequently, the investor's expectations cannot be met. If the PV module performance is underestimated, the produced energy cannot be handled, and the inverter will cut the excess energy to maintain its designated alternative current (AC) output. The technical term for this behavior is "clipping," and it also causes losses in revenue to the investor. Additionally, the degradation of the PV module is accelerated as excessive power is dissipated in the form of heat. In addition, degradation of the PV module will have an influence on the economic and environmental benefits [42]. The grid operator also needs to have reliable data on the installed capacity for the proper operation of the electric grid. As the electric grid always needs to be balanced, he has to provide for sufficient reserve capacity to level out energy fluctuations. For an accurate rating of PV module performance, precise measurement procedures under STC conditions have been established. For more convenient handling of the measurement procedures, the standard cell temperature is defined as 25°C. Other industry standards that define certain conditions for module performance testing are PVUSA test condition (PTC), high-temperature conditions (HTC), low-temperature conditions (LTC), and nominal operating cell temperature (NOCT). As there is no system that might set standards on a global scale, the solar industry has adopted IEC norms as de facto standards. Through Committee 82, which relies on volunteers drawn from academia, government, and industry, the International Electrotechnical Commission provides for the constant actualization of these standards. The primary standard for performance testing is IEC 60904: PV devices, which describes the measurement principle and was first released in 1987. It specifies how measurements are conducted, methods for correcting temperature, and spectral mismatch. Further, few additional norms on specific topics have been developed for the treatment of other specific performance issues. For commercially sold PV modules IEC 61215-2005, IEC 61646-2008, and IEC 62108-2007, consider the special characteristics of silicon, thin-film, and CPV modules, respectively, for performance measurements.

3.4.1 Indoor testing

Several PV module test procedures have been developed to achieve reliable performance measurements. These procedures simulate different stress conditions PV modules might suffer from in the field. Thereby, the natural degradation process is imitated and conclusions on real degradation behavior can be drawn. Statistics laboratories that conduct IEC certification must be accredited according to IEC/ISO 17025. Box 3.1 shows a brief about the visual inspection method that is generally chosen for testing PV modules.

3.4.2 Module power measurement

Module power measurement is the most commonly followed testing method. Here, various parameters that affect the overall performance of the PV module are monitored and their role in efficiency conversion is quantified. Recently, sun simulators have been used for this measurement. In Table 3.4, sun simulators according to IEC standards are given under different classes that are classified based on spectral fit. Box 3.2 briefs the power measurement method using IEC 61215 and 61646 10.2 standards; this is more of an experimental testing with parameter measurement.

Testing of a PV module considering different conditions is considered. Briefs on the use, test conditions, procedure, and requirements are highlighted in Box 3.3, Box 3.4, Box 3.5, and Box 3.6.

BOX 3.1

To identify the defects in the PV module by visual inspection

Testing method: Visual inspection.
Use: Identify defects visually.
Check for:

- Cracked, bent, misaligned, torn external surfaces
- Faulty interconnections, joints
- Voids in, visible corrosion of active layers
- Visible corrosion of output connections, interconnections, bus bars
- Failure of adhesive bonds
- Bubbles or delamination from edge to cell
- Tacky surfaces of plastic materials
- Faulty terminations, exposed electrical parts
- Any other conditions that may affect performance
- Number plate not present or readable

TABLE 3.4 Sun simulators according to IEC 60904-9.

Sun Simulator Classes IEC 60904-9 Ed.2				
			Temporal instability	
Class	**Spectral fit**[a]	**Max. inhomogenity**[b]	**LTI**	**STI**
A	0.75–1.25	2.0%	0.5%	2.0%
B	0.60–1.40	5.0%	2.0%	5.0%
C	0.40–2.00	10.0%	10.0%	10.0%

[a] *Given in IEC 60904-9 Ed.2.*
[b] *Nonhomogeneity of irradiance.*

3.5 Novel PV module concepts

Research in PV modules is ongoing and continuously comes up with a few novel concepts. Some of the novel PV modules are shown in Fig. 3.8. In this section, special focus is on exploring such novel PV modules and some of them are briefly discussed below.

BOX 3.2

To identify the defects in the PV module by power measurement.

Testing method: Power measurement.
Standards: IEC 61215 and 61646 10.2.
Use: Power output at STC:

- Module temperature: 25°C
- Irradiance level: 1000 W/m^2
- Spectrum air nass 1.5 (PI: calibrated class A+ simulator as IEC 60904-9

Insights:

- Electronic load to measure *I-V* curve

Reference device (IEC 60904-2)

- Accuracy of repetitive tests, before/after treatments: ±0.5%
- Total accuracy (any technology) ±2–3%.

BOX 3.3

To identify the defects in the PV module by an isolation test.

Use: Sufficient insulation between electrical parts and frame (or outside world).
Procedure:

- Wrap the module entirely with a copper foil (foil has to stick on the surface)
- Connect the short-circuited terminal-cables of the module with "+" of DC source
- Connect the foil/frame of the module with the "−" of the DC source
- Increase DC voltage slowly (<500 V/s) until 1 kV + 2 max. System voltage is reached, then keep for 1 min.
- Measure the isolation resistance.

Requirements: (isolation resistance) × (module area) ≥40 MΩ m^2.

BOX 3.4

To identify the defects in the PV module by wet leakage.

Use: To evaluate the insulation of the module under wet operating conditions.
Setup: A water tank/pool to contain the module horizontally covered with water.
Procedure:

- Connect the short-circuited cables of the module with "+" of DC source
- Connect the fluid or the water with the "−" of the DC source
- Increase DC voltage slowly (<500 V/s) until 0.5 kV or max. System voltage is reached, then keep for 1 min.
- Measure the isolation resistance:

Requirements: (isolation resistance) × (module area) ≥40 MΩ m^2.

BOX 3.5

To identify the defects in the PV module by weak light performance.

Use: Determination of module power at an irradiance level of 200 W/m^2 at 25°C.

- Very relevant for yield; average irradiance in Germany during daytime: 230 W/m^2.
- Same spectrum and temperature as STC
- PI uses an A + AA simulator (instead of BBB as required by IEC 60904-9)

PI offers additional irradiance levels: 100, 400, 800 W/m^2.

Flexible module: The name itself suggests that these solar PV modules are flexible, which means they can fit any kind of mounting surface. Originally, a mono c-Si material is used for manufacturing these modules. But recently, the CIGS-type thin-film solar material is used. Modules with c-Si material are semiflexible in nature and with CIGS are flexible. These modules are lightweight in nature and mostly thin in size, which makes them suitable

BOX 3.6

To identify the defects in the PV module by thermal cycling.

Test method: Thermal cycling.
Damp Heat.
Use: Ability of the module to withstand the long-term penetration of humidity.
Test conditions:

- Temperature: 85°C (± 2°C)
- Relative Humidity: 85% (±5%)
- Duration of treatment: 1000h

Requirements (tests after 2–4h recovery time):

- No visual defects
- Isolation resistance ≥40 MΩ m^2
- Power loss less than 5%

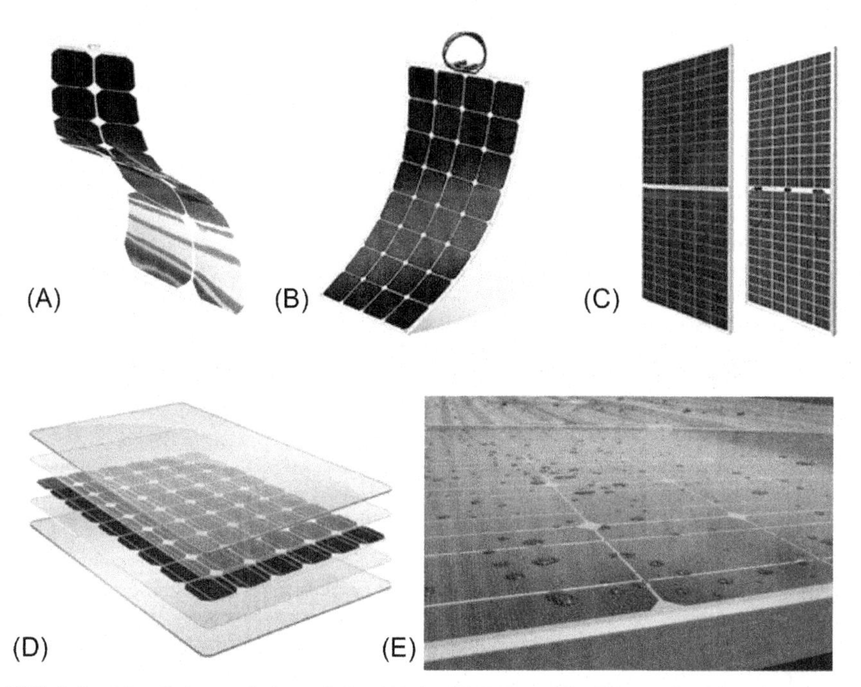

FIG. 3.8 Novel photovoltaic modules: (A) flexible PV module; (B) semiflexible PV module; (C) bifacial PV module; (D) double-glass PV module; (E) antireflection coated PV module.

for rolling and bending and more suitable for applications such as car roofing and window curtains. The commercial application of flexible modules is very limited due to their lower power conversion efficiencies [43, 44]. Like this, there are advantages and disadvantages in using flexible PV modules and these are briefly given in Table 3.5.

Bifacial PV module: The existing PV modules are classified under monofacial as they are capable of generating electricity when sunlight is incident on the surface. The other category of PV module is bifacial, and the name itself suggests that the PV module can generate electricity from both sides. The electrical power output mainly depends on the installation configuration. For example, with monofacial PV modules under open rack installation configurations, there is a huge amount of energy loss in terms of incident solar irradiation. Here, the direct and diffused solar irradiation components are only useful, but the reflected component of solar irradiance is lost. In such an open rack configuration, the use of a bifacial PV module could help in improving energy as they are capable of using a reflective component of solar irradiance for

TABLE 3.5 Advantages and disadvantages of flexible PV modules.

List of advantages	List of disadvantages
Portable in nature and can be mounted to any surface.	As they are mounted directly to the mounting surface, there won't be any gap between the module and surface. This does not allow any air circulation, thereby chances of overheating are very high. This ultimately affects the overall power conversion efficiency.
Compared to the other PV modules, power loss is a bit lower. The main reason for this is the use of thin films for making flexible modules. Thin films are thermally stable and the overall efficiency drops due to temperature is low.	They are not well suitable for positing with respect to the sun path. These modules are mostly fixed to a mounting surface. Hence, efficient angling and tilt angle optimization specific to the installation site are not possible.
Performace of thin-film flexible modules is quite promising as they absorb sunlight even in the infrared and UV ranges.	Damage due to external activities is very high with PV flexible modules
Most thin-film flexible modules are available in semitransparent options. They can be used for a few applications such as building facades and windows.	The warranty by the manufacturers is very limited. At present, the offered warranty is less than 18 months.
Light weight and low cost are other favorable conditions. As they have less weight, they can be fixed onto low weight-bearing structures.	Not suitable for grid integration and other commercial applications

electricity generation. The improved power with the use of bifacial modules is termed bifacial gain [45, 46].

Double-Glass PV Module: In the current PV market, the most-seen PV modules have a back sheet that is made of polymer. Also, they have a typical frame that is made of aluminum. But in recent years, the use of PV modules for various applications has encouraged manufacturers to develop frameless products. PV modules that do not have a frame are called double-glass PV modules. In these modules, heat-strengthened glass is used on either side of the solar cell. This makes the sandwich structure where the layer of solar cells is laminated between the two glasses. This eliminates the polymer back sheet. The application of double-glass PV modules includes daylight maximization in buildings. Due to the transparent nature of glass, the incident solar irradiation on the PV module passes through it to illuminate the indoor areas [47].

Antireflection Coated Glass: In PV modules, solar irradiance and temperature are the most important factors affecting the overall performance. Coming to solar irradiance, the albedo is one of the components responsible for energy loss, due to which the overall energy production is less. Antireflection-coated (ARC) glasses are those that reduce the reflected component of solar irradiance from the solar panels, thereby boosting the overall efficiency as well as energy generation. Thickness and refractive index (RI) are the two major factors in ARC glasses [48].

The thickness of the ARC glass is calculated using the formula:

$$\text{Optimal ARC thickness} = \frac{\text{Wavelength}}{4 \times \text{RI of ARC}}$$

The refractive index of the ARC material is given as the geometric mean of the RI of the surrounding material and the RI of the semiconductor:

$$\text{ARC RI} = \sqrt{(\text{RI of surrounding material} \times \text{RI of semiconductor})}$$

3.6 Conclusion

Photovoltaic technologies are the vital components in solar power generation. The role of PV technologies in sustainable power generation is realized. In this chapter, a study has been carried out on photovoltaics. The efficiency of the PV module using the performance characteristics, especially based on *I-V* and *P-V* curves, is also presented. Photovoltaic modules and their types are presented. On the other side, recent trends in PV technologies are explored. The advantages of using thin-film PV modules over the crystalline module are also presented. The service life and reliability of PV modules with a focus on faults and their identification methods are briefly explained. The testing of PV modules considering different experimental scenarios is presented along with the procedure. Overall, we believe the chapter presents the overview of PV module technologies that would serve as useful material for academicians and researchers.

References

[1] Mints P. SPV market research market update April 2018. 2018
[2] Glunz SW, Preu R, Biro D. Crystalline Silicon Solar Cells: State-of-the-Art and Future Developments. vol. 1. Elsevier Ltd; 2012.
[3] Mathew M, Kumar NM, i Koroth, R.P. Outdoor measurement of mono and poly c-Si PV modules and array characteristics under varying load in hot-humid tropical climate. Mater Today Proc 2018;5(2):3456–64.
[4] Ajitha A, Kumar NM, Jiang XX, Reddy GR, Jayakumar A, Praveen K, Kumar TA. Underwater performance of thin-film photovoltaic module immersed in shallow and deep waters along with possible applications. Results Phys 2019;15:102768.
[5] Kumar NM, Sudhakar K, Samykano M. Performance comparison of BAPV and BIPV systems with c-Si, CIS and CdTe photovoltaic technologies under tropical weather conditions. Case Stud Therm Eng 2019;13:100374.
[6] Duflou JR, Peeters JR, Altamirano D, Bracquene E, Dewulf W. Demanufacturing photovoltaic panels: comparison of end-of-life treatment strategies for improved resource recovery. CIRP Annals 2018;67(1):29–32.
[7] Parida B, Iniyan S, Goic R. A review of solar photovoltaic technologies. Renew Sustain Energy Rev 2011;15(3):1625–36.
[8] Romeo N, Bosio A, Romeo A. An innovative process suitable to produce high-efficiency CdTe/CdS thin-film modules. Sol Energy Mater Sol Cells 2010;94(1):2–7.
[9] Green MA, Dunlop ED, Levi DH, Hohl-Ebinger J, Yoshita M, Ho-Baillie AWY. Solar cell efficiency tables (version 54). Prog Photovoltaics Res Appl 2019;27(7):565–75.
[10] Powalla M, Paetel S, Ahlswede E, Wuerz R, Wessendorf CD, Magorian Friedlmeier T. Thin-film solar cells exceeding 22% solar cell efficiency: an overview on CdTe-, Cu(In,Ga)Se 2-, and perovskite-based materials. Appl Phys Rev 2018;5(4):041602.
[11] Slaoui A, Collins RT. Photovoltaics: advanced inorganic materials. In: Ref. Modul. Mater. Sci. Mater. Eng. 2016.
[12] Metaferia W, Schulte KL, Simon J, Johnston S, Ptak AJ. Gallium arsenide solar cells grown at rates exceeding 300 μm h − 1 by hydride vapor phase epitaxy. Nat Commun 2019;10(1):3361.
[13] Zou H, Du H, Ren J, Sovacool BK, Zhang Y, Mao G. Market dynamics, innovation, and transition in China's solar photovoltaic (PV) industry: a critical review. Renew Sustain Energy Rev 2017;69(November 2016):197–206.
[14] Simon P. Photovoltaics report. Freiburg: Fraunhofer ISE; 2019.
[15] Bett AW, Burger B, Dimroth F, Siefer G, Lerchenmüller H. High-concentration PV using III-V solar cells. In: Conference Record of the 2006 IEEE 4th world conference on photovoltaic energy conversion, WCPEC-4, vol. 1; 2007. p. 615–20.
[16] Philipps SP, Bett AW, Horowitz K, Kurtz S. Current status of concentrator photovoltaic (CPV) technology; 2017.
[17] Sellami N, Mallick TK. Optical efficiency study of PV crossed compound parabolic concentrator. Appl Energy 2013;102:868–76.
[18] Belkasmi M, Bouziane K, Akherraz M, Sadiki T. Optimization of a solar tracker for CPV system principles and pratical approach. In: Proceedings of 2014 international renewable and sustainable energy conference, IRSEC 2014; 2014. p. 169–74.

[19] Burhan M, Ernest CKJ, Choon NK. Electrical rating of concentrated photovoltaic (CPV) systems: long-term performance analysis and comparison to conventional PV systems. Int J Technol 2016;7(2):189–96.
[20] Pérez-Higueras P, Muñoz E, Almonacid G, Vidal PG. High concentrator photovoltaics efficiencies: present status and forecast. Renew Sustain Energy Rev 2011;15(4):1810–5.
[21] Köntges M, et al. Review of failures of photovoltaic modules; 2014.
[21a] Renganathan RMV, Natarajan V, Jahaagirdar T. Failure mechanism in PV systems. In: Grad. course MAE 598 Sol. Commer. 2016. p. 1–17.
[21b] Rajput P, Malvoni M, Kumar NM, Sastry OS, Tiwari GN. Risk priority number for understanding the severity of photovoltaic failure modes and their impacts on performance degradation. Case Stud Therm Eng 2019;16:100563.
[22] Munoz MA, Alonso-García MC, Vela N, Chenlo F. Early degradation of silicon PV modules and guaranty conditions. Sol Energy 2011;85(9):2264–74.
[23] Aghaei M. Novel methods in control and monitoring of photovoltaic systems. Politecnico di Milano; 2016.
[24] Andrews R. Aerial inspections for solar asset optimization. In: Solar Asset Management; 2018.
[25] Tsanakas I, Botsaris PN. On the detection of hot spots in operating photovoltaic arrays through thermal image analysis. Mater Eval 2013;71(April):457–65.
[26] Weinreich B, Haas R, Zehner M, Becker G. Optimierung thermografischer Fehleranalyseverfahren auf Multi-MW-PV-Kraftwerke, In: 26th PV-Symposium Bad Staff., no. 1; 2011. p. 10.
[27] Breitenstein O, Warta W, Langenkamp M. Lock-in thermography: basics and use for evaluating electronic devices and materials; 2005.
[28] Buerhop C, Schlegel D, Niess M, Vodermayer C, Weißmann R, Brabec CJ. Reliability of IR-imaging of PV-plants under operating conditions. Sol Energy Mater Sol Cells 2012;107:154–64.
[29] Buerhop C, Jahn U, Hoyer U, Lercher B, Wittmann S. Abschlussbericht Machbarkeitsstudie Überprüfung der Qualität von Photovoltaik- Modulen mittels Infrarot-Aufnahmen; 2007.
[30] VATh. Electrical infrared inspections ▪ low voltage. February, p. 17, 2016.
[31] International Electrotchnical Commission (IEC). IEC TS 62446-3—photovoltaic (PV) systems—requirements for testing, documentation and maintenance—part 3: Photovoltaic modules and plants—outdoor infrared thermography. Geneva, 2017.
[32] Kirsten A, et al. Aerial infrared thermography of a utility—scale PV power plant after a meteorological tsunami in Brazil, In: WCPEC, no. 20; 2016. p. 1–3.
[33] Maldague X. Introduction to NDT by active infrared thermography. Mater Eval 2002; 60(9):1–22.
[34] Kropp T, Berner M, Stoicescu L, Werner H. Self-sourced daylight electroluminescence from photovoltaic modules. IEEE J Photovolt 2017;7(5):1184–9.
[35] Frazão M, Silva JA, Lobato K, Serra JM. Electroluminescence of silicon solar cells using a consumer grade digital camera. Measurement 2017;99:7–12.
[36] Djordjevic S, Parlevliet D, Jennings P. Detectable faults on recently installed solar modules in Western Australia. Renew Energy 2014;67:215–21.
[37] Muehleisen W, et al. Outdoor detection and visualization of hailstorm damages of photovoltaic plants. Renew Energy 2018;118(November):138–45.
[38] Eder G, Knöbl K, Voronko Y, Grillberger P, Kubicek B. UV-fluorescence measurements as tool for the detection of degradation effects in PV-modules. In: 8th Eur. Weather. Symp., no. September; 2017. p. 1–7.
[39] van Sark W, Nemet G, Schaeffer GJ, Alsema E. Photovoltaic solar energy; 2010.
[40] Kurtz S, Wohlgemuth J, Hacke P, Kempe M, Sample T, Amano J. Ensuring quality of PV modules preprint, In: IEEE photovoltaic specialists conference; 2011, [June].

[41] Gambogi WJ, et al. Assessment of PV module durability using accelerated and outdoor performance analysis and comparisons. In: 2014 IEEE 40th Photovolt. Spec. Conf; 2014. p. 2176–81.
[42] Rajput P, Malvoni M, Manoj Kumar N, Sastry OS, Jayakumar A. Operational performance and degradation influenced life cycle environmental–economic metrics of mc-Si, a-Si and HIT photovoltaic arrays in hot semi-arid climates. Sustainability 2020;12(3):1075.
[43] Trapani K, Millar DL. The thin film flexible floating PV (T3F-PV) array: the concept and development of the prototype. Renew Energy 2014;71:43–50.
[44] Planes E, Yrieix B, Bas C, Flandin L. Chemical degradation of the encapsulation system in flexible PV panel as revealed by infrared and Raman microscopies. Solar Energy Mater Solar Cells 2014;122:15–23.
[45] Guerrero-Lemus R, Vega R, Kim T, Kimm A, Shephard LE. Bifacial solar photovoltaics—a technology review. Renew Sustain Energy Rev 2016;60:1533–49.
[46] Baumann T, Nussbaumer H, Klenk M, Dreisiebner A, Carigiet F, Baumgartner F. Photovoltaic systems with vertically mounted bifacial PV modules in combination with green roofs. Solar Energy 2019;190:139–46.
[47] Tang J, Ju C, Lv R, Zeng X, Chen J, Fu D, Jaubert JN, Xu T. The performance of double glass photovoltaic modules under composite test conditions. Energy Procedia 2017;130: 87–93.
[48] Bauer G. Absolutwerte der optischen Absorptionskonstanten von Alkalihalogenidkristallen im Gebiet ihrer ultravioletten Eigenfrequenzen. Ann Phys 1934;411(4):434–64.

CHAPTER

4

Solar photovoltaic thermal (PVT) module technologies

Mahdi Shakouri[a], *Hossein Ebadi*[b], *and Shiva Gorjian*[c,d]

[a]School of Environment, College of Engineering, University of Tehran, Tehran, Iran, [b]Biosystems Engineering Department, Faculty of Agriculture, Shiraz University, Shiraz, Iran, [c]Biosystems Engineering Department, Faculty of Agriculture, Tarbiat Modares University (TMU), Tehran, Iran, [d]Leibniz Institute for Agricultural Engineering and Bioeconomy (ATB), Potsdam, Germany

4.1 Introduction

From the early solar PV cells to the very recent designs, the operating temperature is considered to be one of the critical parameters in photovoltaic (PV) modules. Exceeding from the optimum value causes a significant decrease in the semiconductor bandgap, which consequently affects the open-circuit voltage negatively. In order to diminish the temperature effect in the PV cells, scientists have proposed cooling methods to maintain the cells at low temperatures and hurdle any temperature growth. Showing promising results, this technique has become more common, which alternatively increases the lifespan of the PV modules by preserving their optimum condition.

The dissemination of energy recovery has triggered developers to take advantage of this removed thermal power for other applications and has formed the concept of the *photovoltaic* and *thermal* (PVT) energy harvesting technique [1]. Most of the PV cells convert a small fragment of the received solar radiation into electricity and the rest of the energy is wasted as heat. Therefore, the PVT is a new configuration of a solar system that has been formed based on the combination of a PV module with a thermal collector. This is capable of exploiting solar radiation for the simultaneous generation of electricity and heat. In a PVT collector, the PV module is fixed to a heat exchanger that contains either air or water as a cooling medium,

Photovoltaic Solar Energy Conversion
https://doi.org/10.1016/B978-0-12-819610-6.00004-1

which converts more solar energy per area than the corresponding cooperation of a separate PV module and a solar thermal collector. Therefore, this causes a reduction in manufacturing and installation costs [2].

In recent years, serious steps have been taken toward the implementation of PVT modules to meet domestic needs, especially in places with space limitations for both PV and thermal collector installation. One of the merits of this technology is the aesthetic advantage compared to conventional thermal collectors with robust designs. The other enticing feature is the interesting niche market that can be formed from either the utility or autonomous systems [3]. However, there is still low industrial involvement in PVT mass production and the overall general knowledge needs to be improved through recent developments.

4.1.1 Working principle of PVT modules

As has been shown in Fig. 4.1, the thermal collector in a PVT module captures the excess generated heat by means of a fluid to either recover heat or cool the PV module [2]. In this process, solar radiation reaches the module at G intensity where a fraction is lost to the ambient as Q_{loss} and the remaining portion empowers the PV module (Q_e) with a given electric efficiency (η_e). The accumulation of solar energy increases the temperature of the PV module and generates the thermal power of Q_{th}, which is transferred to the thermal module through a heat transfer mechanism depending on the fluid medium and module design with a thermal efficiency of η_{th}. Finally, thermal insulation makes the entire system more efficient by reducing and eliminating the sides and back heat losses.

The general energy equation in a simple PVT module can be defined as the following [2]:

$$\eta_e = \frac{Q_e}{GA} \tag{1}$$

$$\eta_{th} = \frac{Q_{th}}{GA} \tag{2}$$

$$\eta_o = \eta_e + \eta_{th} \tag{3}$$

where G (W/m^2) is the solar radiation, A (m^2) is the aperture area of the module, and η_o is the overall efficiency.

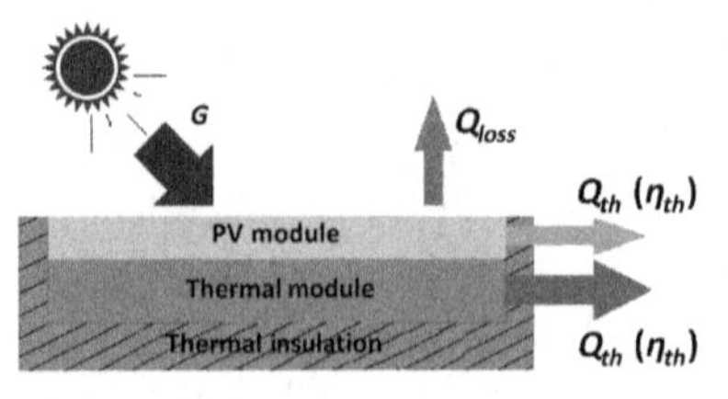

FIG. 4.1 The fundamental energy flow in a typical PVT module [2].

4.1.2 Historical development of PVT modules

The very first experimental investigations on the viability of PVT technology were initiated in the mid-1970s when Wolf [4] analyzed the annual performance of a PVT system for a single-family residence and concluded that the proposed system has the capability to work in a hybrid mode. Two years later, Kern et al. [5] developed a program to evaluate a PVT collector for different designs with heating and cooling applications. After a decade of intense research, scientists published design features of a PVT module that included single-crystal silicon cells and an air collector, where the optimum combination was introduced for the studied configuration [6]. The 21st century started with a focus on collector designs and novel configurations for higher thermal absorption. In this regard, Sopian et al. [7] studied a double-pass solar PVT module and demonstrated that air-based PVT modules enjoy a high potential to be integrated into a dryer. Another study was also conducted over the arrangement of the air passages, where four different designs were compared and the results revealed promising information for air-based PVTs [8].

The performance of the air-based PVT modules connected in series was also studied by researchers, and the results indicated a high potential for this technology to be integrated with buildings [9]. In another attempt, a PVT module including air channels modified with single-row oblique plates to assist in flow direction was studied for a comprehensive understanding of the system's thermal behavior [10]. A major enhancement in module design was achieved as fins were introduced to the PVT configuration. In an experiment, the researchers investigated the effects of the vertical fins inside a double-pass air-based PVT module and found that the presence of fins considerably decreased the cell temperature, which also improved the exchange of heat to the thermal module [11]. A multioperational ventilated PVT module was implemented for integration with a commercial building as a roof-mounted module. It was evaluated under a 6-month operation to find the optimum performance in relation to the net required energy for certain buildings [12]. Agriculture drying is a salient area for PVT integration as numerous researchers have addressed the findings obtained in this field [13].

In the literature, different types of solar cells have been used in PVT modules. The crystalline solar cells were tested in an air-based PVT system connected to a typical residential building as a stand-alone system [14]. Othman et al. [15] introduced a double-pass air-based PVT module using transparent solar cells. In this research, they utilized two parallel transparent PV modules for electricity generation. Thin-film PV cell technology was also employed for PVT development in the way that scientists were able to develop an ultralight and ultrathin unglazed PVT system [16]. In another study conducted by Nualboonrueng et al. [17], polycrystalline and amorphous silicon solar cells were compared to find the best option for a PVT module working in a tropical climate. The PVT modules can be classified based on different criteria such as heat extraction mechanism, application, solar collection design, and heat exchange medium. Fig. 4.2 represents the detailed classification of PVT modules.

4.1.3 Flat-plate PVT modules

A flat-plate PVT (FP-PVT) module usually consists of a flat absorber sheet beneath the PV module, which can include a cover glass to form a glazed type while unglazed types are in use for specific designs [19]. In this type, the main layouts are glass cover, air ducts, water channels, the PV module, the adhesive layer, the flat-plate or tubular absorber, and the insulation. A white Tedlar, a transparent layer of ethylene-vinyl acetate (EVA) resin, is usually used to enhance the PV radiation absorption efficiency and reduce the PV reflection [20]. The selection of the working medium is very crucial because it has a remarkable effect on the overall performance where higher operating temperatures lead to lower electricity efficiency. Thus, a working fluid (usually air or water) is used in contact with the PV module in the form of a heat exchanger to carry away the excess heat from the module, resulting in higher electricity and thermal efficiency.

4.1.3.1 Water-based FP-PVT modules

Compared to air, the liquid form of working fluids in a PVT collector is able to carry more heat from the PV module, which results in higher electric efficiency owing to the higher density and greater heat capacity. However, the manufacturing costs of the heat exchanger are also added to the PVT module [21, 22]. Several liquids have been studied by scientists to develop different PVT modules; among them, water has been the most commonly used working fluid. In an attempt by Chow et al. [23], a flat-box PVT module consisting of a water storage tank, a polycrystalline silicon PV module with 14.5% electric efficiency, and an aluminum alloy thermal absorber attached to the box-structure modules with two transverse headers was developed to perform under the thermosyphon effect. The illustration of the main components and a photograph of the proposed PVT module are depicted in Fig. 4.3. From the experimental tests, it was concluded that the average thermal efficiency ranges from 37.6% to 48.6% with higher obtained values in summer and the average electric efficiency between 10.3% to 12.3% with the higher recorded values in winter.

Nanofluids are the other type of working fluid recently utilized in liquid-based FP-PVT modules. They have attained a significant fraction of progress in PVT technology. The introduction of nanofluids was aimed

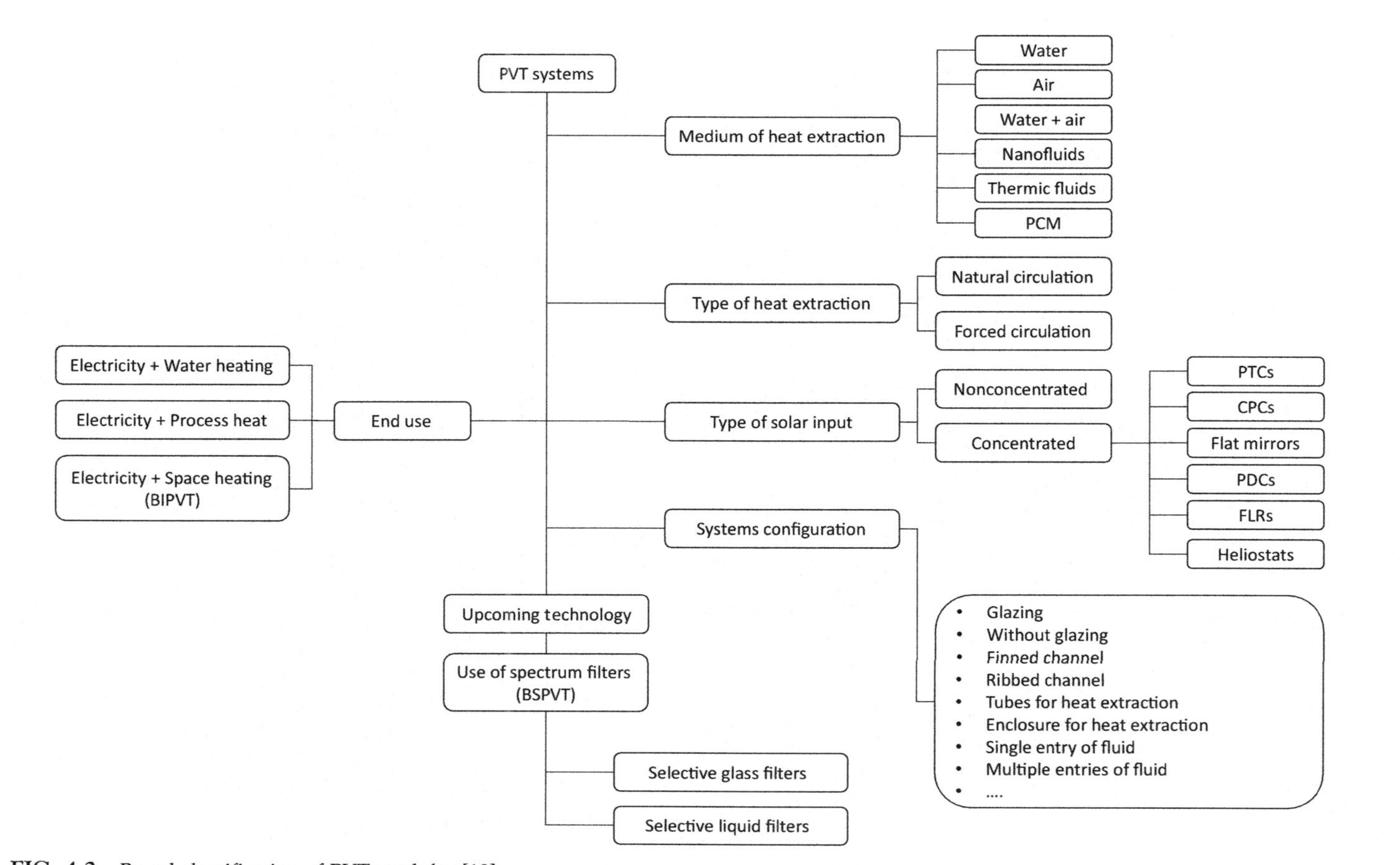

FIG. 4.2 Broad classification of PVT modules [18].

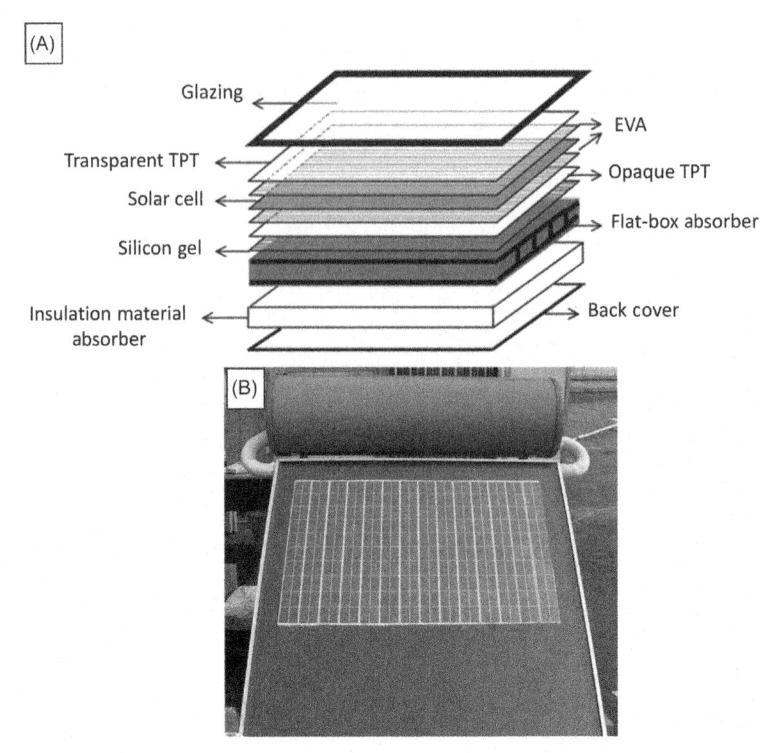

FIG. 4.3 A thermosyphon flat-box PVT module; (A) The main layout of the module, (B) photograph of the constructed model [23].

to enhance the thermal conductivity of conventional fluids, making them a high-potential option for utilization in PVT modules.

Sardarabadi et al. [24] studied the use of silica/water nanofluid in an FP-PVT module and reported that the addition of the nanofluid increased the overall efficiency up to 7.9% where the total energy reached 24.31%. Therefore, nano-based liquid fluids not only can work as the coolant medium, but also can have an effect as an optical filter, which depends on the base fluid, particle size, and other factors [22]. In addition to the fluid type, absorber design and flow passage configurations can also be distinctive factors for categorizing different liquid-based PVTs.

Generally, the water-based PVT modules are categorized based on the flow configuration and their components, as shown in Fig. 4.4. The first pattern is the *sheet and tube* design (Fig. 4.4A), in which a flat absorber is usually employed with tubes arranged in series or parallel connections where a PV module is mounted or laminated above the absorber. The alternative designs based on this flow pattern can be attained by changing the number of glass covers where researchers have not suggested using more than two covers [27]. The second flow pattern is the *channel type*

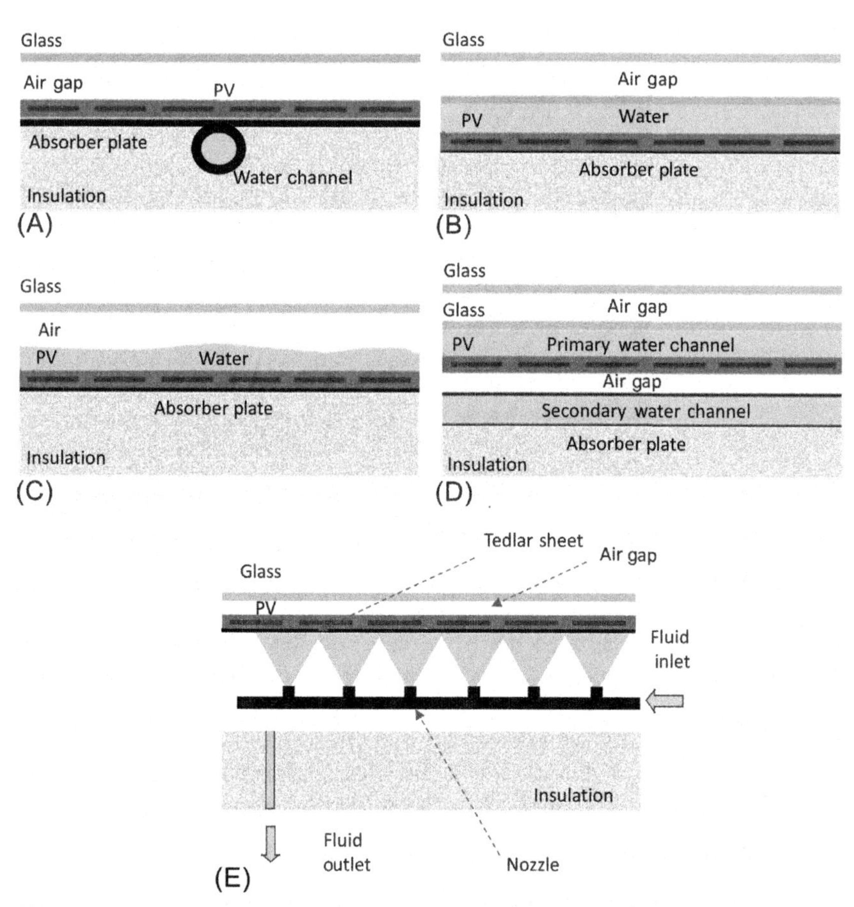

FIG. 4.4 Water-based FP-PVT designs based on the flow configuration: (A) sheet and tube type, (B) channel type, (C) free-flow type, (D) dual-flow type [25], and jet impingement [26].

(Fig. 4.4B) in which the fluid flows inside a channel above the PV module where glass covers are employed to transmit the sunlight. In this type, the selection of the fluid is very critical because the absorption spectrum of PV cells should not be covered by the fluid to yield the maximum electric efficiency [27]. The design indicated in Fig. 4.4C is known as the *free-flow* pattern in which water moves freely above the PV module with only one glass cover compared to the previous design. In this type, water evaporation can be a problem when the PVT module works at high temperatures while the condensation of the liquid on the cover can also affect the absorption efficiency. The next design (Fig. 4.4D) is devoted to the *dual-channel* types in which water flows above a semitransparent module and gets heat in the way considered by the design presented in Fig.4.4B. Placing an air gap below the PV module and using a blackened metallic absorber, the secondary channel is formed as the running water absorbs incoming sunlight through the channel walls [25]. The last design employs a jet array impingement, which was proposed by Hasan et al. [26]. In this design, jet impingement is applied to the back of the PV module via a series of nozzles. The cooling fluid is pumped through the lines and after heat extraction, it is collected by the collection line placed at the bottom of the frame enclosure.

In an experimental study [28], scientists compared the performance of the two proposed designs, including sheet and tube and rectangular channel (fully wetted) absorbers, for an unglazed FP-PVT module. They concluded that the fully wetted design results in higher thermal performance because in this concept, the heat transfer area is improved more so than in the sheet and tube design. It was also found that the electric efficiency of the collector with a fully wetted absorber is nearly 2% more than an individual PV module.

A passive FP-PVT collector integrated with reflectors and a finned absorber was also developed by Ziapour et al. [29]. They studied the effect of an efficient way in covering the outer glazing during night time for thermal storage purposes and collecting reflected sunlight during the day time operation. Results indicated that this novel design could decrease night heat losses and increase daily solar absorption. The spiral flow direction was also tested for a hybrid flat-plate solar collector using Al_2O_3/water nanofluid [30]. Results demonstrated that the use of nanofluid with the spiral flow direction can increase the electrical, thermal, and overall efficiencies by 13%, 45%, and 58%, respectively.

4.1.3.2 *Air-based FP-PVT modules*

In an air-based FP-PVT module, the air is used as the medium for cooling the PV module through circulation to extract heat from the PV module. There are a number of different arrangements such as with/without glazing, natural/forced flow of air, fined channels, etc. In order to classify different PVT modules, several criteria can be employed ranging from design factors to the material shape of the module [31–35]. As a typical air-based FP-PVT collector, the PV module is laminated on the black plate absorber, which can be covered by a glass cover. Although there are many different designs, especially based on absorber designs, the most common models based on the flow patterns are proposed by Hegazy [8] and are presented in Fig. 4.5.

In the first design, air passes through the space between the absorber plate and the glazing (Fig. 4.5A). However, if the air flows under the absorber plate, the second design is formed (Fig. 4.5B). In the third design, air passes through two parallel channels, one over and one under the absorber, causing a dual-channel configuration (Fig. 4.5C). In the last design, air first gets preheated by passing over the absorber and then

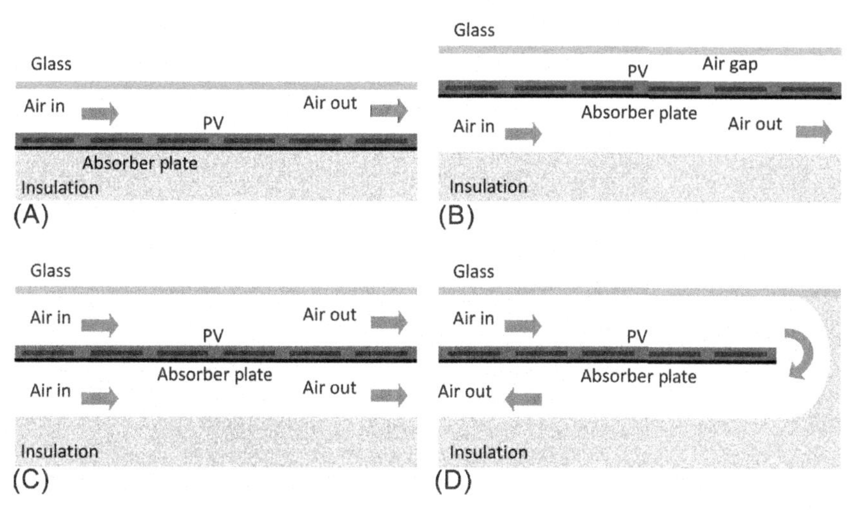

FIG. 4.5 Air-based FP-PVT designs based on the flow configuration: (A) single-pass over the absorber, (B) single-pass under the absorber, (C) single-pass dual channel, and (D) dual pass [8].

absorbs the remaining heat from the channel underneath the flat plate (Fig. 4.5D). Comparing the performance of different designs indicates that the highest daily thermal efficiency refers to the second design, which is followed by the third and fourth designs. The first design has a distinctive lower value, especially in higher mass flow rates. In the case of electricity generation, all the designs have a close performance where the highest efficiency is devoted to the second design. Moreover, based on the amount of electric power consumed by the fan in each design, it is concluded that the first two designs demand the most energy, followed by the fourth and third designs [8].

A typical solar air-based FP-PVT that provides hot air as well as electricity can be employed in different applications such as drying, distillation, and space heating. Farshchimonfared et al. [36] developed an air-based FP-PVT module to deliver thermal power to a building associated with the air distribution system. In their study, an unglazed solar collector covered with a crystalline PV module was installed on a building roof. They found that the optimum depth for a constant temperature rise of 10°C varies between 0.09 to 0.026m. In another study carried out by Slimani et al. [37], a double-pass air-based FP-PVT module, including a monocrystalline PV module, was integrated into a drying chamber. As shown in Fig. 4.6, the PVT collector received solar radiation that reached the aluminum sheet and increased its temperature. Cool air enters the module from the air channel and absorbs heat from the environment while cooling down the PV. As a double-pass configuration, the flow is directed to the upper channel between the PV module and glazing, and leaves the

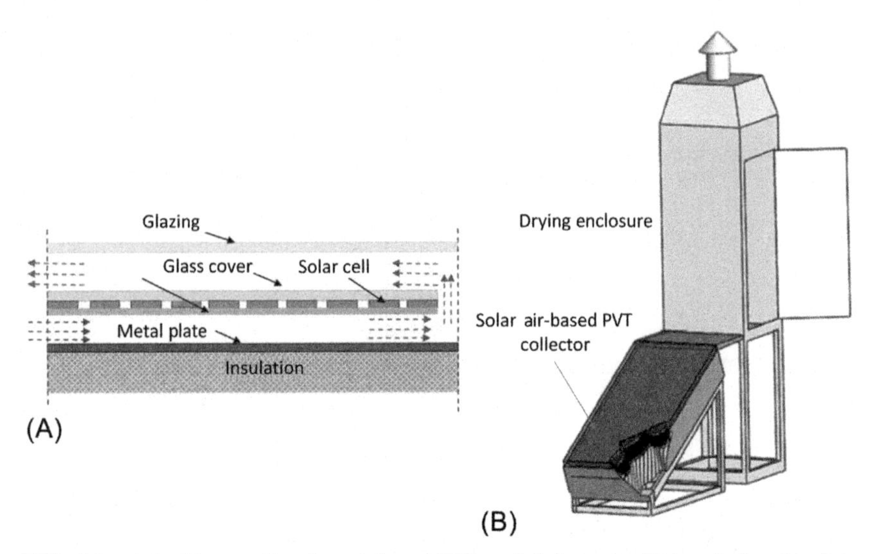

FIG. 4.6 A double-pass flat-plate air-based PVT module integrated into a drying chamber: (A) cross-sectional view of the PVT collector, and (B) three-dimensional modeling of the entire system [37].

PVT module to the drying enclosure through an air duct. Numerical results indicated that the proposed system can be a promising design for indirect solar drying with the desired air temperature. Moreover, the energy analysis revealed that the electric, thermal, and overall efficiencies were 10.5%, 70%, and 90%, respectively.

A building-integrated PVT (BIPVT) module is one of the interesting types of air-based FP-PVT modules that has gained attention. A prototype of an open-loop hybrid unglazed transpired collector (UTC) (Fig. 4.7) was designed for integration with a building façade by Athienitis et al. [38]. In their study, a full-scale model was constructed with 70% PV covering and mounted on an office building in Montreal, Canada. In their design, it was aimed to provide heat ventilation for cold months by the combination between the PV array installation and the UTC designs. Experimental data demonstrated the viability of the proposed system for preheating the fresh air and generating electricity.

4.1.3.3 *Bifluid-based FP-PVT modules*

Bifluid-based FP-PVT modules are defined as FP-PVT modules that incorporate two fluids (usually water and air) to enhance the heat extraction phenomenon in the thermal module. The main scope in this model is the cogeneration of hot water and hot air, in addition to electricity. This configuration is able to pose several advantages compared to two other designs. One of the promising merits that comes from this hybrid module

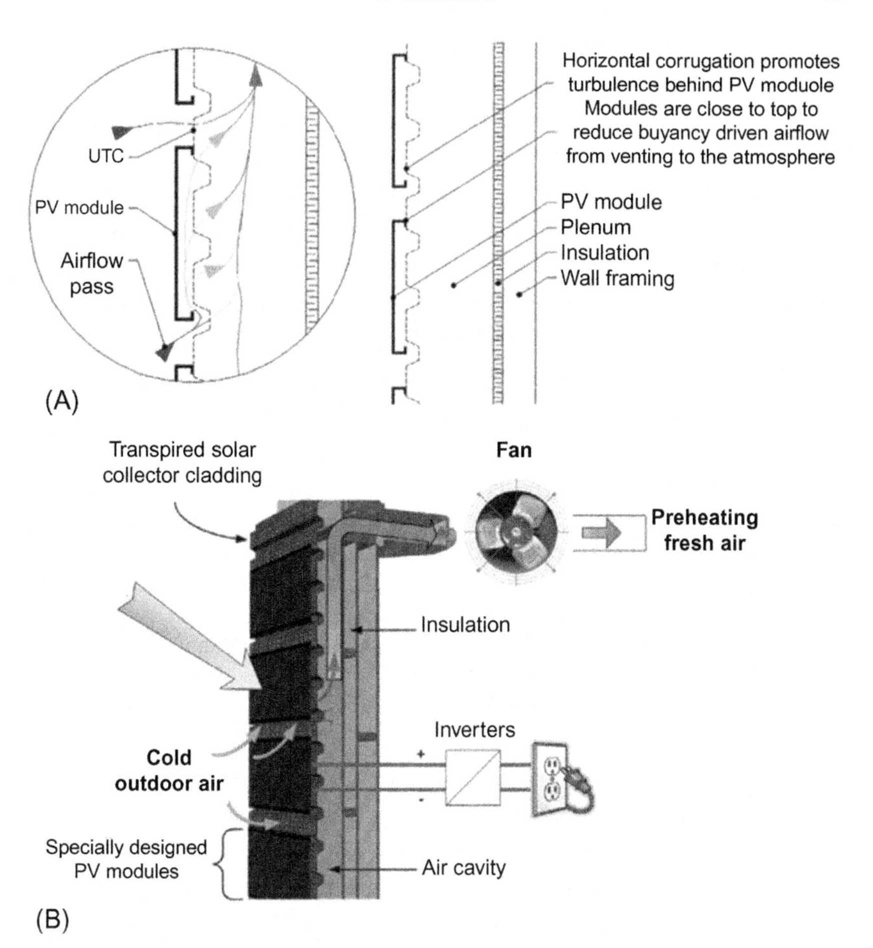

FIG. 4.7 A BIPVT model: (A) schematic view of the proposed PVT design, and (B) PV attachments and airflow pathways [38].

is the high potential of the overall efficiency and better uniformity, which can meet more thermal needs than single-fluid designs. Another advantage is that a major cost reduction can be achieved when a versatile collector provides power in one device [15]. In research conducted by Othman et al. [15], a transparent PV module was placed at the top surface and water tubes were attached to the above and bottom side of the black-painted absorber plate, as shown in Fig. 4.8A. The transparent PV module not only generated electricity but also allowed solar radiation to reach the absorber plate in a way that acted as glazing to trap sunlight for more thermal absorption. Air as the second heat transfer medium also passed through the collector to remove the fraction of heat that cannot be absorbed by the water. The results indicated that the proposed PVT

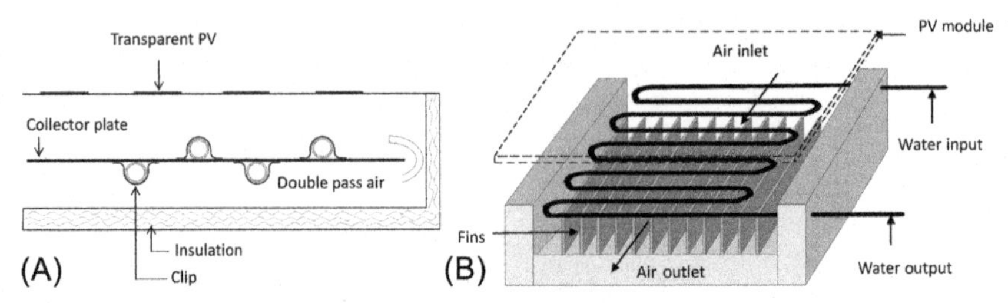

FIG. 4.8 Bifluid-based PVT concepts using water and air simultaneously: (A) Double-pass airflow [39], and (B) Single-pass airflow with fins [15].

module can be both economically and functionally more favorable than the other designs. In another study, a bifluid-based FP-PVT collector was developed based on the concept visualized in Fig. 4.8B. A sheet-and-tube heat exchanger as the absorber plate was used for water circulation and fins were implemented to enhance the air heat transfer inside the PVT module. Indoor testing revealed that thermal efficiencies for the air and water fluids were 58.10% and 62.31%, respectively, while both efficiencies are related and have direct interactions [39].

4.1.4 Concentrating PVT modules

The simultaneous generation of electric and thermal powers in a PVT module, using tracking devices and concentrating reflectors, boosts the overall efficiency of the whole device. It also brings more financial benefits to electricity conversion than power generation that comes from a standalone PV module [40]. At present, the use of the concentrating PVT (CPVT) modules is only applied to large-scale facilities in which the system constitutes components with considerable size, usually as solar towers, parabolic trough concentrators (PTCs), compound parabolic concentrators (CPCs), parabolic dish concentrators (PDCs), and large Fresnel reflectors assisted with single- or two-axis tracking systems. In the case of sun tracking, single-axis mechanisms have higher viability for building integration. Moreover, the building-integrated CPVT systems can be mounted on a roof or located on the building façade [41]. In general, the PTCs commonly are integrated with flat roofs, which is preferable to be hidden from public view. In this design, solar tracking is usually achieved through a tracking system in which the concentrator/receiver structure is rotated on a single axis. Alternatively, the linear Fresnel lenses are utilized to split the beam fraction from the diffuse solar radiation and concentrate it on the PVT module to improve solar illumination per absorbed area. [42]. In the case of employed solar cells, some common types such as GaInP, GaInP/GaAs, AlGaP/GaP, GaP, GaN, and SiC can be the choice for integration with high-temperature concentrating modules [43]. Multijunction cells have also been used in CPVT systems; in an experiment, a dual-junction GaInP/GaAs cell was used to work at 400°C [44].

Kandilli et al. [45] developed and evaluated a CPV module performing with a novel concept of spectral decomposition (Fig. 4.9). The proposed

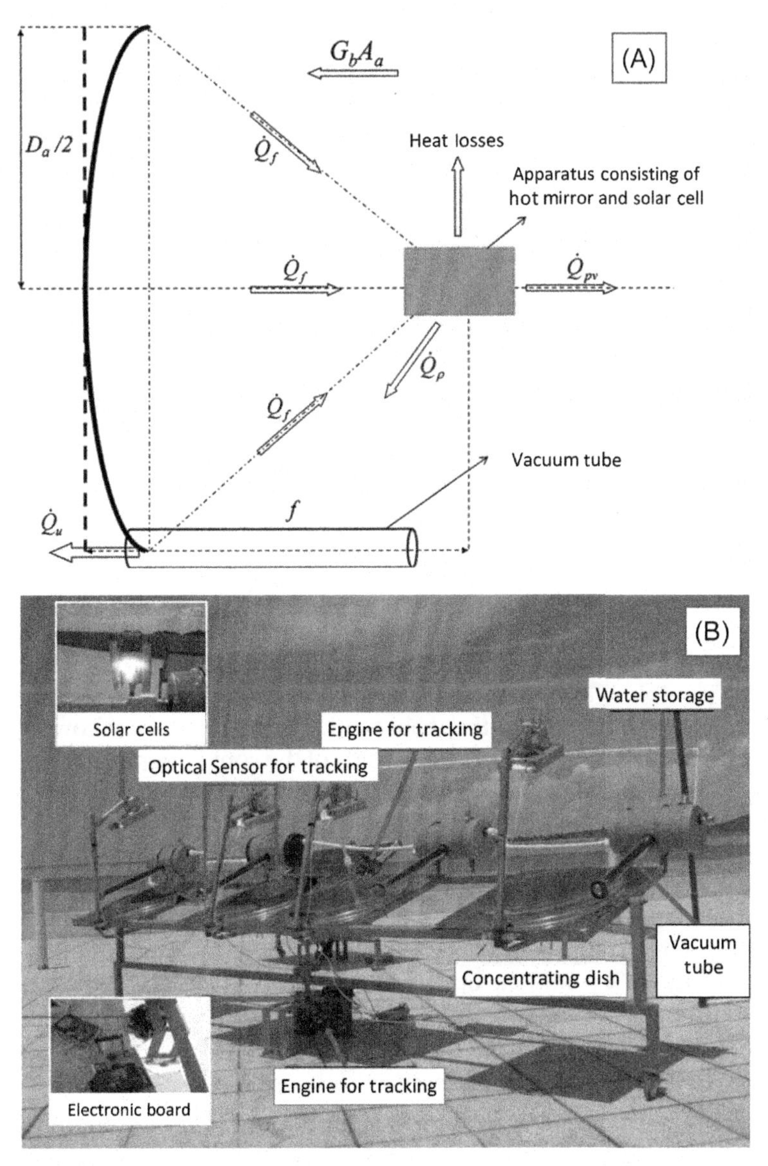

FIG. 4.9 Illustration of a CPV system: (A) schematic view of the proposed concept, and (B) the experimental test setup [45].

system comprised a hot mirror located in the focal area to pass the fraction of solar energy with the wavelength beyond the visible region, and hamper the infrared (IR) and ultraviolet (UV) radiation. The rejected portion, including IR and UV rays, was transferred through a proper optical design to a vacuum tube in which it was collected as the thermal power. The experimental data indicated that the amount of electric power generated per solar cells was 4.6 W while the thermal power of 141.21 W can be obtained from the vacuum tubes and through a cogeneration process. Furthermore, the energy efficiencies derived from the concentrator, vacuum tube, and overall system were 15.35%, 49.86%, and 7.3%, respectively.

Calise et al. [46] also proposed a novel concentrating solar power system including a CPVT system equipped with a PDC. The premise was to utilize the high-temperature collector for providing hot diathermic oil while electricity was also generated by the PV module. Additionally, the heated oil was supposed to produce extra electric power via an organic Rankine cycle (ORC). A prototype was constructed and tested using a double-axis tracking system in which triple-junction silicon cells were placed inside a planar receiver. Comparing two functional modes such as only-CPVT and CPVT+ORC revealed that the first system (only CPVT) exhibited more economic profits where the net present value (NPV) was 20 years and 15% higher than the second system. However, the CPVT +ORC mode was able to provide 6% more electric power.

4.1.4.1 CPVT modules integrated with PTCs

Xu et al. [47] investigated the viability of an innovative low concentrating PVT module incorporated into a heat pump system. The PVT module was installed with fixed truncated parabolic concentrators (employed as the heat pump's evaporator) and reflected the incoming radiation toward the surface of the PV cells without absorption. The test results showed that the electrical output increases as 1.36 folds of the conventional system. In a similar work, scientists developed a hybrid PTC equipped with 32 laminated solar cells connected in series and mounted on two sides of a V-shape absorber (Fig. 4.10). The water was used to transfer heat from the PV cells and the system was designed to provide hot water. A one-axial tracking system was implemented to track the sun around the east-west axis. Analysis indicated that the electric production efficiency reached only 6.4% while the optical efficiency was 0.45%, which are poor values compared to the conventional FP-PVT modules [41].

4.1.4.2 CPVT modules integrated with Fresnel lenses

The integration of Fresnel lenses with PV modules is another technique deemed by CPVT developers to be a promising solution for higher production efficiency. Xu et al. [48] presented a highly concentrating PVT system (HCPVT) with a 30 kW power production capacity. As

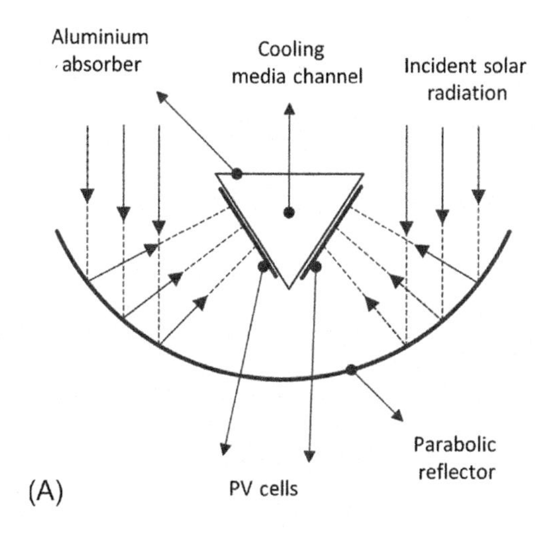

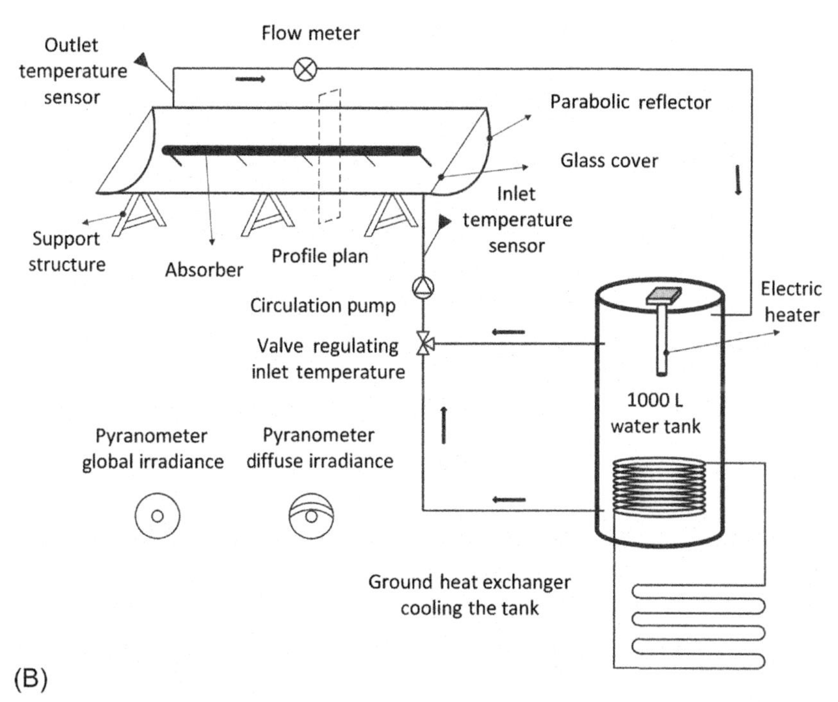

FIG. 4.10 (A) Profile of the absorber and reflector, (B) Diagram of the experimental setup used for a parabolic trough PVT system [41].

shown in Fig. 4.11, the system was composed of 32 point-focus Fresnel lens modules installed on two-axial tracking devices. The entire connection was made of two strings, each composed of 16 series-connected modules. Running water through the receivers collected the

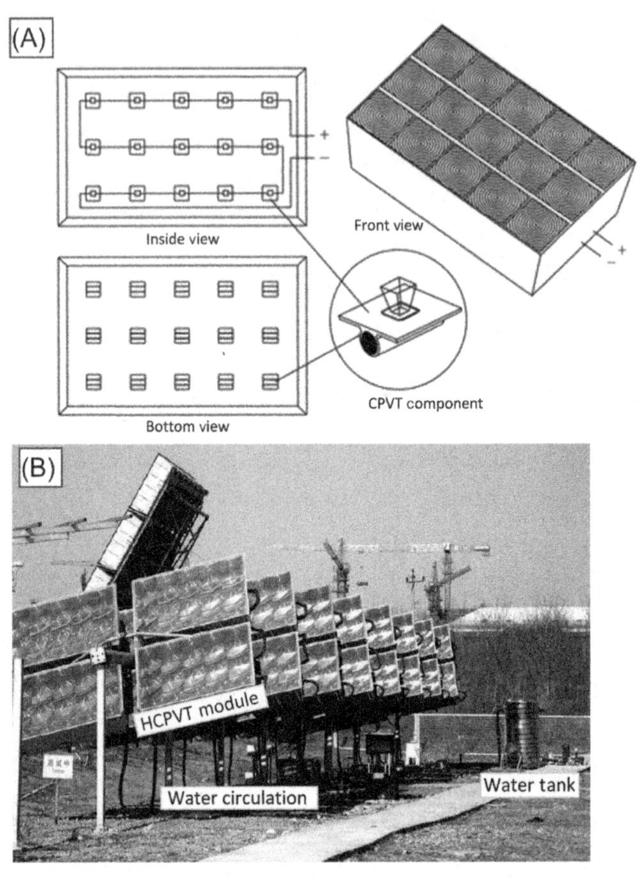

FIG. 4.11 A 30kW HCPVT system integrated with Fresnel lens: (A) schematic of the HCPVT module, and (B) photograph of the 32 modules installed on dual-axial tracking structures [48].

concentrated solar energy, and it was conveyed toward the storage tank. A typical module presented in this work consisted of 15 Fresnel lenses connected side by side on an aluminum frame. The heat transfer equipment was composed of high-temperature solar cells such as InGaP/GaAs/Ge laminated on a frame in the way that water tubes were in perfect contact to cool them down. Moreover, an optical prism was pasted on top of the solar cells to have uniform radiation distribution on the PV surface. These 15 solar cells were also connected in series and tubes were connected relatively by pipes. Experimental results showed that under outdoor conditions, the electricity and thermal efficiencies were 30 and 30%, respectively, indicating that the total solar energy conversion efficiency reached 60%.

4.1.4.3 CPVT modules integrated with CPCs

The CPC collectors emerged as the low-temperature collectors with a low concentration ratio. The CPC design originates from the combination of two parabolas that results in multireflection absorption and reduces the need for tracking the sun's hourly angle. The integration of the CPCs with PVT collectors has been studied by many scientists.

Tripathi et al. [49] designed a PVT-CPC collector to generate both electricity and hot water, as shown in Fig. 4.12. In this work, they utilized semitransparent PV cells and tested the module with manual tracking at 3h intervals. Data analysis demonstrated that this configuration can make 19.60kWh of electric power and 150.40kWh of thermal power. Moreover, the comparison between tracking and nontracking operations showed that the annual net thermal power would be 1.25 higher in the tracking mode.

From the literature review, conventional PVT designs, including the flat-plate and concentrating types, have reached a reasonable maturity in technology and can be applied experimentally in several applications. General conclusions from the studied models reveal that air-based PVT modules are simpler in design compared to the water-based modules with lower construction costs while the thermal and electrical efficiencies of water-based types are quite higher. Although bifluid-based modules are able to yield higher electric efficiency through an enhanced cooling mechanism, the final temperature of the fluids (water and air) is not comparable to the water- or air-based modules, which results in their integration with fewer applications [22]. Additionally, CPVT modules have significant potential for high thermal and electric efficiencies, but their working temperature requires more advanced solar cells, which may lead to higher construction costs. Table 4.1 presents some experimental investigations with the most focus on FP-PVT and CPVT designs.

4.1.5 Innovative PVT modules

4.1.5.1 Heat pipe-based PVT modules (HP-PVT)

Heat pipes are appraised as efficient heat transfer mechanisms that are based on both thermal conductivity and phase transition. A conventional heat pipe is composed of three main units of evaporation, adiabatic, and condensing, which are shown in Fig. 4.13. The heat pipe-based PVT (HP-PVT) module provides a plausible way for heat removal and transmission while there would also be a possibility to use a PV cell layer in conjunction with a flat-plate heat pipe, which includes a series of microchannel arrays working as the evaporation section of the heat pipes. The flat-plate structure offering better thermal conduction is introduced as the more practical

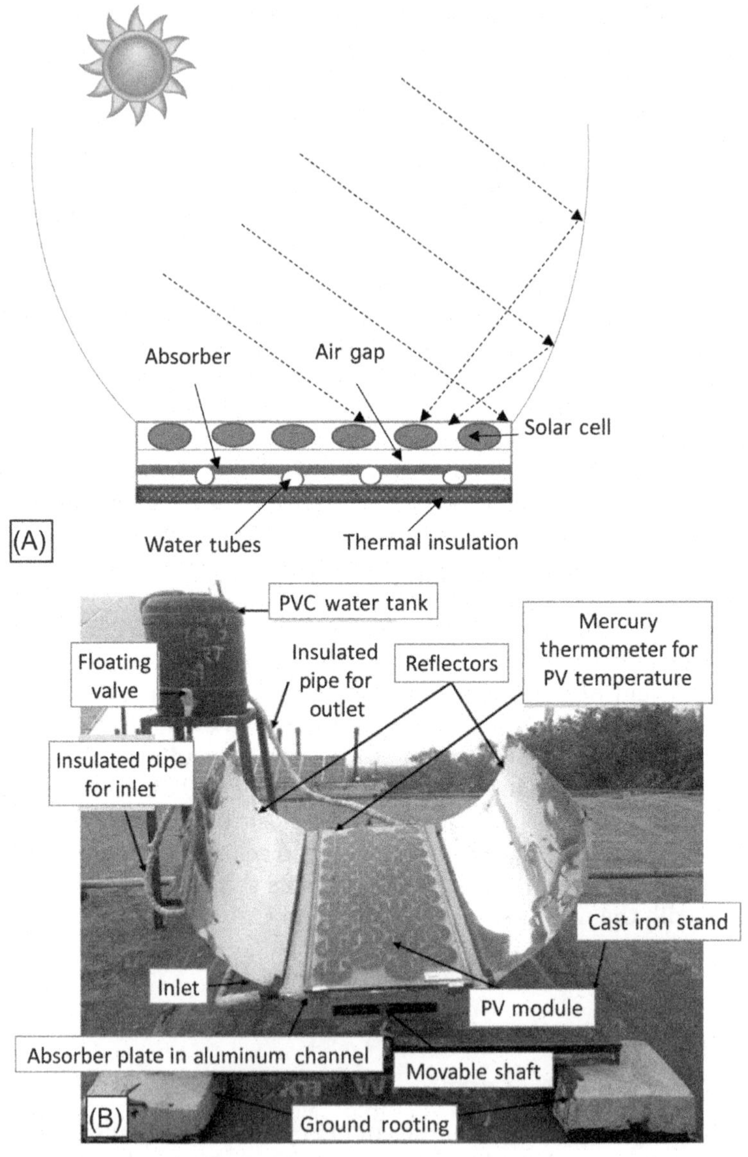

FIG. 4.12 A water-based PVT collector integrated with a CPC: (A) cross-sectional view of the PVT-CPC, and (B) photograph of experimental setup [49].

design because the heat transferred between the PV cell layer and the component for heat extraction is improved [64, 65].

Nowee et al. [66] conducted an experimental study on the performance of a heat pipe integrated with a novel PVT module. The performance comparison between the spring and summer seasons revealed that the average

TABLE 4.1 Summary of some experimental works studying different FP-PVT and CPVT modules.

PVT module	Working fluid	Location	Main findings	Ref.
FP-PVT	Water	Algeria	• The hybrid thermal efficiency was improved in comparison with the conventional thermal collectors; • Providing 37°C hot water with thermal efficiency varying from 50% to 80%.	Touafek et al. [50]
FP-PVT	Air	Iran	• Fans inducing forced convection were powered by electricity generated by the PVT module; • Setting a glass cover above the PV module increased the thermal efficiency and decreased electric efficiency.	Ameri et al. [51]
FP-PVT	Water	Malaysia	• The PVT module was mounted on a dual-axis tracking system; • The overall efficiency was significantly improved, varying from 70.53% to 81.5%.	Rahou et al. [52]
FP-PVT	Water and air	India	• The primary energy-saving efficiency of 35.32% was obtained while electric efficiency was 20.87%; • PVT collector with glazing resulted in more thermal efficiency than without glazing, where the average value reached 16.72%; • Glazing reduced the electric efficiency, where 7.72% was achieved by an unglazed PVT collector.	Michael et al. [53]
FP-PVT	Water	Italy	• PVT collector was able to produce 40°C water for domestic heat applications in summer days;	Buonomano et al. [54]

Continued

TABLE 4.1 Summary of some experimental works studying different FP-PVT and CPVT modules—cont'd

PVT module	Working fluid	Location	Main findings	Ref.
			• The experimental electric and thermal efficiencies were 15% and 13%, respectively; • The payback period was determined as 4 years.	
FP-PVT	Nanofluids (SiO_2, TiO_2, and SiC)	Malaysia	• When SiC was employed, the overall efficiency reached the maximum value of 81.73% where the electric efficiency was 13.52%; • The best performance was followed by TiO_2 and SiO_2 nanofluids, respectively.	Al-Shamani et al. [55]
FP-PVT	Air	Korea	• The heat exchanger was modified using bending round-shaped heat-absorbing plates; • Thermal efficiency was obtained at about 26–45%, depending on the test conditions while the average electric efficiency was measured as 15.7%.	Kim et al. [56]
FP-PVT	Nanofluids (CuO/water and Al_2O_3/water)	Korea	• CuO/water increased both thermal and electric efficiencies by 21.30% and 0.07% than water-based PVT. • Al_2O_3/water enhanced thermal efficiency by 15.14% than water-based PVT; however, electric efficiency remained constant.	Lee et al. [57]
PVT-PTC	Water	Tunisia	• The maximum electric efficiency was obtained as 10.02%; • More electric efficiency was achieved in higher water mass flow rates; • Overall efficiency was 26% where thermal efficiency was about 16%.	Chaabane et al. [58]

TABLE 4.1 Summary of some experimental works studying different FP-PVT and CPVT modules—cont'd

PVT module	Working fluid	Location	Main findings	Ref.
PVT-V-Trough	Water	New Zealand	• The overall efficiency reached 32% at solar noon where the electric efficiency was about 15%. • The fraction of heat loss was high and improved efficiency can be achieved by reducing convective and radiative heat losses.	Künnemeyer et al. [59]
PVT-LFR	Water	Australia	• The overall efficiency can reach 70% while the average thermal and electric efficiencies were 50% and 8%, respectively.	Vivar et al. [60]
PVT-CPC	Water	China	• The final water temperature varied seasonally from 43.6°C to 56.9°C. • The maximum thermal and electric efficiencies were achieved as 37.2% and 10.6% respectively.	Li et al. [61]
PVT-Linear Fresnel Lens	Water	China	• The performance of the Fresnel lens was only limited to direct radiation. • The average total efficiency of the PVT module was measured as 53%. • The electric efficiency was improved by 9.3%, reaching 16.2% when cooling was applied to the PV module.	Karimi et al. [62]

improvements for the electrical and thermal efficiencies were 5.67% and 16.35% during spring and 7.7% and 45.14% during summer, respectively, while the temperature of the module surface can be declined by 15°C. Gang et al. [67] developed another HP-PVT module with water as the working fluid and conducted different experimental tests to evaluate the system's performance. The experimental results demonstrated that heat gain, electric gain, and total PVT efficiencies were improved when

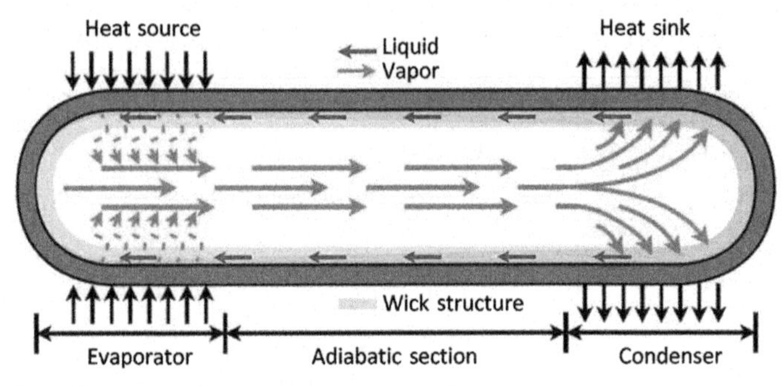

FIG. 4.13 Schematic diagram of a heat pipe [63].

the water flow rate increased. It was also concluded that the effect of the water flow rate is more significant on heat gain than on electric gain. Moreover, when tube spacing of the heat pipes decreases, the heat and electrical gain as well as PVT efficiencies are increased.

In another attempt, Gang at al. [68] developed an HP-PVT module and tested it under outdoor conditions. As shown in Fig. 4.14, an aluminum plate was used as the mainframe sheet embodying nine water-copper heat pipes. Each pipe was designed to have a 1000 mm evaporator section and a 120 mm condenser section. Heat pipes were attached to the back of the aluminum sheet through the evaporator part whereas the condenser sections were immersed in a water box. The monocrystalline silicon solar cells were laminated on the front side of the aluminum sheet where a layer of black Tedlar-Polyester-Tedlar (TPT) was used between them to act as the electrical insulation and to increase solar absorption. In the operation procedure, solar radiation was collected by the PV arrays and the black TPT layer. The heat was transferred from the PV to the aluminum sheet and consequently to the evaporator section of the heat pipes. Then, water conveyed the thermal energy to the running water inside the flow channel and via the condenser section. Experimental data revealed that the system has the potential to generate both electricity and heat with efficiencies of 9.4% and 41.9%, respectively.

4.1.5.2 *Refrigerant-based PVT modules (RB-PVT)*

The use of refrigerants as the working medium in solar collectors opens the opportunity to utilize a lower-temperature phase transition, improving the energy conversion efficiency of the PVT modules. As shown in Fig. 4.15, Tsai [69] developed a system comprised of (I) PVT modules (working as the evaporator), (II) a compressor, (III) a condenser that is immersed inside a water tank, and, (IV) an electronic expansion valve. PVT modules were configured in such a way that they were thermally connected in parallel and electrically in series modes. Low-pressure R134a liquid, coming

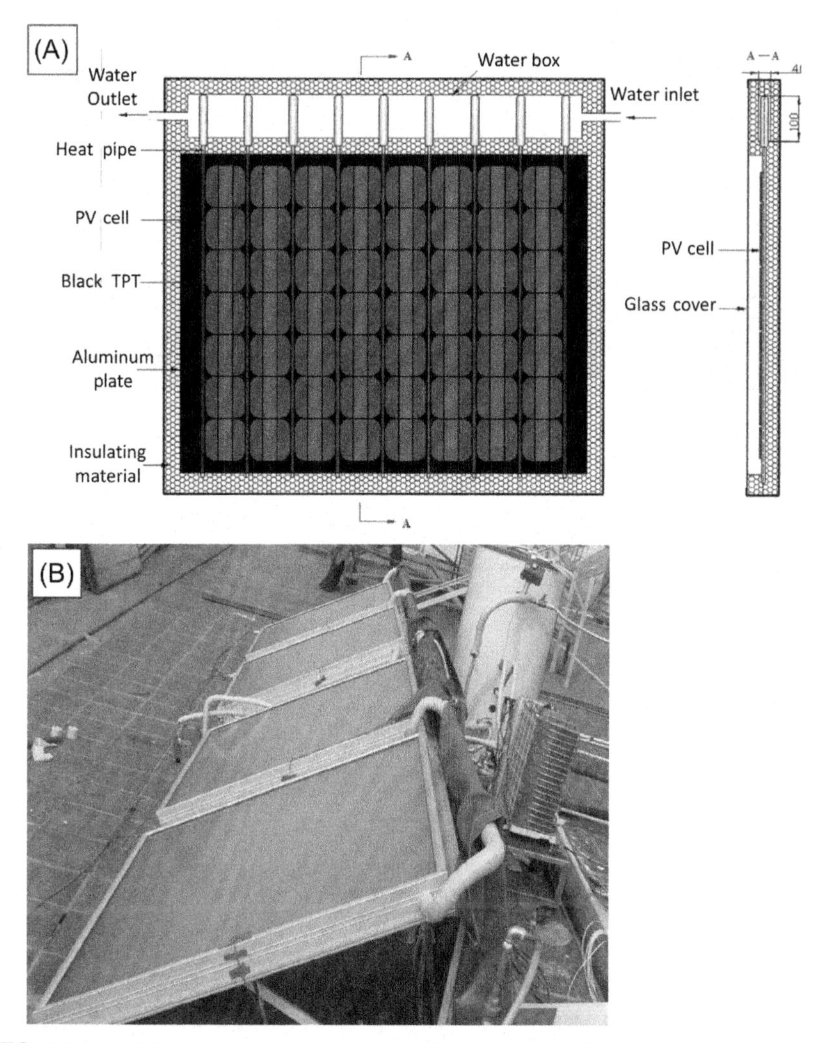

FIG. 4.14 A HP-PVT system: (A) the schematic view of the collector's main components, and (B) photograph of the experimental setup [68].

from stage 4 to 1, collects the sensible heat and becomes vapor while its temperature increased. Then the vaporized refrigerant went from stage 1 to stage 2 and becomes compressed and warmer through the compressor. Directed toward the condenser (from stage 2 to 3), the refrigerant leaves the useful heat in the water tank for further heating purposes. After its condensation, the high-pressured refrigerant flows toward the electronic expansion valve experiencing a throttling process and becomes a low-pressure liquid, ending the refrigeration cycle (stage 3 to 4). The results proved that the compressor electricity consumption is lower than the generated electric power by the PVT module. In this study, it was premised that

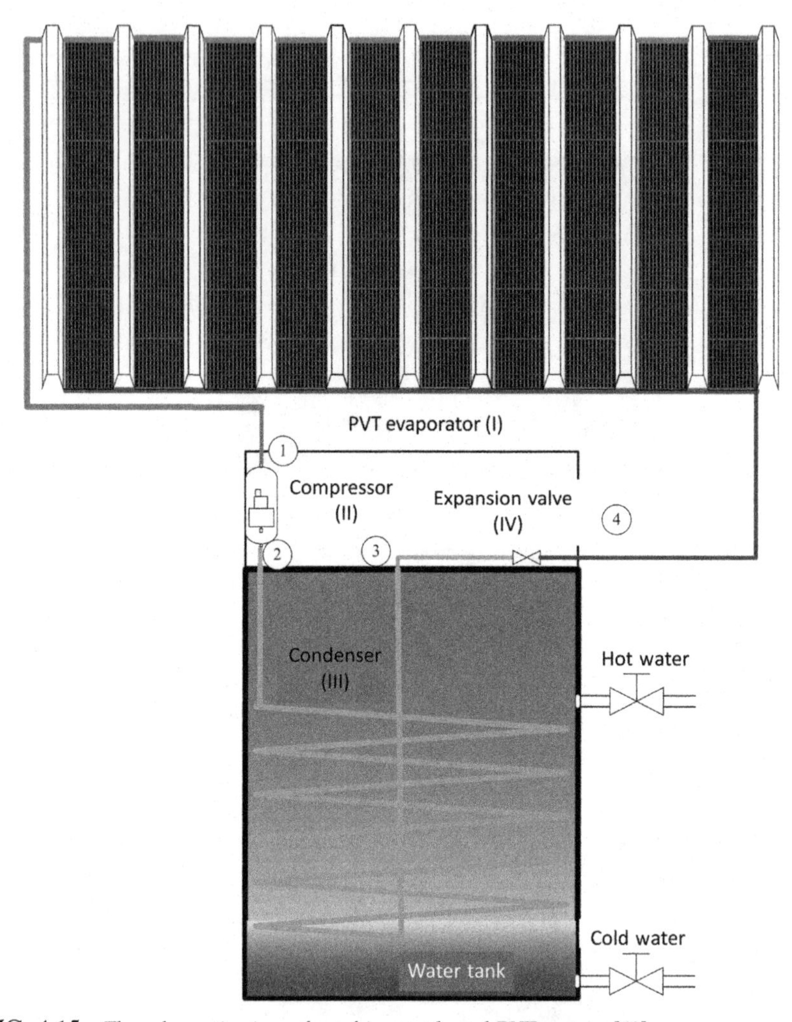

FIG. 4.15 The schematic view of a refrigerant-based PVT system [69].

the electric power generated by the PV module was utilized to drive the compressor and electronic devices.

Fu et al. [70] derived results from different experiments. The achievements showed that the system is able to yield daily average energy efficiency varying from 61.1% to 82.1% when it is performed in a solar-assisted heat-pump mode. Also, they claimed that an average coefficient of performance of 4.01 can be obtained if the high solar radiation is present. Ji et al. [71] evaluated the performance of a PV-based solar-assisted heat pump system using R22 as a refrigerant at four different operating modes. Results revealed that the average PV is improved and efficiency ranges from 5.4% to 13.4%, based on the condenser inlet temperature.

4.1.5.3 Thermoelectric-based PVT modules (TE-PVT)

There are many efforts to enhance the electric power generated from solar energy and one of the very promising solutions is to use thermoelectric (TE) modules. In the case of PVT, TE modules can be placed in the space between the PV modules and the thermal collector in the way that the heat that comes from the back of the solar cells passes through the TE modules, which produces electric power and then reaches the heat extractor medium for heating purposes. The basic arrangement of both FP-PVT and CPVT collectors integrated with TE modules is depicted in Fig. 4.16.

Chávez-Urbiola et al. [72] developed a FP-PVT module incorporated with TE generators (Fig. 4.16A). In their proposed design, a high thermally conductive back electrode was used to enhance the contact between the PV module and the TE generators. When solar energy first reaches the PV arrays, the electric power is generated and excess energy results in temperature growth in solar cells. As the heat is delivered to the TE modules, more electric power is generated from these modules, which improves the total generated electric power. Consequently, the heat reaches the thermal collector and is extracted with water. On its way toward a tank for thermal storage, water passes through a plane flat-plate thermal collector that has been designed only to enhance the thermal power gained from the entire PVT module. As Fig. 4.16B illustrates, the same principle is applied to the CPVT modules in which the PV panel must be free of any surface roughness for good thermal contact with other components [73].

4.1.5.4 PCM-based PVT modules (PCM-PVT)

One of the emerging solutions that improves the thermal efficiency of a heat extraction process is the integration of phase-change materials (PCM) with PVT modules. Having a thermal storage effect, PCMs are able to store the extra heat and release it when it is necessary, especially for nighttime operations. The use of PCMs in PVT modules has been proved as an efficiency enhancement technique because both thermal and electrical efficiencies are boosted after its integration [74]. Fig. 4.17 illustrates the main layouts of the two different configurations of the PCM-PVT modules. As Fig. 4.17A depicts, when solar radiation is delivered on the surface of a typical PCM-PVT module, the visible and ultraviolet fractions are converted to electric power by the PV module placed as the first layer.

Silicon oil may be used to ensure that there is no air gap between the PV module and the PCM tank [75]. Then the excess heat is transferred to the PCM through the upper sheet and is absorbed as sensible heat, which increases the PCM temperature until it reaches the melting point. When the PCM is completely melted, the incoming heat will be stored as sensible heat. In this process, a fraction of heat is transferred from PCM to the copper

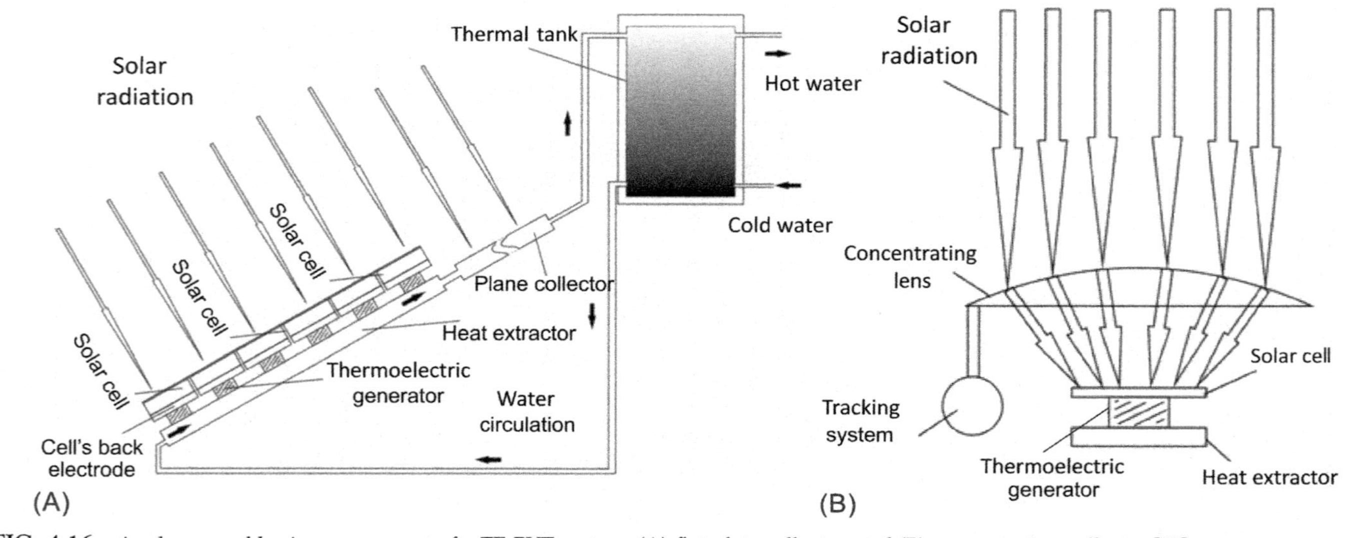

FIG. 4.16 A scheme and basic arrangement of a TE-PVT system: (A) flat-plate collector, and (B) concentrating collector [72].

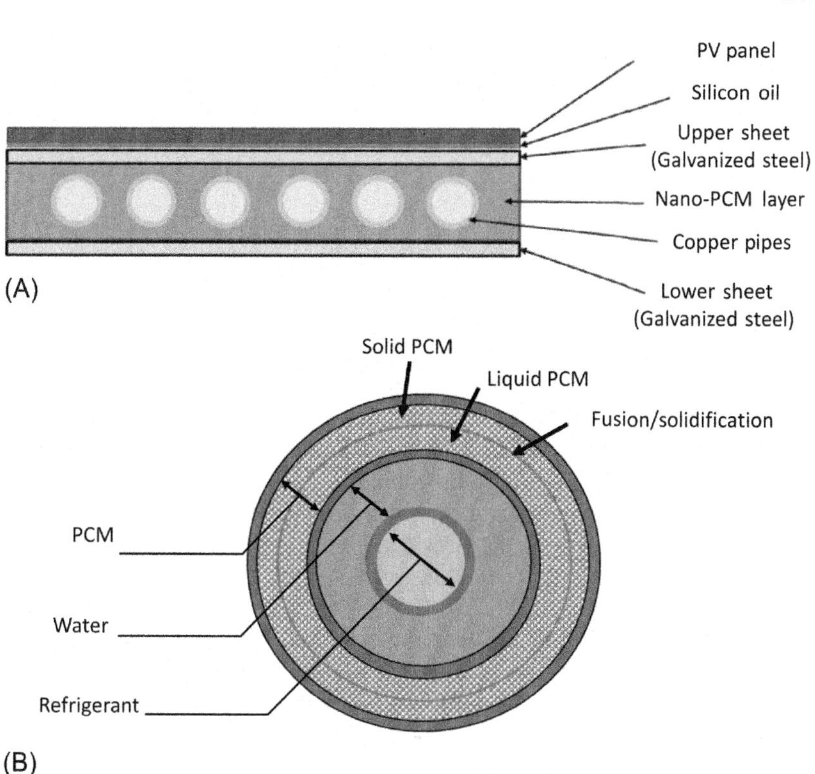

FIG. 4.17 The main layout of a PCM-PVT module: (A) the absorber enclosure is filled by the PCM [75], and (B) the tubular filling of the PCM [76].

pipes through conduction and the working fluid absorbs the heat via convection. The heated fluid will be used for different heating applications. Therefore, the state of the PCM and the system heat removal efficiency will be functions of a variety of operating factors. Afterward, when the solar incident declines, the upper sheet becomes cool and the PCM releases the heat to the copper tubes, which maintains the fluid temperature constant while it starts solidification. As a result, the PCM heat supply can be a boon to the PVT modules that work in locations with intermittent availability of sunlight. In addition, during the operation, PCM absorbs more energy from the PV module and reduces its temperature more than the conventional PVT modules. This fact also proves that PCM integration will increase electric efficiency [74].

Diallo et al. [76] proposed a novel PVT module integrated with heat pipes and the PCM. In their design, it was assumed that as the heat pipe fluid absorbs heat, the refrigerant in the evaporation section will be evaporated. As Fig. 4.17B shows, the heat is transferred from the central tube (contains refrigerant) to the water while the excess heat reaches the PCM in the outer tube and is stored as the PCM starts melting/fusion. Consequently, when heat flux from the heat pipe decreases, the PCM releases the heat to the water and maintains the hot water temperature constant. Results reported the thermal and electric efficiencies as 55.6% and 12.2%, respectively.

4.1.5.5 Beam split PVT system (BSPVT)

Most common solar cells are not able to exploit the full incoming terrestrial solar spectrum and their performance is limited to the fraction of photons that have equal energy with their bandgap. Therefore, solar radiation can be split into two parts, desirable and undesirable, for electricity generation. The undesirable part is usually visualized as losses in the PV modules. However, if a beam filter is implemented to the system and placed before the PV module, it can only transmit the desirable portion of incoming radiation, which will significantly maintain the operating temperatures of the solar cells at the desired level and improve the electric efficiency [77]. The idea of the beam split PVT (BSPVT) module has been developed for this purpose.

The BSPVT module also utilizes the unwanted fraction of solar radiation by using it for heating purposes. In research conducted by Joshi et al. [1], the liquid-based beam filters were extensively studied and at the end, selective fluids displaying viable applications were introduced for utilization in BSPVT systems. Prior to this research, there were numerous studies that concerned the spectral beam-splitting techniques beginning in 1955 and continuing through recent years that can be found in the literature [78]. Experimentally testing a liquid spectrum filter for a BSPVT system is not very abundant and is at the early stages. According to the study conducted by Joshi et. al. [79], the fundamental principle of a BSPVT module is depicted in Fig. 4.18.

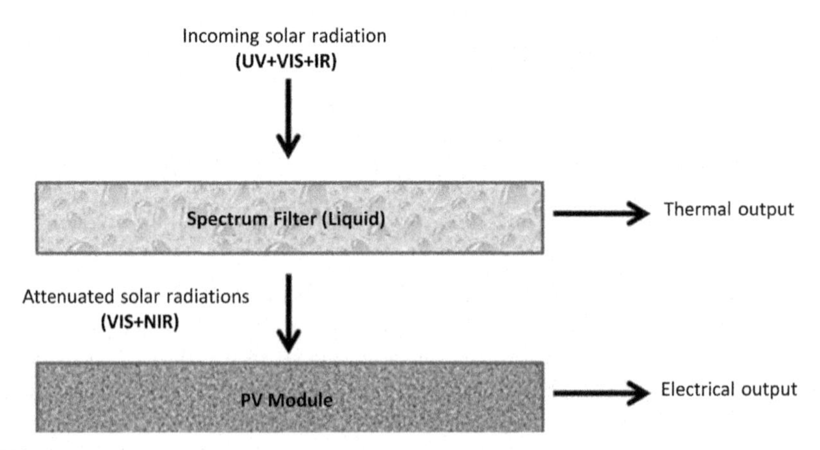

FIG. 4.18 Energy flow takes place in a BSPVT system [79].

In this process, as the solar radiation passes through the spectrum filter, UV and IR fractions are absorbed and used for thermal applications, where the rest of the energy (VIS and NIR) is delivered to the PV module. As a typical design, the filter is usually composed of two glasses and liquid, thus the attenuation of the solar beam is the product of all refraction, reflection, absorption, and transmission mechanisms taking place through these components [79]. A novel linear concentrating BSPVT module was proposed by Stanley et al. [80]. In their design, two independent heat extractors were used to provide high-grade thermal power in addition to electricity generated by the PV modules. A glass optical filter was used that only allows light with a wavelength between 700 to 1100 nm to reach the silicon solar cells and collects the reaming light as heat. A PTC with a 42 × concentration ratio was also integrated with the system to achieve the high intensity of solar energy. Two fluid channels such as a high-temperature stream and a mid-temperature stream were designed to enhance the heat absorption mechanism and increase the total thermal power.

Fig. 4.19 shows the schematic view of a BSPVT module acting as the receiver of the PTC. In this design, a semiconductor doped glass (SDG) filter was mounted in an axial direction inside the thermal receiver tube to filter out the undesirable wavelengths and transmit the desirable portion toward the PV. Reaching high temperatures, propylene glycol extracts the heat stored on the SDG and conveys it to the thermal storage facilities. In the space between the receiver channel and the PV module, sunlight passes through a transparent silicon layer, which is placed to reduce reflective losses. A thin side reflector is also mounted to the internal side of the aluminum casing to improve the concentration of transmitted light. As the cells receive desirable photons, electricity is generated. The integration of the cooling mechanism as the second heat extractor helps to harvest the dissipated heat from PV cells while keeping them at desirable operating temperature. After experimental data collection, researchers found out that the system is able to produce 120°C hot water with 31% thermal efficiency and a total system efficiency of 50%.

From the literature review, novel designs of PVT modules are increasing with a wider range of applications and integrations. Recent studies show promising results where BSPVT modules in cooperation with nanofluids as the optical filter have resulted in an exclusive performance [81]. A heat pipe integrated with a PVT module is also an interesting design that accelerates the heat transfer; however, there are some restrictions in designs such that the flat-plate and oscillating heat pipes are more favorable in this regard [22]. The utilization of PCM with PVT modules brings several advantages and can be introduced as a low-cost and practical technique for performance augmentation [82]. Table 4.2 summarizes some of the studies carried out on innovative PVT modules with the most focus on the novelty and main findings.

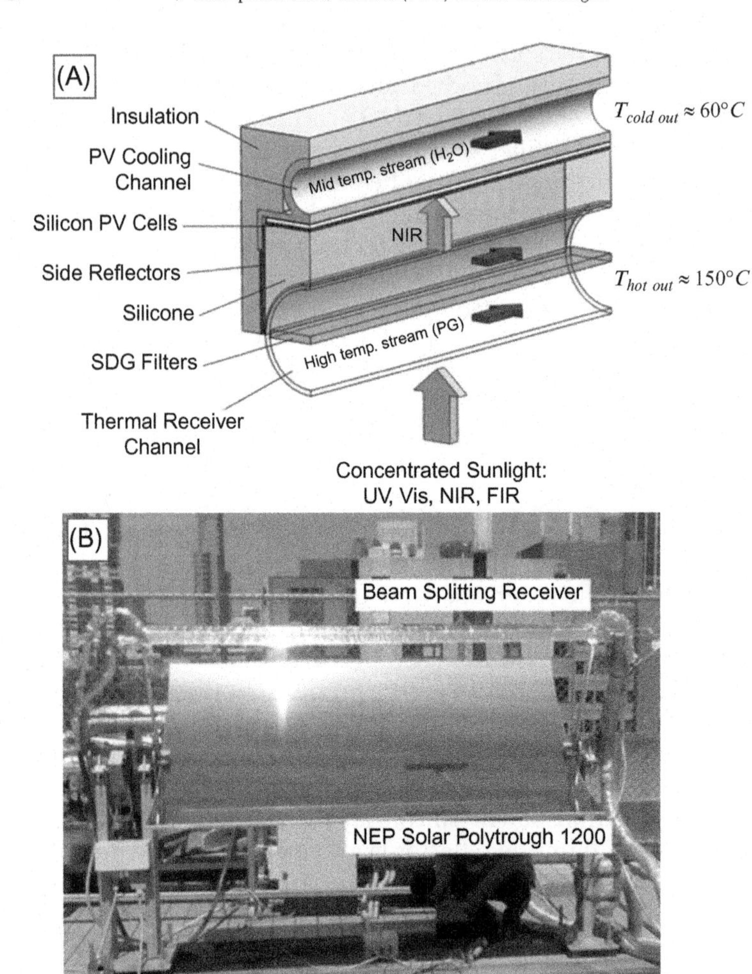

FIG. 4.19 A BSPVT module: (A) cross-section view of the receiver unit, and (B) the experimental setup [80].

TABLE 4.2 Summary of some experimental works studying different innovative PVT modules.

PVT module	Working fluid	Location	Main findings	Ref.
HP-PVT	Water/Glycol	UK	• The value of the solar thermal energy conversion efficiency of the PVT module was obtained as 50%; • The electric efficiency was improved by 15% compared to the uncooled PV.	Jouhara et al. [83]

TABLE 4.2 Summary of some experimental works studying different innovative PVT modules—cont'd

PVT module	Working fluid	Location	Main findings	Ref.
BSPV	Water, silicone oil, coconut oil	India	• Filter thickness was selected to be 1 cm; • The average electric and thermal efficiencies were determined as 12.53 and 47%; • All selected fluids are suitable for implementation with C-Si solar cells.	Joshi et al. [84]
BSPV	Nanofluid (Ag-SiO2/ Water)	Australia	• 4.5 L nanofluid was used as the selective absorbing fluid; • 12% more weighted energy output was obtained compared with standalone PV;	Crisostomo et al. [85]
TE-PVT	Air	China	• The system was implemented as a ventilator to provide fresh air for a room; • The average thermal and electric efficiencies were 26.7% and 10% for a sunny day in winter.	Liu et al. [86]
RB-PVT	Refrigerant	China	• The average value of the thermal-based overall efficiency of the unit was obtained as 150%; • The average electric efficiency of 11.8% was achieved.	Zhou et al. [87]
PCM-HP-PVT	R-134	China	• The overall efficiency was improved by 17.20% and 33.31% compared to the existing PVT and BIPVT, respectively; • The highest thermal efficiency was calculated as 71.67%.	Yu et al. [88]
HP-PVT	Water	China	• The electric and thermal efficiency increases with growth in the number of heat pipes and water mass flow rate; • The optimum length of the heat pipe condenser section was reported as 12 mm; • Water and R-134a were suggested as the working fluid for the mild and cold climate regions, respectively.	Zhang et al. [89]

Continued

TABLE 4.2 Summary of some experimental works studying different innovative PVT modules—cont'd

PVT module	Working fluid	Location	Main findings	Ref.
PCM-PVT	Water	Malaysia	• The maximum cell temperature reduction was 8.1°C for the conventional PVT system and 12°C for the PCM-PVT system; • The electric efficiencies of 13.56%, 13.74%, and 13.87% were measured for PV, PVT, and PCM-PVT, respectively.	Fayaz et al. [90]
PCM-PVT	Water	Egypt	• Nano-Al_2O_3 was used as the PCM mixture; • Testing different occupation ratios for water and PCM revealed that although 100% PCM does not yield the maximum efficiency, it can be a plausible solution for PV cooling when water usage is a problem; • 100% of water occupation resulted in the highest electric and thermal efficiencies	Salem et al. [91]
PCM-PVT	Nanofluid (SiC)	Malaysia	• Nano-SiC-paraffin was used as the PCM; • The electrical and thermal efficiencies were measured as 13.7 and 72% respectively; • The maximum temperature of the nanofluid reached 36.5°C	Al-Waeli et al. [92]

4.2 Conclusion

From the study presented in this chapter, the most commonly investigated PVT modules are FP-PVT and CPVT types, which have reached a reasonable maturity of technology and applications. Domestic water heating, agricultural drying, and space heating are among the most applied areas for the implementation of PVTs. The advent of novel designs has broadened the horizon of this technology by introducing innovative integrations with nanofluids, PCMs, spectra splitting filters, heat pipes, refrigerants, and TE generators. Recent developments in solar cell technology have also been influential on the major progress in CPVT designs, where new multijunction solar cells can endure high-temperature working conditions and last for long periods. After three decades of research and development on PVT modules, their share in the market is still small and beyond expectations. The PVT module costs are not explored well and more studies are necessary to reduce the uncertainties around this technology. Moreover, the PVT modules should be able to provide higher average temperatures, getting closer to the heat demand in the market, in order to play a more significant role in people's lives. Thus, future studies around the advanced modules with higher efficiencies, longer durability, and improved compatibility are required.

Acknowledgment

The authors acknowledge the advisory support of the Renewable Energy Research Institute and the financial support [grant number IG/39705] for the Renewable Energy Research Group of Tarbiat Modares University.

References

[1] Joshi SS, Dhoble AS, Jiwanapurkar PR. Investigations of Different Liquid Based Spectrum Beam Splitters for Combined Solar Photovoltaic Thermal Systems. J Sol Energy Eng Trans ASME 2016;138. https://doi.org/10.1115/1.4032352.
[2] Ramos CAF, Alcaso AN, Cardoso AJM. Photovoltaic-thermal (PVT) technology: review and case study IOP Conference Series: Earth and Environmental Science n.d. doi: https://doi.org/10.1088/1755-1315/354/1/012048.
[3] Zondag HA, Van Helden WGJ, Bakker M, Affolter P. PVT Roadmap: a European guide for the development and market introduction of PVT technology; 2005.
[4] Wolf M. Performance analyses of combined heating and photovoltaic power systems for residences. Energy Convers 1976;16:79–90. https://doi.org/10.1016/0013-7480(76)90018-8.
[5] Kern Jr EC, Russell MC. Combined photovoltaic and thermal hybrid collector systems. In: 13th Photovolt. Spec. Conf; 1978. p. 1153–7.
[6] Cox CH, Raghuraman P. Design considerations for flat-plate-photovoltaic/thermal collectors. Sol Energy 1985;35:227–41. https://doi.org/10.1016/0038-092X(85)90102-1.
[7] Sopian K, Liu HT, Kakac S, Veziroglu TN. Performance of a double pass photovoltaic thermal solar collector suitable for solar drying systems. Energy Convers Manag 2000;41:353–65. https://doi.org/10.1016/S0196-8904(99)00115-6.
[8] Hegazy AA. Comparative study of the performances of four photovoltaic/thermal solar air collectors. Energy Convers Manag 2000;41:861–81. https://doi.org/10.1016/S0196-8904(99)00136-3.
[9] Dubey S, Solanki SC, Tiwari A. Energy and exergy analysis of PV/T air collectors connected in series. Energy Build 2009;41:863–70. https://doi.org/10.1016/j.enbuild.2009.03.010.
[10] Ali AHH, Ahmed M, Youssef MS. Characteristics of heat transfer and fluid flow in a channel with single-row plates array oblique to flow direction for photovoltaic/thermal system. Energy 2010;35:3524–34. https://doi.org/10.1016/j.energy.2010.03.045.
[11] Kumar R, Rosen MA. Performance evaluation of a double pass PV/T solar air heater with and without fins. Appl Therm Eng 2011;31:1402–10. https://doi.org/10.1016/j.applthermaleng.2010.12.037.

[12] Cartmell BP, Shankland NJ, Fiala D, Hanby V. A multi-operational ventilated photovoltaic and solar air collector: application, simulation and initial monitoring feedback. Sol Energy 2004;76:45–53. https://doi.org/10.1016/j.solener.2003.08.037.
[13] Tiwari S, Agrawal S, Tiwari GN. PVT air collector integrated greenhouse dryers. Renew Sustain Energy Rev 2018;90:142–59. https://doi.org/10.1016/j.rser.2018.03.043.
[14] Farshchimonfared M, Bilbao JI, Sproul AB. Full optimisation and sensitivity analysis of a photovoltaic–thermal (PV/T) air system linked to a typical residential building. Sol Energy 2016;136:15–22. https://doi.org/10.1016/j.solener.2016.06.048.
[15] Othman MY, Hamid SA, Tabook MAS, Sopian K, Roslan MH, Ibarahim Z. Performance analysis of PV/T Combi with water and air heating system: an experimental study. Renew Energy 2016;86:716–22. https://doi.org/10.1016/j.renene.2015.08.061.
[16] Hischier I, Hofer J, Gunz L, Nordborg H, Schlüter A. Ultra-thin and lightweight photovoltaic/thermal collectors for building integration. Energy Procedia 2017;122:409–14. https://doi.org/10.1016/j.egypro.2017.07.457.
[17] Nualboonrueng T, Tuenpusa P, Ueda Y, Akisawa A. Field experiments of PV-thermal collectors for residential application in Bangkok. Energies 2012;5:1229–44. https://doi.org/10.3390/en5041229.
[18] Joshi SS, Dhoble AS. Photovoltaic-thermal systems (PVT): technology review and future trends. Renew Sustain Energy Rev 2018;92:848–82. https://doi.org/10.1016/j.rser.2018.04.067.
[19] Ibrahim A, Othman MY, Ruslan MH, Mat S, Sopian K. Recent advances in flat plate photovoltaic/thermal (PV/T) solar collectors. Renew Sustain Energy Rev 2011;15:352–65. https://doi.org/10.1016/j.rser.2010.09.024.
[20] El Amrani A, Mahrane A, Moussa FY, Boukennous Y. Solar module fabrication. Int J Photoenergy 2007;2007. https://doi.org/10.1155/2007/27610.
[21] Tripanagnostopoulos Y. Aspects and improvements of hybrid photovoltaic/thermal solar energy systems. Sol Energy 2007;81:1117–31. https://doi.org/10.1016/j.solener.2007.04.002.
[22] Sathe TM, Dhoble AS. A review on recent advancements in photovoltaic thermal techniques. Renew Sustain Energy Rev 2017;76:645–72. https://doi.org/10.1016/j.rser.2017.03.075.
[23] Chow TT, He W, Ji J. Hybrid photovoltaic-thermosyphon water heating system for residential application. Sol Energy 2006;80:298–306. https://doi.org/10.1016/j.solener.2005.02.003.
[24] Sardarabadi M, Passandideh-Fard M, Zeinali HS. Experimental investigation of the effects of silica/water nanofluid onPV/T (photovoltaic thermal units). Energy 2014;66:264–72. https://doi.org/10.1016/j.energy.2014.01.102.
[25] Aste N, del Pero C, Leonforte F. Water flat plate PV–thermal collectors: a review. Sol Energy 2014;102:98–115. https://doi.org/10.1016/j.solener.2014.01.025.
[26] Hasan HA, Sopian K, Jaaz AH, Al-Shamani AN. Experimental investigation of jet array nanofluids impingement in photovoltaic/thermal collector. Sol Energy 2017;144:321–34. https://doi.org/10.1016/j.solener.2017.01.036.
[27] Zondag HA, de Vries DW, van Helden WGJ, van Zolingen RJC, van Steenhoven AA. The yield of different combined PV-thermal collector designs. Sol Energy 2003;74:253–69. https://doi.org/10.1016/S0038-092X(03)00121-X.
[28] Kim J-H, Kim J-T. The experimental performance of an unglazed PVT collector with two different absorber types. Int J Photoenergy 2012;2012. https://doi.org/10.1155/2012/312168.
[29] Ziapour BM, Palideh V, Mokhtari F. Performance improvement of the finned passive PVT system using reflectors like removable insulation covers. Appl Therm Eng 2016;94:341–9. https://doi.org/10.1016/j.applthermaleng.2015.10.143.

[30] Gangadevi R, Vinayagam BK, Senthilraja S. Experimental investigations of hybrid PV/ Spiral flow thermal collector system performance using Al2O3/water nanofluid. In: IOP Conf. Ser. Mater. Sci. Eng., vol. 197. Institute of Physics Publishing; 2017. https://doi.org/10.1088/1757-899X/197/1/012041.

[31] Zhang X, Zhao X, Smith S, Xu J, Yu X. Review of R&D progress and practical application of the solar photovoltaic/thermal (PV/T) technologies. Renew Sustain Energy Rev 2012;16:599–617. https://doi.org/10.1016/j.rser.2011.08.026.

[32] Dubey S, Sandhu GS, Tiwari GN. Analytical expression for electrical efficiency of PV/T hybrid air collector. Appl Energy 2009;86:697–705. https://doi.org/10.1016/j.apenergy.2008.09.003.

[33] Tonui JK, Tripanagnostopoulos Y. Performance improvement of PV/T solar collectors with natural air flow operation. Sol Energy 2008;82:1–12. https://doi.org/10.1016/j.solener.2007.06.004.

[34] Agrawal S, Tiwari GN. Energy and exergy analysis of hybrid micro-channel photovoltaic thermal module. Sol Energy 2011;85:356–70. https://doi.org/10.1016/j.solener.2010.11.013.

[35] Othman MY, Hussain F, Sopian K, Yatim B, Ruslan H. Performance study of air-based photovoltaic-thermal (PV/T) collector with different designs of heat exchanger. Sains Malaysiana 2013;42.

[36] Farshchimonfared M, Bilbao JI, Sproul AB. Channel depth, air mass flow rate and air distribution duct diameter optimization of photovoltaic thermal (PV/T) air collectors linked to residential buildings. Renew Energy 2015;76:27–35. https://doi.org/10.1016/j.renene.2014.10.044.

[37] Slimani MEA, Amirat M, Bahria S, Kurucz I, Aouli M, Sellami R. Study and modeling of energy performance of a hybrid photovoltaic/thermal solar collector: configuration suitable for an indirect solar dryer. Energy Convers Manag 2016;125:209–21. https://doi.org/10.1016/j.enconman.2016.03.059.

[38] Athienitis AK, Bambara J, O'Neill B, Faille J. A prototype photovoltaic/thermal system integrated with transpired collector. Sol Energy 2011;85:139–53. https://doi.org/10.1016/j.solener.2010.10.008.

[39] Jarimi H, Abu Bakar MN, Othman M, Din MH. Bi-fluid photovoltaic/thermal (PV/T) solar collector: experimental validation of a 2-D theoretical model. Renew Energy 2016;85:1052–67. https://doi.org/10.1016/j.renene.2015.07.014.

[40] Brogren M, Karlsson B. Low-concentrating water-cooled PV-thermal hybrid systems for high latitudes, In: Conf. Rec. IEEE Photovolt. Spec. Conf; 2002. p. 1733–6. https://doi.org/10.1109/pvsc.2002.1190956.

[41] Bernardo LR, Perers B, Håkansson H, Karlsson B. Performance evaluation of low concentrating photovoltaic/thermal systems: a case study from Sweden. Sol Energy 2011;85:1499–510. https://doi.org/10.1016/j.solener.2011.04.006.

[42] Chemisana D. Building integrated concentrating photovoltaics: a review. Renew Sustain Energy Rev 2011;15:603–11. https://doi.org/10.1016/j.rser.2010.07.017.

[43] Daneshazarian R, Cuce E, Cuce PM, Sher F. Concentrating photovoltaic thermal (CPVT) collectors and systems: theory, performance assessment and applications. Renew Sustain Energy Rev 2018;81:473–92. https://doi.org/10.1016/j.rser.2017.08.013.

[44] Perl EE, Simon J, Friedman DJ, Jain N, Sharps P, McPheeters C, et al. (Al)GaInP/GaAs tandem solar cells for power conversion at elevated temperature and high concentration. IEEE J Photovoltaics 2018;8:640–5. https://doi.org/10.1109/JPHOTOV.2017.2783853.

[45] Kandilli C. Performance analysis of a novel concentrating photovoltaic combined system. Energy Convers Manag 2013;67:186–96. https://doi.org/10.1016/j.enconman.2012.11.020.

[46] Calise F, Daccadia MD, Vicidomini M, Ferruzzi G, Vanoli L. Design and dynamic simulation of a combined system integration concentrating photovoltaic/thermal solar collectors and organic rankine cycle. Am J Eng Appl Sci 2015;8:100–18. https://doi.org/10.3844/ajeassp.2015.100.118.
[47] Xu G, Zhang X, Deng S. Experimental study on the operating characteristics of a novel low-concentrating solar photovoltaic/thermal integrated heat pump water heating system. Appl Therm Eng 2011;31:3689–95. https://doi.org/10.1016/j.applthermaleng.2011.01.030.
[48] Xu N, Ji J, Sun W, Huang W, Jin Z. Electrical and thermal performance analysis for a highly concentrating photovoltaic/thermal system. Int J Photoenergy 2015;2015. https://doi.org/10.1155/2015/537538.
[49] Tripathi R, Tiwari GN. Annual performance evaluation (energy and exergy) of fully covered concentrated photovoltaic thermal (PVT) water collector: an experimental validation. Sol Energy 2017;146:180–90. https://doi.org/10.1016/j.solener.2017.02.041.
[50] Touafek K, Haddadi M, Malek A. Experimental study on a new hybrid photovoltaic thermal collector. Appl Sol Energy (English Transl Geliotekhnika) 2009;45:181–6. https://doi.org/10.3103/S0003701X09030104.
[51] Ameri M, Mahmoudabadi MM, Shahsavar A. An experimental study on a photovoltaic/thermal (PV/T) air collector with direct coupling of fans and panels. Energy Sources Part A Recovery Util Environ Eff 2012;34:929–47. https://doi.org/10.1080/15567031003735238.
[52] Rahou M, Othman MY, Mat S, Ibrahim A. Performance study of a photovoltaic thermal system with an oscillatory flow design. J Sol Energy Eng Trans ASME 2014;136. https://doi.org/10.1115/1.4024743.
[53] Michael JJ, Selvarasan I, Goic R. Fabrication, experimental study and testing of a novel photovoltaic module for photovoltaic thermal applications. Renew Energy 2016;90:95–104. https://doi.org/10.1016/j.renene.2015.12.064.
[54] Buonomano A, Calise F, Vicidomini M. Design, simulation and experimental investigation of a solar system based on PV panels and PVT collectors. Energies 2016;9:497. https://doi.org/10.3390/en9070497.
[55] Al-Shamani AN, Sopian K, Mat S, Hasan HA, Abed AM, Ruslan MH. Experimental studies of rectangular tube absorber photovoltaic thermal collector with various types of nanofluids under the tropical climate conditions. Energy Convers Manag 2016;124:528–42. https://doi.org/10.1016/j.enconman.2016.07.052.
[56] Kim S-M, Kim J-H, Kim J-T. Experimental study on the thermal and electrical characteristics of an air-based photovoltaic thermal collector. Energies 2019;12:2661. https://doi.org/10.3390/en12142661.
[57] Lee JH, Hwang SG, Lee GH. Efficiency Improvement of a photovoltaic thermal (PVT) system using nanofluids. Energies 2019;12:1–16.
[58] Chaabane M, Charfi W, Mhiri H, Bournot P. Performance evaluation of concentrating solar photovoltaic and photovoltaic/thermal systems. Sol Energy 2013;98:315–21. https://doi.org/10.1016/j.solener.2013.09.029.
[59] Künnemeyer R, Anderson TN, Duke M, Carson JK. Performance of a V-trough photovoltaic/thermal concentrator. Sol Energy 2014;101:19–27. https://doi.org/10.1016/j.solener.2013.11.024.
[60] Vivar M, Everett V, Fuentes M, Blakers A, Tanner A, Le Lievre P, et al. Initial field performance of a hybrid CPV-T microconcentrator system. Prog Photovoltaics Res Appl 2013;21:1659–71. https://doi.org/10.1002/pip.2229.
[61] Li G, Pei G, Ji J, Yang M, Su Y, Xu N. Numerical and experimental study on a PV/T system with static miniature solar concentrator. Sol Energy 2015;120:565–74. https://doi.org/10.1016/j.solener.2015.07.046.
[62] Karimi F, Xu H, Wang Z, Chen J, Yang M. Experimental study of a concentrated PV/T system using linear Fresnel lens. Energy 2017. https://doi.org/10.1016/j.energy.2017.02.028.

[63] Makki A, Omer S, Sabir H. Advancements in hybrid photovoltaic systems for enhanced solar cells performance. Renew Sustain Energy Rev 2015;41:658–84. https://doi.org/10.1016/j.rser.2014.08.069.

[64] Quan ZH, Li NJ, Zhao YH. Experimental study of solar photovoltaic/thermal (PV/T) system based on flat plate heat pipe. In: Asia-Pacific Power Energy Eng. Conf. APPEEC; 2011. https://doi.org/10.1109/APPEEC.2011.5749155.

[65] Xiao T, Zhen HQ, Yao HZ. Experimental investigation of solar panel cooling by a novel micro heat pipe array. IEEE Computer Society: Asia-Pacific Power Energy Eng. Conf. APPEEC; 2010. https://doi.org/10.1109/APPEEC.2010.5449518.

[66] Moradgholi M, Nowee SM, Abrishamchi I. Application of heat pipe in an experimental investigation on a novel photovoltaic/thermal (PV/T) system. Sol Energy 2014;107:82–8. https://doi.org/10.1016/j.solener.2014.05.018.

[67] Gang P, Huide F, Huijuan Z, Jie J. Performance study and parametric analysis of a novel heat pipe PV/T system. Energy 2012;37:384–95. https://doi.org/10.1016/j.energy.2011.11.017.

[68] Gang P, Huide F, Tao Z, Jie J. A numerical and experimental study on a heat pipe PV/T system. Sol Energy 2011;85:911–21. https://doi.org/10.1016/j.solener.2011.02.006.

[69] Tsai H-L. Modeling and validation of refrigerant-based PVT-assisted heat pump water heating (PVTA–HPWH) system. Sol Energy 2015;122:36–47. https://doi.org/10.1016/j.solener.2015.08.024.

[70] Fu HD, Pei G, Ji J, Long H, Zhang T, Chow TT. Experimental study of a photovoltaic solar-assisted heat-pump/heat-pipe system. Appl Therm Eng 2012;40:343–50. https://doi.org/10.1016/j.applthermaleng.2012.02.036.

[71] Ji J, Pei G, Chow TT, Liu K, He H, Lu J, et al. Experimental study of photovoltaic solar assisted heat pump system. Sol Energy 2008;82:43–52. https://doi.org/10.1016/j.solener.2007.04.006.

[72] Chávez-Urbiola EA, Vorobiev YV, Bulat LP. Solar hybrid systems with thermoelectric generators. Sol Energy 2012;86:369–78. https://doi.org/10.1016/j.solener.2011.10.020.

[73] Zakharchenko R, Licea-Jiménez L, Pérez-García SA, Vorobiev P, Dehesa-Carrasco U, Pérez-Robles JF, et al. Photovoltaic solar panel for a hybrid PV/thermal system. Sol Energy Mater Sol Cells 2004;82:253–61. https://doi.org/10.1016/j.solmat.2004.01.022.

[74] Gaur A, Ménézo C, Giroux-Julien S. Numerical studies on thermal and electrical performance of a fully wetted absorber PVT collector with PCM as a storage medium. Renew Energy 2017;109:168–87. https://doi.org/10.1016/j.renene.2017.01.062.

[75] Al-Waeli AHA, Sopian K, Chaichan MT, Kazem HA, Ibrahim A, Mat S, et al. Evaluation of the nanofluid and nano-PCM based photovoltaic thermal (PVT) system: an experimental study. Energy Convers Manag 2017;151:693–708. https://doi.org/10.1016/j.enconman.2017.09.032.

[76] Diallo TMO, Yu M, Zhou J, Zhao X, Shittu S, Li G, et al. Energy performance analysis of a novel solar PVT loop heat pipe employing a microchannel heat pipe evaporator and a PCM triple heat exchanger. Energy 2019;167:866–88. https://doi.org/10.1016/j.energy.2018.10.192.

[77] Joshi SS, Dhoble AS. Photovoltaic-thermal systems (PVT): technology review and future trends. Renew Sustain Energy Rev 2018;92:848–82. https://doi.org/10.1016/j.rser.2018.04.067.

[78] Imenes AG, Mills DR. Spectral beam splitting technology for increased conversion efficiency in solar concentrating systems: a review. Sol Energy Mater Sol Cells 2004;84:19–69. https://doi.org/10.1016/j.solmat.2004.01.038.

[79] Joshi SS, Dhoble AS. Analytical approach for performance estimation of BSPVT system with liquid spectrum filters. Energy 2018;157:778–91. https://doi.org/10.1016/j.energy.2018.05.204.

[80] Stanley C, Mojiri A, Rahat M, Blakers A, Rosengarten G. Performance testing of a spectral beam splitting hybrid PVT solar receiver for linear concentrators. Appl Energy 2016;168:303–13. https://doi.org/10.1016/j.apenergy.2016.01.112.

[81] Ahmed A, Baig H, Sundaram S, Mallick TK. Use of nanofluids in solar PV/thermal systems. Int J Photoenergy 2019;2019:1–17. https://doi.org/10.1155/2019/8039129.

[82] Sarafraz MM, Safaei MR, Leon AS, Tlili I, Alkanhal TA, Tian Z, et al. Experimental investigation on thermal performance of a PV/T-PCM (photovoltaic/thermal) system cooling with a PCM and nanofluid. Energies 2019;12:2572. https://doi.org/10.3390/en12132572.

[83] Jouhara H, Milko J, Danielewicz J, Sayegh MA, Szulgowska-Zgrzywa M, Ramos JB, et al. The performance of a novel flat heat pipe based thermal and PV/T (photovoltaic and thermal systems) solar collector that can be used as an energy-active building envelope material. Energy 2016;108:148–54. https://doi.org/10.1016/j.energy.2015.07.063.

[84] Joshi SS, Dhoble AS. Experimental investigation of solar photovoltaic thermal system using water, coconut oil and silicone oil as spectrum filters. J Brazilian Soc Mech Sci Eng 2017;39:3227–36. https://doi.org/10.1007/s40430-017-0802-0.

[85] Crisostomo F, Taylor RA, Rosengarten G, Everett V, Hawkes ER. Performance of a linear Fresnel-based concentrating hybrid PV/T collector using selective spectral beam splitting, In: Proceedings of the 52nd Annual Conference, Australia Solar Energy Society (Australian Solar Council); 2014.

[86] Liu Z, Zhang L, Luo Y, Zhang Y, Wu Z. Performance evaluation of a photovoltaic thermal-compound thermoelectric ventilator system. Energy Build 2018;167:23–9. https://doi.org/10.1016/j.enbuild.2018.01.058.

[87] Zhou C, Liang R, Zhang J, Riaz A. Experimental study on the cogeneration performance of roll-bond-PVT heat pump system with single stage compression during summer. Appl Therm Eng 2019;149:249–61. https://doi.org/10.1016/j.applthermaleng.2018.11.120.

[88] Yu M, Chen F, Zheng S, Zhou J, Zhao X, Wang Z, et al. Experimental investigation of a novel solar micro-channel loop-heat-pipe photovoltaic/thermal (MC-LHP-PV/T) system for heat and power generation. Appl Energy 2019;256:113929. https://doi.org/10.1016/j.apenergy.2019.113929.

[89] Zhang T, Yan ZW, Xiao L, Fu HD, Pei G, Ji J. Experimental, study and design sensitivity analysis of a heat pipe photovoltaic/thermal system. Appl Therm Eng 2019;162. https://doi.org/10.1016/j.applthermaleng.2019.114318.

[90] Fayaz H, Rahim NA, Hasanuzzaman M, Rivai A, Nasrin R. Numerical and outdoor real time experimental investigation of performance of PCM based PVT system. Sol Energy 2019;179:135–50. https://doi.org/10.1016/j.solener.2018.12.057.

[91] Salem MR, Elsayed MM, Abd-Elaziz AA, Elshazly KM. Performance enhancement of the photovoltaic cells using Al2O3/PCM mixture and/or water cooling-techniques. Renew Energy 2019;138:876–90. https://doi.org/10.1016/j.renene.2019.02.032.

[92] Al-Waeli AHA, Chaichan MT, Sopian K, Kazem HA, Mahood HB, Khadom AA. Modeling and experimental validation of a PVT system using nanofluid coolant and nano-PCM. Sol Energy 2019;177:178–91. https://doi.org/10.1016/j.solener.2018.11.016.

CHAPTER

5

Solar PV systems design and monitoring

Mohammadreza Aghaei[a,b], *Nallapaneni Manoj Kumar*[c], *Aref Eskandari*[d], *Hamsa Ahmed*[e], *Aline Kirsten Vidal de Oliveira*[f], *and Shauhrat S. Chopra*[c]

[a]Eindhoven University of Technology (TU/e), Design of Sustainable Energy Systems, Energy Technology, Department of Mechanical Engineering, Eindhoven, The Netherlands, [b]Department of Microsystems Engineering (IMTEK), University of Freiburg, Solar Energy Engineering, Freiburg im Breisgau, Germany, [c]School of Energy and Environment, City University of Hong Kong, Kowloon, Hong Kong, [d]Electrical Engineering Department, Amirkabir University of Technology (Tehran Polytechnic), Tehran, Iran, [e]Physics Department, Oldenburg University, Oldenburg, Germany, [f]Federal University of Santa Catarina, Florianopolis, Brazil

5.1 PV system components

A solar PV system is a combination of numerous subcomponents with specific functionality. However, the overall function of the PV system is to generate electricity from incoming solar radiation. Depending upon the installation type, a few additional components might be required; however, the basic components required for electricity generation would be more or less similar for all types of PV installations. The required components are grouped into four categories: PV cells/modules/strings, power electronic components, energy storage devices, and electrical and mechanical components [1–3].

5.1.1 PV modules

A PV cell is one of the specialized semiconductor diodes. It works on the principle of the photovoltaic effect. PV cells convert light energy into electricity. Most of the PV cells help in electricity generation by

Photovoltaic Solar Energy Conversion
https://doi.org/10.1016/B978-0-12-819610-6.00005-3

absorbing light from the visible range of the solar spectrum. A few solar cells are used for the same functionality by light absorption from infrared and ultraviolet radiation. The output generated by the PV cells is very low, so such cells are grouped together to increase the electricity output. The grouping of these solar cells is done based on the voltage and current requirements. Series and parallel combinations are preferred, where the series combination of two or more PV cells adds up the voltages to a desired value. Similarly, the parallel combination of such series connected PV cells adds up the current levels to a desired level. In order to get favorable voltage levels, several PV cells are connected, making a PV module. These PV modules are further grouped based on the voltage requirements. In most solar power plant installations, a wide range of system voltage is chosen (typically ranging from 48 to 600 V). For increasing voltage levels, the modules are grouped in series connection forming a string. Fig. 5.1 shows the string connection of PV modules [4].

5.1.2 Power electronic components

Power electronic components are widely used in PV systems. The most common types of power electronic devices are power conditioning units, for example, the inverter. The PV array generally produces direct current (DC) electricity, which has to be further converted to alternative current (AC). Inverters are used for converting DC electricity to AC. In a PV system, the inverter selection is more crucial, and this generally decides the DC system operating voltage. There are wide ranges of inverters on the market, and the selection can be made based on the system voltage and required peak power rating. The selection of inverter should be very effective as it may have an effect on the overall performance of the PV system [4]. There is a wide range of inverter sizes and types for grid-connected PV systems. Most of the

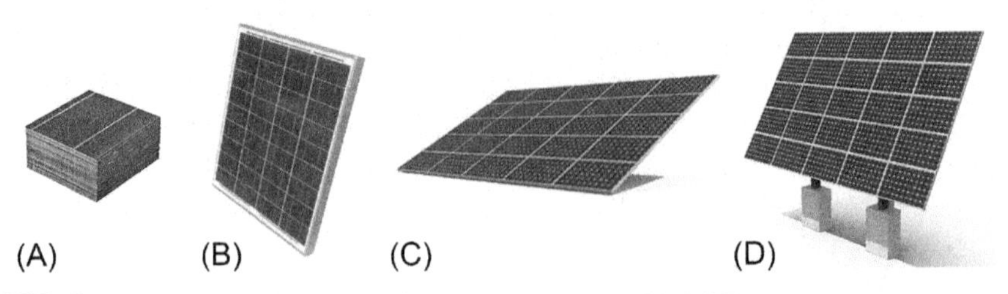

FIG. 5.1 (A) PV cells; (B) PV module; (C) PV array; and (D) PV system.

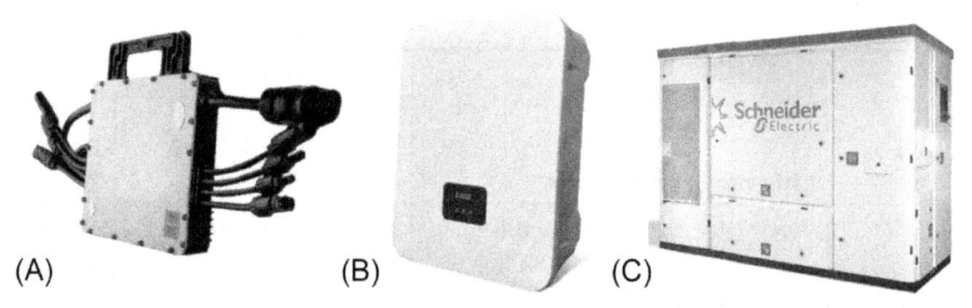

FIG. 5.2 (A) Microinverter; (B) string inverter; and (C) central inverter. *(A) Source: Hoymiles, (B) Source: EAST Group, (C) Source: Schneider electric.*

grid-connected PV systems use one of the following types of inverters, shown in Fig. 5.2.

- Microinverters.
- String inverters.
- Central inverters.

This classification is based on the size and output power ratings of the inverters. Among these three types, the microinverters have a low output power rating and appear to be in smaller sizes. The other two types (string and central) have higher output power ratings and are available in a wide range of sizes. In terms of connection, the microinverters are connected to a single PV module, where they convert the DC output of that particular module to AC output. The string and central inverter are connected to the PV strings based on the current and voltage ratings [5].

5.1.3 Energy storage

In a PV system, energy storage devices are used. Depending on the type of PV plant, energy storage can be planned. In a standalone PV system, an energy storage option is commonly used whereas in the grid, a connected energy storage system may or may not be used. There exist numerous energy storage options for PV systems; however, the most widely used are batteries and pumped energy storage. A brief description of these energy storage options is given below and the basic comparison is given in Table 5.1.

Battery energy storage: These are the most commonly used energy storage options for PV systems. Batteries store the energy generated by the PV array in the form of chemical energy. There are a wide range of batteries on the market and depending upon the PV plant capacity, batteries can be selected. Also, based on load condition and the required days of autonomy, the battery size can be decided.

TABLE 5.1 Most appropriate energy storage options for solar PV plants.

Energy storage	Efficiency	Drawback	Feasibility		
			Stand-alone PV	Grid-connected PV	Hybrid PV
Battery	Very high	Energy loss due to system operating and weather conditions	Feasible	Feasible	Feasible
Pumped energy	Very high	Loss of stored water due to evaporation	Not feasible	Feasible	Feasible

Pumped energy storage: This is also one of the energy storage option used in many of the solar PV plants in coordinated operation with hydro-power plants. In this storage, the excess energy from the solar PV plant is stored in the form of the gravitational potential energy of water. Here, the water from low elevated heights is pumped to an overhead storage tank that is at a higher elevation. Whenever there is a need for energy use, the water from the overhead tank is allowed to generate power using a water turbine.

5.1.4 Electrical and mechanical components

In PV systems, there are a few electrical and mechanical components. The electrical components include the cables, junction boxes, etc., and the mechanical components include the support structures used for installation. When it comes to tracking PV systems, the components are slightly more in number when compared to fixed mode installation. Here, mostly electrical motors are used for tilt angle adjustment [5].

5.2 Types of PV systems

Broadly, PV systems are classified into three types: grid-connected PV systems, stand-alone PV systems, and hybrid PV systems.

5.2.1 Grid-connected PV systems

PV systems whose power is directly fed into the utility or electric grid are generally known as grid-connected PV systems. These are also called on-grid or grid-tied PV systems. These PV systems are capable of only feeding energy into the grid. A typical grid-connected PV system consists of components of PV modules, an inverter, a transformer, and a utility meter. The schematic view of a grid-connected PV system is shown in Fig. 5.3 [5].

5.2.2 Stand-alone PV systems

PV systems that generate electricity to be used locally at the generation center without being injected into a utility grid are called stand-alone PV systems. Here, mostly the energy generated is consumed and any available excess will be stored in batteries. A few examples of such systems are solar streetlights, solar water pumping, and rooftop home solar PV systems. Fig. 5.4 shows the schematic view of a stand-alone PV system.

5.2.3 Hybrid PV systems

In PV systems, we have already shown two different classifications in the above sections. The hybrid solar PV systems typically represent the combination of both grid-tied and stand-alone. In the hybrid configuration,

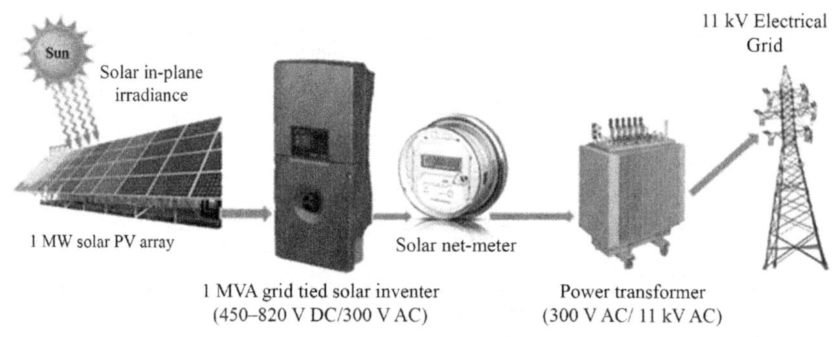

FIG. 5.3 Schematic view of a 1MWp solar PV system connected to a utility grid [5].

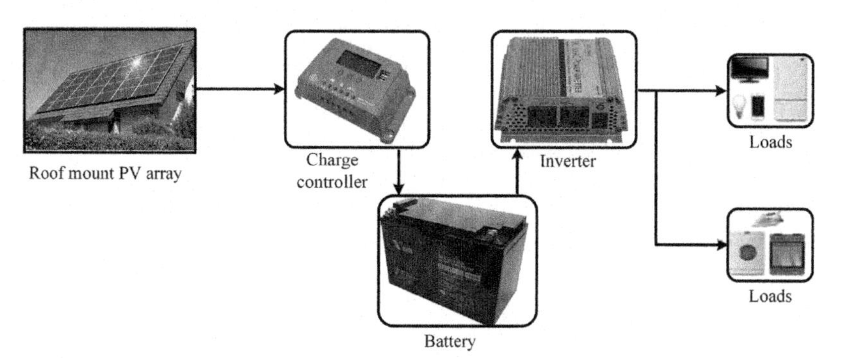

FIG. 5.4 Schematic view of a stand-alone PV system [6].

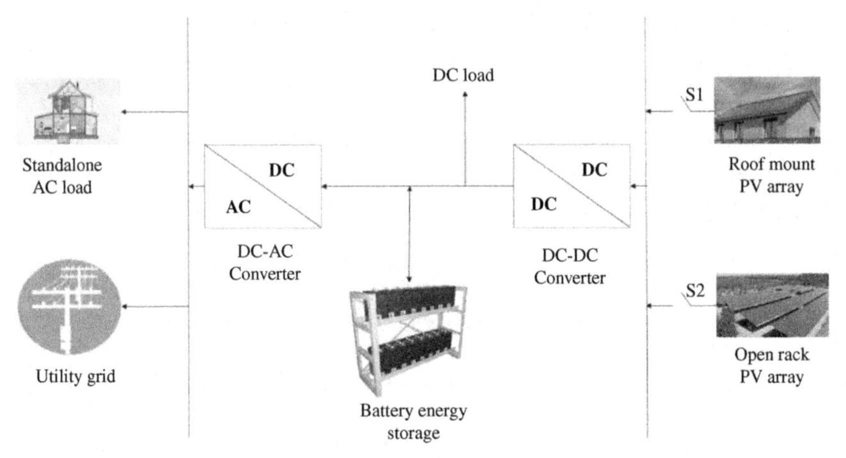

FIG. 5.5 Schematic view of a hybrid PV system [7].

the PV system is capable of generating electrical power, which is locally consumed or stored in batteries, and the excess is injected into the grid by means of net metering. Compared to the other two classifications, the stability in hybrid PV systems is quite high. The main reason is having a sufficient energy backup in the form of batteries. Fig. 5.5 depicts the schematic view of a hybrid PV system.

5.3 Design of PV systems

Designing and sizing PV systems is the most crucial stage in PV implementation. This involves a systematic approach where the collective efforts of multidisciplinary teams should be needed. A five-step procedure for designing a solar PV system includes [5, 8]:

Step 1 Planning and site survey.
Step 2 Assessment of energy requirements.
Step 3 Assessment of solar resource availability.
Step 4 Sizing of the main components of the PV system.
Step 5 Selection of the main components of the PV system.

In the three PV system types, the above-mentioned five-step procedure is adapted to design the system. Almost all the steps are followed in a similar way in each classification; however, Step 4 is different for the grid-connected and stand-alone PV systems.

5.3.1 Grid-connected PV systems

While sizing the grid-connected PV system, Step 1 and Step 3 are chosen initially to understand the location suitability for PV plant installation and to assess the solar resource potential. After the clearance of these two steps, the sizing of the PV system and its components is considered. In grid-connected PV systems, the array capacity is generally a chosen value depending upon the area availability. Once the PV array capacity is chosen, the next major step is sizing the inverter. In grid-connected PV systems, the inverter power sizing is a very delicate problem, where many installers would recommend having an inverter with a PV array power ratio of 1.0–1.1. However, the inverter sizing should be made by considering the overload condition where the energy loss is high during the operation phase of the PV plant. Hence, while sizing the grid-connected PV system's inverter, two main conditions are considered [9]:

Overload behavior: In PV systems, overload behavior is commonly seen and it is the issue of improper system component sizing. At present, with all existing modern inverters, this behavior can be avoided by ensuring that the Pmpp of the array overcomes its Pnom DC limit. This condition allow the inverter to operate as expected at its nominal power condition. This condition ensures that there won't be any overheating of the system component.

Loss evaluation: It is a method that gives clear understanding on inverter sizing for a grid-connected PV. In this condition, power distribution diagrams are plotted considering the operating conditions, which are then used for energy loss evaluation. The difference of the point where maximum power potential is obtained to the nominal DC power results in energy loss. While sizing an inverter, low power or energy losses are ensured in the PV system [9].

5.3.2 Stand-alone PV systems

A stand-alone PV system design follows the five-step procedure mentioned in the previous section. In the first step, planning and site survey are performed. Step 2 is followed by conducting a questionnaire that is related to energy consumption. Step 3 is to check the climatic conditions and to confirm whether sufficient solar resources are available. Once the solar resource potential is identified, the panel generation factor is evaluated. It generally determines the peak watt rating of the PV plant. It also gives information on the numbers of PV panels required and the actual peak sunshine available at the installation site [4].

$$\text{Panel generation factor} = \frac{I_{POA,t} \times \text{Sunshine}}{I_o} \tag{5.1}$$

Once the panel generation factor is estimated, the energy required from the PV modules can be estimated by using the data from the load assessment step. Based on the energy requirement value, the total peak rating of the PV is estimated by using the following relation [10]:

$$\text{Total Watts rating} = \frac{\text{Energy required}}{\text{Panel generation factor}} \tag{5.2}$$

Based on the obtained total watt rating, the number of PV modules required to meet the load is estimated as follows [10]:

$$\text{No. of PV modules} = \frac{\text{Total Watts rating}}{\text{Panel peak capacity}} \tag{5.3}$$

Once the number of PV modules is estimated, the inverter sizing is done. In most cases, the inverter size is considered as 25–30% higher peak rating of the PV array. Say, for example, a 100 kW solar PV system is required and 130 kW of inverter. A simple relation for identifying the inverter size is given as follows [10]:

$$\text{Inverter size} = \text{Total Watt rating} \times 1.3 \tag{5.4}$$

The next component to be chosen is the battery, and battery sizing is done as follows [10]:

$$\text{Battery rating} = \frac{\text{Total Watt} \times \text{Hours of operation} \times \text{Days of atomony}}{\text{Efficency} \times \text{Depth of discharge} \times \text{System Voltage}} \tag{5.5}$$

5.3.3 Hybrid PV systems

As discussed in the previous section, the hybrid PV system consists of both stand-alone operations as well as grid-connected modes. The selected sizing procedure in the hybrid design is almost the same as the procedure given above. However, in hybrid PV systems, load is considered as the first preference, followed by battery storage and grid connection.

5.4 Failures and faults of PV systems

Among the most common failures that affect PV systems are junction box failures, bypass diode failures, and broken glasses, which most likely happen and cause a high loss of energy production, in addition to high electrical shock and fire hazard risks. Problems with soiling or shading are also very common and cause a significant loss of efficiency. They can also cause mismatch faults that interfere in the maximum power point (MPP) tracking system, reducing power production and producing hot spots. Mismatch faults can also be generated by encapsulant degradation, antireflection coating deterioration, and manufacturing defects [9, 11]. Most of the other possible failures are rare and less hazardous to energy production and safety [12].

However, there are other components of the PV system that are sources of failure . The inverters, for instance, are considered the leading cause of PV system failure. The inverters are likely to fail because they are also the most complex and active component of the PV systems [3]. However, their faults are also easy to detect because their malfunction causes high energy production loss [13]. Inverter problems can be classified into three categories [14, 15]:

- Manufacturing and design problems: not designed to certain environmental conditions, poor component or cooling mechanism sizing.
- Control problems: interaction of inverter with the grid and with the PV panels.
- Electrical component failures: mostly due to thermal and electrical stress, with electrolytic capacitors, the IGBT, and cooling and circulating fans cited as common sources of failures in inverters.

Other faults can occur in other components of PV systems such as fuses, switches, DC contactors, surge arrestors, cables and other protections, and the so-called BoS. The failures of BOS components can suppress the production of several PV modules at once, as one component failure can affect the entire string or even an entire string box [15]. They also affect the safety of the PV system, as undersizing or failures of components and cables can cause fire hazards. Also, poor connections and wiring as well as connection insulation rupture and protective device failure can also be related to severe problems such as electric arcs and short circuits, as listed below [16]:

- Ground faults: an electrical short circuit that creates a low impedance path between an active cable and the ground. Ground fault detection and interruption devices are installed to mitigate ground fault effects.
- Line-to-line faults: electrical short circuit between two active points in a PV system, occurring between two points on the same string or between two adjacent strings. Depending on the current flow in the short circuit, current protective devices can detect those faults.
- Arc faults: current path in the air caused by any discontinuity in the conductors or insulation breakdowns in adjacent current-carrying conductors. Arcs can cause fire hazards and can be caused by solder disjoints, cell damage, corrosion of connectors, abrasion from different sources, etc. Most of the arcs can be extinguished by disconnecting the PV system.

PV energy generation can also be reduced by problems in the tracker system, mainly because of failures in the engine or the gearbox [12]. The choice for central trackers or decentralized trackers directly affects the effect of this kind of failure on a PV plant.

In off-grid systems or hybrid systems where the installation of a battery is required, faults in this equipment can also be observed, such as unusual degradation of the batteries, connection faults, high self-consumption, increase of series resistance, and degradation due to high-temperature operation or charge controller failures.

5.5 PV monitoring system

A PV monitoring system is very essential to assure the expected and permanent performance of a PV system. The monitoring system collects the required data in a PV system and transmits it to the control center that lets users evaluate and control the system to decrease maintenance costs, monitor the performance indicators of power generation, and keep track of fault events. In recent years, different PV monitoring systems have been presented. In this part, we present an overview of different PV monitoring systems. This consists of a detailed review of different technologies and techniques in PV monitoring systems [2, 9, 17–27].

5.5.1 Overview of PV monitoring systems

Fig. 5.6 illustrates the general scheme of a PV monitoring system. The basic components used in PV monitoring systems are sensors that measure the parameters in a PV system in actual conditions. The signal processing unit is another significant unit. This unit amplifies and clears signals for subsequent processing. Also, this unit includes a processor that sends the signal processing unit outputs to a PC through a dedicated

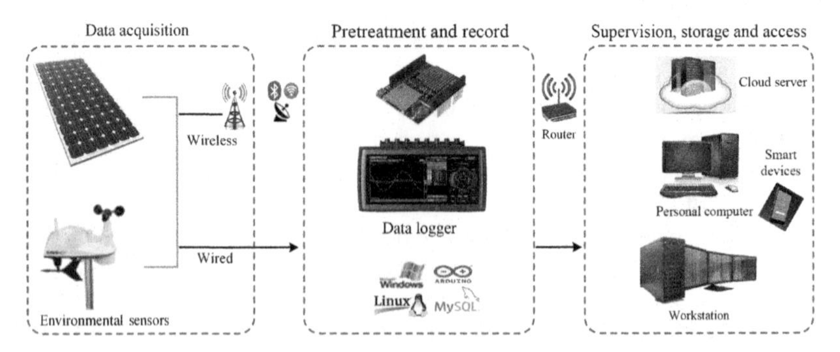

FIG. 5.6 General scheme of a PV monitoring system [28].

protocol in real time. The PC is applied for analyzing, saving, and showing data. According to data analysis and user commands, information is transmitted to the control unit for subsequent operation [28, 29].

5.5.2 Sensors and measured parameters

The power output of solar energy is intermittent and unpredictable. The PV output power may change drastically due to dependence on environmental conditions. Also, high PV penetration levels lead to fluctuations in the utility grid. These aforementioned problems can cause instability in the utility grid. Hence, monitoring the PV system parameters is essential to ensure safe operation and integration of the utility grid with high PV penetration. A significant duty in the PV monitoring system is measuring the parameter selection. The guidelines for this parameter selection are presented in accordance with standard IEC 61724 [30]. Table 5.2 shows a list of measured parameters.

Sensor selection depends on the monitoring objectives and location. The main sensors used in the PV monitoring system to evaluate the aforementioned parameters are current sensors, voltage sensors, solar irradiance

TABLE 5.2 Parameters to be measured [29].

Parameters	Specific parameters	Symbol
Meteorology	Irradiance	G
	Air speed	S_w
	Ambient temperature	T_{am}
PV array	Module temperature	T_m
	Output power	P_a
	Output current	I_a
	Output voltage	V_a
Storage system	Current from storage	I_{fs}
	Current to storage	I_{rs}
	Operating voltage	V_s
	Power from storage	P_{fs}
	Power to storage	P_{rs}
Load	Load power	P_L
	Load current	I_L
	Load voltage	V_L
Electrical grid	Power from the utility grid	P_{fu}
	Power to the utility grid	P_{ru}
	Current from the utility grid	I_{fu}
	Current to the utility grid	I_{ru}
	Utility voltage	V_u

sensors, temperature sensors, anemometer wind speed sensors, hygrometer sensors, and barometer pressure sensors.

5.5.3 Data acquisition system (DAS)

The data acquisition system (DAS) plays an important role in any monitoring system and is used to collect data from different sensors of a PV system. Then, this data is digitalized for storage and the DAS sends data to the control center for processing and presentation [31]. The basic scheme of DAS is illustrated in Fig. 5.7.

The controller is one of the most important components in the DAS. It manages data from the sensors to the control center. Hence, the proper choice of controller is of great importance. The microcontroller, DAS, data-logger, and module are used as the controller. Compared to the DAS and modules, microcontrollers and data-loggers are cheaper and can be easily programmed [32].

5.5.4 Data transmission methods

Data transmission is the transfer of data from point to point. The components of any communication system include a transmitter that prepares the taken data from sensors, a communication channel that carries the data to its destination, and a receiver that records, presents, and controls the data. The communication channels used to send data are listed below:

5.5.4.1 Wired communication

Wired communication consists of two subcategories. Coaxial cables have good bandwidth, a low error rate, and low resistance. The main advantage of coaxial cable is the low leakage of magnetic and electric fields. However, the covered distance is its major limitation [33].

Second, optical fiber cable allows data transmission over long distances with high bandwidth in comparison to electrical cables. Also, optical fiber presents a high speed of data transfer from 100 to 200 Mbps. But it is fragile, and expensive [34].

5.5.4.2 Wireless communication

Wireless communication technologies have been introduced widely in the literature. Drews et al. [35] demonstrated that using satellite produces a slow and expensive method for data transmission. In ref. [36], a global

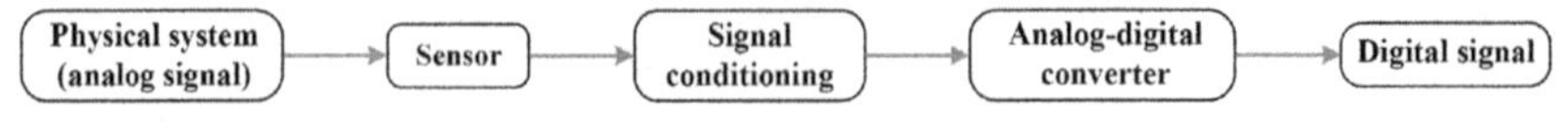

FIG. 5.7 Basic scheme of a DAS [29].

system mobile (GSM) mechanism using short message service (SMS) and general packet radio service (GPRS) were proposed to transmit data with high reliability and precision. This mechanism transmits data at a high speed and large volume. GPRS devices allow high-speed data transfer. Due to the payment for these services, this mechanism has a high operating expense.

Another option for wireless communication is the radio frequency. It is an excellent choice in regions without phone lines. This mechanism is low cost in sending and receiving data. However, this mechanism requires transfer frequency permission and also costs a lot. Therefore, it is hard to implement [14].

Wireless local area networking (WLAN) is another choice for data transfer. This mechanism covers a vast area [37]. WLAN has flexibility in data transfer and also can communicate without limitations in the future. Nevertheless, this mechanism has a few drawbacks such as less bandwidth and poorer service quality because of interference [38].

Hua et al. [39] presented a simple network called Bluetooth. The main disadvantage of this method is that it does not communicate over long distances. Also, wi-fi has good speed in data transfer, but it is costly in comparison with other mechanisms such as Bluetooth and ZigBee [40].

The ZigBee module is a low-cost and low-power wireless control and monitoring application. It is intended to be simpler and less expensive than other wireless mechanisms. It is typically integrated with microcontrollers and wireless mechanisms (nodes), which leads to supporting a big network capacity. Last, TCP/IP is commonly known as internet protocol. It is intended as the most appropriate mechanism for real-time monitoring applications [41].

5.5.4.3 *Power line communication (PLC)*

The only wired communication technology at a comparable cost to wireless communication technologies is the power line communication (PLC). This mechanism works at a high frequency with data transfer speeds around 200 Mbps [42].

5.5.5 Data storage methods

Data storage is required for performing any analysis. The full set of collected data can be accumulated in an SD (secure digital) card. The SD card is a dedicated nonvolatile memory card format, which keeps the data stable even though the electricity is turned off. SD cards allow lower electromagnetic interference and prepare an easy solution for data storage and tracking. Moreover, they have more storage capacity than other storage techniques. SD cards have a few disadvantages such as fragility, inappropriate placement, and being easily influenced with viruses [29].

A digital signal oscilloscope (DSO) is employed that can display and store electrical signals in ref. [43]. A database system called My Sequel is also used for storing data. Due to storage in the HTML format, this method yields better performance and data storage. Also, users can easily access the data through any web browser [44].

5.5.6 Data analysis methods

Data analysis is very crucial for conclusions about system performance. Various techniques have been introduced for data analysis in order to assess the performance of different PV systems. It is not easy to evaluate the output performance of a PV system because many variables affect it. The major problems are the dependence of the system output on environmental and climate factors [45].

Nowadays, standard guidelines are used to evaluate the performance of a PV system. These guidelines explain energy generation, system losses, and solar sources [30]. There are two main disadvantages in these guidelines: (1) they do not present accurate information about the system performance, and (2) they do not allow the evaluation of the performance of devices individually in a PV system. A two-step approach has been proposed to enhance the specifications of existing software [46]. This approach has used MATLAB software with statistical analysis for implementation. The first step describes the performance criteria with using the collected offline data of the PV monitoring system. The second step examines system performance by using real-time data. These two approaches provide a suitable solution for diagnosing and locating malfunctions that was not possible through standard guidelines.

Another approach for assessing the performance of the PV system is the LABVIEW interface. LABVIEW is a graphical user interface built by a virtual instrument (VI). LABVIEW supports the automatic control of system parameters and data acquisition. It is also able to communicate with MATLAB. Therefore, the system designer is given the ability to optimally control and monitor the PV systems. Also, LABVIEW uses a high pass filter to eliminate noise existing in the collected signals [47]. The LABVIEW interface was developed to monitor a solar PV plant in Bogota [48]. Microsoft Excel is used to store solar radiation data and analyze data using various statistical methods. Then, LABVIEW calculates the efficiency of the PV system and power conditioning unit (PCU). In addition, LABVIEW uses Fourier analysis for harmonic AC signal which produced through the PCU.

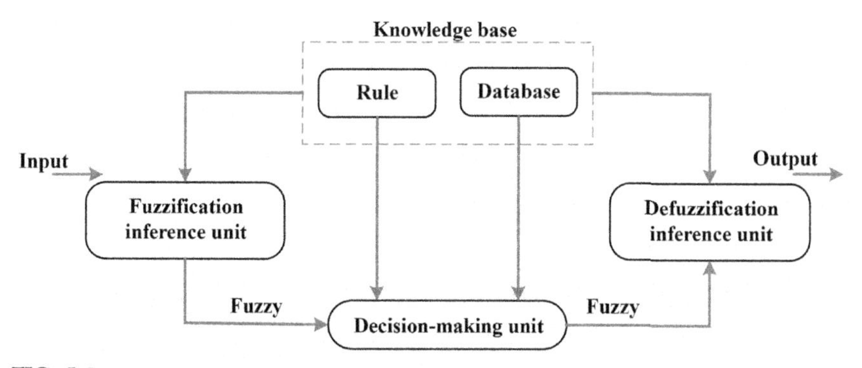

FIG. 5.8 Fuzzy inference system diagram [29].

Artificial intelligence approaches are used to analyze collected data from PV plants. An artificial neural network (ANN) and fuzzy logic are also the most important AI approaches [49]. ANN is a suitable approach for time-series forecasting. Neural networks may create various results by parameter tuning, but they can be designed using environmental and electrical parameters in order to identify unpleasant conditions with appropriate accuracy. This method can locate faults by system performance deviation from the expected [50].

Fig. 5.8 shows a fuzzy inference system diagram. It uses multiple-input and multiple-output (MIMO) methods for data analysis. This method decreases the effect of the outliers on an appropriate function [51].

5.5.6.1 Predictive monitoring

Predictive models for the energy generation of PV systems are used as another approach for PV monitoring systems. Various studies have used this approach.

Spataru et al. [52] presented a monitoring system that accurately predicts the energy generation of the PV system. This approach monitors PV array conditions applying the Sandia Implemented Model. Normal operation is introduced using the predicted output energy of the PV array by the implemented model. The proposed monitoring system detects energy losses over 5% in the PV module through a comparison between the predicted and measured energies.

Moreover, in ref. [53], the specifications of a PV module were simulated under various weather conditions to track the performance degradation of the PV module. Then, the actual generated energy was compared with simulated energy in the MATLAB/Simulink program. Any deviation between these two values is known as a malfunction in the PV system.

The measured AC energy generation was compared with the predictive model to detect online malfunctions. A considerable difference between these two parameters is presented as a malfunction. In order to achieve a better and more accurate model of a PV system, environmental parameters such as solar radiation and temperature are considered in the model implementation [54].

5.5.7 Perspectives and issues in PV monitoring systems

In recent years, significant progress has been made in PV monitoring systems. The main reasons for this progress are the availability of the required data and the development of appropriate novel algorithms. However, the operation of PV monitoring systems is accompanied by challenges. A summary of these challenges is discussed below:

1. Environmental conditions, typical wired communication in PV monitoring system to transmit data with its associated restrictions. They are constantly exposed to sunlight, which reduces their lifespan or these restrictions are caused by reduction in the reliability of the sensors. This reduction is due to radio frequency interference, a high moisture rate, dust, corrosive environments, and vibrations. Such situations lead to damaged cables or weakness in sensor performance [55].
2. Yield degradation, the lifetime of a complete PV system equipment is nearly equivalent. Also, there is no experimental protocol to confirm the PV monitoring system's lifespan. Moreover, the degradation prediction of different pieces of equipment in PV systems is difficult under various environmental pressures. Wohlgemuth et al. [56] presented a degradation rate in PV panels of about 80% per year. This degradation is affected by some factors such as corrosion of joints, series resistance increase, and changes in the color of bus bars. These factors have progressive effects and certainly affect power loss estimation.
3. Resource limitation, implementation of a PV monitoring system has three main restrictions that include data processing, storage system, and energy yield. Due to limited battery power in sensors and storage systems, communication systems should be set up to offer high energy yield [57].
4. Device calibration, one of the main characteristics of the PV monitoring system is the existence of various integrated commercial products. These products must be calibrated in laboratories, but this is difficult because these products are integrated. In addition, the performance of these products may change after calibration because of ignoring the effects of software. A typical calibration method can solve these problems. However, calibration expenses increase with this solution because this requires specialist technicians in the workplace [58].

To understand the challenges of the PV monitoring system, it is essential to detect the expected features of PV monitoring systems in the future. These features include online reporting, precision measurements, suitable storage services, safe access, unmanned action capability, triggers/alerts, and process prediction. These features can be achieved through fulfilling the goals below:

- Ensuring the accuracy of meters.
- Efficient data acquisition system, storage methods, and transfer methods.
- Applying automation.
- Quick fault detection and elimination.
- Online presentation of parameters.
- Estimation of system performance.

5.6 Environmental impacts on PV systems

Climatic conditions have always been of great interest for discussion as they have a considerable influence on the electricity coming out of the vast majority of PV systems. Different weather statuses in different countries have to be taken into account for the selection of the solar module type to be installed in that area depending on the standard test condition (STC) value that it gives, as various kinds of solar cells can give different values of power under the same atmospheric conditions. Hence, the solar module type chosen for a certain area can provide a higher output power than the one selected for another geographical location. The atmospheric behaviors that can degrade the solar module performance include temperature, irradiation level, and other metrological reasons.

This section will emphasize the previously mentioned weather terms affecting thin-film solar cell performance. The known paths for solving these issues will also be stated correspondingly.

5.6.1 Effect of temperature

PV thin-film modules are influenced by different atmospheric conditions such as temperature, dust, light intensity, and wind.

One of the main advantages of thin-film modules is the high-efficiency values at high temperatures. All solar module parameters are influenced by different temperatures; the higher the temperature, the worse the module performance. As a consequence, this can lead to fast degradation in cell power and efficiency [59]. The main reason behind this degradation at high temperatures is that the open-circuit voltage (V_{oc}) decreases linearly with increasing the temperature due to the reduction in the bandgap. There is a direct relation between the V_{oc} and the bandgap. Therefore, V_{oc} will be reduced in addition to the fill factor (*FF*) and the maximum power point. On the other hand, the short-circuit current (J_{sc}) will slightly increase, but the increase in the current will not compensate for the decrease in the other parameters.

The term temperature coefficient (T_k) is used to define how the module is sensitive to the temperature: the less sensitive the module is, the higher the output power and hence the better the module performance. Each module type has a different temperature coefficient. For thin-film modules, the temperature coefficient has a value around −0.0984%, which means that thin-film modules are less sensitive to temperature compared to the mono- and polycrystalline PV modules [60].

There are a few methods to deal with the high-temperature impact on the modules such as active cooling, passive cooling, and the wind method.

Active cooling is done by supplying the module with equipment to control the temperature; nevertheless, the cost has to be taken into consideration. Fig. 5.9 shows the efficiency of the PV panel with and without cooling; the cooling keeps the efficiency of the panel above 14%, whereas the PV panel exhibits lower efficiency without cooling. Therefore, a solar-driven cooling system reduces the operating temperature of the module [61].

Passive cooling can be done by double-face façades, floating PVs, and submerged PV installations. Although the system complexity is an issue, it has several advantages, including cost-effectiveness, low noise,

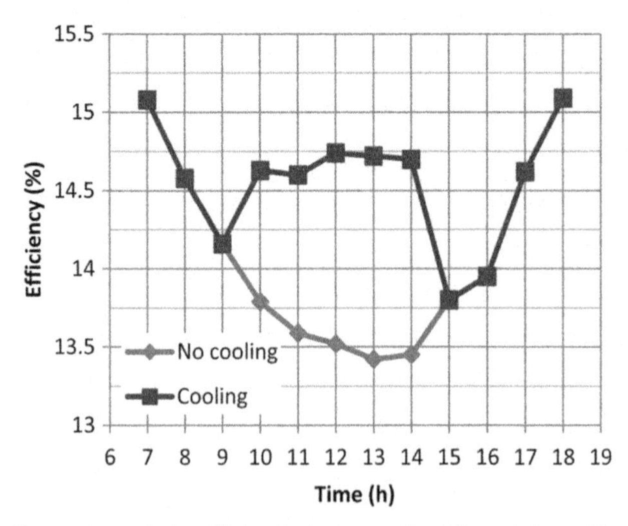

FIG. 5.9 Comparison of the efficiencies between the PV modules with and without cooling [61].

and long-term stability. For more detail about cooling methods, readers are encouraged to check out the following reference: [62].

The wind cooling method is based on normal wind. This method can play an effective role not only in cooling the solar modules, but also in helping to remove dust particles on the modules [59].

5.6.2 Irradiation levels

As in the case of the temperature effect, the irradiation strength has a direct impact on the solar parameters: the higher the radiation intensity, the more the potential for providing more power by the module [59].

Similar to the temperature coefficient at the STC, the low light behavior has to be addressed in the datasheet for researchers and designers. This is important as it helps them to specifically decide about the best module for a particular geographical site [63]. Therefore, the need for a solar concentrator is sometimes of great benefit to strengthen the incident light on the modules and hence improve the performance of the modules.

5.6.3 Spectral effect

The change in total intensity and spectral distribution has a considerable effect on the short-circuit current and the efficiency of the PV module. The performance variation depends strongly on the module type. This is related to the material bandgap, which determines the upper wavelength limit of the spectral response. The outdoor performance of the PV module can be increased between the seasons up to 15% [64], even under the same condition of irradiation level and operating temperature.

The change of solar irradiance will affect the spectral response because it is the ratio of the current generated through the PV module to the incident solar energy on its surface. Also, it is directly proportional to the performance.

For clear-sky days for low bandgap materials such as crystalline Si solar cells, the spectrum has little effect (from 4% to 5% change in efficiency between the seasons). On the other hand, large bandgap materials such as a-Si are severely affected by the spectral change (−10% to +15% change in efficiency can be observed between the seasons) [64].

Finally, one can sum up that all types of solar modules are significantly influenced by the atmospheric fluctuation behavior locally as well as the weather nature differences globally. These environmental factors are temperature, irradiation, and spectral irradiance.

Depending on the STC, the power obtained by each panel will be different and could be statistically chosen where the maximum yield can be the optimum for every sort of solar module.

5.7 Performance assessment

In general, the performance of solar PV systems will vary according to their configurations. Mostly, the performance is dependent on site location, the PV-specific technology of the PV modules, and the specific type of PV system installation. For example, the performance of ground-mount PVs may not be the same as the performance of building-integrated PVs. The method of performance evaluation is more or less similar, irrespective of the chosen PV installation configuration, PV technology, and system components. As per the International Energy Agency (IEA) under the Photovoltaic Power Systems Program (PVPS) project, a detailed performance approach along with numerous indicators are given. The indicators include array yield, final yield, reference yield, capture loss, performance ratio, and system component efficiencies. Here, the system component efficiencies include the PV array efficiency, inverter efficiency, battery efficiency, and overall plant efficiency. With these indicators, one can readily evaluate and understand the performance of the PV system and its influencing factors. Based on these indicators, a decision toward PV plant operation and maintenance can be performed [9]. The detailed performance evaluation method along with various indicators are explained below.

5.7.1 Array, final, reference, and corrected reference yield

The three indicators (array and final and reference yield) are generally evaluated specific to the nominal power rating of the installed capacity. These three indicators are defined for the specific time period (maybe a day, month, and year), and in some situations where the need for understanding the PV system performance based on season can be considered [65–68].

Array yield (Y_A): It is the indicator used to evaluate the energy output from a PV array at its nominal peak power rating. The energy generated at this stage is DC energy. It is shortly defined as the ratio of DC energy output ($E_{dc,t}$) of the PV array to the peak rating of the installed PV capacity (P):

$$Y_A = \frac{E_{dc,t}}{P} \tag{5.6}$$

Units: kWh/kWp/day, kWh/kWp/month, kWh/kWp/year, kWh/kWp/season.

Final yield (Y_F): It is the indicator used to evaluate the energy output from a PV system at its nominal peak power rating. The energy generated at this stage is AC energy. It is shortly defined as the ratio of AC energy output ($E_{ac,t}$) of the PV system to the peak rating of the installed PV capacity (P) (5.5)–(5.8).

$$Y_F = \frac{E_{ac,t}}{P} \quad (5.7)$$

Units: kWh/kWp/day, kWh/kWp/month, kWh/kWp/year, kWh/kWp/season.

Reference yield (Y_R): It is the indicator used to measure the number of peak sun hours at the reference irradiance. It is shortly defined as the ratio of irradiance on the plane of the PV array ($I_{poa,t}$) to the reference irradiance at the standard testing conditions ($I_o = 1$ kW/m^2) [65–68].

$$Y_R = \frac{I_{poa,t}}{I_o} \quad (5.8)$$

Units: kWh/kWp/day, kWh/kWp/month, kWh/kWp/year, kWh/kWp/season.

Corrected reference yield (Y_{CR}): This indicator can be used in the PV system performance evaluation when the influence of module temperature is considered. The module temperature is varied for different configurations of PV plant installation [65–68]. The corrected reference yield is defined as:

$$Y_{CR} = Y_R(1 - \alpha(T_M - T_{STC})) \quad (5.9)$$

For evaluating the corrected reference yields, data related to temperature coefficient (α, whose units %/°C), module temperature (T_M, whose units °C), and ambient temperature (T_{STC}, whose units °C) are needed. Here, the temperature coefficient values will be varied as per the chosen PV technology, and the module temperature must be measured using the temperature measuring instruments. Otherwise, the module temperature can be estimated using the nominal operating cell temperature (NOCT) method, where the PV module is said to be operated at that temperature (T_{noct}) [69, 70].

$$T_M = T_{STC}\left(\frac{T_{noct} - 20}{800}\right) \times I_{poa,t} \quad (5.10)$$

Most researchers use the aforementioned equations, however, there are some limitations that are related to wind condition, ground clearance, and installation configuration. The above NOCT method may give the best results for low wind conditions. Considering the installation types and wind conditions, a simple and generalized method is given here to estimate the module temperature for different installation types of PV plants. The relation shown in Eq. (5.6) uses the mounting coefficient (M_c), which is different for different PV installation types. The mounting coefficient values are given as 1, 1.2, 1.8, and 2.4 for freestanding, flat roof, sloped roof, and façade integrated installation types, respectively [65–69].

$$T_M = T_{STC} + M_c\left(\frac{0.32}{8.91 + 2W_s}\right) \times I_{poa,t} \tag{5.11}$$

5.7.2 Efficiency indicators

The efficiency indicators reveal information about the performance quality and whether the PV system components individually or the PV system overall is performing as per rated. Indicators such as PV module efficiency, inverter efficiency, system efficiency, capacity factors, and performance ratio come under this category.

Module efficiency (η_{pv}): This is an indicator that readily informs the operating condition of PV module where it is defined as the ratio of the DC energy generated at the output terminals of the PV module or array to the energy available at the plane of the array due to incident solar irradiance [65–69].

$$\eta_{pv} = \left(\frac{E_{dc,t}}{I_{poa,t} \times PV\,\text{array area}}\right) \times 100 \tag{5.12}$$

Inverter efficiency (η_{inv}): This is an indicator that readily informs the operating condition of the inverter where it is defined as the ratio of the AC energy generated at the output terminals of the inverter to the DC energy at the input terminals of the inverter [65–69].

$$\eta_{inv} = \left(\frac{E_{ac,t}}{E_{dc,t}}\right) \times 100 \tag{5.13}$$

System efficiency (η_{system}): This is an indicator that readily informs the overall PV system efficiency where it is defined as the ratio of the AC energy injected to the grid or load to the energy available at the plane of the array due to incident solar irradiance [65–69].

$$\eta_{pv} = \left(\frac{E_{\text{grid or load},t}}{I_{poa,t} \times PV\,\text{array area}}\right) \times 100 \tag{5.14}$$

Capacity factor (CF): This is an indicator proposed for power generating units with respect to specific operating time. In a PV system, this indicator is used to evaluate the overall performance and is given as the ratio of the amount of generated energy by the PV plant to the operating time, which is generally considered as 24h per day or 8760h per year at its nominal peak installed capacity [65–69].

$$CF = \left(\frac{E_{\text{grid or load},t}}{8760 \times P}\right) \times 100 \tag{5.15}$$

Although the capacity factor is most commonly used, it may not be an effective indicator to assess the performance of the PV plant. The main reason for this is the limited operating time of the PV plant, which is approximately 12h. Hence, a new indicator called performance ratio is used.

Performance ratio (PR): This represents the overall performance quality of the PV system considering all the effects that are involved in various stages of energy conversion in PV systems [65–69].

$$PR = \left(\frac{Y_F}{Y_R}\right) \times 100 \tag{5.16}$$

5.7.3 Indicators for energy loss

The energy conversion process in the PV system involves various losses. These losses occur at various system components used in PV systems. Four indicator or loss categories are considered: array capture loss, thermal capture loss, miscellaneous capture loss, and system losses. The energy losses are location- and component-specific as they might operate in harsh weather conditions. So, evaluating them individually will inform decisions toward the maintenance of PV plants [65–69].

Array capture loss (L_{AC}): This is an indicator that readily informs the variation in observed energies between the actual and theoretical. It is generally estimated by taking the difference in energy levels between the reference and array yields [65–69].

$$L_{AC} = Y_R - Y_A \tag{5.17}$$

Thermal capture loss (L_{TC}): This is an indicator that readily informs the variation in observed energies due to the effect of temperature. The energy loss at the module level due to observed temperature deviation from STC value is considered. The difference of reference yield with the corrected reference yields is considered as thermal capture loss [65–69].

$$L_{TC} = Y_R - Y_{CR} \tag{5.18}$$

Miscellaneous capture loss (L_{MC}): This is an indicator that is used to measure the energy loss associated with the DC side of the PV systems. This occurs due to various conditions such as low irradiance, shading condition, soiling effects, mismatch, snow, etc. It is evaluated as the difference between the corrected reference yield to the array yield [65–69].

$$L_{MC} = Y_{CR} - Y_A \tag{5.19}$$

System loss (L_S): This is an indicator that is used to measure the energy loss associated with the electrical and electronic components of the PV system. Generally, it is associated with the AC side of the system and includes components such as the inverter, cables, transformers, etc. It is evaluated as the difference between the array yield and the final yield [65–69].

$$L_S = Y_A - Y_F \tag{5.20}$$

5.7.4 Degradation and risk assessment

Degradation and risk due to various faults are the two other indicators that affect the performance of the PV system. The lifetime of the PV system is excepted to be 25 years, hence during its lifetime, several challenges could result in fault conditions and finally lead to energy degradation [71].

Degradation: This is an indicator that readily informs the system performance in terms of energy outputs. Over a period of time, a gradual reduction in the generated output is observed, and to measure that, an indicator called the degradation rate is used. It is generally evaluated by means of statistical methods that show the relation between the performance ratio of the PV plant over the period of time [72].

$$DR = 100 \times \left(\frac{12 \times m}{l}\right) \tag{5.21}$$

$$y = mx + l \tag{5.22}$$

where DR is the degradation rate (%/year), l is the y-intercept of the trend line,and m is the slope of the line. The degradation rate will vary with respect to PV technology and location weather conditions.

Risk: This is an indicator that is considered for understanding the risk associated with various failure modes and fault conditions of the PV system. Failure mode and effect analysis (FMEA) based on the occurrence, severity, and detectability specify the effect of each failure state and its reasons on the system. Risk priority number (RPN) is used for understanding the FEMA, and it is as follows [73]:

$$RPN = S \times O \times D \tag{5.23}$$

where "S" represents severity, which has no dimensions. Severity specifies a single failure state, which strongly affects the performance of the system. "O" defines the occurrence, this dependence on the probability of the system failing over exposure time. "D" specifies the detection, which technology (tool) is capable of detecting failures in a system during its exposure time [74].

5.8 Conclusion

PV systems consist of modules, inverter, converters, energy storage, and electrical and mechanical equipment to generate AC and DC power. Generally, PV systems are classified into three types: grid-connected PV systems, stand-alone PV systems, and hybrid PV systems. Designing and sizing PV systems is the most crucial stage in a PV project. Among the most common failures that affect PV system performance are junction box failures, bypass diode failures, and broken glasses. Inverter problems can be classified into three categories: manufacturing and design problems, control problems, and electrical component failures. However, failures due to poor connections, wiring, connection insulation rupture, and protective devices can cause severe problems in PV systems such as electric arcs, short circuits, etc. PV systems should be monitored through a set of comprehensive techniques in accordance with IEC standards. PV system monitoring is very essential to ensure the expected and stable performance of the PV systems. The monitoring system collects the required data in the PV system and transmits it to the control center, which lets user evaluate and control this system to decrease maintenance costs, monitor the performance indicators of power generation, and keep track of fault events. Moreover, the predictive models for the energy generation of PV systems are used as another approach for PV monitoring systems. The climatic conditions have a considerable influence on the reliability and durability of PV systems. This chapter has outlined the main components of a PV system. In addition, designing different types of PV systems was discussed. However, the main scope of this chapter was to summarize the typical defects and failures in PV systems and also to present an overview of existing PV monitoring systems reported in the literature. In this chapter, the most important instruments and systems required for PV monitoring have been discussed. Moreover, various aspects of environment conditions on PV system performance were presented. The main indicators such as the array, final, and reference yield as well as the performance ratio for the assessment and comparison of PV systems have been addressed according to the IEC.

References

[1] Aghaei M, Thayoob YHM, Imamzai M, Piyous P, Amin N. Design of a cost-efficient solar energy based electrical power generation system for a remote Island - Pulau Perhentian Besar in Malaysia. In: Proceedings of the 2013 IEEE 7th International Power Engineering and Optimization Conference, PEOCO 2013; 2013.

[2] Leva S, Aghaei M. Failures and defects in PV systems review and methods of analysis; 2018 p. 56–84.

[3] Aghaei M. Novel methods in control and monitoring of photovoltaic systems. Politecnico di Milano; 2016.

[4] Kumar NM, Subathra MSP, Moses JE. On-grid solar photovoltaic system: components, design considerations, and case study. In: Proceedings of the 4th International Conference on Electrical Energy Systems, ICEES 2018; 2018.

[5] Manoj Kumar N, Sudhakar K, Samykano M. Techno-economic analysis of 1 MWp grid connected solar PV plant in Malaysia. Int J Ambient Energy 2019.

[6] Kumar NM, Reddy PRK, Sunny KA, Navothana B. Annual energy prediction of roof mount PV system with crystalline silicon and thin film modules. In: *Spec. Sect. Curr. Res. Top. Power, Nucl. Fuel Energy, SP-CRTPNFE*, no. October; 2016. p. 24–31.
[7] Kumar NM, Reddy PRK, Praveen K. Optimal energy performance and comparison of open rack and roof mount mono c-Si photovoltaic Systems. Energy Procedia 2017;117:136–44.
[8] van Sark W, Reinders A, Verlinden P, Freundlich A. Photovoltaic solar energy from fundamentals to applications; 2017.
[9] Reinders A, Verlinden P, van Sark W, Freundlich A. Photovoltaic solar energy: from fundamentals to applications. Wiley; 2017.
[10] Chandel M, Agrawal GD, Mathur S, Mathur A. Techno-economic analysis of solar photovoltaic power plant for garment zone of Jaipur city. Case Stud Therm Eng 2014.
[11] Meyer EL, Van Dyk EE. Assessing the reliability and degradation of photovoltaic module performance parameters. IEEE Trans Reliab 2004;53(1):83–92.
[12] Renganathan RMV, Natarajan V, Jahaagirdar T. Failure mechanism in PV systems. In: Grad. course MAE 598 Sol. Commer; 2016. p. 1–17.
[13] Golnas A. PV system reliability: an operator's perspective. IEEE J Photovolt 2013; 3(1):416–21.
[14] Triki-Lahiani A, Bennani-Ben Abdelghani A, Slama-Belkhodja I. Fault detection and monitoring systems for photovoltaic installations: a review. Renew Sustain Energy Rev 2018;82(July 2017):2680–92.
[15] Cristaldi L, Faifer M, Lazzaroni M, Khalil MMAF, Catelani M, Ciani L. Diagnostic architecture: a procedure based on the analysis of the failure causes applied to photovoltaic plants. Meas J Int Meas Confed 2015.
[16] Alam MK, Khan FH, Johnson J, Flicker J. PV faults: Overview, modeling, prevention and detection techniques. In: 2013 IEEE 14th Workshop on Control and Modeling for Power Electronics, COMPEL 2013; 2013.
[17] Vidal de Oliveira AK, Amstad D, Madukanya UE, Rafael L, Aghaei M, Rüther R. Aerial infrared thermography of a CdTe utility-scale PV power plant. In: 46th IEEE PVSC; 2019.
[18] Sizkouhi AMM, Esmailifar SM, Aghaei M, Vidal de Oliveira AK, Rüther R. Autonomous path planning by unmanned aerial vehicle (UAV) for precise monitoring of large-scale PV plants. In: 46th IEEE PVSC; 2019.
[19] Vidal De Oliveira AK, Aghaei M, Rüther R, Aghaei M. Automatic fault detection of photovoltaic arrays by convolutional neural networks during aerial infrared thermography. In: 36th EU PVSEC—European PV solar energy conference and exhibition; 2019.
[20] Aghaei M, Grimaccia F, Gonano CA, Leva S. Innovative automated control system for PV fields inspection and remote control. IEEE Trans Ind Electron 2015;62(11).
[21] Aghaei M, Dolara A, Grimaccia F, Leva S, Kania D, Borkowski J. Experimental comparison of MPPT methods for PV systems under dynamic partial shading conditions. In: EEEIC 2016—international conference on environment and electrical engineering; 2016.
[22] Grimaccia F, Aghaei M, Mussetta M, Leva S, Quater PB. Planning for PV plant performance monitoring by means of unmanned aerial systems (UAS). Int J Energy Environ Eng 2015;6(1).
[23] Kirsten A, et al. Aerial infrared thermography of a utility—scale PV power plant after a meteorological tsunami in Brazil. In: *WCPEC*, no. 20; 2016. p. 1–3.
[24] Quater PB, Grimaccia F, Leva S, Mussetta M, Aghaei M. Light unmanned aerial vehicles (UAVs) for cooperative inspection of PV plants. IEEE J Photovolt 2014;4(4).
[25] Grimaccia F, Leva S, Dolara A, Aghaei M. Survey on PV modules' common faults after an O&M flight extensive campaign over different plants in Italy. IEEE J Photovolt 2017;7(3).

[26] Aghaei M, Quater PB, Grimaccia F, Leva S, Mussetta M. Unmanned aerial vehicles in photovoltaic systems monitoring applications. In: *29th Eur. Photovolt. Sol. Energy Conf. Exhib. (EU PVSEC 2014)*, no. 22–26 September; 2014. p. 2734–9.
[27] Aghaei M, Leva S, Grimaccia F. PV power plant inspection by image mosaicing techniques for IR real-time images. In: 2017 IEEE 44th Photovoltaic Specialist Conference, PVSC 2017; 2017.
[28] Rahman MM, Selvaraj J, Rahim NA, Hasanuzzaman M. Global modern monitoring systems for PV based power generation: a review. Renew Sustain Energy Rev 2018; 82(October 2017):4142–58.
[29] Madeti SR, Singh SN. Monitoring system for photovoltaic plants: a review. Renew Sustain Energy Rev 2017;67:1180–207.
[30] IEC. Photovoltaic system performance monitoring—guidelines for measurement, data exchange, and analysis. *IEC Standard 61724*; 1998.
[31] Benghanem M, Maafi A. Data acquisition system for photovoltaic systems performance monitoring. IEEE Trans Instrum Meas 1998;47(1):30–3.
[32] Koutroulis E, Kalaitzakis K. Development of an integrated data-acquisition system for renewable energy sources systems monitoring. Renew Energy 2003;28(1):139–52.
[33] Wang W, Xu Y, Khanna M. A survey on the communication architectures in smart grid. Comput Netw 2011;55(15):3604–29.
[34] Kribus A, Zik O, Karni J. Optical fibers and solar power generation. Sol Energy 2000; 68(5):405–16.
[35] Drews A, et al. Monitoring and remote failure detection of grid-connected PV systems based on satellite observations. Sol Energy 2007;81(4):548–64.
[36] Ahmad T, Hasan QU, Malik A, Awan NS. Remote monitoring for solar photovoltaic systems in rural application using GSM network. Int J Emerg Electr Power Syst 2015;16(5):413–9.
[37] Tseng CL, et al. Feasibility study on application of GSM-SMS technology to field data acquisition. Comput Electron Agric 2006;53(1):45–59.
[38] Bucci G, Fiorucci E, Landi C, Ocera G. Architecture of a digital wireless data communication network for distributed sensor applications. Meas J Int Meas Confed 2004; 35(1):33–45.
[39] Hua J, Lin X, Xu L, Li J, Ouyang M. Bluetooth wireless monitoring, diagnosis and calibration interface for control system of fuel cell bus in Olympic demonstration. J Power Sources 2009;186(2):478–84.
[40] Mahmood A, Javaid N, Razzaq S. A review of wireless communications for smart grid. Renew Sustain Energy Rev 2015;41:248–60.
[41] López MEA, Mantinan FJG, Molina MG. Implementation of wireless remote monitoring and control of solar photovoltaic (PV) system. In: *2012 Sixth IEEE/PES Transmission and Distribution: Latin America Conference and Exposition (T&D-LA)*; 2012. p. 1–6.
[42] Han J, Choi CS, Park WK, Lee I, Kim SH. PLC-based photovoltaic system management for smart home energy management system. IEEE Trans Consum Electron 2014; 60(2):184–9.
[43] Sinton RA, Cuevas A, Stuckings M. Quasi-steady-state photoconductance, a new method for solar cell material and device characterization. In: *Conference record of the IEEE photovoltaic specialists conference*; 1996. p. 457–60.
[44] Ding J, Zhao J, Ma B. Remote monitoring system of temperature and humidity based on GSM. In: *2009 2nd international congress on image and signal processing*; 2009. p. 1–4.
[45] Xiao W, Lind MGJ, Dunford WG, Capel A. Real-time identification of optimal operating points in photovoltaic power systems. IEEE Trans Ind Electron 2006;53(4):1017–26.
[46] Vergura S, Acciani G, Amoruso V, Patrono GE, Vacca F. Descriptive and inferential statistics for supervising and monitoring the operation of PV plants. IEEE Trans Ind Electron 2009;56(11):4456–64.

[47] Boquete L, et al. A portable wireless biometric multi-channel system. Meas J Int Meas Confed 2012;45(6):1587–98.
[48] Forero N, Hernández J, Gordillo G. Development of a monitoring system for a PV solar plant. Energy Convers Manag 2006;47(15–16):2329–36.
[49] Mellit A, Kalogirou SA, Hontoria L, Shaari S. Artificial intelligence techniques for sizing photovoltaic systems: a review. Renew Sustain Energy Rev 2009;13(2):406–19.
[50] Chao KH, Ho SH, Wang MH. Modeling and fault diagnosis of a photovoltaic system. Electr Power Syst Res 2008;78(1):97–105.
[51] Hu X, Sun F. Fuzzy clustering based multi-model support vector regression state of charge estimator for lithium-ion battery of electric vehicle. In: 2009 International Conference on Intelligent Human-Machine Systems and Cybernetics, IHMSC 2009, vol. 1; 2009. p. 392–6.
[52] Spataru S, Sera D, Kerekes T, Teodorescu R. Photovoltaic array condition monitoring based on online regression of performance model. In: Conference record of the IEEE photovoltaic specialists conference; 2013. p. 815–20.
[53] Hamdaoui M, Rabhi A, El Hajjaji A, Rahmoun M, Azizi M. Monitoring and control of the performances for photovoltaic systems. In: Int. Renew. Energy Congr., no. January; 2009. p. 69–71.
[54] Platon R, Martel J, Woodruff N, Chau TY. Online fault detection in PV systems. IEEE Trans Sustain Energy 2015;6(4):1200–7.
[55] Gungor VC, Hancke GP. Industrial wireless sensor networks: Challenges, design principles, and technical approaches. IEEE Trans Ind Electron 2009;56(10):4258–65.
[56] Wohlgemuth JH, Kurtz S. Using accelerated testing to predict module reliability. In: Conference record of the IEEE photovoltaic specialists conference; 2011. p. 003601–5.
[57] Gungor VC, Lambert FC. A survey on communication networks for electric system automation. Comput Netw 2006;50(7):877–97.
[58] Gungor VC, Lu B, Hancke GP. Opportunities and challenges of wireless sensor networks in smart grid. IEEE Trans Ind Electron 2010;57(10):3557–64.
[59] Mussard M, Amara M. Performance of solar photovoltaic modules under arid climatic conditions: a review. Sol Energy 2018;174(July):409–21.
[60] Adeeb J, Farhan A, Al-Salaymeh A. Temperature effect on performance of different solar cell technologies. J Ecol Eng 2019;20(5):249–54.
[61] Wu S, Xiong C. Passive cooling technology for photovoltaic panels for domestic houses. Int J Low-Carbon Technol 2014;9(2):118–26.
[62] Dupré O, Vaillon R, Green MA. Thermal behavior of photovoltaic devices; 2017.
[63] Sağlam Ş. Meteorological parameters effects on solar energy power generation. WSEAS Trans Circuits Syst 2010;9(10):637–49.
[64] Eke R, Betts TR, Gottschalg R. Spectral irradiance effects on the outdoor performance of photovoltaic modules. Renew Sustain Energy Rev 2017;69(October 2016):429–34.
[65] Malvoni M, Leggieri A, Maggiotto G, Congedo PM, De Giorgi MG. Long term performance, losses and efficiency analysis of a 960 kWP photovoltaic system in the Mediterranean climate. Energy Convers Manag 2017;145:169–81.
[66] Li C, et al. Performance of off-grid residential solar photovoltaic power systems using five solar tracking modes in Kunming, China. Int J Hydrogen Energy 2017;42 (10):6502–10.
[67] Yadav SK, Bajpai U. Performance evaluation of a rooftop solar photovoltaic power plant in Northern India. Energy Sustain Dev 2018.
[68] Shiva Kumar B, Sudhakar K. Performance evaluation of 10 MW grid connected solar photovoltaic power plant in India. Energy Reports; 2015.
[69] Kumar NM, Kumar MR, Rejoice PR, Mathew M. Performance analysis of 100 kWp grid connected Si-poly photovoltaic system using PVsyst simulation tool. Energy Procedia 2017.

[70] Gökmen N, Hu W, Hou P, Chen Z, Sera D, Spataru S. Investigation of wind speed cooling effect on PV panels in windy locations. Renew Energy 2016.
[71] Silvestre S, Kichou S, Guglielminotti L, Nofuentes G, Alonso-Abella M. Degradation analysis of thin film photovoltaic modules under outdoor long term exposure in Spanish continental climate conditions. Sol Energy 2016.
[72] Kumar NM, Malvoni M. A preliminary study of the degradation of large-scale c-Si photovoltaic system under four years of operation in semi-arid climates. Results Phys 2019.
[73] Tatapudi S, Sundarajan P, Libby C, Kuitche J, TamizhMani G. Risk priority number for PV module defects: influence of climatic condition; 2018.
[74] Rajput P, et al. Risk priority number for understanding the severity of photovoltaic failure modes and their impacts on performance degradation. Case Stud Therm Eng 2019;(October):100563.

CHAPTER

6

On-farm applications of solar PV systems

Shiva Gorjian[a,b], *Renu Singh*[c], *Ashish Shukla*[d], *and Abdur Rehman Mazhar*[d]

[a]Biosystems Engineering Department, Faculty of Agriculture, Tarbiat Modares University (TMU), Tehran, Iran, [b]Leibniz Institute for Agricultural Engineering and Bioeconomy (ATB), Potsdam, Germany, [c]Centre for Environment Science and Climate Resilient Agriculture, Indian Agricultural Research Institute, New Delhi, India, [d]Centre for Built and Natural Environment, Faculty of Engineering, Environment and Computing, Coventry University, Coventry, United Kingdom

6.1 Introduction

The agricultural sector is certainly the only provider of human food. With a projected global population increase of 25% in the next 20 years, the demand for food and energy is expected to be significantly raised as a consequence. Most agricultural activities are directly or indirectly powered by fossil fuels that cause greenhouse gas (GHG) emissions, leading to climate change as a consequence of global warming. In this regard, sustainable alternative energy resources are required to reduce growing concerns about the environmental impacts of nonrenewable fuels [1]. Photovoltaic (PV) systems offer a promising solution to run agricultural activities in a sustainable manner. It has been reported that PV technology is among the fastest-growing energy technologies [2, 3]. Solar PV systems can be employed to either supply electricity or both heat and electricity (through the use of PVT systems[a]) to supply the energy demands of various agricultural activities. In small-scale farms and the protected

[a]PVT modules are discussed in Chapter 4.

Photovoltaic Solar Energy Conversion
https://doi.org/10.1016/B978-0-12-819610-6.00006-5

cultivation environments of greenhouses, PV systems are preferable to be installed as distributed power generation systems. However, in farms with large available areas, there would definitely be a need for large centralized PV power plants [4–6]. PV systems can be installed either in grid-connected (grid-tied) or stand-alone configurations.[b] The main benefit of grid-connected PV systems is that they do not need any energy storage units (ESU), causing a significant reduction in total costs. In contrast, stand-alone PV systems are more ideal for distant farms and regions where access to the power grid is sophisticated or impossible.

In terms of capacity, grid-connected PV systems are generally classified into small-scale (1–5 kW), medium-scale (5–250 kW), and large-scale PV systems (more than 10 kW) [7]. One of the necessities for installing PV systems is the need for large areas, so agricultural farms can fulfill this requirement with the additional advantage of financial gain through electricity generation [8, 9]. Farm PV systems contribute toward the reduction of energy supplied from the grid, reducing peak load demand and mitigating transmission losses as generation is done onsite [10, 11]. The cost-effectiveness of PV systems depends on the availability of solar radiation in a specific region as well as electricity prices. Legal regulations can also prominently affect the profitability of PV installations, even in regions with the same solar energy potential [8, 12]. This chapter provides a holistic overview of the most commonly used applications of PV systems in agriculture.

6.1.1 Agrophotovoltaic (APV)

The idea of *agrophotovoltaic* (APV) was first proposed by Goetzberger and Zastrow [13] in 1982. It revolves around the coproduction of solar PV energy and agricultural products on the same field. Nowadays, this technique is also known as an *agrivoltaic system.* The proposed idea included the installation of PV panels 2 m above the ground to enlarge the space between them, avoiding excessive crop shading. Scientists claimed that if PV panels are mounted sufficiently high above the ground, about two-thirds of the received solar radiation reaches the surface beneath. After three decades of intense research and development, this concept became commercialized and implemented in different projects and pilot plants worldwide [14, 15].

According to the results from a case study by Amaducci et al. [16], the yield production comparison revealed that all agrivoltaic systems working in combined mode are more advantageous than either ground-mounted PV systems installed on farm lands or solely crop production

[b]Design principles of PV systems are presented in Chapter 5.

FIG. 6.1 (A) APV system installed in Italy [14], and (B) APV facility installed above the potato farm (RESOLA project), Heggelbach, Germany [15].

as a regular farming practice. Results from the various studies confirmed that using AVP systems increases land-use efficiency by raising farm revenues more than 30%, but only if yield losses due to shading effects are kept down by a selection of proper crops [15]. Additionally, the agrivoltaic systems decrease water evaporation rates, ensuring that crops have more accessible water than crops cultivated in regular lands fully exposed to sun, which results in stable crop yields [17, 18]. Fig. 6.1 shows two installed APV systems in Italy and Germany. The PV panels may need to be placed farther apart, enabling farm equipment to navigate the rows.

Some APV projects already employ tracking PV modules, allowing for maximizing the electric efficiency and light availability improvements for crops at the same time. Mobile PV panel installation can also result in better rainfall distribution below the APV systems [15]. Fig. 6.2 shows an established APV system using mobile PV panels employing a dual-axis sun-tracking system.

APV systems are also expected to have the most promising potential in regions with arid and semiarid climates due to their capability in

FIG. 6.2 View of an Agrovoltaico plant equipped with a dual-axis sun-tracking system [16].

improving water productivity by decreasing evapotranspiration and the adverse impacts of the excessive reception of solar radiation [15, 19]. Although APV systems offer several benefits with their implementations in farm fields, their installation has been limited by several governments and local councils due to their landscape impacts, soil occupation, and competition with food production systems. Additionally, the successful penetration of agrivoltaic systems depends on farmer acceptance, which is definitely affected by the benefits coming from these systems [20, 21].

6.1.2 Aquavoltaics

Bodies of water offer vital elements for natural ecosystems in addition to life and most human activities. According to the Food and Agriculture Organization (FAO), *aquaculture,* also known as *aquafarming,* is the farming of aquatic organisms consisting of fish, molluscs, crustaceans, and aquatic plants under a controlled environment. Currently, more than 40% of seafood is produced from aquaculture farms [22]. Floatovoltaics (FV) is a PV system floating on any sized water body that mitigates water losses by covering the water and preventing evaporation by up to 70–85%. The combination of FV with aquaculture forms the concept of aquavoltaics, bringing the benefit of efficient water use for both food and energy generation while simultaneously assisting in land-use change. In this regard, FV systems offer benefits to farmers by enabling them to implement a solar power generation system and preserving the use of water bodies, including ponds, canals, reservoirs, and dams more efficiently. The systems can be implemented in both industrial-sized farms as well as small remote villages [23, 24]. Fig. 6.3 shows some FV installations around the world.

FV installation on the surface of an aquaculture system forms a cooling mechanism for PV modules, enhancing the power conversion efficiency (PCE) between 5% and 22% depending on the deployment method. FVs can also perform as fish aggregation techniques which suggest a controlling framework for the behavior of fishes by protecting them, and presumably improve the yield [28]. Kim et al. [29] proposed the concept of a floating PV system for harvesting salt and electricity at the same salt farm for the first time. In this case, a fiber-reinforced plastic (FRP) seawater vessel was prepared where the PV modules were installed. The test rig of the proposed system is shown in Fig. 6.4A. The results from the experimental tests indicated that the water temperature and depth do not have a strong correlation and each can independently affect the performance of the PV modules while the impact of water salinity on the performance of the PV module is insignificant in a short time. They claimed that the electric performance of the floating PV system was higher than the

FIG. 6.3 (A) The FV system installed on an irrigation pond, Niente winery, California's Napa Valley region [25], (B) The FV system installed on a sandpit lake in Zwolle, the Netherlands [26], and (C) The FV system installed at Tengeh Reservoir in Tuas, Singapore [27].

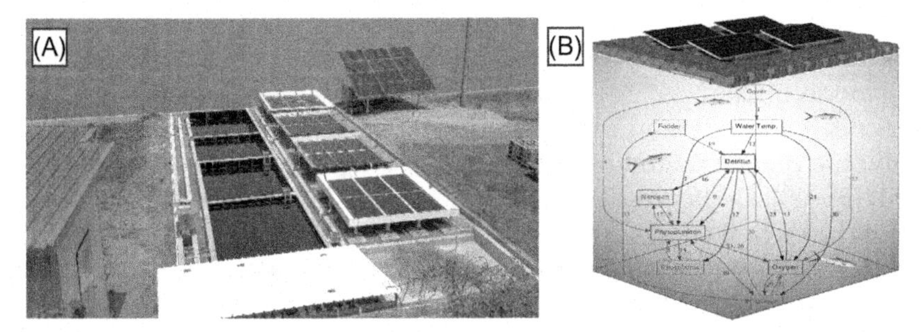

FIG. 6.4 (A) The developed aquavoltaic system for simultaneous salt and electricity harvest [29], and (B) Representation of the developed model simulating the fish pond ecosystem covered with floating PV panels [30].

land-mounted PV modules, mainly due to the cooling effect of the seawater. Château et al. [30] developed a dynamic model to study the principal biochemical processes in a milkfish (*Chanos chanos*) pond covered by the floating PV (FPV) panels. Also, some experiments were conducted and data were recorded from ponds with and without FPV cover within two production seasons of summer and winter. Results indicated that the FV system may display some mild negative consequences for fish production due to the reduction of dissolved oxygen (DO) levels. Also, it was concluded that the energy gained from this integration (with the capacity of about 1.13MW/ha) is significant, which could compensate for the losses in fish production. The ecological model developed in this study is shown in Fig. 6.4B.

In aquavoltaic systems, the mounting method of the PV panels is important as they are usually installed on floating cages, and therefore require specific structures to ensure their endurance under dynamic environments. Currently, there are four distinguished methods for floating PV systems:

thin-film FV, submerged FV, surface-mounted FV, and microencapsulated phase change material (MEPCM). Detailed descriptions of these methods are presented in Ref. [28]. Salinity is the other threat that requires special sealing and cabling for PV systems at use in marine areas. In addition, restrictions in solar tracking systems, power storage facilities, and power distribution are other challenges in aquavoltaic systems that need to be considered [28, 31].

6.2 Applications of solar PV systems in agriculture

Numerous researchers have studied various applications of solar PV systems in agricultural practices broadly categorized as conventional and modern activities.[c] Solar PV energy possesses huge potential to power agricultural facilities and farm equipment, including PV-powered water pumping and irrigation systems [32–37], PV-powered desalination systems [38–40], PV-powered solar dryers [41–45], PV-powered greenhouses [46–48], PV-powered livestock and dairy farming systems [9, 49–53], PV-powered crop protection systems [54, 55], etc. Fig. 6.5 shows an overview of the most common applications of solar PV systems in agriculture, which are discussed in the following sections in detail.

6.2.1 PV-powered solar water pumping systems

PV-powered solar water pumping systems are among the most promising solar-powered farm applications, especially at remote locations with no reliable access to grid electricity and cost-effective diesel fuel. These systems are eco-environment with low maintenance and without any fuel costs.

Solar-powered pumping systems are similar to conventional ones, except that they consume solar electricity generated by the PV modules instead of grid electricity or diesel fuel. There are various possible designs to develop solar water pumping systems, but a general system consists of the main components of a PV array, a DC/AC surface mounted/submersible/floating motor pump set (depending on the application), a power conditioning unit, and a water storage tank (Fig. 6.6). Direct integration of the PV array with the water pumping system was first introduced in the late 1970s when there were some limitations in earlier designs in terms of overall performance due to the lack of appropriate design [56, 57]. Direct-coupled PV DC water

[c]Applications of PV systems in precision agriculture (PA) are presented in Chapter 7.

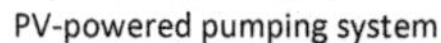

PV-powered pumping system PV-powered solar dryer PV-powered dairy system

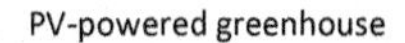

PV-powered greenhouse PV-powered crop protection system

FIG. 6.5 Overview of the most common PV-powered applications in agriculture.

pumping systems are straightforward and authentic but lack the ability to work at the maximum power point of the PV generator due to the high dependency on daily weather conditions. However, using a maximum power point tracker (MPPT) may enhance the performance of the system [56, 58]. In PV-powered pumping systems, the PV array is usually installed on a reliable frame where manual or automatic tracking is used. Because the PV panels generate DC electricity and most of the available water pumps on the market need an AC electrical input, an inverter is needed to convert the DC power to AC power. Additionally, the battery storage units can be replaced by water storage tanks in a way that water is pumped to the elevated tanks in the presence of sunlight and is stored for nocturnal or overcast applications [59, 60].

The first generation of solar PV-powered pumping systems included centrifugal pumps powered by DC/AC motors with hydraulic efficiencies ranging from 25% to 35%. The second generation used positive displacement pumps, progressing cavity pumps, or diaphragm pumps generally attributed by low PV input power demands and hydraulic efficiencies up to 70% [56].

Crop irrigation is one of the most crucial activities in agriculture, and it is quite energy-intensive. The PV-powered pumping systems for irrigation purposes can generally provide the head of between 5 to 200 m at a flow rate of 250 m^3/day [61]. In order to develop a PV-powered pumping

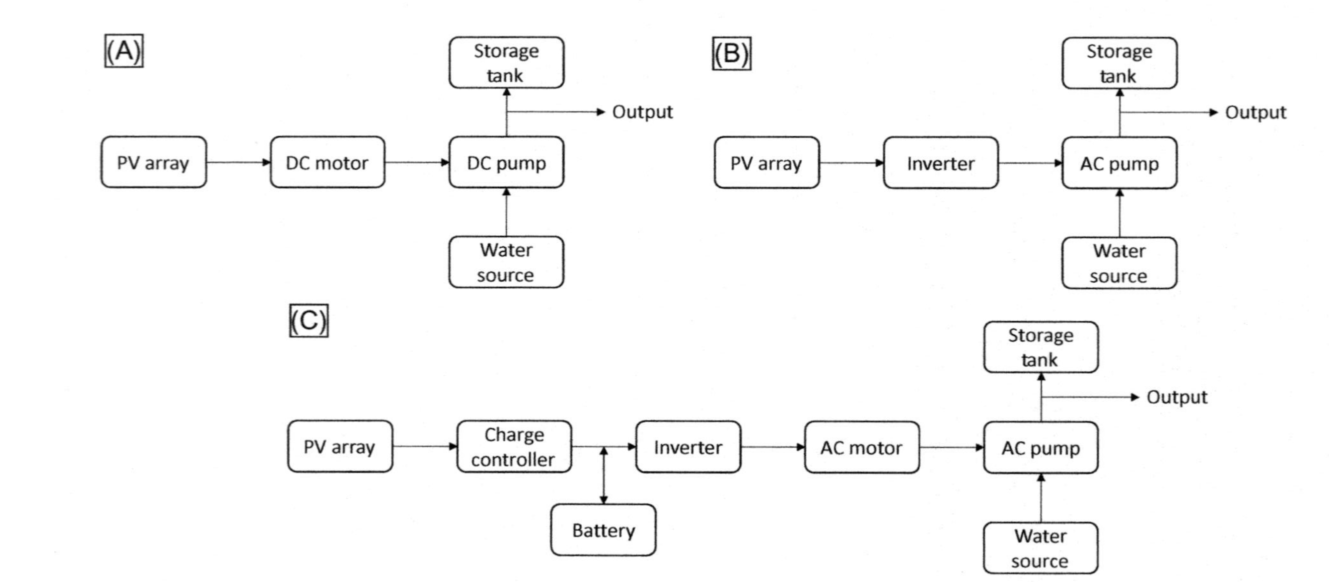

FIG. 6.6 Different configurations of PV-powered water pumping systems [56]: (A) The direct-coupled PV DC water pumping system, (B) The PV AC water pumping system, and (C) The PV-powered water pumping system with battery storage.

irrigation system, the first step is the assessment of the water resources to determine the rational amount of water in the ground and surface water resources. The other important steps are determining the water demand of the crops, studying the weather and solar radiation data, estimating and calculating the harvest of solar power, determining the application rate of irrigation schedules, and choosing the PV modules, motor pump set, other electronic and hydraulic devices, and auxiliary units [57]. The main parameters that affect the performance of a PV-powered water pumping system are [56]:

- Availability of solar radiation and variations of the air temperature at the location of installation.
- Total dynamic head (TDH) (suction head + discharge head + frictional losses).
- The water flow rate.
- The total amount of water requirement.
- The amount of potential energy needed to elevate the water to the discharge level (hydraulic energy).

Also, the efficiency of the PV system has a significant impact on the performance of the PV-powered solar pumping system. The hydraulic energy E_h (kWh/d) needed to provide a volume V (m^3) of water at TDH (m) is presented by [56]:

$$E_h = \rho \times g \times V \times TDH \tag{6.1}$$

where ρ (kg/m^3) represents the water density, g (9.81 m/s^2) is the acceleration of gravity, and TDH (m) is the total dynamic head. The power required by the PV array, P_{pv} (W), is given by:

$$P_{pv} = E_w / \left(I_T \times \eta_{mp} \times F \right) \tag{6.2}$$

where I_T (kWh/m^2day) denotes the average daily solar irradiation incident on the surface of the PV array, F is the array mismatch factor, and η_{mp} is the daily subsystem efficiency. The volume of water pumped V (m^3) is evaluated as:

$$V = \left(P_{pv} \times I_T \times \eta_{mp} \times F \right) \tag{6.3}$$

The efficiency of the motor-pump system η_{mp} can be obtained as follows:

$$\text{Efficiency} = \text{hydraulic energy output/energy input} \tag{6.4}$$

The efficiency of the PV array (%) is expressed as:

$$\eta_{pv} = \frac{P_{pv}(W)}{I_T\left(\frac{W}{m^2}\right) \times A_c(m^2)} \times 100 \tag{6.5}$$

where A_c (m^2) is the surface of the PV array. The overall efficiency of the system is obtained as:

$$\eta_T = \left(\eta_{pv} \times \eta_{mp}\right) \tag{6.6}$$

6.2.1.1 Developments and prospects

Al-Smairan [62] conducted a case study analyzing the Tall Hassan station (Fig. 6.7A) located in Jordan and compared two alternative power sources, common diesel engines and PV systems. In their study, the station was primarily employed as the source of drinking water for domestic use and watering purposes with daily consumption of $45\,m^3$/day. A submersible pump was incorporated to take the water from the groundwater well and feed two $55\,m^3$ storage tanks. In the case of solar-powered mode, 108 PV modules were used with a total peak of 5.94 kW, connected to a DC/AC inverter as a stand-alone PV system. The economic evaluation revealed that the solar-powered mode was the most profitable system when the fuel price was more than 0.4 US$/L.

Dietmar Stuck, Austria [63], developed the NSP solar pump system with a novel design (Fig. 6.7B). The NSP pump was designed in a way to retrieve water from wells at a depth of 100 m and up to 20,000 L/d, which also makes it suitable for use as an irrigation tool. The main advantage of the NSP pump is that it needs no maintenance and is salt-water resistant. In this water pumping system, the required power is provided by the PV panels with no battery energy storage system. As is shown in Fig. 6.7B, an electric motor runs a winder connected to the pumping lever and moves it up and down, allowing water to be pumped to the surface.

FIG. 6.7 (A) PV-powered water pumping system in Tall Hassan station, North Badia, Jordan [62], and (B) The NSP PV-powered solar water pump [63].

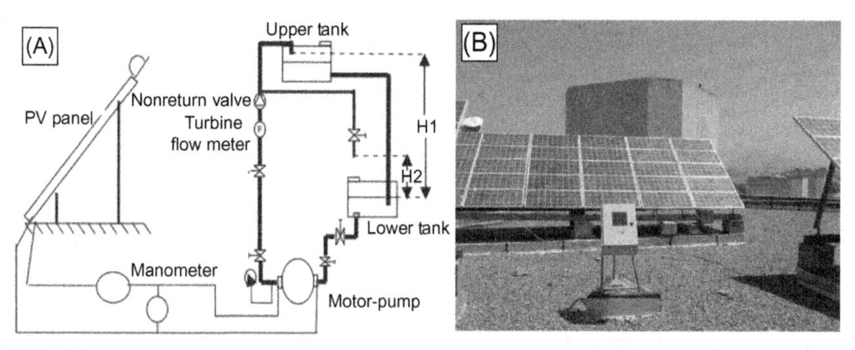

FIG. 6.8 (A) Schematic view of the PV-powered water pumping system, and (B) PV array and insolation measuring equipment at the University of Science Technology, Oran [64].

The operation assessment of a directly coupled DC PV-powered water pumping system was experimentally evaluated by Mokeddem et al. [64]. The system was comprised of a 1.5 kWp PV array, a DC motor, and a centrifugal water pump. A schematic view and photo of the experimental setup are shown in Fig. 6.8. The operation investigation was conducted over a period of 4 months under various climatic conditions and different amounts of isolation with a two static head arrangement. The results indicated that the system is quite compatible with low energy demand irrigation in isolated areas where there is no access to the power grid. They asserted that because the system works with simple electronic control and no battery, both the initial and maintenance costs can be minimized. The efficiency of the system was also reported as less than 30%, which is common for directly coupled PV pumping systems.

Hassan and Kamran [65] proposed an irrigation water pump with a novel architecture powered by PV/utility employed maximum power point tracking with no battery backup. As shown in Fig. 6.9, the single control input is delivered to the pump controller in order to regulate the desired water flow rate and magnify the PV resource employment at the same time. They claimed that in the proposed system, there is the possibility of adding PV panels incrementally to the grid-powered pumping system. This would allow a reduction in the initial investment costs needed for entire solar deployment while the investment for partial solar deployment is paid back in decreased electricity tariffs to the farmer.

Tabaei and Ameri [66] experimentally investigated the performance of a PV-powered water pumping system employing static booster reflectors, as shown in Fig. 6.10, under the climate conditions of Kerman, Iran. The experiments were conducted for two installed stainless steel 304 and aluminum foil reflectors. In this design, the PV panels are connected to the water pump system using one positive displacement surface water pump coupled to a PM (permanent magnet) DC motor. The results indicated the

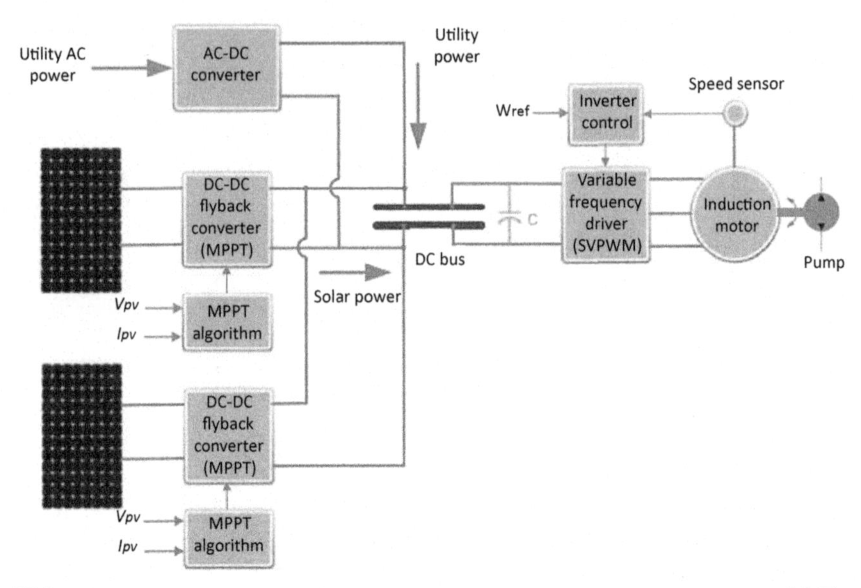

FIG. 6.9 The developed hybrid PV/utility-powered water pumping system in ref. [65].

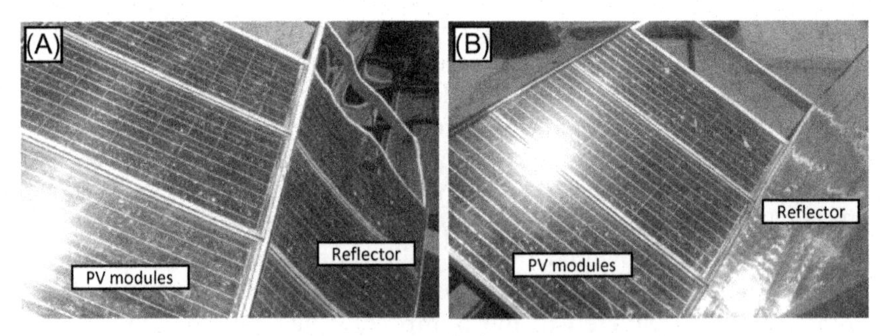

FIG. 6.10 Experimental setup of the PV-power solar pumping system [67]: (A) With stainless steel 304 reflector, and (B) with aluminum foil reflector.

performance enhancement of the PV modules as 14% and 8.5% due to using the aluminum foil and stainless steel 304 reflectors, respectively.

With technical innovations in PV-powered pumping systems, great attention has been paid to the deployment of this technology worldwide. It is obvious from the literature that in recent years, remarkable progress on PV-powered solar pumping technology has been achieved that has overcome the embedded technical limitations of the early developed solar water-pumping systems. With the expected increasing growth in the cost of fossil fuels and the fall in the peak watt cost of solar cells, PV power is going to become more affordable in the future.

6.2.2 PV-powered desalination systems

As a consequence of climate change due to global warming, the number of deserts and unfertile lands has increased in the world, causing acute water shortages. Agriculture accounts for more than two-thirds of global water use, including more than 80% in developing countries (Fig. 6.11). It is estimated that increasing the global population to more than 9 billion people by the end of 2050 will need about a 50% rise in agricultural production and a 15% growth in water withdrawal. The agriculture water comes from various sources of surface water, groundwater from wells, and rainwater [69]. Desalination is currently the only viable solution for increasing the water supply beyond what is available from the hydrological cycle. In a typical desalination process, salt and other impurities are removed from saline water (either seawater or brackish groundwater), which makes it consumable by plants in agricultural applications and by humans as drinkable water. Desalination is beneficial to the agricultural industry because it offers an alternative to brackish water for irrigation, decreasing the adverse effects of the direct use of salt water on crop growth and yield due to the negative impacts of salinity on soil properties. Desalination can also remarkably reduce civil works and labor requirements [70, 71].

Due to the accelerated achievements in solar energy technologies (both PV and thermal), there is a strong tendency to integrate solar systems with

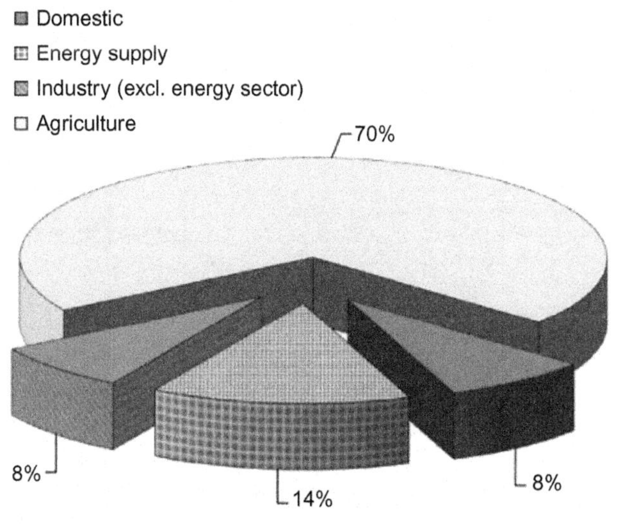

FIG. 6.11 Global use of freshwater in different sections [68].

desalination plants with the goal of improving energy efficiency. Solar desalination technologies are mainly classified as direct and indirect technologies in which direct systems (also known as solar stills) produce freshwater simply in a solar collector while indirect systems use solar thermal or electric power to drive desalination processes [72, 73]. Solar PV systems can be coupled with electricity-driven indirect desalination plants or used to power electromechanical devices in thermally driven desalination plants as well as solar distillation systems. A comprehensive study of PV-powered desalination technologies is presented in Chapter 8.

6.2.3 Solar-powered dryers

The main goal behind the food drying process is to reduce the moisture content to a secure and desirable level, and thus extend the life of the dried products. Sun drying is a conventional technique that is broadly employed to prevent agricultural crops from spoiling, especially in rural areas. The sun-drying method comes with some disadvantages such as spoiling due to rain, wind, moisture, and dust, and deterioration of crops because of decomposition, insect attacks, fungi, etc. In addition, sun drying is labor-intensive and needs a considerable amount of energy as well as land area for spreading the product [74]. Solar drying is one of the most promising technologies among various solar energy applications. Drying can be defined as a mechanism to remove products' moisture involving both heat and mass transfer. Solar drying offers an alternative method for drying agricultural products in a clean and sanitary environment that can reduce crop losses, energy consumption, and drying time in addition to improving the quality of the final products [75]. The basic principle of a solar dryer is shown in Fig. 6.12.

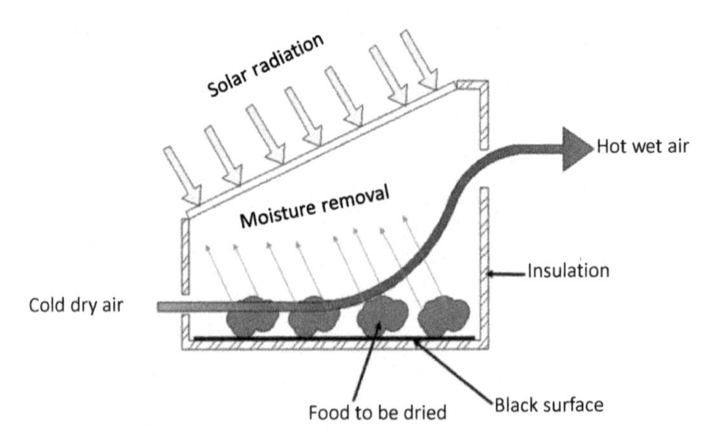

FIG. 6.12 Working principle of a simple solar dryer [76].

Solar drying methods are mainly classified into the open sun-drying method and drying with solar dryers. Additionally, solar dryers can broadly be categorized as passive and active systems. Solar dryer classification is mainly based on the heat transfer mechanisms they employ to withdraw the moisture content from the product. In passive solar dryers, buoyancy forces or wind flow are utilized to circulate the heated air resulting from solar thermal utilization while in active solar dryers, the hot air is circulated by means of mechanical devices such as fans or blowers, which can be driven by the power supplied from PV modules or grid electricity [76, 77]. The passive and active solar dryers can be further grouped as direct, indirect, and mixed solar dryers depending on the way that solar radiation is collected by the space under the drying process. A wide classification of solar dryers is shown in Fig. 6.13. In direct solar drying, heat is produced by the exposure of sunlight on the crop itself and the internal space of the drying cabinet. The indirect solar dryers, also known as conventional dryers, do not receive solar radiation directly because a separate solar-powered unit collects thermal energy, which can then be transferred to the drying chamber. Finally, the mixed solar drying method uses both features of the direct and indirect methods when a faster drying rate is required [75, 77].

6.2.3.1 PV-powered solar dryers

One of the electrical demands of active solar dryers is the need to power a fan to induce the forced convection flow. In this regard, the PV modules can be integrated with an electric circuit to run one or some DC fans, which are placed at the entrance of the air collector [78], the exit of the air collector [79], or the exit of the drying chamber [80]. Goud et al. [78] developed an indirect type of solar dryer using PV modules to maintain the airflow through the system. In their work, three DC fans were placed at the entrance of the solar collector and powered by three PV panels ($10\,W_p$). The schematic view of the developed solar dryer is shown in Fig. 6.14. As can be seen in this figure, a solar air collector is incorporated with the drying chamber, which includes drying products placed on several trays to contact the hot air. The absorbed solar energy by the collector is transferred to the incoming air (running via an electric DC fan), resulting in the air to be heated at the entrance of the drying chamber. As the hot air stream reaches the trays, the heat and mass transfer take place on the products' surface in which water is gradually extracted from the product, leading to dried material production. Analyzing the experimental data, average air velocities of 1.0824 and 1.0774 m/s were recorded in green chili and okra drying processes, respectively. Moreover, it was concluded that the integration of PV modules with the presented fans is appropriate and can be proposed as a reliable system.

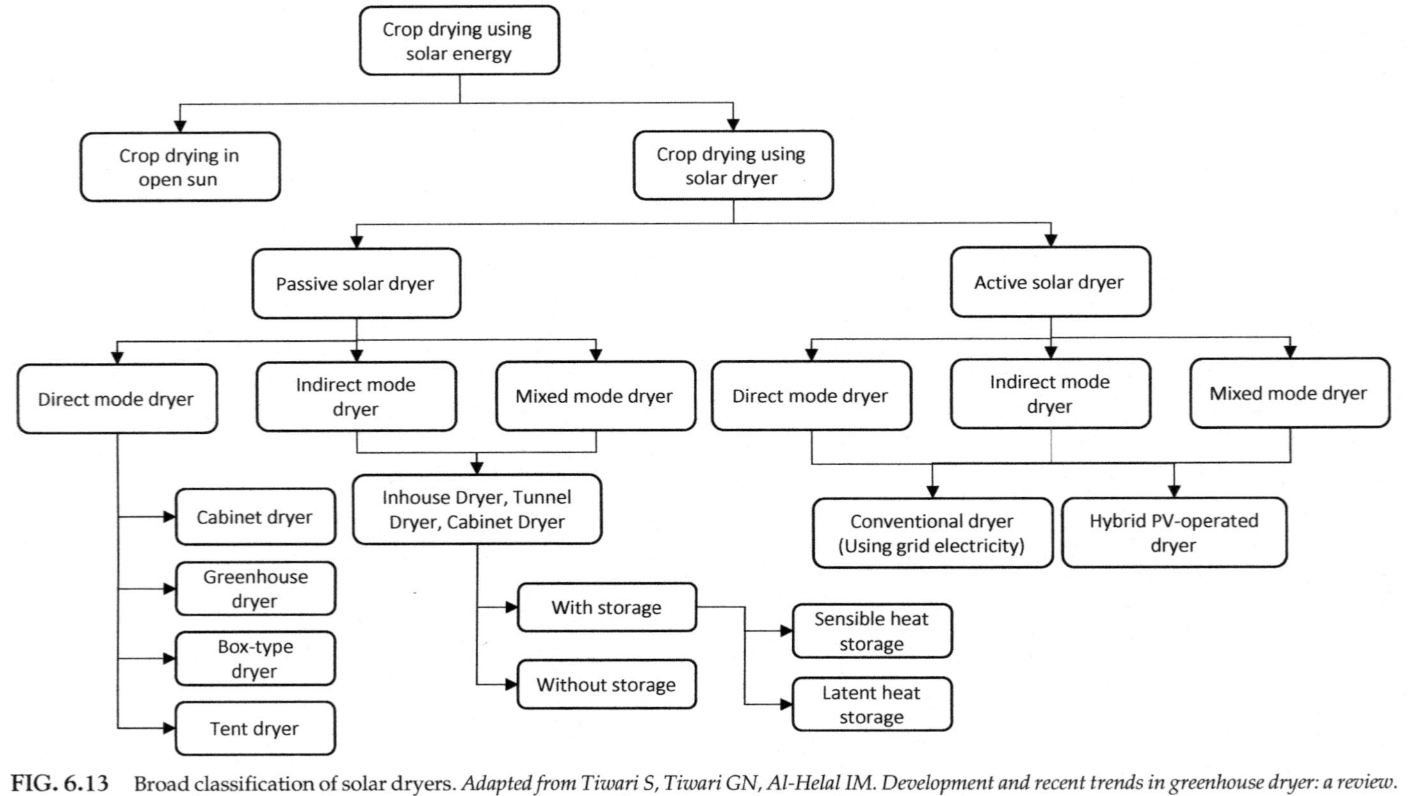

FIG. 6.13 Broad classification of solar dryers. *Adapted from Tiwari S, Tiwari GN, Al-Helal IM. Development and recent trends in greenhouse dryer: a review. Renew Sustain Energy Rev 2016;65:1048–64. https://doi.org/10.1016/j.rser.2016.07.070, Agrawal A, Sarviya RM. A review of research and development work on solar dryers with heat storage. Int J Sustain Energy 2016;35:583–605. https://doi.org/10.1080/14786451.2014.930464.*

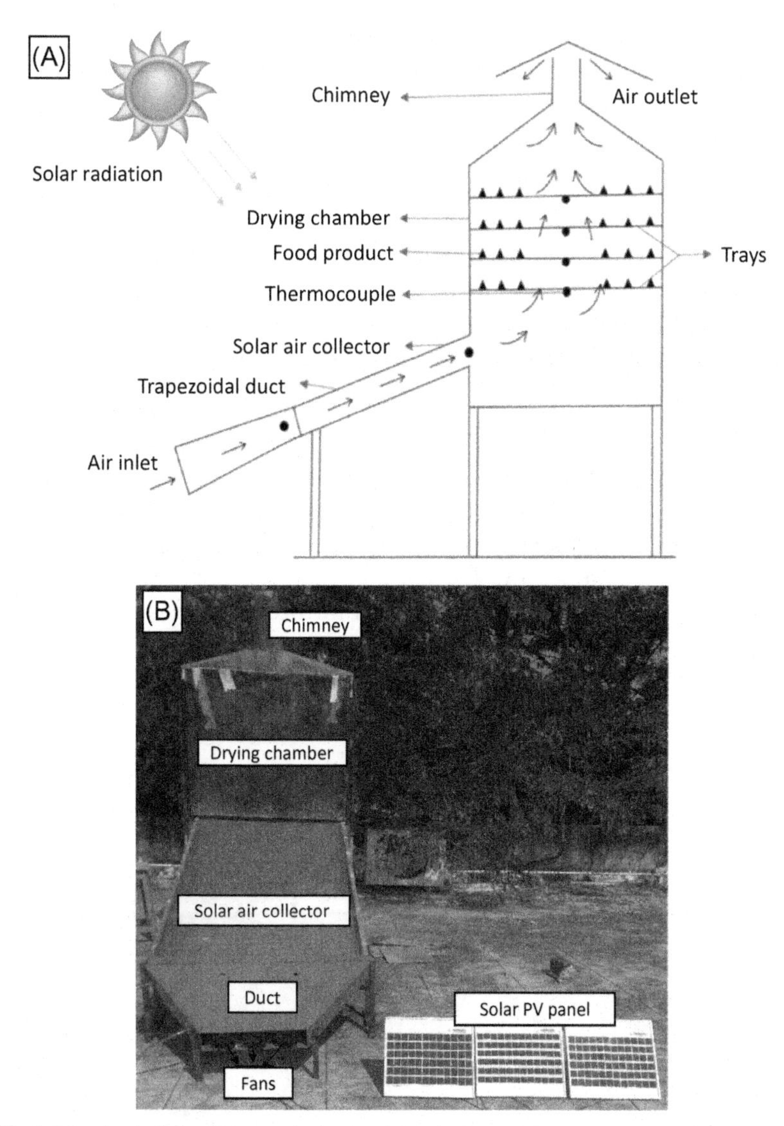

FIG. 6.14 An indirect PV-powered solar dryer [78]: (A) Schematic view of the dryer, and (B) photo of the experimental setup.

Janjai et al. [81] tested a PV-ventilated greenhouse dryer in which three roof-mounted PV panels were used to facilitate the drying process carried out on dry peeled longan and banana. The results indicated that the proposed design is able to reduce drying time and improve product quality compared to conventional sun-drying designs. In a similar study conducted by Eltawil et al. [42], a greenhouse-like tunnel dryer coupled with a solar thermal collector and a PV panel to power a DC fan was investigated for drying mint. They concluded that the developed system results in a reduction of peppermint drying time and improves the quality of the dried product.

Infrared drying is one of the techniques used to enhance the quality of the product and decrease the drying time [82]. Although the combination of the solar dryer with an infrared system has not been extensively studied, those limited works investigating this novel integration suggest significant potential [83, 84], especially when PV panels are incorporated to provide the electrical power needed for the infrared source. In this regard, Ziaforoughi et al. [44] introduced a solar dryer assisted by a PV-powered infrared system to investigate the drying process of potato slices. As Fig. 6.15 shows, in this design, an indirect dryer is used in which heated air is passed through the potato slices laid on the trays inside the drying cabinet and leaves the system from a chimney located on the cabinet topside. An intermittent infrared source, integrated inside the drying area, is used to boost the drying process with the aid of a temperature controller that controls the lamp current and maintains the product temperature at the desired level. In this work, the infrared source is powered with a PV panel installed beside the dryer. Results revealed that the utilization of PV technology culminates in a 40–69% reduction in energy consumption while the drying time can also be decreased by 31–52% with infrared assistance.

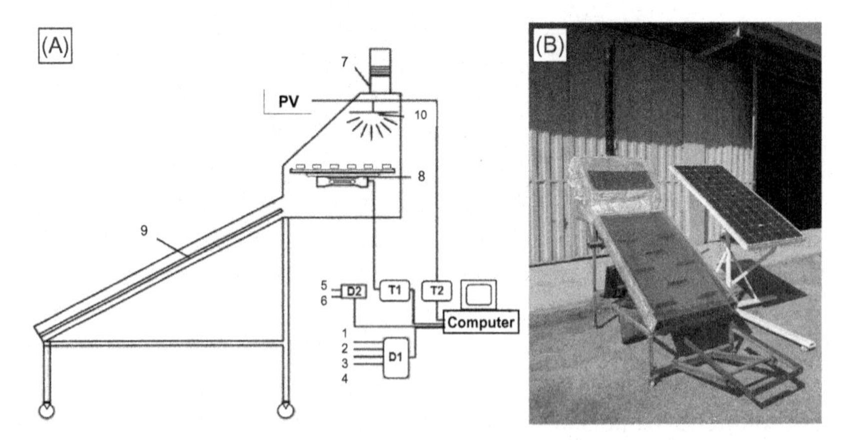

FIG. 6.15 The indirect solar dryer assisted with a PV-powered infrared system [44]; (A) Schematic view of the experimental setup consisting of data loggers (D1 and D2), chimney (7), load cell (8), aluminum absorber (9), infrared lamp (10), and transmitters (T1 & T2); and (B) photo of the developed system.

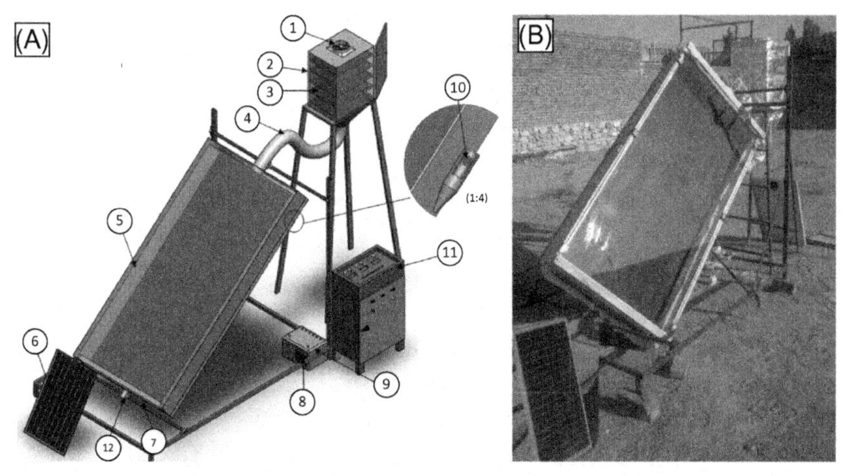

FIG. 6.16 An active solar dryer assisted with a PV-powered solar tracker [85]: (A) Schematic view of the proposed design including axial fan (1), drying cabinet (2), trays (3), connecting tube (4), solar collector (5), PV panel (6), air inlet (7), charge controller (8), battery unit (9), sun-tracking sensor (10), control panel (11), mechanical pivot (12); and (b) photo of the experimental setup.

Samimi-Akhijahani et al. [85] proposed an indirect solar dryer assisted with a PV-powered tracking system to dry tomato slices. As shown in Fig. 6.16, they incorporated a tracking flat-plate solar air collector coupled to a drying cabinet through the transferring tube, in which a one-axial fan powered by the PV panel was implemented to transfer the heated air from the collector to the drying trays in a forced convection mode. Photocell sensors were also coupled with the sun-tracking unit to control a PV-powered 12 V DC motor. The results from the performance evaluation indicated that the proposed arrangement can reduce the drying time by 16.6–36.6% while the quality of the final product was not influenced negatively by the tracking mode.

6.2.3.2 PVT-powered solar dryers

PVT utilization with a solar dryer not only provides the electricity supply for the assisted systems, but also delivers heat to the thermal collector and enhances the total performance consequently. Fig. 6.17 represents a PVT-powered solar dryer developed by Tiwari et al. [86]. As the figure shows, in this design, an air-based PVT module was constructed out of two sections, PV and solar thermal modules. During operation, solar radiation is captured by the entire module, generating electricity and thermal energy, respectively, from the PV and thermal sections. As the air is fed into the module, fresh air passing beneath the PV module makes it cooler while extracting the excess heat. Further, the thermal collector transfers

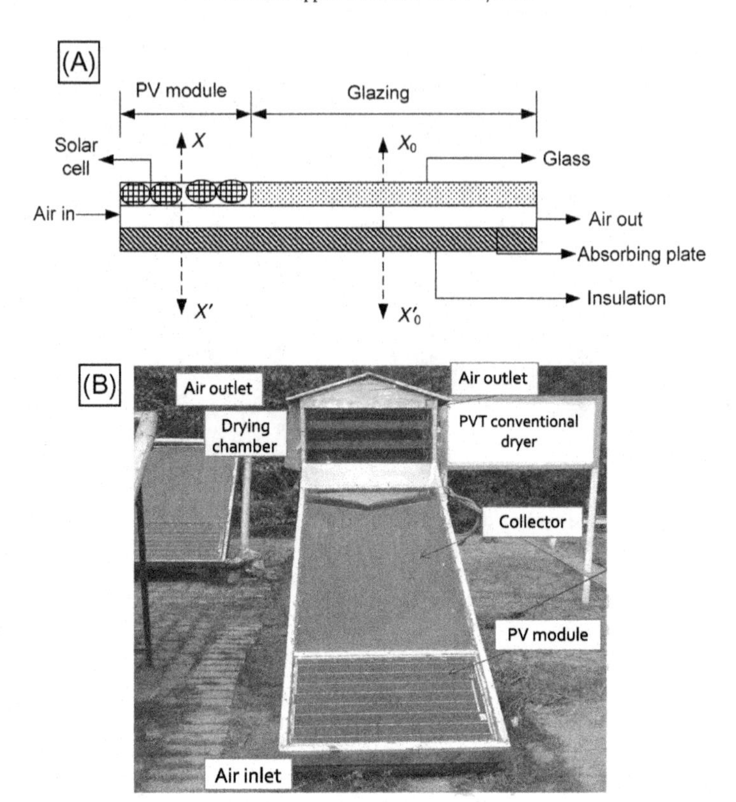

FIG. 6.17 An air-based PVT powered solar dryer [86]; (A) Cross-section of the air-based PVT module, and (B) photo of the proposed system.

the absorbed heat to the preheated air and provides adequate thermal power for the drying process where the hot air passes through the product, absorbing its moisture content and leaving the system from the exhaust gate. In the proposed model, scientists used the generated electricity to power the DC fans for air circulation purposes. They concluded that the developed system including one solar collector was able to yield 61.56% of thermal efficiency.

In another system introduced by Tiwari et al. [87], a mixed-type greenhouse solar dryer assisted with a PVT module for three different purposes was developed. First, the PV module generated electricity to drive DC fans, inducing forced mode operation. Second, the heat absorbed by the module was transferred to the drying cabinet, and third, the PVT module was arranged to block the direct exposure of the drying material to improve the product quality. Experimental results showed that the quality and decoloration of the dried product were promoted and mitigated, respectively. Additionally, the thermal energy and exergy values were determined to be 2.03 and 0.535 kWh, respectively. In the research conducted by Fterich et al. [41], the performance of a mixed-mode forced convection solar dryer equipped with a roof-mounted PVT module was investigated. In the given design, the air is pumped into a drying chamber through a pipe by means of a DC fan powered by the PV module. Getting preheated by the PVT absorber including several aluminum tubes, the air is able to dry tomato slices in a more efficient way compared to other conventional methods. Performance analysis showed that mixed solar drying reduces the tomato's initial moisture content from 91.94% to 22.32% for 44 h.

Dorouzi et al. [88] developed an indirect solar dryer assisted with a PVT system that was based on a desiccant regeneration circle. Closed-loop air circulation was formed using a DC fan, an auxiliary heater regulated the drying temperature, and a desiccant bed absorbed the exhaust air mist. In this work, calcium chloride solution as the desiccant fluid flowed freely over the PV panel to attain the excess heat and became regenerated. It is operated when the air relative humidity (RH) exceeds the set point. The results indicated that the proposed dryer has the capability to meet the entire electric power requirement when the drying temperature varies from 60–65°C and the RH set point for the regeneration circle activation is 28% of drying of the tomato. Fig. 6.18 depicts the schematic view and a photograph of the developed system.

In an experimental investigation, a solar-powered fluidized bed dryer assisted with an infrared system was introduced by Mehran et al. [89] for drying paddy grains (Fig. 6.19). The proposed system included a solar water collector, a PV-powered infrared lamp, a gas-powered water heater, and a desiccant wheel. Two operating conditions of natural gas drying (NGD) and solar-assisted drying (SAD) were considered during the experiments. The results indicated the highest total energy consumption of 1.163 kWh for the dryer under the NGD test mode and the lowest value of 0.314 kWh under the SAD mode. In addition, the specific energy consumption values under the drying in SAD and NGD modes varied from 8.30–22.12 to 16.73–32.62 kWh/kg of the evaporated water, respectively. Moreover, it was concluded that although using an infrared lamp results in an increase in the solar energy fraction to 0.741, its use in the fluidizing chamber has no remarkable impact on the drying speed of the crop.

The use of PV-powered solar dryers brings great advantages considering their technical viability and energy-saving potential. Although recent studies in this field show significant progress, there is still room for improvements in terms of decreasing costs and increasing performance, both thermal and electrical, to make the use of these sustainable facilities widespread.

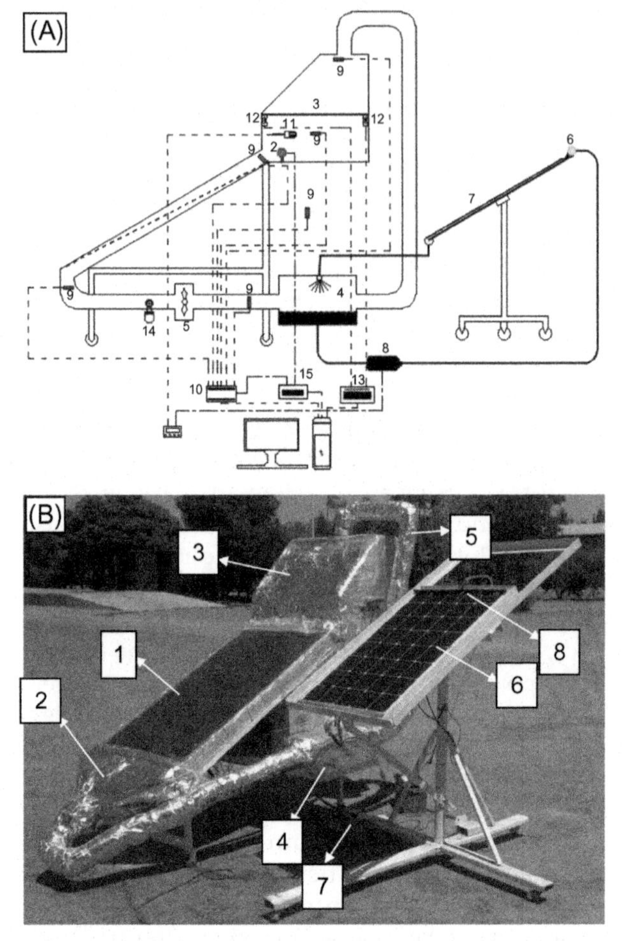

FIG. 6.18 An indirect solar dryer assisted with a liquid desiccant-based PVT collector [88]: (A) Schematic view of the experimental test rig, (B) A photo of the system including a solar collector (1), air entrance to the collector (2), drying chamber (3), liquid desiccant bed (4), connecting tube (5), PV panel (6), regeneration system pump (7), and distribution pipe (8).

FIG. 6.19 The experimental installation of the solar-assisted drying system developed in ref. [89].

6.2.4 PV-powered livestock farming systems

6.2.4.1 PV applications in livestock farms

Dairy farming is an important endeavor in agriculture to promote economic benefits and expand food production around the world. Because most dairy operations are energy-intensive, farms rely heavily either on utility power network access or renewable energy conversion systems to meet their electric demands. The high capacity of the vast available areas is promising and highlights the integration of PV arrays for clean, sustainable, and profitable electricity generation.

Nacer et al. [90] reported that on-farm electricity consumption ranges from 330 to 566 kWh/cow/year and the utilization of a grid-connected PV system can cause a major reduction, such as 80 million tons in annual GHG emissions. In another work, Bazen et al. [91] evaluated the economic potential of PV technology applications in the poultry industry. They stated that solar energy needs more advance incentives from the government to be employed by farmers as the main source of electrification. They also implied that although solar PV technology can ease environmental concerns pertaining to conventional "coal-fired" technology, new legislation offering strike constraints based on environmental factors may reinforce its deployment and make it more financially attractive. In a comprehensive study by Bey et al. [52], the feasibility of using grid-connected PV systems in dairy farms was evaluated. It was found that the use of PV systems not only reduces GHG emissions but also increases milk production by 8%. In addition, Zhang et al. [53] found that PV integration into dairy farms can be another source of income for the farmers when there is a surplus of electricity. In order to electrify a typical dairy farm with PV systems, it is important to estimate different electric loads that can be broken down to lighting, ventilation, water pumping, milking, and milk cooling, as shown in Fig. 6.20.

Lighting load

Artificial lighting has a direct effect on animals kept in livestock buildings and can trigger animals' milk production rate, health, and fertility [92]. Therefore, standard measures must be taken to provide a comfortable environment for livestock, where scientists recommend 200 lx of artificial lighting [93]. Therefore, the daily electrical energy required by the lighting system can be determined as the following [9]:

$$E_L = P_L N_L t_L \tag{6.7}$$

where E_L (Wh/d) denotes the daily electricity consumed by the lighting system, P_L (W) is the lamp power, N_L is the number of lamps required, and t_L (h) is the time duration of lighting per day.

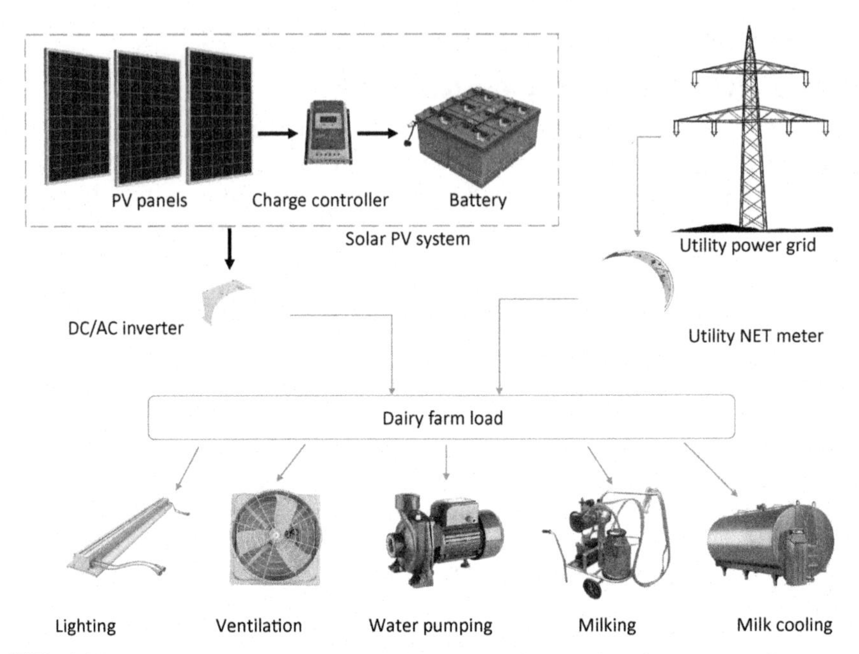

FIG. 6.20 The layout of a grid-connected PV system that supplies electric power for a typical dairy farm.

Ventilation

Ventilation is a key issue in livestock farming, as any failure may cause significant health problems or discomfort in the animals. Passive ventilation is a natural process in which odors and moisture are removed through openings and by natural drafts (convection). The effectiveness of this technique can be satisfactory while there are numerous designs that suggest the utilization of mechanical or forced ventilation in a combined operation mode. In this type, several electric pieces of equipment such as fans, blowers, etc., are incorporated to blow out the air and take fresh air inside the housing. As a result, another load is added to the total electrical power required for a farm, which can be evaluated as the following [9]:

$$E_V = P_F N_F f_F \tag{6.8}$$

where E_V (Wh/d) represents the daily electricity consumed by the ventilation system, P_F (W) is the fan power, N_F is the number of fans required, and f_F is the fan usage frequency (8 h/d for dairy cows).

Livestock watering

Water pumping is one of the primary electricity demands in the livestock farming industry, and it is usually employed for watering systems. The size of the load varies seasonally and is based on the animal species, where the estimated daily water intake for cattle ranges from 3 to 11 gal/d for growing cattle and nursing cows, respectively [94]. In order to obtain the energy needed for pumping, E_P (Wh/d), several factors such as the pump power, P_P (W), the number of animals, N_A, the amount of water needed for each animal per day, q (L/d), and the pump flow rate, Q_V (L/nm), must be considered; these are expressed in the following equation (6.9):

$$E_P = P_P \frac{N_A q}{Q_V} \tag{6.9}$$

Milk cooling and milking system

The milking machine has a significant importance in dairy farm production rate and its daily operation compels operators to pay necessary attention to the rate of energy consumption. Running for hours each day takes a considerable share of the total electricity requirements as well as the power needed for cooling and storing the milk inside the tanks. Therefore, the amounts of energy consumed by the milking machine, E_M (Wh), and the milk cooling tank, E_{MC} (Wh), can be derived from the following expressions (6.9):

$$E_M = P_M N_A t_M \tag{6.10}$$

$$E_{MC} = P_T t_{MS} \tag{6.11}$$

where P_M and P_T (W) are the powers of the milking machine and cooling tank, respectively, while N_A denotes the number of animals, t_M is the milking time, and t_{MS} is the milk storage time (h).

Maammeur et al. [95] studied a pilot farm in Algeria and developed a grid-connected PV system supplying electrical energy for a typical animal farm and irrigation system. In this simulation, the farm's power consumption was estimated for each month and three different periods of the year were distinguished with various energy demands, and a 50 kW PV power system was modeled using HOMER software. The illustration depicted in Fig. 6.21 indicates the results obtained from this study, where it shows the annual electric consumption, monthly grid, solar PV generation, and total energy required for a farm's equipment. At the same time, PV systems have been used in conjunction with other technologies in dairy farm setups [96]. Blanchet et al. [96] designed and installed a designated solar/biomass hybrid energy system resulting in a payback of 3.4 years and an annual greenhouse gas reduction of 275.9 tons of CO_2. Zhang et al. [97] carried out an economic assessment of a PV water pumping system integration with dairy farms and found that this created new sources of revenue with a very lucrative internal rate of return in developed countries. The electricity generation from PV systems is also used in several

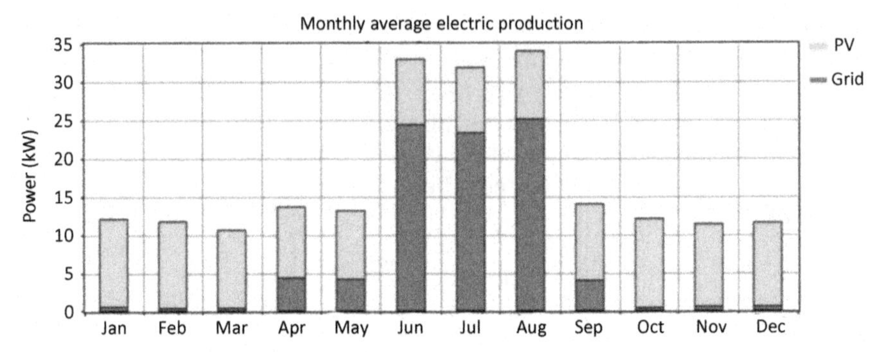

FIG. 6.21 The simulation results of the grid-connected PV system integrated with a farm with an average daily energy consumption of 281 kWh/day [95].

add-on activities in dairy farms such as energizing equipment for monitoring cow activity [98]. Furthermore, PV systems are also utilized for postproduction activities such as cooling of milk and associated produce, integration with anaerobic digesters, and integration with other hybrid energy systems.

6.2.4.2 Utilization of PVT collectors in livestock

Farms require both electricity and hot water for various activities in an integrated consumption pattern. Hybrid water-based PVT collectors are useful for providing both these utilities at the same time. As an example, a dairy cow uses 350 kWh of electricity per year, where 40% of this energy is used for heating water. In a study by Shortall et al. [99], a summary of the daily and seasonal trends for electricity used for milking dairy farms was presented. Upton et al. [100] used a mechanistic model to predict the distribution of the energy consumption in farms and found that on average, 28% of the total electricity production is used toward water heating. Wang et al. [101] performed a thermo-economic evaluation of a spectral splitting hybrid PVT system on a dairy farm. It was reported that there was excellent decarbonization potential and such systems are economically viable. Van Campen et al. [102] presented a comprehensive study of the hybrid utilization of solar technologies for livestock watering and aquaculture. Pringle at al. [28] investigated synergies for dual usage of water production from PV electricity generation in aquaculture. Solar hybrid PVT systems have the potential to reduce the operating and environmental impact if they are used in the farm industry; they can also make the system more flexible and reliable, particularly in cold-climate regions. The report presented by Park et al. [103] on direct energy use in agricultural practices signifies that this sector is one of the major energy consumers with the highest potential for the use of low-carbon technologies. It is estimated that if the current food system sourced all its energy from renewable energies, it could still provide 3000 times the quantity it does at the moment.

6.2.5 PV-powered solar greenhouses

Greenhouses are one of the most integral centralized elements of modern agricultural systems. They are used to protect crops from harsh weather conditions outside while also offering contamination prevention and pest protection. Additionally, greenhouses bring the possibility of growing specified types of crops such as fruits, vegetables, flowers, herbs, etc., throughout the year [104, 105]. The integration of PV systems with greenhouses has recently received remarkable attention because they have great potential to address the main challenges embedded with the greenhouse sector, including the limitation in arable land availability and the obligation for mitigating GHG emissions. The electricity demand of the greenhouses can be supplied either by settling the PV modules on the greenhouse skin (roofs or walls) or installing them as a separate unit alongside, as shown in Fig. 6.22. The integrated PV systems can be grid-connected or stand-alone, where the surplus electricity can easily be fed into the grid in the latter case, providing an additional source of revenue for the farmers.

Until now, integrations of a wide range of various PV module technologies with greenhouses have been studied by researchers, including the integration of them with PV modules [108–110], PVT modules [111–113], and concentrating PVT (CPVT) modules [114, 115].

FIG. 6.22 (A) Solar PV panels mounted as a part of the roofs in commercial greenhouses implemented at Prides Corner Farms [106], and (B) The PV systems installed close to the greenhouses in Hongxingqiao Town, Changxing County, in east China's Zhejiang Province [107].

6.2.5.1 PV-integrated solar greenhouses

According to the literature, the most widely used PV modules in greenhouses are mono- and polycrystalline silicon types. Although the integration of both opaque and semitransparent PV modules with greenhouses has been reported in several studies, the light transparency provided by the second type can mitigate the light-blocking issue inside the greenhouse caused by traditional opaque modules.

Yildiz et al. [110] developed a novel grid-connected PV-powered cooling system and integrated the proposed system with a greenhouse. The main components of the design are: (I) a converter, (II) PV cells, (III) an inverter, (IV) a fan, (V) an earth-to-air heat exchanger (HAHE), and (VI) a greenhouse. As Fig. 6.23 shows, the blower is used to circulate the air through the underground galvanized pipe (HAHE) in a closed-loop circle to maintain the greenhouse temperature at the desired level. A positive displacement blower with a 5300 m^3/h volumetric flow rate and 736 W

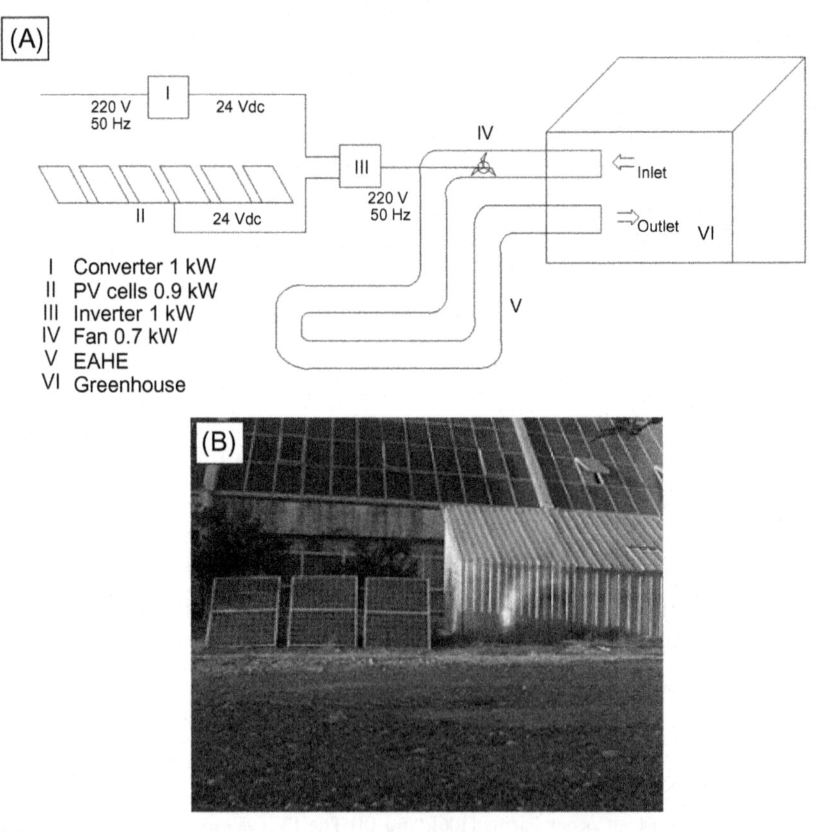

FIG. 6.23 A grid-connected PV-powered cooling system integrated with a greenhouse [110]; (A) Schematic view of the system, and (B) Photo of the experimental setup.

FIG. 6.24 (A) The unshaded studied greenhouse, and (B) the semitransparent BIPV panel [116].

capacity was set to be powered by the grid-connected PV system. Experimental data revealed that the average heat discharge rate was 5.02kW and 11h/day operation yielded 8.10kWh of electricity consumption, in which 34.55% came from photovoltaic conversion and 65.45% was supplied from the grid.

Emam Hassanien et al. [116] investigated the result of semitransparent building-integrated PVs (BIPVs) installed on top of a greenhouse (Fig. 6.24), and studied the growth of tomatoes as well as the behavior of the indoor environment. Also, they estimated the produced energy and payback period of the system. In this research, three modules covering more than 20% of the greenhouse roof area and tilting 30 degrees south were implemented with 0.08m spacing between the plastic cover and the BIPV. Each employed module exhibited 170Wp peak power with 8.25% efficiency. Analyses indicated that the amount of electric energy produced by the BIPV was 637kWh per year. A fall in solar radiation was also observed under the BIPV and measured as 35–40% higher compared to the mode covered by the polyethylene sheets on typical sunny days. They also suggested that the shading effects of the BIPV modules decreased the air temperature by 1–3°C on sunny days with no significant influence on the relative humidity. Additionally, the payback period of the system was determined to be 9 years.

6.2.5.2 PVT-integrated solar greenhouses

Reducing the payback period of the PV systems integrated with greenhouses is very important. In this regard, the PVT systems present a good alternative by providing both thermal and electric power generation, causing a significant reduction in payback time [117, 118]. According to the literature, in greenhouse applications, air and water are among the most widely used working fluids in PVT modules. However, water-based PVT modules are more effective to reduce PV cell temperature, particularly under hot climate conditions. On the other hand, in terms of

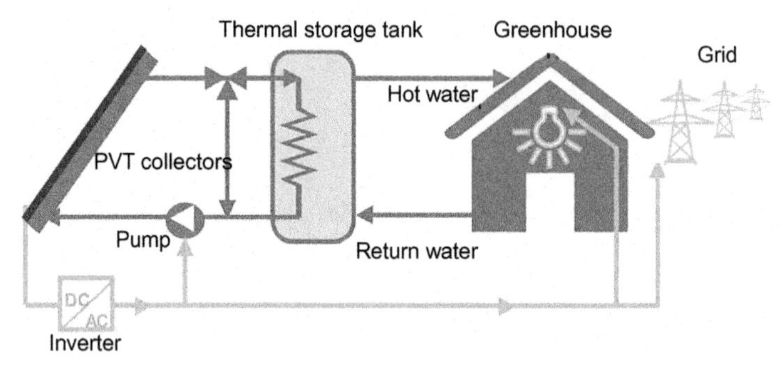

FIG. 6.25 The PVT-CHP system for greenhouses proposed in ref. [121].

simplicity and cost-effectiveness, the air-based PVT modules have added advantages [119, 120].

Wang et al. [121] conducted a technoeconomic assessment of a solar combined heat and power (S-CHP) system assisted with hybrid PVT collectors for dispersed cogeneration in a tomato greenhouse located in Bari, Italy (Fig. 6.25). A water-based PVT S-CHP system was designed and yearly simulations were conducted to anticipate the transient behavior of the S-CHP system and to evaluate the system's energy performance. The simulation data revealed that an area with 30,000 m^2 PV arrays would enable the PVT S-CHP system to make up 73% of the total annual thermal energy needed for the greenhouse, whereas providing a net electrical production of 2.6 was required as the electrical demand per year. They claimed that if the PVT system was implemented, 3010 tCO_2/year can be mitigated from the current CO_2, where the payback time can reach 10.4 years. They concluded that such PVT S-CHP systems have promising technoeconomic viability in the suggested greenhouse applications and could be an alternative for existing systems.

6.2.5.3 CPVT-integrated solar greenhouses

Overheating of solar cells is one of the major setbacks in CPV modules due to their high values of concentration ratios, which can reach higher than 100. Therefore, a permanent cooling mechanism is required to avoid damage to the modules. The integration of thermal modules with CPV units offers significant thermal and electric power as outputs [122–124]. The CPVT modules, using both reflectors or concentrating lenses, bring two main advantages: (1) they are compact and therefore can be easily installed on roofs, and (2) they provide acceptable outputs even under poor weather conditions. A greenhouse with an innovative design and integrated with linear Fresnel lenses was developed by Sonneveld et al. [125]. In their proposed system, a CPVT collector, including a tracing system to position the PVT modules, was mounted on the roof of the greenhouse (Fig. 6.26). The results reported the generated electricity as

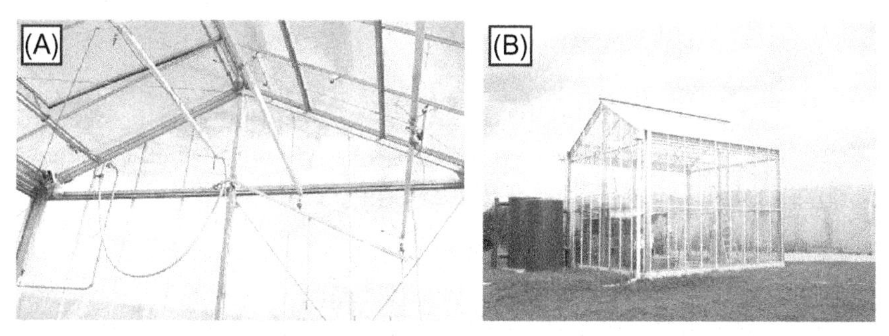

FIG. 6.26 (A) Linear Fresnel lenses and PVT modules integrated into a greenhouse hood, and (B) The solar greenhouse integrated with the CPVT module [125].

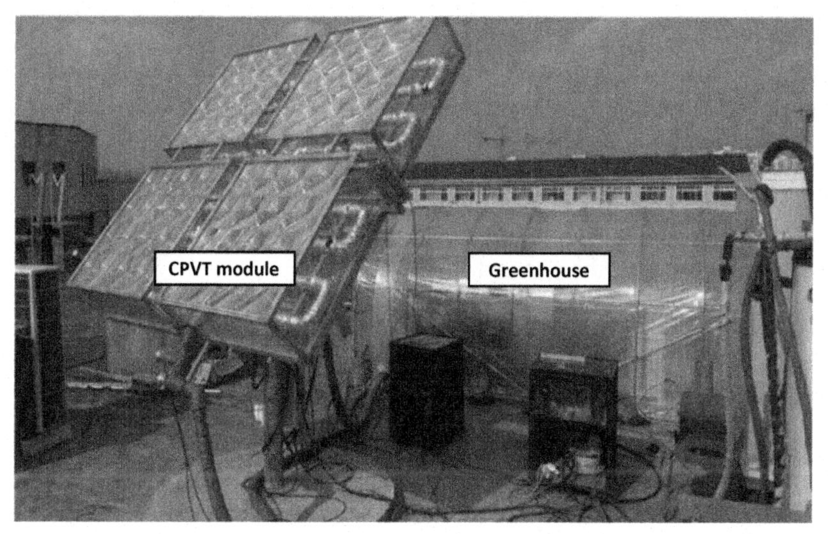

FIG. 6.27 The experimental setup of the greenhouse integrated with a CPVT module [126].

29 kWh/m^2/year with a thermal yield of 518 MJ/m^2. They claimed that the proposed CPVT system can be used as a unit to control the temperature as well as to provide lighting inside the greenhouse.

The energy potential and economic feasibility of a greenhouse integrated with a CPVT module were investigated by Hussain et al. [126]. In this case, two CPVT collectors, one with and the other without a glass-reinforced plastic envelope, were considered (Fig. 6.27). The results indicated that the glass-reinforced CPVT had better efficiency and a lower heat loss compared with the standard one. The results also indicated that using a CPVT collector to supply the required heat for the greenhouse led to a significantly discounted payback period (DPP) and life cycle saving (LCS).

6.2.6 Solar-powered crop protection systems

6.2.6.1 Solar-powered fencing

Each year, there are a number of losses in agricultural products due to animal menace and human interference, which impacts many farmers' revenue and can cause major rescission in the agriculture sector. One of the favorable techniques due to its practicality is the use of solar-powered prohibitive fencing to the farms to protect them from animals and trespassers [127]. This low-cost technology is based on nonlethal electric shocks, in which the object experiences high-voltage low-current shocks supplied in a pulsating pattern. Solar PV technology is usually implemented in this protective system to provide the main power, especially in remote regions where power network access is limited at the farm location. As shown in Fig. 6.28, electricity is generated from solar energy via the PV cells and charges the battery. The battery unit stores electric power for operations at night or cloudy days. The converter unit (energizer) is the heart of the system, and produces a different range of voltage pulses depending on the type of animal, fence length, vegetation load, and power source [129]. In order to induce an effective shock, an earth system is necessary to complete the circuit powered by the energizer where the current passes through the wires reaching the animal and returns back to the energizer via the ground.

Several models have been developed by scientists with different solutions for a wide range of animals. In a new design developed by Gandhimathi et al. [130], GSM (Global System for Mobile Communications) modems were incorporated with a solar fencing system to provide manual operation for shock production based on photoelectric sensor detection.

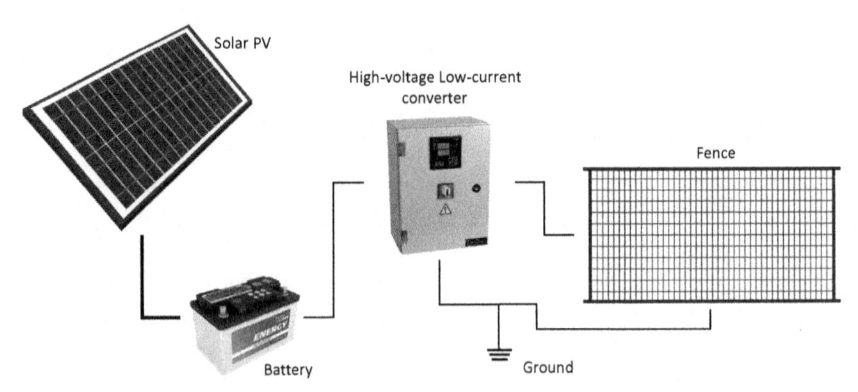

FIG. 6.28 Schematic view of PV-powered fencing used in farms, adapted from [128].

6.2.6.2 Solar-powered bird repeller

In situations where birds cause significant crop damage, this species becomes detrimental to human beings and must be hampered before any catastrophic crop losses. Different techniques have been tested throughout history to control birds, especially in farms where old methods included scarecrows, hawk kites, colored lights, lasers, flashes, chemicals, etc. [54]. However, recent studies suggest new techniques that implement sounds to make birds uncomfortable and avoid the environment. Therefore, solar-powered bird repellent devices are designed to take advantage of available solar energy to power an electric device that makes distress calls through a speaker and irritates birds to keep farms safe [131]. In a typical solar bird repeller, as shown in Fig. 6.29, electricity is generated through the PV panels and then charges the battery, which supplies an amplifier, a speaker, and a convertor in which voltage is downgraded to power an MP3 player. In this process, domestic bird predator calls are loaded to the MP3 player where the amplifier increases the signal level for the speaker [54]. Ogochukwu et al. [132] proposed an ultrasonic bird repeller that worked in the range of 15–25 kHz to disturb birds with four piezo transducers (Fig. 6.29A). They reported that the device was successful in driving birds from the area while it took 5–6 seconds to cover one hectare. In another attempt conducted by Muminov et al. [133], an effective bird repeller system was developed using three sonar sensors and a microcontroller. This configuration has emerged after several experiments reported the fact that pest birds can get used to the same played sounds and not be scared. Therefore, in this design, a special sound (like a gunshot sound) was loaded in addition to bird predator calls. The system was programmed to play the uncommon sound if the first and repeated sound didn't affect the pest bird.

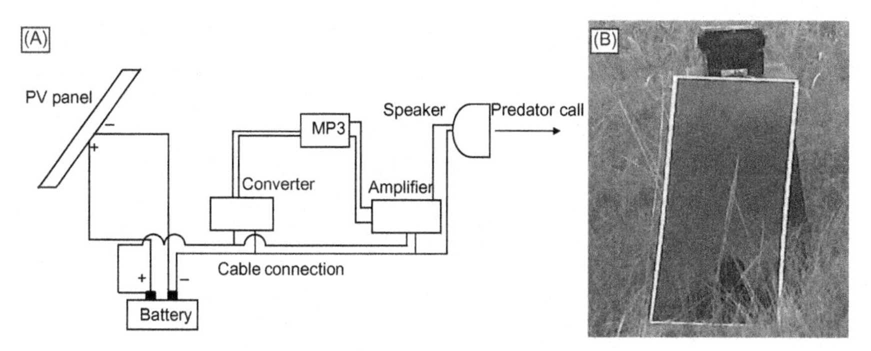

FIG. 6.29 Solar-powered bird repeller device: (A) Schematic view of an audible model [54], and (B) photo of an ultrasonic type [132].

6.2.6.3 Solar-powered light trap

The growing concern over the extreme use of pesticides for many crops drives public opinion toward other integrated pest-management techniques. The application of solar light traps has become a promising method with significant outcomes in pest control [134]. In addition to remarkable portability, this device promotes smooth operation in any field location and can be used for a vast variety of pests [135]. In a typical design, UV LED bulbs are used as the illumination source to attract insects showing sensitivity to the UV wavelength. Solar panels coupled with a battery unit are mainly incorporated to provide supply power with storage ability for at least 8 h. The entire device provides an eco-environment, energy-saving, and effective solution compared to the alternative choices [136]. In most designs, a funnel with collecting tubes is placed below the UV light (Fig. 6.30A) while other models such as the one developed by Sermsri et al. [138] implement an electric trap to shock pests lured to the LED, as shown in Fig. 6.30B.

6.3 Conclusions and future prospects

Similar to other sectors, agriculture has been faced with the problem of growing usage of fossil fuels, along with its rising dependence on energy-intensive activities that limit this sector to be sustainably expanded to meet global food demand, which is affected by climate change and volatility in fuel prices. Considering the high energy demand with the implications of a higher carbon footprint in agricultural activities, the necessity for the adoption of renewable energy sources in farm communities seems vital. The main conclusions of the current chapter on the usage of PV technologies in agriculture are summarized below:

- The application of PV technologies in agricultural farms is quite diverse and has tremendous potential for mitigating carbon emissions and bringing further profitability to the business.
- The development of several hybrid technologies brings further innovation and flexibility in the on-farm use of PV technologies, especially PVT systems for dairy farms, water heating, drying, greenhouses, and desalination systems.
- More cost-efficient solutions with cheaper energy storage options are required to further improve the uptake of PV technologies. This should

FIG. 6.30 Solar-powered light traps: (A) The model (with funnel) used in the Brinjal field [137], and (B) the model (with electric wire mesh) implemented in a coconut farm [138].

be possible with a unified legislative, political, social, and economic will for a sustainable future.

According to the literature, the main benefits and challenges of integrating PV technologies with agricultural activities are presented in Table 6.1. It is clear that further research and development, both technologically and socioeconomically, are required to accelerate PV adoption in the agriculture sector.

TABLE 6.1 Benefits and challenges of PV usage in agriculture.

Benefits	Challenges
• Sustainable and environmentally friendly with a zero-carbon footprint.	• PV cells heat up during energy conversion and due to solar irradiation. This reduces their efficiency, requiring efficient cooling solutions.
• It can easily be used in hybrid with conventional technologies to minimize unreliability and maximize advantages.	• Unpredictable source of energy, requiring thermal and/or electrical storage options.
• The system is literally noise-free in comparison with conventional systems as there are minimal moving parts.	• Compared to conventional fossil-fuel technologies, the investment cost is considerably high.
• No logistical mismatches between the generation and consumption of solar energy.	• High pay-back times with a low benefit-to-cost ratio.
• Especially suited for remote isolated areas typical of agricultural zones with no access to fossil fuels or local power grids.	• Maintenance concerns, especially cleaning PV panels, require water, which is already scarce in arid zones.
• It can be adapted in many versatile combinations with hybrid technologies specifically suited to the application.	• Efficiencies and energy output are low and are highly dependent on many external factors, making the system unreliable.
• With technological advancements and rampant research in this sector, the unit costs are lowering with massive improvements.	• Usually associated with small-scale applications. Operation control and optimization are major research areas at the moment.

Acknowledgment

We would like to thank the Renewable Energy Research Institute (RERI) for advisory support and Tarbiat Modares University for financial support [grant number IG/39705]. Additionally, the collaboration of the Centre for Built and Natural Environment, Coventry University, in this research is duly acknowledged.

References

[1] Bolyssov T, Yessengeldin B, Akybayeva G, Sultanova Z, Zhanseitov A. Features of the use of renewable energy sources in agriculture. Int J Energy Econ Policy 2019;9:363–8. https://doi.org/10.32479/ijeep.7443.

[2] Global Status Report. Renewable 2017. 72; 2017. https://doi.org/10.1016/j.rser.2016.09.082.

[3] Rajagopal D, Sexton SE, Roland-Holst D, Zilberman D. Challenge of biofuel: filling the tank without emptying the stomach? Environ Res Lett 2007;2:44004. https://doi.org/10.1088/1748-9326/2/4/044004.

[4] Caballero F, Sauma E, Yanine F. Business optimal design of a grid-connected hybrid PV (photovoltaic)-wind energy system without energy storage for an Easter Island's block. Energy 2013;61:248–61. https://doi.org/10.1016/j.energy.2013.08.030.

[5] Bojić M, Blagojević M. Photovoltaic electricity production of a grid-connected urban house in Serbia. Energy Policy 2006;34:2941–8. https://doi.org/10.1016/j.enpol.2005.04.024.

[6] Fernández-Infantes A, Contreras J, Bernal-Agustín JL. Design of grid connected PV systems considering electrical, economical and environmental aspects: a practical case. Renew Energy 2006;31:2042–62. https://doi.org/10.1016/j.renene.2005.09.028.

[7] Castro M, Delgado A, Argul FJ, Colmenar A, Yeves F, Peire J. Grid-connected PV buildings: analysis of future scenarios with an example of Southern Spain. Sol Energy 2005;79:86–95. https://doi.org/10.1016/j.solener.2004.09.022.

[8] Piechocki J, Sołowiej P, Neugebauer M, Chen G. Development in energy generation technologies and alternative fuels for agriculture. In: Adv. Agric. Mach. Technol. CRC Press; 2018. p. 89–112. https://doi.org/10.1201/9781351132398-4.

[9] Maammeur H, Hamidat A, Loukarfi L, Missoum M, Abdeladim K, Nacer T. Performance investigation of grid-connected PV systems for family farms: case study of North-West of Algeria. Renew Sustain Energy Rev 2017;78:1208–20. https://doi.org/10.1016/j.rser.2017.05.004.

[10] Soroudi A, Ehsan M. A possibilistic-probabilistic tool for evaluating the impact of stochastic renewable and controllable power generation on energy losses in distribution networks—a case study. Renew Sustain Energy Rev 2011;15:794–800. https://doi.org/10.1016/j.rser.2010.09.035.

[11] Alam Hossain Mondal M, Sadrul Islam AKM. Potential and viability of grid-connected solar PV system in Bangladesh. Renew Energy 2011;36:1869–74. https://doi.org/10.1016/j.renene.2010.11.033.

[12] Mekhilef S, Faramarzi SZ, Saidur R, Salam Z. The application of solar technologies for sustainable development of agricultural sector. Renew Sustain Energy Rev 2013;18:583–94. https://doi.org/10.1016/j.rser.2012.10.049.

[13] Goetzberger A, Zastrow A. On the coexistence of solar-energy conversion and plant cultivation. Int J Sol Energy 1982;1:55–69. https://doi.org/10.1080/01425918208909875.

[14] Trommsdorff M. An economic analysis of agrophotovoltaics: opportunities, risks and strategies toward a more efficient land use. Germany: University of Freiburg; 2016.

[15] Weselek A, Ehmann A, Zikeli S, Lewandowski I, Schindele S, Högy P. Agrophotovoltaic systems: applications, challenges, and opportunities. A review. Agron Sustain Dev 2019;39:35. https://doi.org/10.1007/s13593-019-0581-3.

[16] Amaducci S, Yin X, Colauzzi M. Agrivoltaic systems to optimise land use for electric energy production. Appl Energy 2018;220:545–61. https://doi.org/10.1016/j.apenergy.2018.03.081.

[17] Owley J, Wilson MA. The new agriculture : from food farms to solar farms. Columbia J Environ Law 2019;44(2):409–77.

[18] van der Horst RR. Solar farms on agricultural land: a partial equilibrium analysis MSc in environmental sciences. Wageningen University and Research; 2019.

[19] Borgogno Mondino E, Fabrizio E, Chiabrando R. Site selection of large ground-mounted photovoltaic plants: a GIS decision support system and an application to Italy. Int J Green Energy 2015;12:515–25. https://doi.org/10.1080/15435075.2013.858047.

[20] Dinesh H, Pearce JM. The potential of agrivoltaic systems. Renew Sustain Energy Rev 2016;54:299–308. https://doi.org/10.1016/j.rser.2015.10.024.

[21] Marrou H, Guilioni L, Dufour L, Dupraz C, Wery J. Microclimate under agrivoltaic systems: Is crop growth rate affected in the partial shade of solar panels? Agric For Meteorol 2013;177:117–32. https://doi.org/10.1016/j.agrformet.2013.04.012.
[22] Naylor RL, Goldburg RJ, Primavera JH, Kautsky N, Beveridge MCM, Clay J, et al. Effect of aquaculture on world fish supplies. Nature 2000;405:1017–24. https://doi.org/10.1038/35016500.
[23] Christoffel PF. Development of a GIS-based decision support tool for environmental impact assessment and due-diligence analyses of planned agricultural floating solar systems. University of South Africa; 2019.
[24] Xue J. Photovoltaic agriculture—new opportunity for photovoltaic applications in China. Renew Sustain Energy Rev 2017;73:1–9. https://doi.org/10.1016/j.rser.2017.01.098.
[25] Solaripedia-Green Architecture & Building; 2020. http://www.solaripedia.com/13/147/wineries_and_thieves_go_solar_(california,_usa).html [Accessed 15 January 2020].
[26] BayWa r.e. starts building 27.4MWp floating PV plant on Dutch lake | PV Tech 2020. https://www.pv-tech.org/news/baywa-r.e.-starts-construction-of-27.4mwp-plant-on-dutch-sandpit-lake [Accessed 13 March 2020].
[27] Tengeh Reservoir to house one of world's largest floating solar panel systems; 2020. https://www.straitstimes.com/singapore/tengeh-reservoir-to-house-one-of-worlds-largest-floating-solar-panel-systems [Accessed 15 January 2020].
[28] Pringle AM, Handler RM, Pearce JM. Aquavoltaics: synergies for dual use of water area for solar photovoltaic electricity generation and aquaculture. Renew Sustain Energy Rev 2017;80:572–84. https://doi.org/10.1016/j.rser.2017.05.191.
[29] Kim B, Lee S, Kang S, Jeong M, Gim GH, Park J, et al. Aquavoltaic system for harvesting salt and electricity at the salt farm floor: Concept and field test. Sol Energy Mater Sol Cells 2020;204:110234. https://doi.org/10.1016/j.solmat.2019.110234.
[30] Château P-A, Wunderlich RF, Wang T-W, Lai H-T, Chen C-C, Chang F-J. Mathematical modeling suggests high potential for the deployment of floating photovoltaic on fish ponds. Sci Total Environ 2019;687:654–66. https://doi.org/10.1016/j.scitotenv.2019.05.420.
[31] Menicou M, Vassiliou V. Prospective energy needs in Mediterranean offshore aquaculture: Renewable and sustainable energy solutions. Renew Sustain Energy Rev 2010;14:3084–91. https://doi.org/10.1016/j.rser.2010.06.013.
[32] Rezk H, Abdelkareem MA, Ghenai C. Performance evaluation and optimal design of stand-alone solar PV-battery system for irrigation in isolated regions: a case study in Al Minya (Egypt). Sustain Energy Technol Assessments 2019;36:100556. https://doi.org/10.1016/j.seta.2019.100556.
[33] Osma-Pinto G, Ordóñez-Plata G. Measuring the effect of forced irrigation on the front surface of PV panels for warm tropical conditions. Energy Reports 2019;5:501–14. https://doi.org/10.1016/j.egyr.2019.04.010.
[34] Narvarte L, Almeida RH, Carrêlo IB, Rodríguez L, Carrasco LM, Martinez-Moreno F. On the number of PV modules in series for large-power irrigation systems. Energy Convers Manag 2019;186:516–25. https://doi.org/10.1016/j.enconman.2019.03.001.
[35] Mohammed Wazed S, Hughes BR, O'Connor D, Kaiser CJ. A review of sustainable solar irrigation systems for Sub-Saharan Africa. Renew Sustain Energy Rev 2018;81:1206–25. https://doi.org/10.1016/j.rser.2017.08.039.
[36] Parvaresh Rizi A, Ashrafzadeh A, Ramezani A. A financial comparative study of solar and regular irrigation pumps: case studies in eastern and southern Iran. Renew Energy 2019;138:1096–103. https://doi.org/10.1016/j.renene.2019.02.026.
[37] Almeida RH, Ledesma JR, Carrêlo IB, Narvarte L, Ferrara G, Antipodi L. A new pump selection method for large-power PV irrigation systems at a variable frequency. Energy Convers Manag 2018;174:874–85. https://doi.org/10.1016/j.enconman.2018.08.071.
[38] Jones MA, Odeh I, Haddad M, Mohammad AH, Quinn JC. Economic analysis of photovoltaic (PV) powered water pumping and desalination without energy storage for agriculture. Desalination 2016;387:35–45.

[39] Stuber MD, Sullivan C, Kirk SA, Farrand JA, Schillaci PV, Fojtasek BD, et al. Pilot demonstration of concentrated solar-powered desalination of subsurface agricultural drainage water and other brackish groundwater sources. Desalination 2015;355:186–96.
[40] Dehesa-Carrasco U, Ramírez-Luna JJ, Calderón-Mólgora C, Villalobos-Hernández RS, Flores-Prieto JJ. Experimental evaluation of a low pressure desalination system (NF-PV), without battery support, for application in sustainable agriculture in rural areas. Water Sci Technol Water Supply 2016;17:579–87.
[41] Fterich M, Chouikhi H, Bentaher H, Maalej A. Experimental parametric study of a mixed-mode forced convection solar dryer equipped with a PV/T air collector. Sol Energy 2018;171:751–60. https://doi.org/10.1016/j.solener.2018.06.051.
[42] Eltawil MA, Azam MM, Alghannam AO. Energy analysis of hybrid solar tunnel dryer with PV system and solar collector for drying mint (MenthaViridis). J Clean Prod 2018;181:352–64. https://doi.org/10.1016/j.jclepro.2018.01.229.
[43] Eltawil MA, Azam MM, Alghannam AO. Solar PV powered mixed-mode tunnel dryer for drying potato chips. Renew Energy 2018;116:594–605. https://doi.org/10.1016/j.renene.2017.10.007.
[44] Ziaforoughi A, Esfahani JA. A salient reduction of energy consumption and drying time in a novel PV-solar collector-assisted intermittent infrared dryer. Sol Energy 2016;136:428–36. https://doi.org/10.1016/j.solener.2016.07.025.
[45] Chauhan PS, Kumar A, Nuntadusit C. Heat transfer analysis of PV integrated modified greenhouse dryer. Renew Energy 2018;121:53–65. https://doi.org/10.1016/j.renene.2018.01.017.
[46] Mahdavi S, Sarhaddi F, Hedayatizadeh M. Energy/exergy based-evaluation of heating/cooling potential of PV/T and earth-air heat exchanger integration into a solar greenhouse. Appl Therm Eng 2019;149:996–1007. https://doi.org/10.1016/j.applthermaleng.2018.12.109.
[47] Selmani A, Outanoute M, El Khayat M, Guerbaoui M, Ed-Dahhak A, Lachhab A, et al. Towards autonomous greenhouses solar-powered. Procedia Comput Sci 2019;148:495–501. https://doi.org/10.1016/j.procs.2019.01.062.
[48] Hollingsworth JA. Environmental and economic impacts of solar powered integrated greenhouses; 2019.
[49] Lukuyu JM, Blanchard RE, Rowley PN. A risk-adjusted techno-economic analysis for renewable-based milk cooling in remote dairy farming communities in East Africa. Renew Energy 2019;130:700–13. https://doi.org/10.1016/j.renene.2018.06.101.
[50] Nadjemi O, Nacer T, Hamidat A, Salhi H. Optimal hybrid PV/wind energy system sizing: application of cuckoo search algorithm for Algerian dairy farms. Renew Sustain Energy Rev 2017;70:1352–65. https://doi.org/10.1016/j.rser.2016.12.038.
[51] Nacer T, Hamidat A, Nadjemi O. Feasibility study and electric power flow of grid connected photovoltaic dairy farm in Mitidja (Algeria). Energy Procedia 2014;50:581–8. https://doi.org/10.1016/j.egypro.2014.06.071.
[52] Bey M, Hamidat A, Benyoucef B, Nacer T. Viability study of the use of grid connected photovoltaic system in agriculture: case of Algerian dairy farms. Renew Sustain Energy Rev 2016;63:333–45. https://doi.org/10.1016/j.rser.2016.05.066.
[53] Zhang C, Campana P, Yang J, Zhang J, Yan J. Can solar energy be an alternative choice of milk production in dairy farms? A case study of integrated PVWP system with Alfalfa and milk production in dairy farms in China. Energy Procedia 2017;105:3953–9. https://doi.org/10.1016/j.egypro.2017.03.822.
[54] Suryawanshi VR. Design, manufacture and test of a solar powered audible bird scarer and study of sound ranges used in it. Int J Sci Res 2015;4:1709–11.
[55] Khandar CD, Patil SS, Dindekar ND, Raut PV. Solar power fencing based on Gsm technology for agriculture n.d.

[56] Chandel SS, Nagaraju Naik M, Chandel R. Review of solar photovoltaic water pumping system technology for irrigation and community drinking water supplies. Renew Sustain Energy Rev 2015;49:1084–99. https://doi.org/10.1016/j.rser.2015.04.083.
[57] Gao Z, Zhang Y, Gao L, Li R. Progress on solar photovoltaic pumping irrigation technology. Irrig Drain 2018;67:89–96. https://doi.org/10.1002/ird.2196.
[58] Al-Waeli AHA, El-Din MMK, Al-Kabi AHK, Al-Mamari A, Kazem HA, Chaichan MT. Optimum design and evaluation of solar water pumping system for rural areas. Int J Renew Energy Res 2017;7:12–20.
[59] Dawson FG, Alexander G, Hofmann PL. Solar powered irrigation. vol. 3; 1979.
[60] Aliyu M, Hassan G, Said SA, Siddiqui MU, Alawami AT, Elamin IM. A review of solar-powered water pumping systems. Renew Sustain Energy Rev 2018;87:61–76. https://doi.org/10.1016/j.rser.2018.02.010.
[61] Rathore PKS, Das SS, Chauhan DS. Perspectives of solar photovoltaic water pumping for irrigation in India. Energy Strateg Rev 2018;22:385–95. https://doi.org/10.1016/j.esr.2018.10.009.
[62] Al-Smairan M. Application of photovoltaic array for pumping water as an alternative to diesel engines in Jordan Badia, Tall Hassan station: case study. Renew Sustain Energy Rev 2012;16:4500–7. https://doi.org/10.1016/j.rser.2012.04.033.
[63] Pumpmakers NSP Solar Pump; 2020. https://www.techxlab.org/solutions/pm-pumpmakers-nsp-solar-pump [Accessed 18 January 2020].
[64] Mokeddem A, Midoun A, Kadri D, Hiadsi S, Raja IA. Performance of a directly-coupled PV water pumping system. Energy Convers Manag 2011;52:3089–95. https://doi.org/10.1016/j.enconman.2011.04.024.
[65] Hassan W, Kamran F. A hybrid PV/utility powered irrigation water pumping system for rural agricultural areas. Cogent Eng 2018;5. https://doi.org/10.1080/23311916.2018.1466383.
[66] Tabaei H, Ameri M. The effect of booster reflectors on the photovoltaic water pumping system performance. J Sol Energy Eng 2012;134:1–4. https://doi.org/10.1115/1.4005339.
[67] Sontake VC, Kalamkar VR. Solar photovoltaic water pumping system—a comprehensive review. Renew Sustain Energy Rev 2016;59:1038–67. https://doi.org/10.1016/j.rser.2016.01.021.
[68] Pugsley A, Zacharopoulos A, Mondol JD, Smyth M. Solar desalination potential around the world. In: Renew. Energy Powered Desalin. Handb. Elsevier; 2018. p. 47–90. https://doi.org/10.1016/B978-0-12-815244-7.00002-7.
[69] Morrison J, Morikawa M, Murphy M, Schulte P. Water scarcity & climate change: growing risks for businesses and investors; 2009.
[70] Kumar R, Ahmed M, Bhadrachari G, Thomas JP. Desalination for agriculture: water quality and plant chemistry, technologies and challenges. Water Sci Technol Water Supply 2017;18:ws2017229. https://doi.org/10.2166/ws.2017.229.
[71] Zhang Y, Sivakumar M, Yang S, Enever K, Ramezanianpour M. Application of solar energy in water treatment processes: a review. Desalination 2018;428:116–45. https://doi.org/10.1016/j.desal.2017.11.020.
[72] Ahmed FE, Hashaikeh R, Hilal N. Solar powered desalination—technology, energy and future outlook. Desalination 2019;453:54–76. https://doi.org/10.1016/j.desal.2018.12.002.
[73] Chen C, Jiang Y, Ye Z, Yang Y, Hou L. Sustainably integrating desalination with solar power to overcome future freshwater scarcity in China. Glob Energy Interconnect 2019;2:98–113. https://doi.org/10.1016/j.gloei.2019.07.009.
[74] Fadhel MI, Sopian K, Daud WRW, Alghoul MA. Review on advanced of solar assisted chemical heat pump dryer for agriculture produce. Renew Sustain Energy Rev 2011;15:1152–68. https://doi.org/10.1016/j.rser.2010.10.007.

[75] Tiwari S, Tiwari GN, Al-Helal IM. Development and recent trends in greenhouse dryer: a review. Renew Sustain Energy Rev 2016;65:1048–64. https://doi.org/10.1016/j.rser.2016.07.070.

[76] El Hage H, Herez A, Ramadan M, Bazzi H, Khaled M. An investigation on solar drying: a review with economic and environmental assessment. Energy 2018;157:815–29. https://doi.org/10.1016/j.energy.2018.05.197.

[77] Mustayen AGMB, Mekhilef S, Saidur R. Performance study of different solar dryers: a review. Renew Sustain Energy Rev 2014;34:463–70. https://doi.org/10.1016/j.rser.2014.03.020.

[78] Goud M, Reddy MVV, Chandramohan VP, Suresh S. A novel indirect solar dryer with inlet fans powered by solar PV panels: drying kinetics of Capsicum Annum and Abelmoschus esculentus with dryer performance. Sol Energy 2019;194:871–85. https://doi.org/10.1016/j.solener.2019.11.031.

[79] Mumba J. Development of a photovoltaic powered forced circulation grain dryer for use in the tropics. Renew Energy 1995;6:855–62. https://doi.org/10.1016/0960-1481(94)00088-N.

[80] Barnwal P, Tiwari GN. Grape drying by using hybrid photovoltaic-thermal (PV/T) greenhouse dryer: an experimental study. Sol Energy 2008;82:1131–44. https://doi.org/10.1016/j.solener.2008.05.012.

[81] Janjai S, Lamlert N, Intawee P, Mahayothee B, Bala BK, Nagle M, et al. Experimental and simulated performance of a PV-ventilated solar greenhouse dryer for drying of peeled longan and banana. Sol Energy 2009;83:1550–65. https://doi.org/10.1016/j.solener.2009.05.003.

[82] Wang J, Sheng K. Far-infrared and microwave drying of peach. LWT Food Sci Technol 2006;39:247–55. https://doi.org/10.1016/j.lwt.2005.02.001.

[83] Aktaş M, Şevik S, Amini A, Khanlari A. Analysis of drying of melon in a solar-heat recovery assisted infrared dryer. Sol Energy 2016;137:500–15. https://doi.org/10.1016/j.solener.2016.08.036.

[84] Şevik S, Aktaş M, Dolgun EC, Arslan E, Tuncer AD. Performance analysis of solar and solar-infrared dryer of mint and apple slices using energy-exergy methodology. Sol Energy 2019;180:537–49. https://doi.org/10.1016/j.solener.2019.01.049.

[85] Samimi-Akhijahani H, Arabhosseini A. Accelerating drying process of tomato slices in a PV-assisted solar dryer using a sun tracking system. Renew Energy 2018;123:428–38. https://doi.org/10.1016/j.renene.2018.02.056.

[86] Tiwari S, Tiwari GN. Energy and exergy analysis of a mixed-mode greenhouse-type solar dryer, integrated with partially covered N-PVT air collector. Energy 2017;128:183–95. https://doi.org/10.1016/j.energy.2017.04.022.

[87] Tiwari S, Tiwari GN, Al-Helal IM. Performance analysis of photovoltaic–thermal (PVT) mixed mode greenhouse solar dryer. Sol Energy 2016;133:421–8. https://doi.org/10.1016/j.solener.2016.04.033.

[88] Dorouzi M, Mortezapour H, Akhavan HR, Moghaddam AG. Tomato slices drying in a liquid desiccant-assisted solar dryer coupled with a photovoltaic-thermal regeneration system. Sol Energy 2018;162:364–71. https://doi.org/10.1016/j.solener.2018.01.025.

[89] Mehran S, Nikian M, Ghazi M, Zareiforoush H, Bagheri I. Experimental investigation and energy analysis of a solar-assisted fluidized-bed dryer including solar water heater and solar-powered infrared lamp for paddy grains drying. Sol Energy 2019;190:167–84. https://doi.org/10.1016/j.solener.2019.08.002.

[90] Nacer T, Hamidat A, Nadjemi O. A comprehensive method to assess the feasibility of renewable energy on Algerian dairy farms. J Clean Prod 2016;112:3631–42. https://doi.org/10.1016/j.jclepro.2015.06.101.

[91] Bazen EF, Brown MA. Feasibility of solar technology (photovoltaic) adoption: a case study on Tennessee's poultry industry. Renew Energy 2009;34:748–54. https://doi.org/10.1016/j.renene.2008.04.003.

[92] Reksen O, Tverdal A, Landsverk K, Kommisrud E, Bøe KE, Ropstad E. Effects of photointensity and photoperiod on milk yield and reproductive performance of Norwegian Red cattle. J Dairy Sci 1999;82:810–6. https://doi.org/10.3168/jds.S0022-0302(99)75300-2.
[93] Šístková M, Peterka A, Peterka B. Light and noise conditions of buildings for breeding dairy cows. Res Agric Eng 2010;56:92–8. https://doi.org/10.17221/43/2009-rae.
[94] Buschermohle MJ, Burns RT. Solar-Powered Livestock Watering Systems. n.d.
[95] Maammeur H, Hamidat A, Loukarfi L. Energy intake of a PV system from grid-connected agricultural farm in Chlef (Algeria). Energy Procedia 2013;36:1202–11. https://doi.org/10.1016/j.egypro.2013.07.136.
[96] Blanchet CAC, Pantaleo AM, van Dam KH. A process systems engineering approach to designing a solar/biomass hybrid energy system for dairy farms in Argentina; 2019. p. 1609–14. https://doi.org/10.1016/b978-0-12-818634-3.50269-1.
[97] Zhang C, Campana PE, Yang J, Yu C, Yan J. Economic assessment of photovoltaic water pumping integration with dairy milk production. Energy Convers Manag 2018;177:750–64. https://doi.org/10.1016/j.enconman.2018.09.060.
[98] Tullo E, Fontana I, Gottardo D, Sloth KH, Guarino M. Technical note: validation of a commercial system for the continuous and automated monitoring of dairy cow activity. J Dairy Sci 2016;99:7489–94. https://doi.org/10.3168/jds.2016-11014.
[99] Shortall J, O'Brien B, Sleator RD, Upton J. Daily and seasonal trends of electricity and water use on pasture-based automatic milking dairy farms. J Dairy Sci 2018;101:1565–78. https://doi.org/10.3168/jds.2017-13407.
[100] Upton J, Murphy M, Shalloo L, Groot Koerkamp PWG, De Boer IJM. A mechanistic model for electricity consumption on dairy farms: definition, validation, and demonstration. J Dairy Sci 2014;97:4973–84. https://doi.org/10.3168/jds.2014-8015.
[101] Wang K, Pantaleo AM, Markides CN. Thermoeconomic assessment of a spectral-splitting hybrid PVT system in dairy farms for combined heat and power. Conference; 2019.
[102] Van Campen B, Guidi D, Best G. Solar photovoltaics for sustainable agriculture and rural development; 2000. p. 76. https://doi.org/10.13140/RG.2.1.5056.4962.
[103] Park NACS. Final report to Defra AC0401: direct energy use in agriculture: opportunities for reducing fossil fuel inputs; 2007. p. 52.
[104] Giacomelli GA. Engineering principles impacting high-tunnel environments. Horttechnology 2009;19:30–3. https://doi.org/10.21273/hortsci.19.1.30.
[105] Scognamiglio A, Garde F, Ratsimba T, Monnier A, Scotto E. Photovoltaic greenhouses: a feasible solutions for Islands? Design, operation, monitoring and lessons learned from a real case study. In: 6th World Conf. Photovolt. Energy Convers; 2014. p. 1169–70.
[106] Prides Corner Farms Investment in Solar Panels Pays Off—Greenhouse Grower; 2020. https://www.greenhousegrower.com/technology/prides-corner-farms-investment-in-solar-panels-pays-off/ [Accessed 13 January 2020].
[107] Photovoltaic agricultural greenhouses benefit farmers in Zhejiang, E China-Xinhua; 2020. http://www.xinhuanet.com/english/2018-05/27/c_137210458.htm [Accessed 13 January 2020].
[108] Sgroi F, Tudisca S, Di Trapani A, Testa R, Squatrito R. Efficacy and efficiency of Italian energy policy: the case of PV systems in greenhouse farms. Energies 2014;7:3985–4001.
[109] Carlini M, Villarini M, Esposto S, Bernardi M. Performance analysis of greenhouses with integrated photovoltaic modules. In: Int. Conf. Comput. Sci. Its Appl. Springer; 2010. p. 206–14.
[110] Yıldız A, Ozgener O, Ozgener L. Energetic performance analysis of a solar photovoltaic cell (PV) assisted closed loop earth-to-air heat exchanger for solar greenhouse cooling: an experimental study for low energy architecture in Aegean Region. Renew Energy 2012;44:281–7. https://doi.org/10.1016/j.renene.2012.01.091.

[111] Nayak S, Tiwari GN. Energy and exergy analysis of photovoltaic/thermal integrated with a solar greenhouse. Energy Build 2008;40:2015–21. https://doi.org/10.1016/j.enbuild.2008.05.007.

[112] Tiwari S, Bhatti J, Tiwari GN, Al-Helal IM. Thermal modelling of photovoltaic thermal (PVT) integrated greenhouse system for biogas heating. Sol Energy 2016;136:639–49. https://doi.org/10.1016/j.solener.2016.07.048.

[113] Rocamora MC, Tripanagnostopoulos Y. Aspects of PV/T solar system application for ventilation needs in greenhouses. In: Int. Symp. Greenh. Cool., vol. 719; 2006. p. 239–46.

[114] Sonneveld PJ, Swinkels G, Van Tuijl BAJ, Janssen HJJ, Gieling TH. A Fresnel lenses based concentrated PV system in a greenhouse. In: Int. Symp. High Technol. Greenh. Syst. GreenSys2009, vol. 893; 2009. p. 343–50.

[115] Sonneveld P, Swinkels GJ. A CPV system based on NIR reflecting lamellae integrated into a greenhouse: optimizing of optics. Adv Sci Technol 2010;74:297–302. Trans Tech Publ.

[116] Hassanien R, Hassanien E, Li M, Yin F. The integration of semi-transparent photovoltaics on greenhouse roof for energy and plant production. Renew Energy 2018;121:377–88. https://doi.org/10.1016/j.renene.2018.01.044.

[117] Riffat SB, Cuce E. A review on hybrid photovoltaic/thermal collectors and systems. Int J Low-Carbon Technol 2011;6:212–41. https://doi.org/10.1093/ijlct/ctr016.

[118] Rekha L, Vijayalakshmi MM, Natarajan E. Green house saving potential of a photovoltaic thermal hybrid solar. vol. 2; 2013. p. 370–6.

[119] Agrawal B, Tiwari GN. Optimizing the energy and exergy of building integrated photovoltaic thermal (BIPVT) systems under cold climatic conditions. Appl Energy 2010;87:417–26. https://doi.org/10.1016/j.apenergy.2009.06.011.

[120] Zhang X, Zhao X, Smith S, Xu J, Yu X. Review of R&D progress and practical application of the solar photovoltaic/thermal (PV/T) technologies. Renew Sustain Energy Rev 2012;16:599–617. https://doi.org/10.1016/j.rser.2011.08.026.

[121] Wang K, Pantaleo AM, Mugnozza GS, Markides CN. Technoeconomic assessment of solar combined heat and power systems based on hybrid PVT collectors in greenhouse applications. IOP Conf Ser Mater Sci Eng 2019;609. https://doi.org/10.1088/1757-899X/609/7/072026.

[122] Lamnatou C, Chemisana D. Solar radiation manipulations and their role in greenhouse claddings: Fresnel lenses, NIR- and UV-blocking materials. Renew Sustain Energy Rev 2013;18:271–87. https://doi.org/10.1016/j.rser.2012.09.041.

[123] Lamnatou C, Chemisana D. Solar radiation manipulations and their role in greenhouse claddings: Fluorescent solar concentrators, photoselective and other materials. Renew Sustain Energy Rev 2013;27:175–90. https://doi.org/10.1016/j.rser.2013.06.052.

[124] Chou SK, Chua KJ, Ho JC, Ooi CL. On the study of an energy-efficient greenhouse for heating, cooling and dehumidification applications. Appl Energy 2004;77:355–73. https://doi.org/10.1016/S0306-2619(03)00157-0.

[125] Sonneveld PJ, Swinkels GLAM, van Tuijl BAJ, Janssen HJJ, Campen J, Bot GPA. Performance of a concentrated photovoltaic energy system with static linear Fresnel lenses. Sol Energy 2011;85:432–42. https://doi.org/10.1016/j.solener.2010.12.001.

[126] Imtiaz Hussain M, Ali A, Lee GH. Multi-module concentrated photovoltaic thermal system feasibility for greenhouse heating: model validation and techno-economic analysis. Sol Energy 2016;135:719–30. https://doi.org/10.1016/j.solener.2016.06.053.

[127] A SH. Solar fencing unit and alarm for animal entry prevention; 2017.

[128] Prasanthi G, Harinarayana M. Design of converter for solar power fencing system for an agriculture field. Int J Adv Res Electr Electron Instrum Eng 2014. https://doi.org/10.15662/ijareeie.2014.0311037. [An ISO 2007;3297].

[129] Kadam DM, Dange AR, Khambalkar VP. Performance of solar power fencing system for agriculture; vol. 7. 2011.

[130] Gandhimathi A, Madhumitha P, Kalaivani T. Solar power fencing system for agriculture protection using GSM. Int J Innov Res Comput Commun Eng 2016. https://doi.org/10.15680/IJIRCCE.2016.
[131] Baral SS, Swarnkar R, Kothiya AV, Monpara AM, Chavda SK. Bird repeller—a review. Int J Curr Microbiol Appl Sci 2019;8:1035–9. https://doi.org/10.20546/ijcmas.2019.802.121.
[132] Stella Ogochukwu E, Desmond Okechukwu A, Godfrey NO. Construction and testing of ultrasonic bird repeller. J Nat Sci Res 2012;2.
[133] Muminov A, Jeon YC, Na D, Lee C, Jeon HS. Development of a solar powered bird repeller system with effective bird scarer sounds. In: 2017 Int. Conf. Inf. Sci. Commun. Technol. ICISCT 2017, vol. 2017—December. Institute of Electrical and Electronics Engineers Inc; 2017. p. 1–4. https://doi.org/10.1109/ICISCT.2017.8188587.
[134] Post N, Student G. Ecofriendly management of thrips in capsicum under protected condition Sunitha ND and Narasamma. J Entomol Zool Stud 2018;6.
[135] Brimapureeswaran R, Nivas G, Meenatchi R, Sujeetha ARP, Loganathan M. Development of a new solar light cum glue trap model and its utilization in agriculture. Int J Emerg Technol Innov Eng 2016;2:37–41.
[136] Kumar N. Development and evaluation of eco-friendly solar energy based light trap. Int J Pure App Biosci 2019;7:356–60. https://doi.org/10.18782/2320-7051.5587.
[137] Thangalakshmi S, Ramanujan R. Electronic trapping and monitoring of insect pests troubling agricultural fields. Int J Emerg Eng Res Technol 2015;3:206.
[138] Sermsri N, Torasa C. Solar energy-based insect pest trap. Procedia—Soc Behav Sci 2015;197:2548–53. https://doi.org/10.1016/j.sbspro.2015.07.620.

Further reading

Agrawal A, Sarviya RM. A review of research and development work on solar dryers with heat storage. Int J Sustain Energy 2016;35:583–605. https://doi.org/10.1080/14786451.2014.930464.

CHAPTER

7

Applications of solar PV systems in agricultural automation and robotics

Shiva Gorjian[a,b], Saeid Minaei[a], Ladan MalehMirchegini[c], Max Trommsdorff[d], and Redmond R. Shamshiri[b]

[a]Biosystems Engineering Department, Faculty of Agriculture, Tarbiat Modares University (TMU), Tehran, Iran, [b]Leibniz Institute for Agricultural Engineering and Bioeconomy (ATB), Potsdam, Germany, [c]Mechanical Engineering Department, Alzahra University, Tehran, Iran, [d]Fraunhofer Institute for Solar Energy Systems ISE, Freiburg im Breisgau, Germany

7.1 Precision agriculture (PA)

The origin of precision agriculture (PA) is traced back to the late 1980s with early applications in industrial manufacturing. Based on the definition presented by Blackmore [1], PA is a systems approach with the final goal of decreasing decision uncertainty through better understanding of the reasons for variabilities and the management of uncontrolled variations in agricultural fields.

In today's world, agriculture has faced increasing challenges in several areas such as water availability, resistance to agrochemicals, and environmental protection. In this regard, PA can definitely address a part of these challenges through deploying automation and sensing technologies [2]. Meeting production and sustainability challenges in the future of agriculture entails the integration of existing and to-be-developed PA technologies into crop production systems, including cultural practices, equipment, weather prediction, farm management, etc. Agricultural productivity has

Photovoltaic Solar Energy Conversion
https://doi.org/10.1016/B978-0-12-819610-6.00007-7

been substantially increased over the years through mechanization and automation. Rapid progress in agricultural science and technology has become the main driving force for the deployment of robotics and automation in this sector [3–5].

Research and development in PA over the years has resulted in the adaption of information and communication technologies for farming systems, which makes this approach the technical core of the information-intensive farms of the future [6]. The advent of robotics and autonomous systems (RAS) provides the opportunity to develop a new generation of flexible agricultural equipment and to decrease inputs in more efficient ways by using robotic technologies and intelligent machines [2]. Globally, automated conventional machinery and agricultural robotics are considered key solutions for performing field operations (e.g., planting, spraying, weeding, harvesting) in the most efficient, precise, and productive way to increase farm yields while minimizing environmental impacts [7]. Moreover, robotic platforms with the ability to collect sensory data in the field can further provide information about soils, crops, seeds, livestock, water use, and farm equipment. Additionally, some advanced analytics and low-cost automation techniques such as wireless sensor networks (WSNs) and the Internet of Things (IoT) have already started to assist farmers in analyzing data on weather, soil, temperature, moisture, etc., and give them better insights that can help to optimize yield and to improve planning [2].

7.1.1 Robotics and intelligent machines in agriculture

In recent years, automation techniques, smart sensors, and agricultural robots (ag-robots) have achieved remarkable success in farm applications and have attracted much interest, which has led to the development of more rational and adaptable machinery [8]. Research and development in robotics is still evolving and undergoing changes because of ongoing developments in sensors, reduction in equipment costs, and development of novel control algorithms [9].

The concept of precision autonomous farming (PAF) pertains to automatic agricultural machinery operating safely and efficiently without human interference to properly perform agricultural tasks [10]. For various farm operations, two major tasks must be carried out simultaneously by an autonomous mobile robot (AMR), normally performed by an operator: (1) steering the vehicle, and (2) operating the equipment. Therefore, eliminating the continuous adjustments performed by an operator for steering is the main reason behind the development of autonomous navigation systems (ANSs), which have been employed in agricultural machinery including tractors, cultivators, planters, harvesters, etc. [4]. The safe operation of autonomous vehicles in the field requires real-time risk detection and obstacle avoidance strategies [11, 12]. There are different types of technologies that can assist agricultural robots entering the field; some of them need to be developed while others have already been employed and need to only be adapted for use in agricultural environments. Among them, two major techniques are *artificial intelligence (AI)* and *machine learning* [13].

7.1.2 Autonomous navigation systems (ANSs)

The navigation task in agricultural AMRs is defined as the autonomous and safe guidance of robots in agricultural environments by considering the position of the robot and the detection of existing obstacles in the surrounding area [14]. The autonomous navigation concept revolves around the calculation and implementation of the necessary motions of an autonomous agricultural vehicle covering all targeted crops in a specific cultivation area with the help of task-specific actuation or sensing systems [15]. Four main components of the ANS are *navigation sensors, computational methods, path planning,* and *control strategies.* Fig. 7.1 shows the navigational system based on the interactive communication between the robot's perception that happens in the sensing process and the control process in the actuators.

The most common sensors used in automatic guidance are global navigation satellite systems (GNSS), infrared sensors, machine vision, light detection and ranging (LiDAR), and ultrasonic sensors. The implementation of LiDAR and machine vision can help in positioning the vehicle near the crop (e.g., in harvesting activities) [17, 18]. The use of GPS navigation has become nearly widespread in agriculture with the deployment of RTK

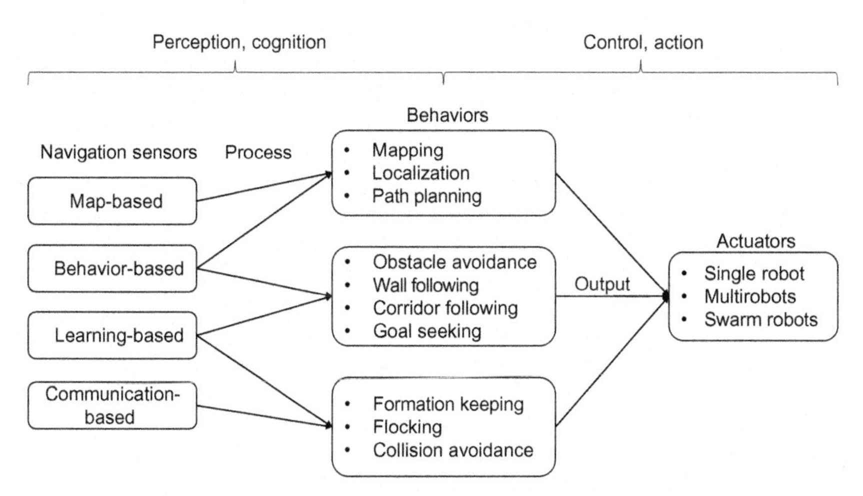

FIG. 7.1 Block diagram of the perception-actuation process in a navigation system [16].

(real-time kinematic) providing accuracy to centimeters for the automated positioning of large farm vehicles such as tractors and combine harvesters. However, in some applications, the accuracy of relative positioning and navigation is more important than absolute positioning [14, 19]. In the situation of simultaneous operation of several fully autonomous vehicles in a shared land, two main challenges in path planning are *optimal routing* and *obstacle avoidance.* In optimal routing, collisions with static and dynamic objects must be avoided while traveled distance and environmental impact should be minimized. In one of the obstacle-avoidance methods, steering control and avoiding obstacles are performed by utilizing local data related to the vehicle's environment. In optimal routing, fiberoptic gyro (FOG) sensors, GNSS, and accelerometers are utilized to compute position and orientation while establishing the steering angle and monitoring the vehicle's clutch and brake position are performed using rotary encoders and proximity sensors, respectively [20, 21]. Different control strategies for steering are based on fuzzy logic (FL), neural networks (NNs), proportional-integral-derivative (PID), feedforward PID (FPID), and genetic algorithms (GA) [4, 14].

Automatic guidance and steering control systems for tractors and farm machinery are quite mature and have been commercialized for about two decades. John Deere has introduced a guidance and steering control technology called AutoTrac that employs the NavCom StarFire GNSS guidance system. The StarFire guidance system provides a range of accuracies for positioning using satellite broadcast correction information or RTK positioning, enabling ±2.5-cm positioning accuracy (Fig. 7.2A). Additionally, a terrain compensation module (TCM) has been employed in guidance systems, in which the precise ground positioning is provided by using sensors that measure the roll, pitch, and yaw of the vehicle. The RTK differential corrections can be broadcast using a mobile RTK modem (Fig. 7.2B) [22].

The Advanced Farming System (AFS) developed by Case IH and Precision Land Management (PLM) of New Holland utilizes guidance systems (AccuGuide, AutoPilot, and Intellisteer) that offer diverse GNSS receiving units and satellite differential correction services available from Trimble and Omnistar (Centerpoint RTX and Rangepoint RTX) [17]. Recently, Case IH announced a proprietary RTK correction service (AFS RTK+) in the United States and Canada that implements an RTK base station network in which corrections can be broadcast through a mobile phone network (Fig. 7.3). The AutoPilot apparatus enables direct integration within the tractor electrohydraulic system to control steering [17].

The hydraulic steering system of the GPS PILOT S3 developed by CLAAS has the potential to be compatible with various kinds of machines such as tractors, combine harvesters, and forage harvesters. This steering

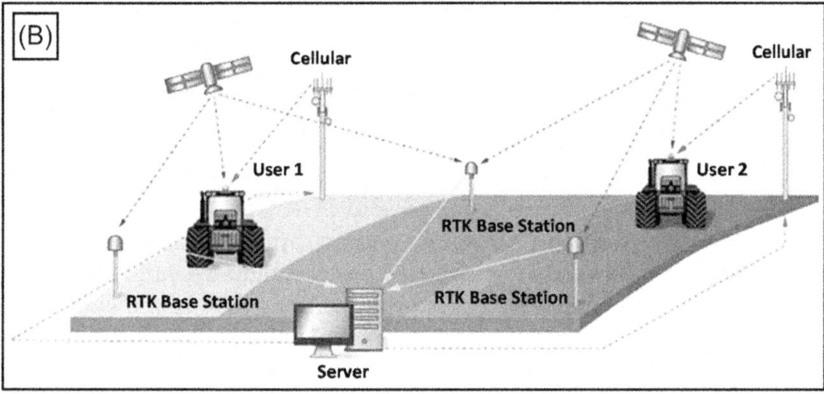

FIG. 7.2 (A) Combine harvester equipped with an AutoTrac system using a GreenStar 2630 Display to run the guidance system and a host of other precision farming applications, developed by John Deere and (B) Mobile RTK corrections using 3G/4G communications [17].

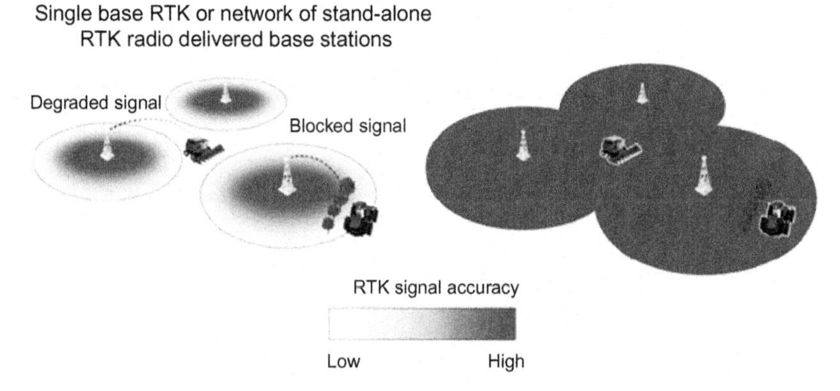

FIG. 7.3 Case IH dealer network RTK correction service [23].

system also offers the ability for use in an automatic steering wheel called GPS PILOT FLEX. This technology reflects precise steering performance and creates good versatility. Being assisted with RTK correction, GPS PILOT FLEX yields flexible and accurate operation [17]. A number of differential GPS correction signal options can be found with CLAAS systems consisting of satellite broadcast signals (EGNOS, OMNISTAR HP/XP/G2); BASELINE HD, which exploits a mobile reference station; and RTK systems utilizing RTK NET, which can provide corrections by means of a mobile phone network [24]. As shown in Fig. 7.4, (1) signals transmitted by GPS satellites are delivered to the machine and the RTK network, (2) a correction signal from networked reference stations is calculated by the central server, (3) the mobile phone network enables the machine to receive the high-precision RTK correction signals, and (4) both signals are converted to the steering signals by the GPS PILOT.

Machine operation and path planning include systems that improve coordinate implement and tractor operations. In this regard, Headland Management Systems (HMS) have been introduced to reduce operator fatigue [17]. Steering assistance within the headland is offered by a few manufacturers with precise alignment in accordance with the next move such as iTEC developed by John Deere and TURN-IN systems developed by CLAAS, which demonstrate path planning capability. One more promising technology compatible with autonomous vehicle operation is known as implement guidance, which properly positions the implement with a proportional response to the variations in loads on slopes [25]. TrueGuide

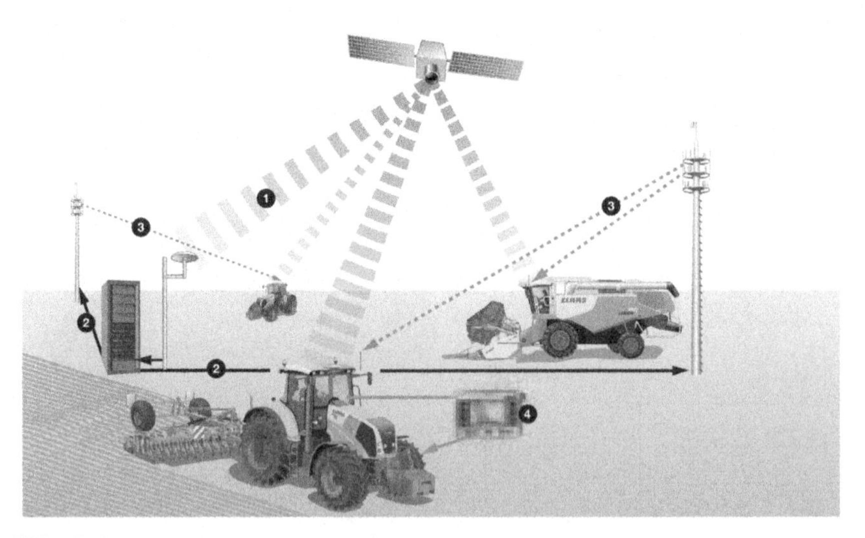

FIG. 7.4 Differential GNSS corrections associated with RTK NET, developed by CLAAS [24].

and TrueTracker implement guidance systems used in Case IH's PLM and New Holland's AFS developed by Trimble are other products. The True-Guide provide a passive implement guidance operating in combination with the tractor guidance system. TrueTracker as an active implement guidance system employs hydraulic mechanisms installed on the implement as well as terrain compensation to form independent implement guidance [17].

7.1.3 Agricultural robots (ag-robots)

The ag-robots move in a challenging, dynamic, and semistructured agricultural environment to perform different field operations. Ag-robots can be classified into two broad categories: *manipulators* and *unmanned ground vehicles* (AGVs). There are two major classifications for ground robots: self-propelled mobile robots and robotic smart implements that are conveyed by a mobile machine [15]. Self-propelled mobile robots come in a wide range of sizes and designs. Conventional self-propelled agricultural machinery, including tractors, combine harvesters, and sprayers, has been robotized using GNSS and autoguidance systems, which were described in the previous section (Fig. 7.2A).

Fig. 7.5A shows an autosteered harvester developed by Kinze and a fully autonomous tractor operating as a grain cart puller for unloading the harvester. Case IH and John Deere have developed autonomous

FIG. 7.5 (A) Autosteered harvester developed by Kinze and a fully autonomous tractor operating as a grain cart puller for unloading [15], (B) Autonomous cab-less tractor, developed by Case IH [26], (C) Autonomous cab-less tractor, developed by John Deere [26], (D) BoniRob: a multipurpose weeding robotic platform for farm applications [27], (E) Lettuce-weeding robot, developed by Blue River [28], and (F) Autonomous seed sower, developed by Small Robot Company [29].

cab-less tractor robots that are matchable with common cultivation implements (Fig. 7.5B). Bear Flag Robotics has also developed a driverless automation kit for tractors and implements that enables farmers to affordably retrofit their existing vehicles with this technology [30]. These robots are primarily designed for energy-intensive farm operations such as plowing, planting, spraying, and harvesting while other smaller and self-propelled robots have been developed for operations that demand low power, including scouting and weeding (Fig. 7.5D). Additionally, robotic smart implements have been developed for various applications, and some of them have been already commercialized for farm applications such as transplanting and mechanical weeding. Blue River Technologies employs computer vision and robotics technologies to build smart agriculture equipment introducing see and spray technology. The company has created farming machines that can zero in on just the crops that need pesticides, decreasing overall chemical usage (Fig. 7.5E) [28]. Small Robot Company presents robots with the ability to seed and supervise every individual plant in crops. These intelligent robots are designed to perform feeding and spraying based on the state of each plant, applying appropriate levels of nutrients and support with minimum waste (Fig. 7.5F) [29].

Manipulator-type ag-robots are mainly used in food processing, dairy, horticulture, and the orchard industry [2]. For instance, soft grippers are used for the selective harvesting of mushrooms, sweet peppers, tomatoes, raspberries, and strawberries. In open fields as well as greenhouses, there are complementary tasks in harvesting where manipulators can play an important role. Currently, researchers are developing robotic arms integrated with cameras to identify the three-dimensional (3D) location of fruits to assist in automated harvest [7].

Automated grasping and manipulation introduce a number of unique challenges compared to other sectors due to the significant variations in natural size and shape between examples of the same product as well as the heterogeneous positions and fragile nature of the agriculture products (e.g., during harvesting) [2]. An apple vacuum harvesting robot has been developed by Abundant Robotics that employs LiDAR to steer along the rows of trees, and machine vision to detect ripened apples, then gently suck and pick them off the trees (Fig. 7.6A). This robot can eventually be adapted to harvest other fruits [31]. Sweeper has developed a sweet pepper harvesting robot for use in commercial greenhouses. It has been aimed to perform in a single stem row cropping system when the crop has nonclustered fruits and little leaf occlusion (Fig. 7.6B) [32]. Soft Robotics has designed and built a new generation of robotic grippers that has flexibility, plug and play, and repeatability, which enables them to be employed in industrial applications (Fig. 7.6C). The *Soft Robotics System* includes soft robotic grippers and a control unit with the ability to adjust variables including size, shape, and weight, all by one device [33].

FIG. 7.6 (A) Apple vacuum-harvesting robot, developed by Abundant Robotics [31], (B) Sweet pepper harvesting robot, developed by Sweeper [32], and (C) Robotic grippers and the control unit, developed by Soft Robotics [33].

In addition to ground ag-robots, unmanned aerial vehicles (UAVs), generally known as drones, with time- and labor-saving features are also used in farm applications. Agriculture drones are semiautonomous because they have to fly according to a specified flight path in terms of waypoints and flight altitude. Therefore, a positioning measurement system must be embedded onboard [34, 35]. The UAVs can carry different sensors required for studying crop-related parameters, including the most common optical sensors such as RGB, multispectral, and hyperspectral cameras. In addition to digital data collection, drones can also perform aerobiological sampling above the farm fields for pest recognition at the early stages of infestation. UAVs are categorized as fixed-wing airplanes and rotary-motor helicopters [35, 36]. An agricultural drone called AgDrone has been developed by HoneyComb. This drone is equipped with an autopilot system called the AgDrone System, allowing the drone to fly itself (Fig. 7.7).

Automation and robotics have changed the face of agriculture at a rapid pace. In this regard, several companies have started transitioning from the conventional farming industry to a modern, advanced, and automated agricultural environment. Despite recent progress, most of the developed agricultural robots have not been commercialized yet due to the specific technical and economic requirements of agricultural tasks. For this purpose, the development of more versatile and robust robots is required.

FIG. 7.7 The AgDrone developed by HoneyComb [37].

7.2 Utilization of PV systems in PA

Productivity in agriculture primarily depends on energy, water, and land resources. In the past, energy was extremely inexpensive and agricultural operations consumed large amounts of energy for their rapid development. Continuous increases in fuel and electricity prices as well as necessities for substantial reductions in greenhouse gas (GHG) emissions have created the need for improvements in farming energy efficiency. In this regard, the exploration of alternative energy sources and their use in agriculture has received remarkable attention.

At present, nearly all agricultural vehicles and equipment run on petroleum products. The progress in agricultural automation and mechanization has increased the energy demand of modern agricultural activities. Therefore, employing renewable energy sources can decrease the energy consumption of this sector without compromising the performance [38]. Solar energy is the most abundant and reliable source of renewable energy that can be considered as a secure and sustainable alternative for use in various industrial and domestic applications. The integration of solar energy with agricultural activities points to the fact that this sector is ready for technological advancements [39]. Photovoltaic (PV) technology is one of the fast-growing power generation methods around the world with the obvious advantages of being sustainable and eco-friendly. Nowadays, PV electricity generation technology has become a high priority option in energy policy strategies at both the national and global scales [40, 41]. Photovoltaic systems are more cost-effective, especially in distributed generation settings compared with electricity grids or diesel generators, with the most operational feasibility in remote locations, ranches, orchards, and other similar agricultural environments [38, 42]. Apart from PV systems being renewable with no carbon emissions, their main benefits in agriculture are their low operation and maintenance costs, modular nature, long life, and reliability.

7.2.1 Solar-powered ag-robots

There are extensive opportunities to implement solar-powered robots in agricultural fields to perform different farm applications, including plowing and seeding, spraying and weeding, fruit harvesting, etc. Today, most ag-robots run on batteries and electric motors. Therefore, the integration of solar PV modules with farm robots would be a feasible option. The use of solar PV energy entails the use of efficient electric motors with the benefits of simpler structure, higher performance, and lower costs [43, 44].

Several companies around the world have focused on the design and development of smart weeding machines using digital cameras to provide

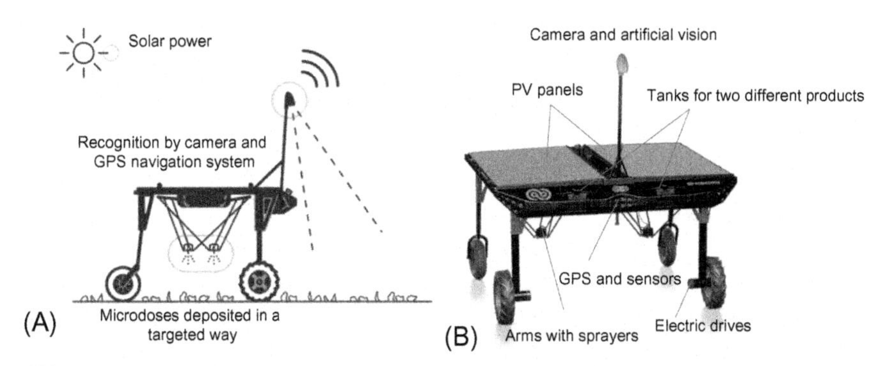

FIG. 7.8 Solar-powered autonomous weeding robot [45]: (A) Schematic view, and (B) Photo of the robot.

large-scale field images and find rows of crops by adopting machine vision techniques to perform weeding operations. In this case, solar-powered spraying and weeding robots would be a sustainable and low-cost solution to fulfill spraying and weeding tasks.

The Swiss company ecoRobotix has developed a solar-powered fully autonomous *spraying robot* for weeding of row crops, meadows, and inter-cropping cultures (Fig. 7.8). The robot positions itself by using a camera and an RTK GPS receiver. The vision system of the robot helps to follow the crop rows and detect weeds in and between the rows. Then, the detected weeds are systemically targeted and a microdose of herbicide is applied by the arms of the robot. The robot's speed is adapted by the concentration of weeds, and therefore, the most performance is achievable in a low to moderate concentration at a reasonable speed. The main advantage of this robot is that it can be fully controlled and configured by using a smartphone. The maximum surface area that can be covered by the robot is 3 ha/day and 10 ha/week. The robot has the best performance in soils that are not too wet or viscous under a wind speed of less than 60 km/h at ground level [45]. The robot's lab testing and its performance in a field are shown in Fig. 7.9. ecoRobotix has also developed the solar-powered *ARA Phenomobile scouting robot* and the *AVO weeding robotic platform*, which are expected to enter the market by the end of 2021. The scouting robot is a lightweight fully autonomous platform for use in scouting and phenotyping applications. The solar power supply makes the robot energy-autonomous even on cloudy days. Like the previous weeding robot, this robot orients and positions itself by RTK GPS navigation equipment (Fig. 7.10). A smartphone application is used to locally control the robot and to load the appropriate navigation path [48]. Also, the robot is equipped with an adjustable payload arm to suspend any sensor at an adjustable height compatible with different crop heights.

FIG. 7.9 (A) Lab testing of the robot's vision and weed control system [46], (B) the robot working in a sugar beet field in Switzerland [47], and (C) The robot arm placing a microdose at the right spot [45].

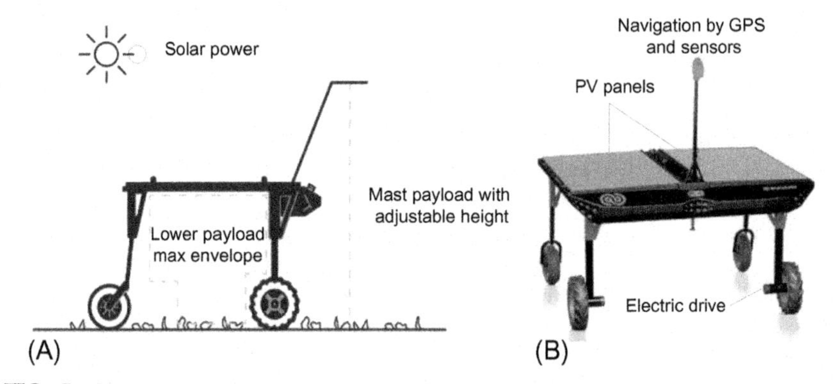

FIG. 7.10 Solar-powered ARA phenomobile scouting robot [48]: (A) schematic view, (B) photo of the robot.

The solar-powered AVO is a precision weed control platform introduced by ecoRobotix equipped with rechargeable batteries and designed for use in planned fields and row crops (Fig. 7.11). Depending on available solar radiation, the battery recharge, and terrain conditions, the robot can treat up to 10 ha/day with more than 90% less herbicide used due to its capability of centimeter-precise detection and spraying. Combining RTK GPS positioning and visual navigation provides a high degree of precision, minimizing crop roll over. The platform is equipped with a four-wheel drive system that enables the robot to operate on slopes commonly encountered in cultivated crops. The platform can be used for both full surface and spot spraying. The spraying tool consists of 52 equidistant nozzles on a height-adjustable spraying ramp. The AVO robot can be controlled and monitored through the ecoRobotix mobile app while its missions are defined via the desktop web app [49].

AgBotII is an autonomous solar-powered ag-robot designed by the Queensland University of Technology (QUT). The AgBotII can detect, classify, and destroy weeds with an accuracy of more than 90% using multiple sensors, software, and electronic devices. It can be controlled by a

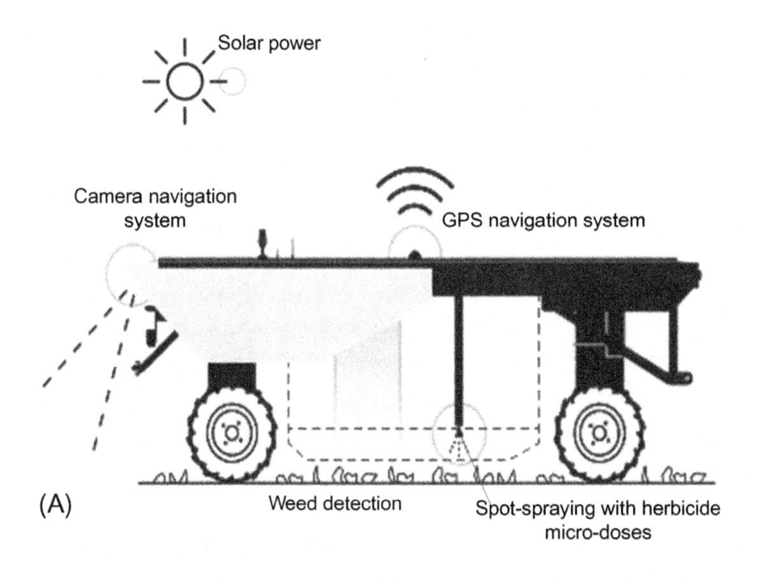

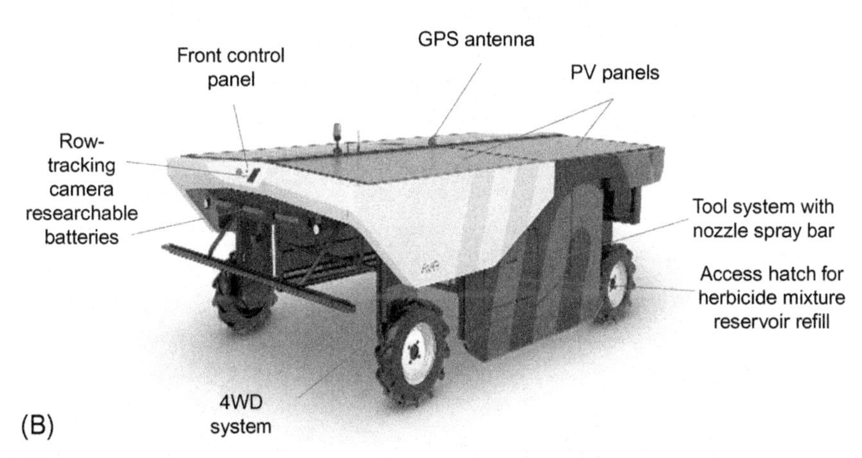

FIG. 7.11 Solar-powered AVO weeding robotic platform [49]: (A) schematic view, and (B) the robot's platform.

computer, tablet, or smartphone app. The robot can also be used to execute fertilization and seeding. The farmer can interact with the robot by a virtual globe software interface and a radiofrequency remote for both autonomous and manual controls according to the need. The key feature of this robot is its capability of weeding with either mechanical or herbicide tools (Fig. 7.12). Its main drive unit contains a 5 kW, 48 VDC electric motor with an efficiency of 75–85%, which is coupled with a two-stage planetary gearbox to induce motion at the favorable speed between 5 and 10 km/h. Two battery packs in parallel arrangement provide a 10.5 kWh electric storage unit that can supply 7–11 h of autonomy for the robot [51]. The batteries can be charged when the robot is docking with the solar charge station, as shown in Fig. 7.12B.

AgBotII employs two quad-core Intel i7 computers, one for navigation and the other for perception purposes. The robot's software has been developed based on the robot operating system (ROS) instruction, which

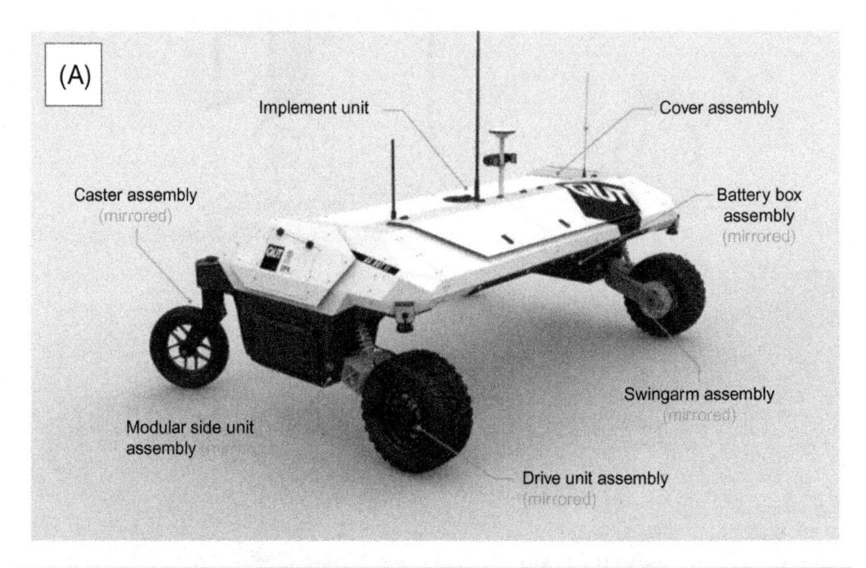

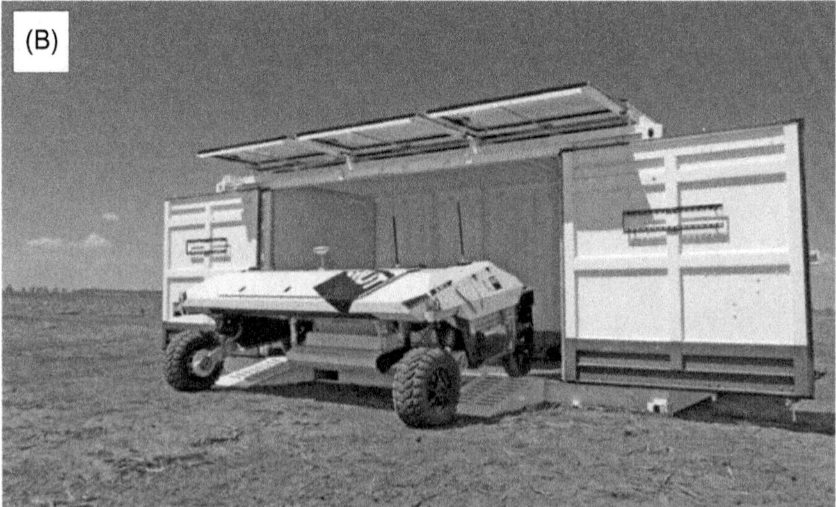

FIG. 7.12 Solar-powered AgBotII weeding robot [50]: (A) Robot's major assemblies, and (B) The robot docking with the solar charge station.

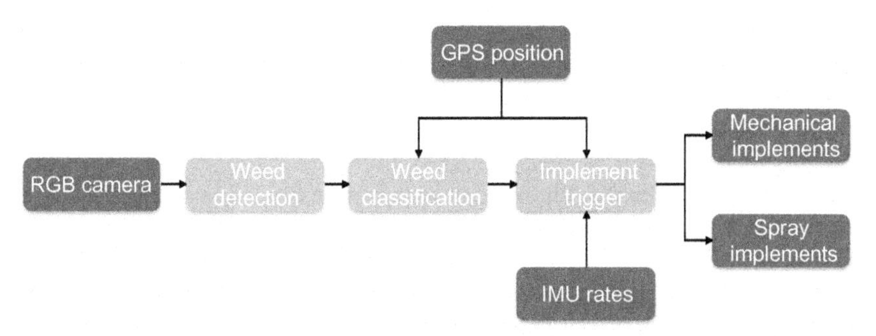

FIG. 7.13 The simplified block diagram of the AgBotII weeding system software. *Adapted from Bawden O, Kulk J, Russell R, McCool C, English A, Dayoub F, et al. Robot for weed species plant-specific management. J Field Robot 2017;34:1179–99. https://doi.org/10.1002/rob.21727.*

is divided over two computers. During weeding, the weed is sensed using the RGB camera and an image-processing computer. The weed classification determines the species from the image of the weed, and then the system chooses and operates the appropriate weeding implement at the right time [51]. The main stages of weed treatment performed by the AgBotII are shown in Fig. 7.13.

Inspired by the ladybird beetle, researchers at the Australian Centre for Field Robotics (ACFR) at the University of Sydney proposed Ladybird as a versatile and modular robot for use in the vegetable industry. Being an omnidirectional robot, Ladybird has the power requirements provided by an embedded bank of batteries and solar energy (Fig. 7.14A). The rechargeable batteries can perform for 7–9h a day while solar panels can operate persistently on clear days. The platform has a multimodal sensor suite consisting of hyperspectral, thermal, infrared, panoramic vision, stereovision with strobe, LiDAR, and GPS, allowing for autonomous navigation and crop perception [53] A spherical camera along with forward- and rear-facing LiDAR enables the robot to avoid obstacles and detect

FIG. 7.14 (A) Ladybird on a beetroot crop in Cowra, NSW, and (B) RIPPA on a lettuce crop in Lindenow, VIC [52].

crop rows, whereas an RTK GPS/INS assists in map-based farm traversal. The hyperspectral line-scanning, stereo vision, and thermal infrared cameras provide crop sensing, where a Nuvo-3005EI7QC computer employing an Intel Core i7-3610QE CPU and 16GB RAM is used for computing and processing [54].

RIPPA (Robot for Intelligent Perception and Precision Application) is another solar-powered autonomous weeding robot developed in ACFR at Sydney University, Australia, for use in the vegetable production industry. In comparison with the Ladybird, the RIPPA's platform arrangement has been revised to provide a lighter design while it is also more robust and convenient to operate [53]. Solar energy is collected by PV panels mounted on top of the machine and a 7h battery life ensures the RIPPA's 24h autonomous operation 7 days a week (Fig. 7.14B). VIIPA (Variable Injection Intelligent Precision Applicator) has been mounted on RIPPA to be applied as a weed control technique by autonomous spot spraying with a high operating rate and direct application of fluid at the microdose level [55]. Mechanical weeding is performed using RGB camera imagery and deep learning detection algorithms for identifying weeds and applying a force using a steel tine to disrupt weeds [52].

Vitirover, a small solar-powered autonomous mower robot, is the output of the Vineyard Vigilant & INNovative Ecological Rover (VVINNER) project cofunded by the European Commission within the framework of the Competitive and Innovation Program. The robot has been designed for permanent performance in vineyards not only to mow but also to monitor the vineyard by recording appropriate data, which can then assist in risk management and decision making (Fig. 7.15A). The Vitirover operates at a speed of less than 500m/h and very close to the vine stocks (distances less than 2cm) with no risk of damaging the plants on slopes up to 30% [59]. All the Vitirover robots are equipped with GPS and sensors to manage a predefined landscape. The robot can operate in all large grassy plots such as PV farms, orchards, parks, and large gardens. It can be controlled by the user through a smartphone app or a computer interface. Due to integration with a PV panel with optimized power management, it can work

FIG. 7.15 (A) Vitirover in New Zealand [56], (B) VineScout trialed at Quinta do Ataíde [57], and (C) Tertill: the solar-powered weeding robot for home gardens [58].

in the vines during all the growing seasons with no requirement for a recharging base [60]. VineScout is a solar-powered robot developed to monitor key vineyard parameters. By using the GPS and mounted sensors, it can find its track among the rows of vines with no need for a human operator [57]. The VineScout is empowered by electric batteries while the onboard sensors and other software are supported by the electricity supplied from the PV panels integrated into the vehicle (Fig. 7.15B). The robot's autonomous improved navigation system enables moving faster with more safety through the rows. The robot can also generate maps at night, thus broadening its work capacity. Solar electricity is able to energize the batteries designed for propelling the robot while on the move, enabling VineScout to navigate a significant range in the field [61].

Tertill is another small solar-powered autonomous weatherproof robot developed for home gardens. It searches for small weeds and attacks them in two ways; (1) as it moves and turns in the garden, the robot's wheels scrub the soil, damaging weed sprouts and preventing them from emerging, and (2) weeds that do emerge are cut down by a spinning string trimmer. Tertill spends its time permanently in the garden for either weeding or collecting energy from the sun (Fig. 7.15C). Charging the battery is possible by using the USB charging port on the robot's underside. The robot uses this power to drive around the garden and avoid rocks, plants, and obstacles taller than 2.5 cm. The robot is equipped with a four-wheel drive system to help navigate relatively smooth terrains such as soft soil, sand, and mulch. Tertill's smartphone app lets the user check its status remotely and provides a way to update the software on the robot. Bluetooth technology provides a wireless connection between Tertill and the app running on the smartphone [58].

FarmBot is a 100% open-source CNC (computer numerical control) autonomous precision farming tool used to take care of a garden plot. FarmBot can simultaneously cultivate a variety of crops in the same area. All FarmBots are driven by four NEMA 17 stepper motors, a microcontroller called Farmduino, and a Raspberry Pi 3 computer. The three-axis robot employs linear guides in three X, Y, and Z directions, allowing seed injectors, watering nozzles, sensors, and weed removal equipment to be positioned precisely (Fig. 7.16A) [62]. The robot can be controlled from a computer, tablet, or smartphone through a web app, giving the capability for remote management of the garden. By using the drag-and-drop web-based interface, 33 different common crops can be chosen to be grown and therefore, gardens can easily be designed and customized. An advantage of FarmBot is that it can alternatively be derived using solar power (Fig. 7.16B). The FarmBot comes in two different models, Genesis and Express. The FarmBot Genesis is designed to be used for experimentation, prototyping, and hacking while FarmBot Express is designed for those who want to get a setup with customization options [63].

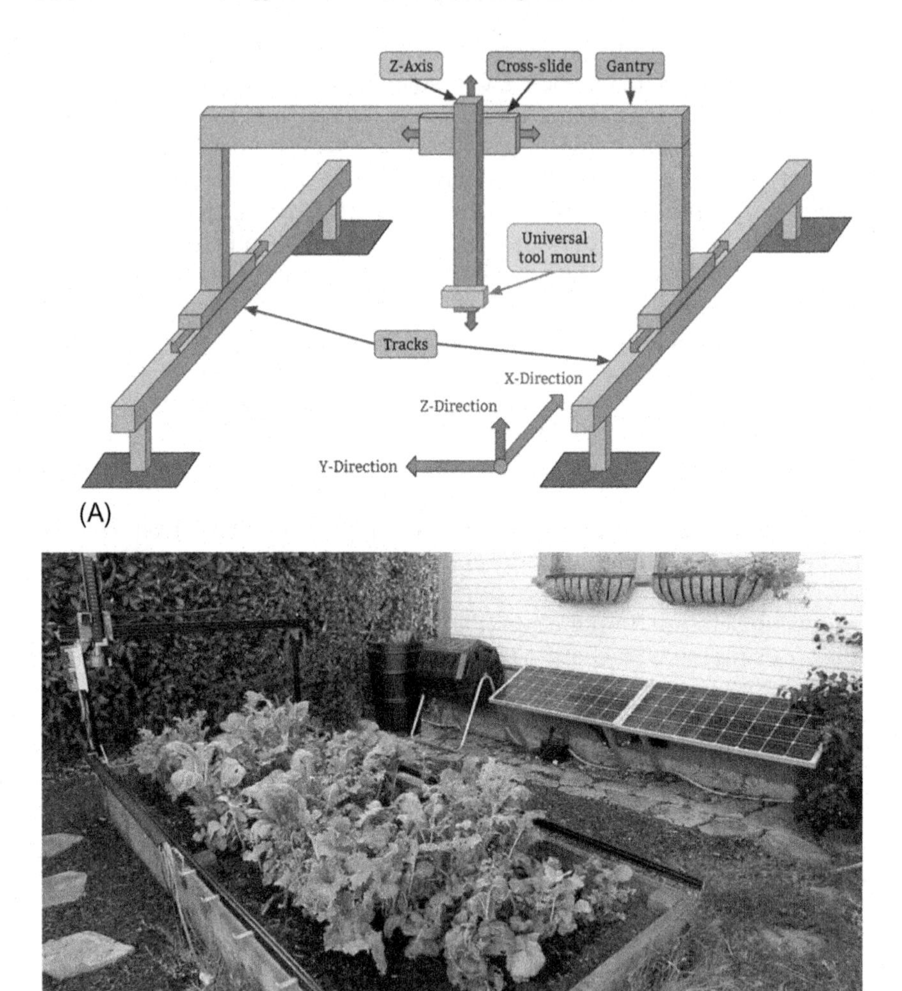

FIG. 7.16 FarmBot precision farming tool [62]: (A) the physical structure, and (B) FarmBot powering with PV modules.

7.2.1.1 Solar-powered robotics in agrophotovoltaics (APV)

Agrivoltaics or agrophotovoltaics (APV)[a] refer to the simultaneous use of agricultural lands for food production and PV power generation (Fig. 7.17). Enabling a large increase in land-use efficiency, APV opens the door for substantially expanding PV capacity while preserving fertile

a Details of the APV concept are presented in Chapter 6.

FIG. 7.17 APV system of Fraunhofer ISE at Lake Constance, Germany [64].

arable land for agriculture [65–67]. In recent years, APV technology has experienced dynamic development, spreading to almost all regions of the world. The installed APV capacity increased from almost 5 MWp in 2012 to at least 2.1 GWp in 2019, with government subsidy programs in Japan, China, France, the United States, and South Korea [65, 66].

In 2019, the Fraunhofer Institute for Solar Energy Systems (ISE) proposed a concept to integrate methods of artificial intelligence and robotics for automatic field analysis and processing into an APV system (APV-BOT). The concept is based on FarmBot, with the difference being that the FarmBot approach focuses on small-scale applications for domestic use without simultaneous PV power generation [68], but the APV-BOT takes advantage of the mounting structure of an APV system that allows for scaling up the FarmBot approach by facilitating the installation of a rail-fixed CNC system (Fig. 7.18A). According to the proposed concept, a gantry robot enables the cultivation of various types of crops, allowing for a maximum level of intercropping (Fig. 7.18B). Through an integrated analysis of plant growth and environmental parameters, the concept also includes dynamic tracking of the PV modules to adjust to the light demands of cultivated crops. An algorithm allows for optimizing the orientation of the PV modules regarding agricultural and electrical yields.

Unlike the above-mentioned solar-powered robotic approaches, the APV-BOT concept represents a fixed and typically on-grid installation that does not rely on any autonomous navigation. In contrast, all planting, cultivating, spraying, harvesting, and weeding activities are steered by the automated control of the CNC system. Further, the APV-BOT produces much more electricity than those needed for the CNC steering system.

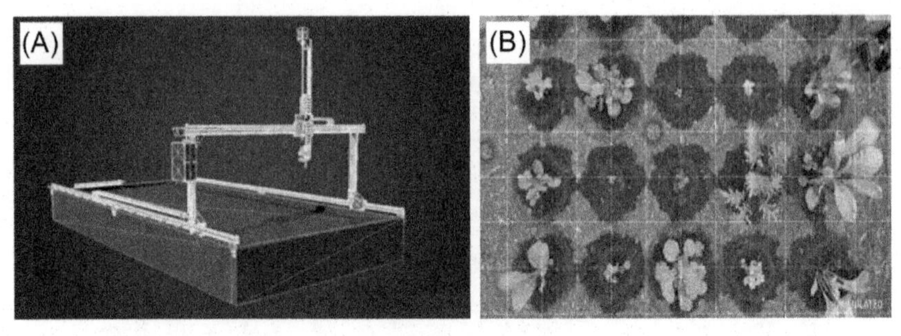

FIG. 7.18 (A) Concept of CNC-controlled field cultivation, and (B) plant classification from a bird's-eye view [69].

From the perspective of APV technology, the APV-BOT represents an appealing option to reduce the cost of the mounting structure because apart from the gantry robot, no machinery employment is required, which allows for lowering the vertical clearance of the APV system.

7.3 Solar-powered electric tractors

7.3.1 Electrification of farm tractors

Currently, most agricultural vehicles are powered by internal combustion engines (ICEs), which heavily depend on fossil fuels. Electric engines, as alternatives, can be employed in farm vehicles in which the electricity demand can be supplied from various renewable energy sources such as wind, solar, and hydro [70]. ICEs are among the most economical ones in the world but have lower efficiency and more complex maintenance in comparison with electric motors, so that the efficiency of ICEs can reach a maximum of 45% while an electric motor can achieve an efficiency about 95% at the same operating condition. The use of electric powertrains allows for higher fuel efficiencies and flexible torque-speed control over the mechanical and hydraulic ones, with more adaptability to PA activities. The ISOBUS, a communication protocol for high-voltage power electronics controlling networks on agricultural machinery, also enables advanced torque and speed controls in these machines [71, 72].

According to the Society of Automotive Engineers (SAE), the hybrid electric vehicle (HEV) is a vehicle in which the propulsive power is provided by both fossil fuels and rechargeable electricity storage systems [72]. Depending on the hybridization range, there are varieties of electric propulsion systems such as micro, mild, full, or plug-in hybrid electric vehicles, and according to the architecture of the drivetrain, they are classified into serial, parallel, and combined (power-split) hybrid configurations [73]. In other designs, in order to save energy, there is no need for ICEs and the EV uses electric power as the only source to move. Both battery electric vehicles (BEVs) and fuel cell vehicles (FCVs) lie in this category. In FCVs, hydrogen can be the primary energy stored in a specific tank and the fuel cell system operates as an energy conversion unit inducing the electric motor. Fig. 7.19 shows a range of electric propulsion systems.

A full hybrid EV demands electric power around 30–50 kW with a battery capacity ranging from 1 to 2 kWh. In addition, plug-in hybrid electric vehicles (PHEVs) usually employ an electric motor demanding higher power, which lies between 30 and 80 kW, where the battery capacity varies between 3 and 10 kWh [73]. Detailed descriptions for all the above-mentioned configurations are presented in Refs. [72–75]. Because there are large varieties of propulsion systems, the most appropriate ones must be selected according to the requirements for specified applications. Three main infrastructures for widespread use of EVs are the *source of electricity*, *batteries*, and *charge stations*. Solar electricity generation from PV systems can provide security against interruptions of limited supply and instabilities of power grids.

Despite the advantages of EVs, their utilization is still beyond expectations. One reason is the energy storage because the specific power of gasoline is around 10,000 Wh/kg, which is not comparable with the 150 Wh/kg reported in the best Li-ion battery. The high values of specific power lead to desirable acceleration, efficient regenerative braking, and fast charging. The charging time for batteries is the most important factor; it must be short enough in EVs to make them compatible with conventional ICE vehicles. Another important reason is battery cost, where one-third of the EV's cost is devoted to the battery [75, 76]. At present, the most widely used energy storage systems are electrochemical batteries with the most

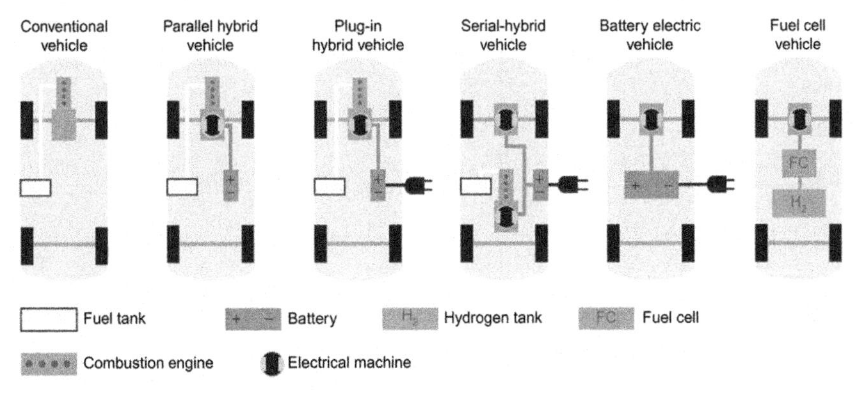

FIG. 7.19 Various electric propulsion systems in comparison with a conventional one [73].

dominant and reliable technologies of Pb-acid, Ni-MH, and Li-ion. The Li-ion batteries are more preferable because they are more energy-efficient, powerful, and lighter. The other types of Na-NiCl (ZEBRA) and $Zn-O_2$ can also be utilized in which the $Zn-O_2$ employs mechanical charging as an alternative to common electric charging. However, the three main drawbacks of batteries are: (1) limited energy capacity, (2) lengthy charging time, and (3) finite charging cycles [77]. Other alternative energy storage systems are mechanical energy storage that uses flywheels, electrostatic storage that is possible by using ultracapacitors, and hydrogen tanks, but none of them can be cost-competitive with batteries [72, 75]. Table 7.1 presents a comparison of the above-mentioned energy storage systems.

7.3.2 History of emerging electric drives in farm tractors

In 1954, International Harvester Company introduced the Electrall feature as an offer on Farmall 400 tractors to electrify operating farm equipment and accessories. The Electrall implemented a device to electrocute insects in the field at night [78]. The first fuel cell tractor in the world was built in 1959 by Allis Chalmers. The tractor was equipped with 1008 alkaline fuel cells, connected in 112 units containing nine cells each arranged in four banks, which could supply the required power to run a 20 horsepower direct current (DC) motor [79]. In the late 1990s, Schmetz, a Dutch farm equipment dealer, proposed a tractor equipped with an

TABLE 7.1 Comparison between various energy storage systems used in EVs[a].

Energy source	Specific energy (Wh/kg)	Specific power (W/kg)	Cycle life	Cost (US$/kWh)
Lead-acid	35	150	700	150
Ni-MH	70	220	1500	1500
Li-ion	130	350	1000	2000
ZEBRA	110	150	1500	700
$Zn-O_2$	200	100	1 (electric fuel)	5000[b]
Flywheels	40	3000	5000	20,000
Ultracap	5	2000	500,000	25,000

[a] *The fuel cells are excluded because they depend on hydrogen tank characteristics.*
[b] *It needs an expensive system to recover Zn anodes.*
Adapted from Dixon J. Energy storage for electric vehicles. Proc. IEEE Int. Conf. Ind. Technol.; 2010. p. 20–25. https://doi.org/10.1109/ICIT.2010.5472647.

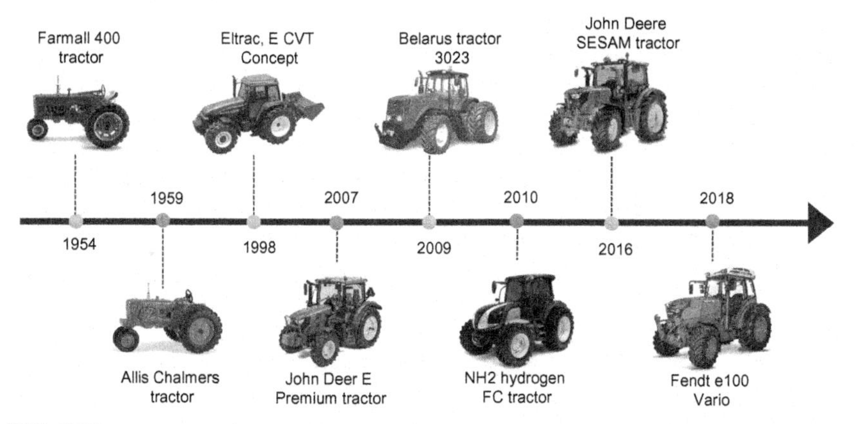

FIG. 7.20 Timeline of hybrid electric farm tractors. *Adapted from Ghobadpour A, Boulon L, Mousazadeh H, Malvajerdi AS, Rafiee S. State of the art of autonomous agricultural off-road vehicles driven by renewable energy systems. Energy Procedia 2019;162:4–13. https://doi.org/10.1016/j.egypro.2019.04.002.*

electric generator-motor located between the diesel engine and the transmission. The tractor exhibited the ability of infinitely variable speed transmission using electrical drives that employed no hydraulic mechanism. In early 2000, the universities and institutions in Germany, in association with Fendt (AGCO), also called MELA (Mobile Elektrische Leistungs- und Antriebstechnik), presented electric driven auxiliaries as a response to tractor electrification challenges [78]. In 2002, the Agricultural Industry Electronics Foundation (AEF) established the ISO11783 (ISObus) standard to boost the adjustable interaction between tractor and implements of any producer [80]. The NH2, a hydrogen-powered tractor, was released by New Holland in 2010. The fuel cell of the NH2 generates about 106 HP and its electric motor is powered to run on a hydrogen tank [79]. Fendt, the German manufacturer of agricultural equipment, showed its first electric tractor, model e100 Vario, at the Agritechnica fair in Hanover in 2018 [81]. A chronology of hybrid electric farm tractors is depicted in Fig. 7.20. In recent years, a number of farm machinery manufacturers have developed several hybrid and fully electric tractors. Recently, advancements in electrification and digitalization have emerged in farm tractors and some companies such as AVL, John Deere, Case IH, Fendt, Autonomous Tractor Corporation (ATC), Ztractor, Electric Tractor Inc. (ETi), etc., have proposed some technical solutions in the development of fully automated electric tractors. Over the next few years, if new technologies are leveraged, the basic layout of conventional farm tractors is expected to be significantly changed.

(A) (B)

FIG. 7.21 (A) A solar-powered tractor with onboard PV integration [82], (B) The electric tractor (Solectrac) charged from an installed 10kW solar array (offboard PV integration) [83].

7.3.3 PV-integrated charging methods

Chargers can be classified into two main groups, *inductive* and *conductive*. The inductive charger has no contacting surface and a magnet is used to transfer the power. Although this coupling brings convenience to drivers, it has not achieved a highly efficient level yet. The conductive charger is a conventional device that induces power through contacting. The two most widely used methods to supply power for electric vehicles are *onboard* and *offboard* integrations [75]. In the PV-integrated offboard charging method, the PV panels are mounted at fixed locations on specified charging stations or roofs of buildings while in the PV-integrated onboard charging method, PV modules are mounted on the vehicle either to assist in propulsion or to run a specific EV application. Fig. 7.21A shows a solar-powered tractor with onboard PV integration. Some constraints to the onboard PV charging systems are related to size, space, weight, and cost while off-board PV charging stations are less complex in design with fewer weight constraints and more available installation spaces and tilting options (Fig. 7.21B) [84, 85].

The PV-integrated offboard charging systems can be grid-connected or stand-alone. In grid-connected charging stations, the system shifts to the utility power network if PV power is not available. In these systems, a bidirectional inverter is employed that works as an inverter if the power flow is from the PV array to the grid, and acts as a rectifier if the power flow is from the grid to the battery embedded in the EV [86]. In the stand-alone configuration, the electricity generated by the PV array is utilized as the power source of the batteries because no utility power is used. A battery bank or energy storage unit (ESU) is used to eliminate the intermittent availability of solar energy. In this way, the excess power generated by the PV array is stored in the batteries, and can then be utilized to charge the batteries embedded in the EVs under low-insolation conditions [85]. Typical setups for stand-alone and grid-connected

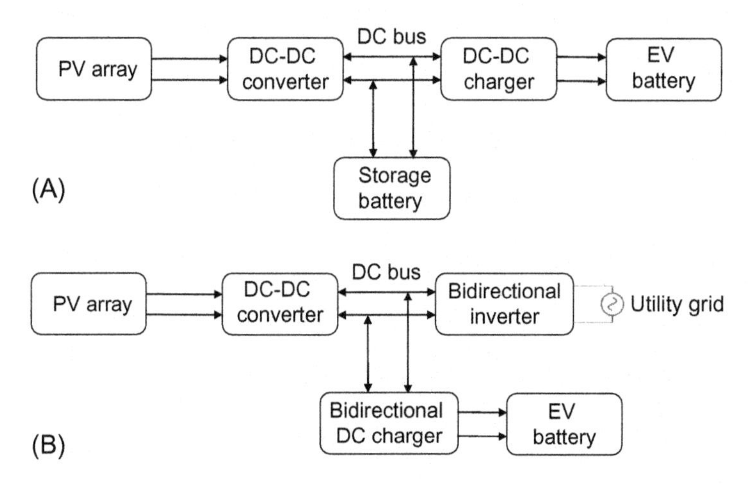

FIG. 7.22 Offboard PV-integrated charging system [85]: (A) stand-alone configuration, and (B) grid-connected configuration.

PV-integrated charging systems are shown in Fig. 7.22. It should be mentioned that batteries can also be employed in grid-connected PV systems in the way that the charging system empowers the battery and transfers the surplus power from the PV array to the grid simultaneously within peak sunshine periods while the battery can be charged from the grid when solar radiation is not sufficient or available [85].

7.3.4 Development of solar-powered farm vehicles

A solar-assisted plug-in hybrid electric tractor (SAPHT) was developed by Mousazadeh et al. [87] for the farm's low power demand applications. In their project, nearly 18% of the daily power utilized by the SAPHT was provided by an onboard PV array while the remaining portion was supplied by the power grid. The six standard farm implements, including the moldboard plow, the row crop planter, the trailer, the boom-type sprayer, the grain spreader, and the cutter-bar mower, were mounted to the SAPHT to determine the operating range of each implement (Fig. 7.23). The results of the field tests indicated that a single-bottom moldboard plow and a two-row planter can work consistently for 4h/day using a 16.5kWh valve-regulated lead-acid (VRLA) battery pack while a standard mower, sprayer, and fertilizer can perform for 3.2h/day, 3.6h/day, and 3.4h/day, respectively. They also reported that with using this battery pack, the SAPHT can pull a two-ton weighted trailer with a speed of 25km/h for 2h on asphalt, at a speed of 18km/h for 2.5h on a sand road,

FIG. 7.23 Field operations of the SAPHT [87]: (A) spaying, (B) grain spreading, (C) pulling, and (D) plowing soil.

and at a speed of 9km/h for 3h in a farm field. It was concluded that the operating hours can be extended with the implementation of batteries with higher capacity and energy density with the same weight.

In the RAMseS (Renewable Energy Agricultural Multipurpose System for Farmers) project, the potential of using a prototype battery-powered electric vehicle (BPEV) charged by a solar PV array (10kWp) in the Southern Mediterranean region was investigated (Fig. 7.24) [88]. In this project, a 10 kW_p solar PV array was installed in a 50-ha agricultural area located at Achkout, Lebanon. The PV array was composed of 72 PV panels where the maximum power point, maximum power voltage, and maximum power current were 138 Wp, 18.2 VDC, and 7.59 A, respectively. The battery storage unit, with a total capacity of 112.8 kWh, constituted 24 lead-acid batteries with a capacity of 2350 Ah and a cell voltage of 2 V. The BPEV had a mass of 1750 kg, a load capacity of 1000 kg, a main motor power of 12 kW, an auxiliary motor power of 12 kW, and a maximum torque of 200 N m, and could easily be switched between two-wheel drive and four-wheel drive. The results from the technical test undertaken by IBMER (Institute for Building, Mechanization, and Electrification of Agriculture) in Poznan, Poland, indicated that the PV-powered BPEV can only be used

(A) (B)

FIG. 7.24 (A) Solar BPEV operates spraying, and (B) Installed PV array with an average daily power rate of 4.5 kWh [88].

for light-duty agricultural tasks, making full replacement of the fossil fuel-powered tractors with PV-BPEVs impossible. They concluded that electric energy storage systems with faster charging and extensive lifespans must be offered to assist in the development of PV-powered EVs.

Solectrac has developed 100% battery-powered electric tractors, providing an opportunity for farmers to supply the required power from solar, wind, and other sources of renewable energy. The Solectrac tractors can accept all Category 1–540 rpm PTO (power take-off) implements, incorporating a 26-kWh onboard battery pack that provides 3–6 h of run time depending on the load [89]. The integrated battery management system (BMS) automatically protects the batteries during charging and discharging. Two types of electric tractors, eUtility and eFarmer, are presented by this company. The eUtility tractors are ideal for use in vineyards, livestock operations, commercial greenhouses, hobby farms, etc. In this type, inefficient hydraulic actuators have been replaced by linear ones tolerating 1000 lbs. of dynamic load and 3000 lbs. of static loads (Fig. 7.25A). The eFarmer tractors employ a simple joystick to control steering and speed, allowing for easy navigation between row crops (Fig. 7.2B) [89].

7.4 Solar-powered wireless sensor networks

7.4.1 Wireless communication technologies

The PA encompasses an interdisciplinary concept of agriculture integration with information technology, assisting in crop yield and quality increase. Crop yield monitoring with a share of 61.4% is the most widely used PA technique among the others. In this regard, different technologies such as information and communications technology (ICT), sensor and

FIG. 7.25 Solar-powered electric tractors (Solectracs) [89]; (A) eUtility model, and (B) eFarmer model.

information processing technology, and geographic information systems (GIS) have been employed in agricultural activities to increase profitability and decrease the labor force [90]. The adoption of sensors in agriculture can assist in real-time data collection by measuring several important parameters, including ambient temperature, soil moisture, soil nutrition, atmospheric humidity, luminance, water level, etc. The collected data can then be transferred through a wireless sensor network (WSN) and analyzed. Farmers can also remotely monitor their crops and equipment by using their computers or smartphones and take immediate action [91]. In this regard, WSN, as the second largest network after the Internet, can help to improve overall efficiency. The use of Internet technology and WSN will create a basis for usage of the Internet of Things (IoT) [92].

7.4.2 Wireless sensor network (WSN)

A wireless sensor network (WSN) consists of a wireless ad-hoc[b] network including several sensor nodes (motes) that are used to sense, measure, and collect information from the surroundings, where they can be deployed to control actuators. A sensor node is comprised of four major components: the *power unit*, the *sensing unit*, the *computing unit*, and the *communication unit* [91, 93]. The nodes sense the ecological phenomena in real,time, which are then converted to digital signals and processed and stored in the computing unit. The processed and stored information is shared with other nodes or the end user through the communication unit [94]. The sensor nodes have the ability to communicate to the server/base station or together as centralized and decentralized architectures regarding different topologies such as mesh, star, etc. The schematic diagram of a typical WSN and its components is shown in Fig. 7.26. As

b In ad-hoc networks, there are no fixed routers and the nodes can be positioned arbitrarily.

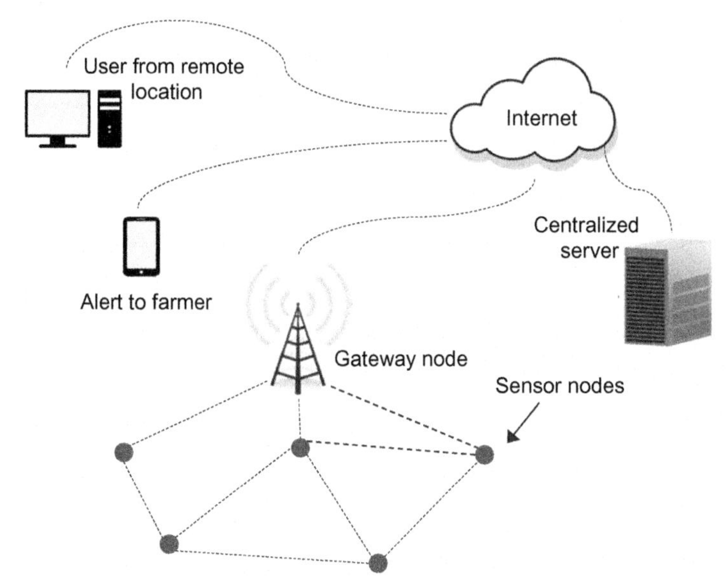

FIG. 7.26 The general architecture of a WSN.

shown in the figure, a WSN usually consists of multiple sensor nodes and a gateway node.

The features of the WSNs are their convenient application, scalability to large sizes, node mobility, and resilience. These networks are more receptive to information from a physical phenomenon than from a single sensor. Therefore, the failure of a node will not be influential on the whole network [95]. In agricultural applications, WSNs are generally classified into two main types, terrestrial and underground. In terrestrial WSNs (TWSNs), sensor nodes are deployed above the land surface to help in evaluating land conditions. In underground WSNs (UWSNs), sensor nodes are deployed inside the soil in an agriculture field for real-time sensing and monitoring of soil conditions [91]. According to the application, power consumption, band frequency, distance, and data rate, there are different technologies in WSNs for efficient data communication, including ZigBee, Bluetooth, Wibree, Wifi, GPRS, and WiMAX (Table 7.2).

7.4.3 Internet of things (IoT)

The IoT refers to a low-power and low-cost network of physical objects instrumented with embedded electronics, sensors, software, and network connectivity, enabling data transmission between the objects. In these networks, the processed information is transferred to a cloud computing unit though a gateway to be processed and stored, and can then be utilized for decision making (Fig. 7.27) [96, 97]. The main goals of using IoT networks in agriculture are improving efficiency as well as increasing accuracy and

TABLE 7.2 Different available communication technologies for WSNs [91].

	Technologies				
Specifications	**Zigbee**	**Bluetooth**	**Wi-Fi**	**GPRS 2G/ 3G/4G**	**WiMax**
Standard	IEEE 802.15.4	IEEE 802.15.1	IEEE 802.11a, b,g,n	–	IEEE 802.16. a,e
Frequency band	2.4GHz	2.4GHz	2.4GHz	2.4GHz	2.4GHz
Range	10–100m	100m	32m	Coverage area of GSM	30miles
Data rate	20Kbit/s to 250Kbit/s	1Mbps	11–54mbps	5–100kbps/ 200kbps/ 0.1–1GB	50–100Mbps
Power consumption	Low	Medium	High	Medium	Medium
Cost	Low	Low	High	Medium	High
Modulation/ protocol	DSSS, CSMA/ CA	FHSS, GFSK	DSSS/ CCK, OFDM	SNDCP, LLC	OFDM
Security	128bits	64 or 128bits	128bits	128bits	160bits

profitability. The most common agricultural applications of IoT systems are livestock monitoring, automated irrigation, soil sensing, and weather monitoring [98].

As an example, IoT sensors can be installed to collect the ambient and soil moisture data to be applied in an automated irrigation system to prevent over- and underwatering of crops. Another example is the use of an IoT-based livestock monitoring system to ensure that animals are being appropriately fed and cared for [99]. The main communication technologies in IoT along with the frequencies and distances that they can cover are presented in Table 7.3. More details about these communication technologies can be found in Ref. [100].

7.4.4 Energy harvesting techniques for WSNs

A main constraint of WSNs is the limited battery capacity of the sensor nodes. Until now, several techniques have been developed to address this problem, which is mainly based on two strategies of using energy-efficient schemes and energy-harvesting methods. In this regard, various

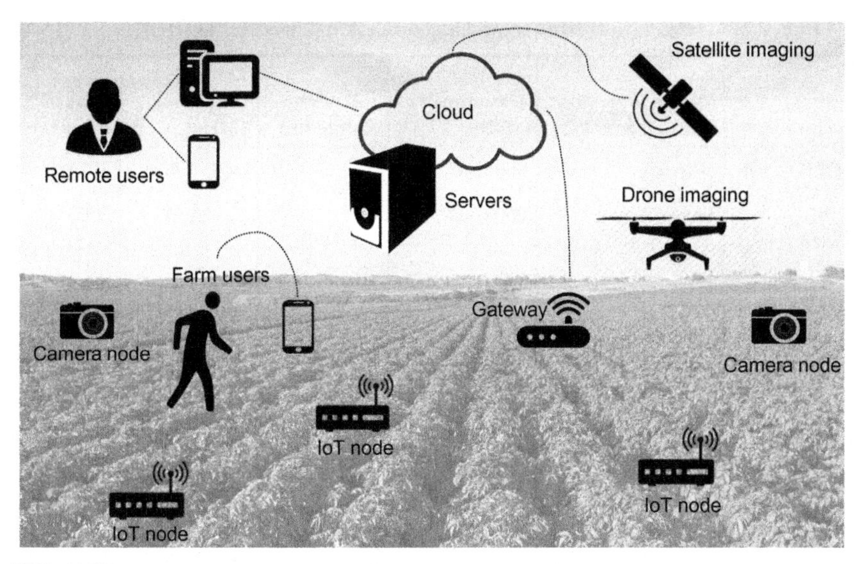

FIG. 7.27 Schematic view of IoT deployment in the field.

techniques for harvesting energy have been introduced that allow sensor nodes to capture a number of energy sources such as solar energy, mechanical vibration energy, wind energy, etc. (Fig. 7.28). These types of energies are directly converted to electricity to provide the required power for the sensor nodes [94]. The harvested energy can be stored in batteries for later use. For example, the batteries of the sensor nodes can be charged by solar energy during the day when adequate sunlight is available and then the stored energy can be utilized during the night. To increase energy efficiency, the sensor nodes can enter restricted sleep periods to decrease power consumption when the batteries have low residual energy [94, 101].

7.4.5 Applications of solar-powered WSNs in agriculture

Solar energy through PV power generation technology can be harnessed in agricultural practices using WSNs. In this method, solar cells are utilized to provide extended, clean, and sustainable energy for sensor nodes [94]. Although solar energy is intrinsically time- and season-dependent, it remains one of the most feasible sources of energy for implementation in WSNs through a power management mechanism [101]. The most common devices in WSNs for storing solar energy are disposable lithium batteries, which limit the widespread use of these sensors. At the moment, the novel power supply techniques in autonomous wireless sensors are supercapacitors and rechargeable batteries. The three main

TABLE 7.3 Different available communication technologies for IoT [100].

Technology	Standard	Downlink/ uplink	Range (m)	Operating frequency (MHz)	Year of discovery
RFID[a]	Wireless	100 kbps	2	0.125–5876	1973
IEEE 802.15.4	6loWPAN	250 kbps	30	826 and 915	2003
Z-Wave[b]	Wireless	100 kbit/s	30	868.42 and 908.42	2013
LTE[c]	3GPP, LTE, and 4G	100 Mbps	35	400–1900	1991
LoRa[d]	Wireless	0.3 37.5 (kb/s)	3000–5000	169, 433, and 868 (Europe) and 915 (North America)	2012
NFC[e]	ISO 18092	106, 212, or 424 Kbits	> 0.2	13.56	2004
UBW[f]	IEEE 802.15.3	11–55 Mbps	10–30	2400	2002
M2M[g]	Open to All Communication Protocols	50–150 Mbps	5–20	1–20	1973
6loWPAN[h]	Wireless	250 Kbps	30	915	2006

[a] *Radio-frequency identification.*
[b] *Zensys wave.*
[c] *Long-term evolution.*
[d] *Longrange.*
[e] *Near field communication.*
[f] *Ultrawide band.*
[g] *Machine to machine.*
[h] *IPv6 low-power wireless personal area network.*

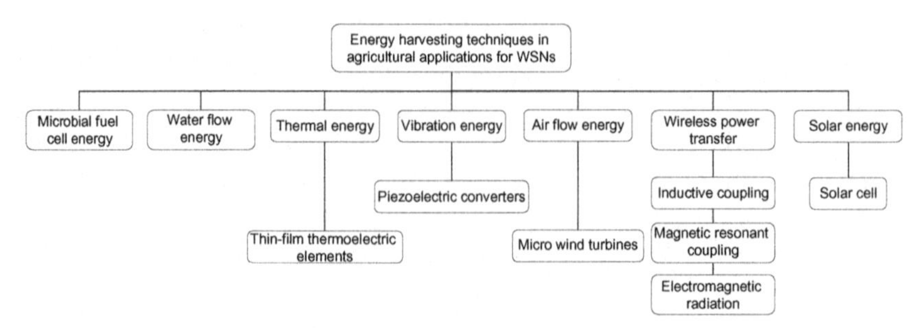

FIG. 7.28 Energy-efficient schemes in agriculture for WSNs [94].

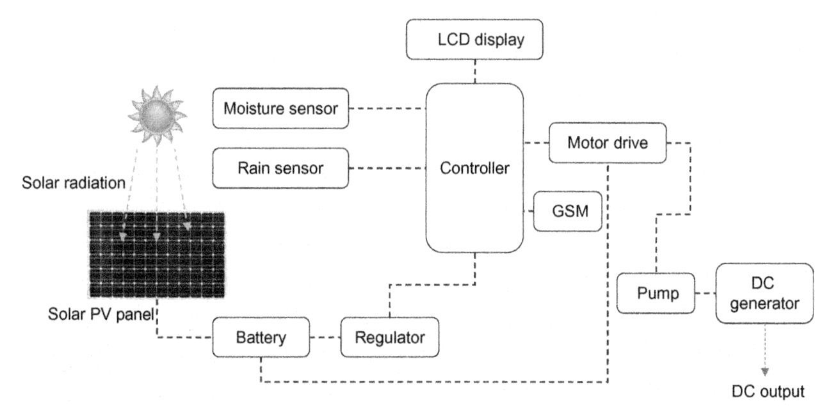

FIG. 7.29 Block diagram of a solar-powered irrigation system based on WSNs [101].

components of a solar PV system to power WSNs are solar cells/modules, control circuits, and batteries [102, 103].

A solar-power automated irrigation system was proposed by Pooja et al. [101] for use in remote areas. The system consists of two main parts, a solar pumping system and an automatic irrigation unit. Solar panels were utilized to charge a battery as a power supply for a motor, which was controlled by a sensing circuit (Fig. 7.29). In this study, a network of sensor nodes was used to collect soil humidity and temperature data, which were transmitted to a remote station. The authors concluded that PV power for irrigation is more affordable in comparison with traditional energy sources for small and remote field applications.

Zhang et al. [104] developed a wireless monitoring node using solar PV panels equipped with an automated tracker to be applied to a paddy field environment. In their work, some parameters including moisture, temperature, pH, water level, and light intensity were captured by the nodes from the field environment. When processing was done, the collected information was uploaded to remote monitoring software via a GPRS module (Fig. 7.30). The use of an automatic tracker for solar PV panels made it possible to achieve high efficiency. The solar panel and battery were coupled to a controller where the battery was charged by the PV panel during the day. The authors claimed that the proposed node could accomplish precise data transmission while the solar power supply met the power demand of the system.

In another study carried out by Vasisht et al. [106], a low-cost IoT platform called FarmBeats was developed. The system could perform data collection from various types of sensors including cameras, drones, and soil sensors. The system was comprised of a solar-powered IoT base station and an intelligent gateway to confirm the availability of services in the cloud (Fig. 7.31). The authors claimed that the developed platform

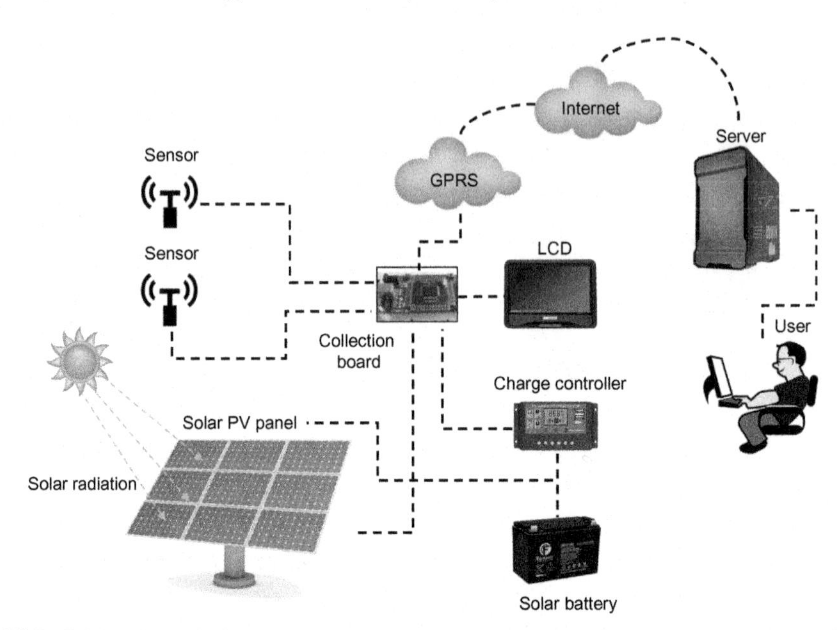

FIG. 7.30 The overall framework for a solar-powered WSN system developed in Ref. [105].

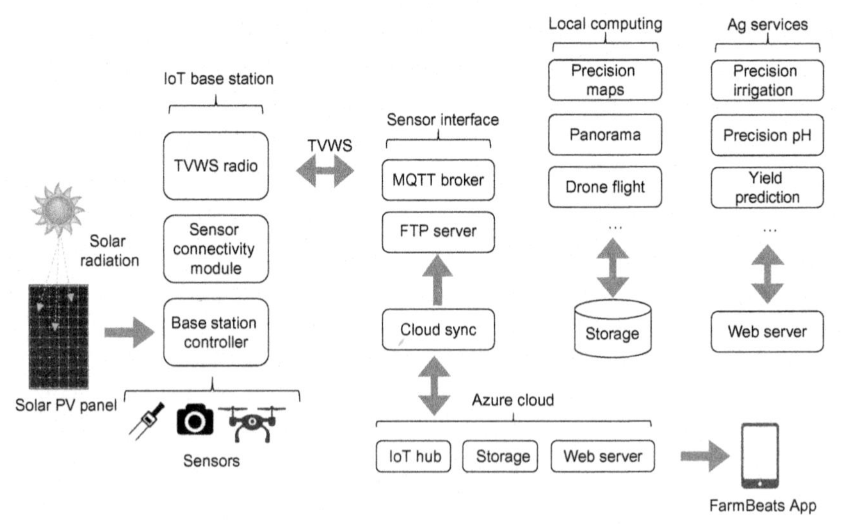

FIG. 7.31 Schematic diagram of the FarmBeats IoT platform [106].

supported high bandwidth sensors using TV white space (TVWS) as a low-cost and long-range technology. They deployed the system in two farms and reported that it was already used by farmers for precision agriculture, including animal and storage monitoring.

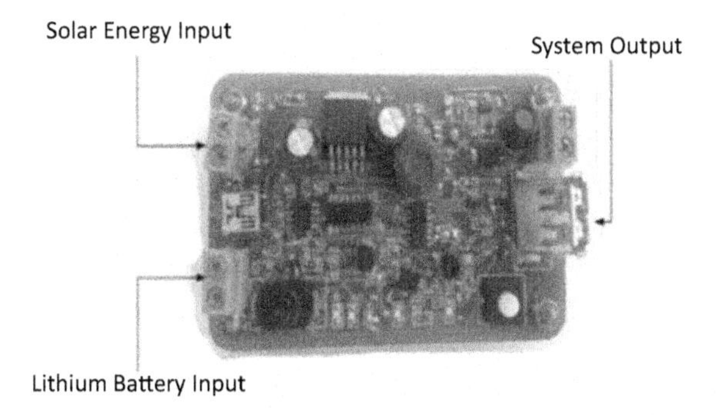

FIG. 7.32 Circuit board of the developed solar-powered ISEH system [103].

A novel intelligent solar energy-harvesting (ISEH) system was proposed by Li and Shi [103] for WSNs. The system was composed of three main components, a solar PV panel, a lithium battery, and a control circuit (Fig. 7.32). The system employed a maximum power point tracking (MPPT) circuit to achieve the highest efficiency for the PV panels. The harvested solar energy could be stored in a battery to operate the IoT nodes when sunlight was not available. They claimed that the proposed system was a low-power device that could be implemented using low-power equipment, which made it suitable for outdoor-based wireless sensor nodes in the IoT.

Chieochan et al. [107] developed a small IoT-based off-grid solar cell system as an auxiliary power unit in a smart-scale Lingzhi mushroom farm. For this purpose, an IoT system with voltage and current sensors was employed to measure and monitor the solar cell charging voltage, the charging current of the battery from the solar cells, and the battery current loading transferred to the fog and sprinkler pumps, as shown in Fig. 7.33. All solar cell information was collected in a Blynk cloud service. The Blynk app assisted in real-time monitoring of the current and voltage variations by creating a time-series graph. The results revealed that employing a solar system is economically viable only in remote areas with no access to grid electricity. In addition, the authors claimed that the integration of the IoT with voltage and current sensors promoted the performance of the solar cell system as an alternative power resource for smart farming applications.

Table 7.4 summarizes some recent studies carried out on WSN applications in agriculture using PV technology as the energy-harvesting method. The literature shows that solar-powered WSNs have been used in diverse agricultural activities, including automated irrigation, soil sensing, weather monitoring, etc. Farmers prefer to integrate PV systems with WSNs due to their simple installation and efficient performance when

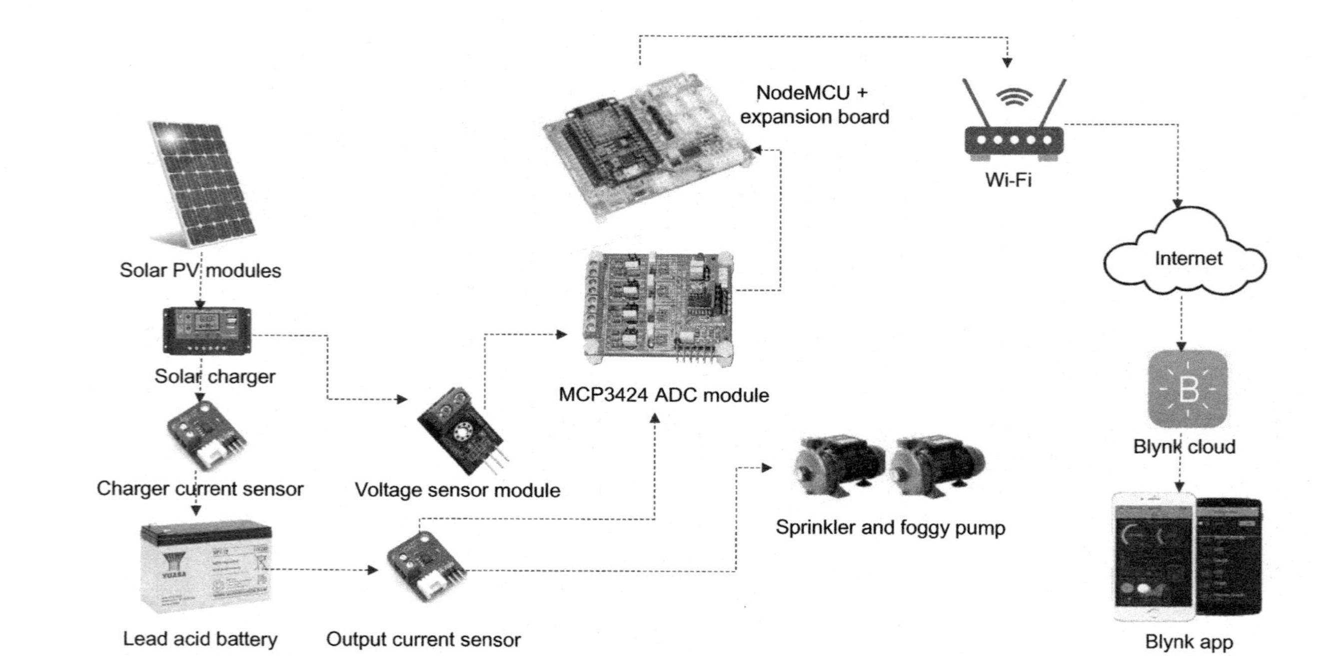

FIG. 7.33 Concept layout of the IoT for a solar cell system with Blynk. *Adapted from Chieochan O, Saokaew A, Boonchieng E. Internet of things (IOT) for smart solar energy: a case study of the smart farm at Maejo University. 2017 Int. Conf. Control. Autom. Inf. Sci., vol. 2017- Janua, IEEE; 2017. p. 262–7. https://doi.org/10.1109/ICCAIS.2017.8217588.*

TABLE 7.4 PV-powered WSNs utilized in agricultural activities.

Wireless "device	Sensors/actuators	Application	Main findings	Reference
GPRS/3G	Wind speed and direction, temperature, humidity, rain gauge, water, and pH levels	Weather monitoring	• The battery support for sensor node operation was limited to 7 days.	Nguyen et al. [108]
nRF24L01 IEEE 802.11b/g/n (Wi-Fi) and cloud computing	Air temperature, wind speed and direction, leaf wetness, soil moisture, air humidity, rain volume/fertilizers or spraying chemicals and watering system	Precision agriculture applications	• The proposed system has an acceptable efficiency; • The system's structure is a little complex; • The architecture is suitable for use in a wide range of PA activities.	Khattab et al. [109]
GSM module/LoRa	Soil temperature and moisture, air temperature and humidity, and light intensity/alert messages	Automation in greenhouse	• The system has been optimized for large-scale applications; • To measure environmental parameters, the module empowers the module; • Multiple parameters can be shown on a mobile application.	Ilie-Ablachim et al. [110]
Wi-Fi/GSM modem	Soil moisture, humidity, and temperature	Automated irrigation	• PV power is more affordable for microirrigation systems; • PV power is cost-competitive for small, remote irrigation applications; • Precise utilization timing makes the system more efficient and low-cost.	Pooja et al. [101]

Continued

TABLE 7.4 PV-powered WSNs utilized in agricultural activities—cont'd

Wireless "device	Sensors/actuators	Application	Main findings	Reference
Wi-Fi/ GPRS module	Temperature, soil moisture, pH, illumination, and water levels	Monitoring in a paddy field environment	• The proposed node can perform accurate data transmission; • The PV power supply can meet power requirements.	Zhang et al. [104]
Wi-Fi and cloud computing	Soil humidity, pH, and water levels	Automation system for soil sensing	• The developed system is affordable and easy to install in the field; • The system can deliver accurate results with good prediction capability.	Navulur et al. [111]
Wi-Fi/4G modem	Soil moisture and temperature, light intensity, relative humidity, ambient temperature, carbon dioxide, solar radiation	Automation of agricultural activities	• The network exhibited an average 83% enhanced lifetime with reduced costs; • The system benefited from optimized sleep time, consuming less power for sensor nodes; • Solar-powered sensor nodes can provide a longer lifetime for the network.	Heble et al. [96]
Wi-Fi/ GPRS module	Soil moisture and temperature	Automated irrigation and remote farm monitoring	• The developed Agribot has a better performance compared to fixed automation devices; • The Agribot requires less hardware for farm monitoring and irrigation compared to a fixed system.	Rahul et al. [112]

the availability of solar radiation is sufficient. In addition, solar PV systems provide a long-term source of power for sensor nodes, which makes the whole network self-sustained.

7.5 Conclusion

PV power technologies facilitate energy access, especially in rural areas and remote communities. In recent years, the implementation of PV systems in agriculture has shown rapid progress, but still more theoretical studies and practical explorations considering both technical and economical aspects are required to make this integration a success. Although PV systems are confirmed to be employed as an alternative for supplying power requirements in smart and precision farming activities, the high investment costs remain a barrier to the widespread deployment of PV technology in this sector. This situation becomes worse when an off-grid PV system is employed because utilizing electric energy storage systems is mandatory in this configuration. Particularly, in some applications, depending on the types and configurations of the PV systems, using hybrid power generation units containing both PV modules and wind turbines or other available renewable-based subsystems can assist in making the whole integration more feasible and affordable.

Acknowledgment

The authors would like to thank the advisory support from the Fraunhofer Institute for Solar Energy Systems ISE and financial support [grant number IG/39705] from Tarbiat Modares University for doing this research. Additionally, the technical advice of Matthias Demant and Stephan Schindele is duly acknowledged.

References

[1] Blackmore BS, Blackmore S. Developing the principles of precision farming. Int Conf Precis Agric 2014;5.
[2] UK RAS Network. UK-RAS White papers. In: The Future of Robotic Agriculture; 2018.
[3] Ren D, Martynenko A. Guest editorial: robotics and automation in agriculture. Int J Robot Autom 2018;33:215–8. https://doi.org/10.2316/Journal.206.2018.3.206-0001.
[4] Zaleski MS. Automation, the workforce, and the future of the laboratory. MLO Med Lab Obs 2011;43:59.
[5] Zhang Q. Precision technology agriculture for crop farming. CRC Press; 2015.
[6] Blackmore S, Apostolidi K, Fountas S. FutureFarm: addressing the needs of the European farm of the future: findings of the first two years. IFAC Proc 2010;43:1–17. https://doi.org/10.3182/20101206-3-JP-3009.00003.
[7] Pitla SK. Agricultural robotics. In: Adv. Agric. Mach. Technol. CRC Press; 2018. p. 157–77. https://doi.org/10.1201/9781351132398-7.

[8] Zhang B, Zhou J. Introductory chapter: recent development and applications of agricultural robots. In: Agric. Robot.—Fundam. Appl., vol. i. IntechOpen; 2019. p. 13. https://doi.org/10.5772/intechopen.81149.
[9] Pedersen SM, Fountas S, Have H, Blackmore BS. Agricultural robots—system analysis and economic feasibility. Precis Agric 2006;7:295–308. https://doi.org/10.1007/s11119-006-9014-9.
[10] Mehta P. Automation in agriculture: agribot the next generation weed detection and herbicide sprayer, a review. J Basis Appl Eng Res 2016;3:234–8.
[11] Pota H, Eaton R, Katupitiya J, Pathirana SD. Agricultural robotics: a streamlined approach to realization of autonomous farming. In: ICIIS 2007—2nd Int Conf Ind Inf Syst 2007, Conf Proc; 2007. p. 85–90. https://doi.org/10.1109/ICIINFS.2007.4579153.
[12] Li M, Imou K, Wakabayashi K, Yokoyama S. Review of research on agricultural vehicle autonomous guidance. Int J Agric Biol Eng 2009;2:1–16. https://doi.org/10.3965/j.issn.1934-6344.2009.03.001-016.
[13] Lakota M, Stajnko D, Vindiš P, Berk P, Kelc D, Rakun J. Automatization and digitalization in agriculture. Poljopr Teh 2019;44:13–22. https://doi.org/10.5937/PoljTeh1902013L.
[14] Shalal N, Low T, McCarthy C, Hancock N. A review of autonomous navigation systems in agricultural environments. In: Innov. Agric. Technol. a Sustain. Futur. Western Australia: Barton; 2013.
[15] Vougioukas SG. Agricultural robotics. Annu Rev Control Robot Auton Syst 2019;2:365–92. https://doi.org/10.1146/annurev-control-053018-023617.
[16] Nurmaini S, Tutuko B. Intelligent robotics navigation system: problems, methods, and algorithm. Int J Electr Comput Eng 2017;7:3711. https://doi.org/10.11591/ijece.v7i6.pp3711-3726.
[17] Thomasson JA, Baillie CP, Antille DL, Lobsey CR, Mccarthy CL. In: Autonomous technologies in agricultural equipment: a review of the state of the art. 2019 Agric. Equip. Technol. Conf., American Society of Agricultural and Biological Engineers; 2019. p. 1–17. https://doi.org/10.13031/913.
[18] Ehlers SG, Field WE. Determining the effectiveness of mirrors and camera systems in monitoring the rearward visibility of self-propelled agricultural machinery. J Agric Saf Health 2017;23:183–201. https://doi.org/10.13031/jash.12034.
[19] Auat Cheein FA, Carelli R. Agricultural robotics: unmanned robotic service units in agricultural tasks. IEEE Ind Electron Mag 2013;7:48–58. https://doi.org/10.1109/MIE.2013.2252957.
[20] De Simone M, Rivera Z, Guida D. Obstacle avoidance system for unmanned ground vehicles by using ultrasonic sensors. Machines 2018;6:18. https://doi.org/10.3390/machines6020018.
[21] Liu J, Jayakumar P, Stein JL, Ersal T. A nonlinear model predictive control formulation for obstacle avoidance in high-speed autonomous ground vehicles in unstructured environments. Veh Syst Dyn 2018;56:853–82. https://doi.org/10.1080/00423114.2017.1399209.
[22] John Deere n.d. https://www.deere.com/assets/publications/index.html?id=004d03e7#2 [Accessed 1 January 2020].
[23] AFS Accuracy | Advanced Farming Systems | Case IH; n.d. https://www.caseih.com/northamerica/en-us/innovations/advanced-farming-systems/afs-accuracy [Accessed 1 January 2020].
[24] Automatic steering wheel for the GPS PILOT S3—Press releases | CLAAS Group—2020. https://www.claas-group.com/press-corporate-communications/press-releases/gps-lenkrad/158138 [Accessed 1 January 2020].
[25] Yisa MG, Terao H. Dynamics of tractor-implement combinations on slopes (part I): state-of-the-art review. J Fac Agric Hokkaido Univ 1995;66:240–62.

[26] Case IH Autonomous Concept Vehicle | Case IH; n.d. https://www.caseih.com/northamerica/en-us/Pages/campaigns/autonomous-concept-vehicle.aspx [Accessed 2 January 2020].
[27] Multipurpose farm robot / weeding—BoniRob—Bosch Deepfield Robotics; n.d. https://www.agriexpo.online/prod/bosch-deepfield-robotics/product-168586-1199.html [Accessed 2 January 2020].
[28] See & Spray Agricultural Machines—Blue River Technology; n.d. http://www.bluerivertechnology.com/ [Accessed 16 October 2019].
[29] Small Robot Company; n.d. https://www.smallrobotcompany.com/ [Accessed 16 October 2019].
[30] Autonomous Tractor—autonomous tractor, auto-steer, self-driving tractor, bear flag robotics, agtech startup; n.d. https://bearflagrobotics.com/#section-0 [Accessed 16 October 2019].
[31] AbundantRobotics.com; 2019. https://www.abundantrobotics.com/ [Accessed 16 October 2019].
[32] Sweeper; 2020. http://www.sweeper-robot.eu/ [Accessed 2 January 2020].
[33] Soft Robotics; n.d. https://www.softroboticsinc.com/ [Accessed 16 October 2019].
[34] Daponte P, De Vito L, Glielmo L, Iannelli L, Liuzza D, Picariello F, et al. A review on the use of drones for precision agriculture. IOP Conf Ser Earth Environ Sci 2019;275. https://doi.org/10.1088/1755-1315/275/1/012022.
[35] Puri V, Nayyar A, Raja L. Agriculture drones: a modern breakthrough in precision agriculture. J Stat Manag Syst 2017;20:507–18. https://doi.org/10.1080/09720510.2017.1395171.
[36] Norasma CYN, Fadzilah MA, Roslin NA, Zanariah ZWN, Tarmidi Z, Candra FS. Unmanned Aerial Vehicle Applications in Agriculture. 506; 2019. https://doi.org/10.1088/1757-899X/506/1/012063.
[37] Agricultural Drone Solution for Scouting & Mapping; 2020. http://www.honeycombcorp.com/agdrone-system/ [Accessed 2 January 2020].
[38] Piechocki J, Sołowiej P, Neugebauer M, Chen G. Development in energy generation technologies and alternative fuels for agriculture. In: Adv. Agric. Mach. Technol. CRC Press; 2018. p. 89–112. https://doi.org/10.1201/9781351132398-4.
[39] Mekhilef S, Faramarzi SZ, Saidur R, Salam Z. The application of solar technologies for sustainable development of agricultural sector. Renew Sustain Energy Rev 2013;18:583–94. https://doi.org/10.1016/j.rser.2012.10.049.
[40] Xue J. Photovoltaic agriculture—new opportunity for photovoltaic applications in China. Renew Sustain Energy Rev 2017;73:1–9. https://doi.org/10.1016/j.rser.2017.01.098.
[41] Gorjian S, Zadeh BN, Eltrop L, Shamshiri RR, Amanlou Y. Solar photovoltaic power generation in Iran: development, policies, and barriers. Renew Sustain Energy Rev 2019;106:110–23. https://doi.org/10.1016/j.rser.2019.02.025.
[42] Chel A, Kaushik G. Renewable energy for sustainable agriculture. Agron Sustain Dev 2011;31:91–118. https://doi.org/10.1051/agro/2010029.
[43] Katole DN, Wankar AM, Buradkar PS, Mahajan AP, Ritu R, Vaidya PM. A review: solar powered multipurpose agricultural robot; 2018. p. 742–4.
[44] Shinde VN, Sharma NS, Kasar PMS. Solar based multi-tasking agriculture robot. 4; 2018. p. 185–8.
[45] ecoRobotics: Weeding Robot 2019.
[46] Farm Robots: Ch3 Switzerland, France and England | LandWISE; n.d. https://www.landwise.org.nz/precision-agriculture/in-search-of-farm-robots-ch3-switzerland-france-and-england/ [Accessed 6 January 2020].
[47] Robots fight weeds in challenge to agrochemical giants; 2019. https://steemit.com/environment/@dulaldulal/robots-fight-weeds-in-challenge-to-agrochemical-giants [Accessed 6 January 2020].

[48] ecoRobotics: Phenomobile Scouting Robot 2019.
[49] ecoRobotics: Weeding Robotic Platform 2019.
[50] Agriculture: leader in autonomous mobility | IDTechEx Research Article; 2020. https://www.idtechex.com/ja/research-article/agriculture-leader-in-autonomous-mobility/11643 [Accessed 25 January 2020].
[51] Bawden O, Kulk J, Russell R, McCool C, English A, Dayoub F, et al. Robot for weed species plant-specific management. J Field Robot 2017;34:1179–99. https://doi.org/10.1002/rob.21727.
[52] Implications of robotics and autonomous vehicles for the grains industry-GRDC; 2020. https://grdc.com.au/resources-and-publications/grdc-update-papers/tab-content/grdc-update-papers/2018/02/implications-of-robotics-and-autonomous-vehicles [Accessed 7 January 2020].
[53] Agriculture | ACFR Confluence; 2020. https://confluence.acfr.usyd.edu.au/display/AGPub/Our+Robots [Accessed 7 January 2020].
[54] Underwood JP, Calleija M, Taylor Z, Hung C, Nieto J, Fitch R, et al. In: Real-time target detection and steerable spray for vegetable crops. Proceedings, Work Robot Agric Int Conf Robot Autom; 2015.
[55] Rippa robot takes farms forward to the future—The University of Sydney; 2020. https://sydney.edu.au/news-opinion/news/2015/10/21/rippa-robot-takes-farms-forward-to-the-future-.html [Accessed 7 January 2020].
[56] Sujaritha DM, Lakshminarasimhan M, Fernandez CJ, Chandran M. Greenbot : a Solar Autonomous Robot to Uproot Weeds in a Grape Field. Int J Comput Sci Eng Commun 2016;4:1351–8.
[57] VineScout: Vineyard Monitoring Robot Trialed at Quinta do Ataíde; 2017.
[58] Franklin Robotics | Home of Tertill, the robotic garden weeder that is powered by the sun; 2020. https://www.tertill.com/ [Accessed 7 January 2020].
[59] Keresztes B, Germain C, Da Costa J-P, Grenier G, Beaulieu XD. Vineyard Vigilant & INNovative Ecological Rover (VVINNER): an autonomous robot for automated scoring of vineyards. In: Int. Conf. Agric. Eng., Zurich; 2014. p. 6–10.
[60] Robot | Grassing management | Automated Mower Robot | Vitirover; 2019. https://www.vitirover.fr/en-robot [Accessed 6 January 2020].
[61] News & Gallery | VineScout; 2019. http://vinescout.eu/web/ [Accessed 13 November 2019].
[62] FarmBot | Open-Source CNC Farming; 2020. https://farm.bot/ [Accessed 7 January 2020].
[63] Axbom S, Ralsgård L. Design of an autonomous weeding vehicle used in the agricultural industry. Lund University; 2018.
[64] Agrophotovoltaics: High harvesting yield in hot summer of 2018—Fraunhofer ISE; 2020. https://www.ise.fraunhofer.de/en/press-media/press-releases/2019/agrophotovoltaics-hight-harvesting-yield-in-hot-summer-of-2018.html [Accessed 25 January 2020].
[65] Dupraz C, Marrou H, Talbot G, Dufour L, Nogier A, Ferard Y. Combining solar photovoltaic panels and food crops for optimising land use: Towards new agrivoltaic schemes. Renew Energy 2011;36:2725–32. https://doi.org/10.1016/j.renene.2011.03.005.
[66] Trommsdorff M. An economic analysis of agrophotovoltaics: opportunities, risks and strategies towards a more efficient land use. Germany: University of Freiburg; 2016.
[67] Weselek A, Ehmann A, Zikeli S, Lewandowski I, Schindele S, Högy P. Agrophotovoltaic systems: applications, challenges, and opportunities. A review. Agron Sustain Dev 2019;39:35. https://doi.org/10.1007/s13593-019-0581-3.
[68] The FarmBot Whitepaper; n.d. https://farm.bot/blogs/news/the-farmbot-white-paper [Accessed 26 December 2019].
[69] The FarmBot Whitepaper; 2020. https://farm.bot/blogs/news/the-farmbot-white-paper [Accessed 25 January 2020].

[70] Ertrac, EPoSS, SmartGrids. European Roadmap Electrification of Road Transport; 2012.
[71] ERTRAC. European roadmap infrastructure for green vehicles; 2012.
[72] Khatawkar DS, James PS, Dhalin D. Modern trends in farm machinery-electric drives: a review. Int J Curr Microbiol Appl Sci 2019;8:83–98. https://doi.org/10.20546/ijcmas.2019.801.011.
[73] Herrmann F, Rothfuss F. Introduction to hybrid electric vehicles, battery electric vehicles, and off-road electric vehicles. Elsevier Ltd.; 2015. https://doi.org/10.1016/B978-1-78242-377-5.00001-7.
[74] Kebriaei M, Niasar AH, Asaei B. Hybrid electric vehicles: an overview. In: 2015 Int. Conf. Connect. Veh. Expo, IEEE; 2015. p. 299–305. https://doi.org/10.1109/ICCVE.2015.84.
[75] Tie SF, Tan CW. A review of energy sources and energy management system in electric vehicles. Renew Sustain Energy Rev 2013;20:82–102. https://doi.org/10.1016/j.rser.2012.11.077.
[76] Dixon J. Energy storage for electric vehicles. In: Proc. IEEE Int. Conf. Ind. Technol; 2010. p. 20–5. https://doi.org/10.1109/ICIT.2010.5472647.
[77] Bhatti AR, Salam Z, Aziz MJBA, Yee KP. A critical review of electric vehicle charging using solar photovoltaic. Int J Energy Res 2016;40:439–61. https://doi.org/10.1002/er.3472.
[78] A. Buning E. Electric Drives in Agricultural Machinery-Approach from the Tractor Side; 2010.
[79] Ghobadpour A, Boulon L, Mousazadeh H, Malvajerdi AS, Rafiee S. State of the art of autonomous agricultural off-road vehicles driven by renewable energy systems. Energy Procedia 2019;162:4–13. https://doi.org/10.1016/j.egypro.2019.04.002.
[80] Moreda GP, Muñoz-García MA, Barreiro P. High voltage electrification of tractor and agricultural machinery—a review. Energy Convers Manag 2016;115:117–31. https://doi.org/10.1016/j.enconman.2016.02.018.
[81] Caban J, Vrabel J, Šarkan B, Zarajczyk J, Marczuk A. Analysis of the market of electric tractors in agricultural production. MATEC Web Conf 2018;244. https://doi.org/10.1051/matecconf/201824403005.
[82] Solar Powered Tractor; 2020. https://commons.wikimedia.org/wiki/File:Solar_Powered_Tractor_(5872711433).jpg#filehistory [Accessed 9 January 2020].
[83] wheelbarrow Farm-The electric tractor has arrived; 2020. https://wheelbarrowfarm.com/news/f/its-finally-here-our-electric-tractor-has-arrived [Accessed 10 January 2020].
[84] Abdelhamid M, Pilla S, Singh R, Haque I, Filipi Z. A comprehensive optimized model for on-board solar photovoltaic system for plug-in electric vehicles: energy and economic impacts. Int J Energy Res 2016;40:1489–508. https://doi.org/10.1002/er.3534.
[85] Sujitha N, Krithiga S. RES based EV battery charging system: A review. Renew Sustain Energy Rev 2017;75:978–88. https://doi.org/10.1016/j.rser.2016.11.078.
[86] Youssef C, Fatima E, Najia E, Chakib A. A technological review on electric vehicle DC charging stations using photovoltaic sources. IOP Conf Ser Mater Sci Eng 2018;353. https://doi.org/10.1088/1757-899X/353/1/012014.
[87] Mousazadeh H, Keyhani A, Javadi A, Mobli H, Abrinia K, Sharifi A. Optimal power and energy modeling and range evaluation of a solar assist plug-in hybrid electric tractor (SAPHT). Trans ASABE 2010;53:1025–35. https://doi.org/10.13031/2013.32586.
[88] Redpath DAG, McIlveen-Wright D, Kattakayam T, Hewitt NJ, Karlowski J, Bardi U. Battery powered electric vehicles charged via solar photovoltaic arrays developed for light agricultural duties in remote hilly areas in the Southern Mediterranean region. J Clean Prod 2011;19:2034–48. https://doi.org/10.1016/j.jclepro.2011.07.015.
[89] Solectrac Electric Tractors; 2020. https://www.solectrac.com/ [Accessed 10 January 2020].

[90] Feng X, Yan F, Liu X. Study of wireless communication technologies on internet of things for precision agriculture. Wireless Pers Commun 2019;108:1785–802. https://doi.org/10.1007/s11277-019-06496-7.
[91] Thakur D, Kumar Y, Kumar A, Singh PK. Applicability of wireless sensor networks in precision agriculture: a review. Wireless Pers Commun 2019;107:471–512. https://doi.org/10.1007/s11277-019-06285-2.
[92] Syafarinda Y, Akhadin F, Fitri ZE, Yogiswara, Widiawanl B, Rosdiana E. The Precision Agriculture Based on Wireless Sensor Network with MQTT Protocol. IOP Conf Ser Earth Environ Sci 2018;207. https://doi.org/10.1088/1755-1315/207/1/012059.
[93] Kumar SA, Ilango P. The impact of wireless sensor network in the field of precision agriculture: a review. Wireless Pers Commun 2018;98:685–98. https://doi.org/10.1007/s11277-017-4890-z.
[94] Jawad HM, Nordin R, Gharghan SK, Jawad AM, Ismail M. Energy-efficient wireless sensor networks for precision agriculture: a review. Sensors (Switzerland) 2017;17:1781. https://doi.org/10.3390/s17081781.
[95] Kiani F, Seyyedabbasi A. Wireless sensor network and internet of things in precision agriculture. Int J Adv Comput Sci Appl 2018;9:99–103. https://doi.org/10.14569/IJACSA.2018.090614.
[96] Heble S, Kumar A, Prasad KVVD, Samirana S, Rajalakshmi P, Desai UB. A low power IoT network for smart agriculture. In: IEEE World Forum Internet Things, WF-IoT 2018, Proc., vol. 2018, January. Institute of Electrical and Electronics Engineers Inc.; 2018. p. 609–14. https://doi.org/10.1109/WF-IoT.2018.8355152.
[97] Na A, Isaac W. Developing a human-centric agricultural model in the IoT environment, In: 2016 Int. Conf. Internet Things Appl. IOTAInstitute of Electrical and Electronics Engineers Inc.; 2016. p. 292–7. https://doi.org/10.1109/IOTA.2016.7562740.
[98] Vuran MC, Salam A, Wong R, Irmak S. Internet of underground things in precision agriculture: architecture and technology aspects. Ad Hoc Networks 2018;81:160–73. https://doi.org/10.1016/j.adhoc.2018.07.017.
[99] DATAFLAIR TEAM. IoT Applications in Agriculture—4 Best Benefits of IoT in Agriculture—DataFlair, 1. https://data-flair.training/blogs/iot-applications-in-agriculture/. (Accessed 6 November 2019).
[100] Khanna A, Kaur S. Evolution of internet of things (IoT) and its significant impact in the field of precision agriculture. Comput Electron Agric 2019;157:218–31. https://doi.org/10.1016/j.compag.2018.12.039.
[101] Pooja VN, Pooja PH, Savitri GC, Megha MS, Nirosha H. Design and implementation of agricultural system using solar power. Int J Sci Res Publ 2017;7:236–8.
[102] Zou T, Lin S, Feng Q, Chen Y. Energy-efficient control with harvesting predictions for solar-powered wireless sensor networks. Sensors 2016;16:53. https://doi.org/10.3390/s16010053.
[103] Li Y, Shi R. An intelligent solar energy-harvesting system for wireless sensor networks. EURASIP J Wirel Commun Netw 2015;2015:1–12. https://doi.org/10.1186/s13638-015-0414-2.
[104] Zhang X, Du J, Fan C, Liu D, Fang J, Wang L. A wireless sensor monitoring node based on automatic tracking solar-powered panel for paddy field environment. IEEE Internet Things J 2017;4:1304–11. https://doi.org/10.1109/JIOT.2017.2706418.
[105] Bhangale SY, Bhide AS. A wireless sensor monitoring node based on automatic tracking solar-powered panel for paddy field environment. Int J Adv Res Innov Ideas Educ 2019;5:1522–30.
[106] Vasisht D, Kapetanovic Z, Won J, Jin X, Chandra R, Kapoor A, et al. In: FarmBeats: An IoT platform for data-driven agriculture. Proc 14th USENIX Symp Networked Syst Des Implement; 2017. p. 515–29.

[107] Chieochan O, Saokaew A, Boonchieng E. Internet of things (IOT) for smart solar energy: a case study of the smart farm at Maejo University. In: 2017 Int. Conf. Control. Autom. Inf. Sci., vol. 2017- Janua, IEEE; 2017. p. 262–7. https://doi.org/10.1109/ICCAIS.2017.8217588.

[108] Nguyen T-D, Thanh TT, Nguyen L-L, Huynh H-T. On the design of energy efficient environment monitoring station and data collection network based on ubiquitous wireless sensor networks. In: 2015 IEEE RIVF Int. Conf. Comput. Commun. Technol.—Res. Innov. Vis. Futur., IEEE; 2015. p. 163–8. https://doi.org/10.1109/RIVF.2015.7049893.

[109] Khattab A, Abdelgawad A, Yelmarthi K. Design and implementation of a cloud-based IoT scheme for precision agriculture. In: 2016 28th Int. Conf. Microelectron., IEEE; 2016. p. 201–4. https://doi.org/10.1109/ICM.2016.7847850.

[110] Ilie-Ablachim D, Patru GC, Florea I-M, Rosner D. Monitoring device for culture substrate growth parameters for precision agriculture: acronym: MoniSen. In: 2016 15th RoEduNet Conf. Netw. Educ. Res., IEEE; 2016. p. 1–7. https://doi.org/10.1109/RoEduNet.2016.7753237.

[111] Navulur S, Sastry ASCS, Prasad MNG. Agricultural Management through Wireless Sensors and Internet of Things. Int J Electr Comput Eng 2017;7:3492. https://doi.org/10.11591/ijece.v7i6.pp3492-3499.

[112] Rahul DS, Sudarshan SK, Meghana K, Nandan KN, Kirthana R, Sure P. IoT based solar powered Agribot for irrigation and farm monitoring: Agribot for irrigation and farm monitoring. In: Proc. 2nd Int. Conf. Inven. Syst. Control. ICISC 2018. Institute of Electrical and Electronics Engineers Inc.; 2018. p. 826–31. https://doi.org/10.1109/ICISC.2018.8398915.

CHAPTER

8

Applications of solar PV systems in desalination technologies

Shiva Gorjian[a,b], *Barat Ghobadian*[a], *Hossein Ebadi*[c], *Faezeh Ketabchi*[d], and *Saber Khanmohammadi*[e]

[a]Biosystems Engineering Department, Faculty of Agriculture, Tarbiat Modares University (TMU), Tehran, Iran, [b]Leibniz Institute for Agricultural Engineering and Bioeconomy (ATB), Potsdam, Germany, [c]Biosystems Engineering Department, Faculty of Agriculture, Shiraz University, Shiraz, Iran, [d]Mechanical Engineering Department, Amirkabir University of Technology (Tehran Polytechnic), Tehran, Iran, [e]Mechanical Engineering Department, University of Kashan, Kashan, Iran

8.1 Introduction

Water is the most vital substance on this planet, and it plays a vital role in nearly all aspects of human life and the ecosystem [1]. All water on Earth can be found in the oceans, seas, rivers, underground waters, etc. Water exists in three states, liquid, solid, and vapor, which are commonly found in the soil and top layers of the Earth's crust [2, 3]. Current estimates are that the Earth's hydrosphere contains 1386 million cubic kilometers. However, about 96.54% of this amount is saline water, and only 2.53% is freshwater [2, 4]. More than two-thirds, equal to 68.7%, of the freshwater is frozen in polar ice caps and glaciers in the form of ice and permanent snow cover while 30.1% exists in the form of fresh groundwater. Therefore, only 0.3% of freshwater is directly available to humans in lakes, reservoirs, and river systems. Intensive use of these water resources disturbs the natural equilibrium established over centuries and will need tens or hundreds of years to be restored [3–5].

Water scarcity is defined as the lack of sufficient quantity and quality of available water resources within a region. Water scarcity is a global issue and

https://doi.org/10.1016/B978-0-12-819610-6.00008-9

is considered the primary concern to human societies as well as a barrier to sustainable development [6, 7]. The well-known measure for evaluation of *national water scarcity* in a region is the renewable water per capita per year. As a threshold, a region is considered to be under regular water stress when the water supply is less than 1700 m^3 per capita per year. Currently, more than 4 billion people are living in water scarcity regions in the world while almost 1 billion people lack access to safe potable water in developing countries [8, 9]. Moreover, it is estimated that about half the world's population will live in water-scarce regions by 2025 [7, 10].

8.2 Desalination and water security

Water security is defined as possessing a sufficient supply of water to meet current and future demands. There are several measures to mitigate water scarcity and therefore achieve water security, such as water conservation, mending infrastructure, and improving the distribution and catchment systems [11, 12]. Although these criteria are suggested as requirements, they can only improve the usage of existent water resources. Therefore, the only applicable way for increasing existing water resources is desalination and water reuse [12, 13]. Desalination is considered *water insurance* that provides new levels of security that reduce both the economic and social risks of water scarcity by reducing demands on aquifers and surface waters [11, 14].

Desalination has primarily become the most important source of water in the Middle East, the Gulf Cooperation Council States, and North Africa (MENA), with more than 50% of the global capacity, and some parts of the Caribbean islands, where water is particularly scarce [15–17].

During desalination, all the salts and minerals are removed from saltwater (seawater or brackish water) and freshwater is produced [11, 18]. In recent years, the global capacity of desalination has grown rapidly, with the highest amount in arid nations [11]. According to the *International Desalination Association (IDA)*, currently more than 18,000 desalination plants with a maximum capacity of around 90 mcm[a] of the produced water per day are operating around the world [4, 19]. In these, seawater desalination is predominant (58.9%) while the rest is allocated to the desalination of brackish groundwater (21.2%) and other saline surface waters (19.9%) [20, 21].

8.3 Benefits and challenges of desalination

Although desalination is known as an important alternative and a promising option in water-scarce countries, there are still several issues facing this technology that are independent of the approach or type of

[a]Million cubic meters.

process. The main concerns about desalination technology are related to energy demand as well as environmental and socioeconomic impacts [22, 23]. Being energy-intensive and relying mostly on fossil fuels result in an increase in greenhouse gas (GHG) emissions, which is the primary constraint to the sustainable development of desalination plants [4, 15, 24]. In the case of environmental concerns, the negative consequences of desalination plants can be mitigated by performing an environmental impact assessment (EIA) and implementing environmental management plans (EMPs) to make this technology more sustainable [23]. Therefore, with the rising demand for freshwater, there is significant potential in the market for alternative energy sources to drive operation of desalination plants around the world [15, 25].

8.4 Conventional desalination technologies

The primary traditional desalination methods are classified into two main areas, thermal evaporation and membrane-based separation technologies [26, 27]. Thermal methods utilize heat to separate water from the salt solution through evaporation, where the water vapor is then condensed and freshwater is produced. The most mature thermal desalination technologies are multistage flash distillation (MSF), multieffect distillation (MED), and vapor compression (VC) [28, 29]. In membrane-based technologies, the concentration gradient, electrical potential, or mechanical pressure is performed as a driving force applied across semipermeable membranes to separate the salt from the water stream. The primary membrane processes include membrane desalination (MD), reverse osmosis (RO) and electrodialysis (ED) [30, 31]. There are also other alternative technologies such as freezing and ion-exchange, which have not reached the commercial scale (Fig. 8.1). Regarding energy consumption, SWRO has a specific electric power consumption of about 3–4 $kWh\,m^{-3}$ with energy recovery while MSF and MED have a total specific energy consumption between 10 and 16 $kWh_T\,m^{-3}$ (about 7.5–12 $kWh_{th}\,m^{-3}$ and 2.5–4 $kWh_{el}\,m^{-3}$) and 5.5–9 $kWh_T\,m^{-3}$ (4–7 $kWh_{th}\,m^{-3}$ and 1.5–2 $kWh_{el}\,m^{-3}$), respectively [32–34].

8.5 Solar-powered desalination systems

Solar energy is known as the most abundant renewable energy source. It has intensively been developed because of both governmental support policies and remarkable improvements in the technological aspects of utilization and the development of renewable energies [35, 36].

Solar energy is an abundant and ecofriendly source of renewable energy with the most promising potential to drive desalination plants.

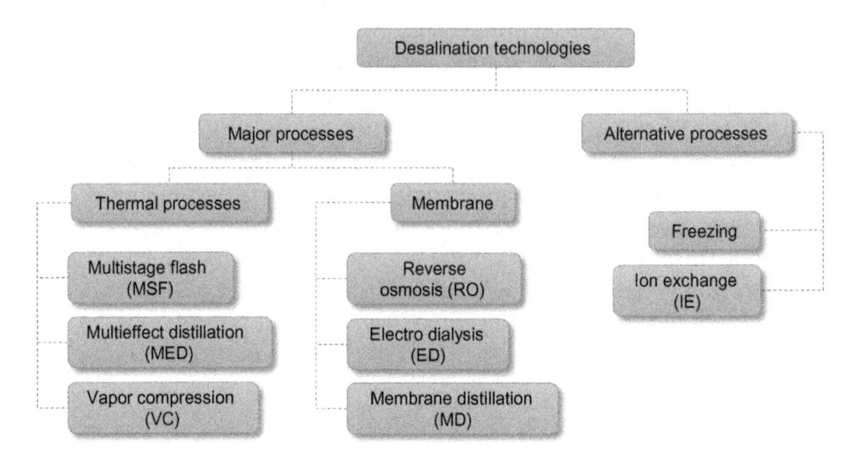

FIG. 8.1 Classification of the most mature desalination technologies [4].

Moreover, solar desalination has been proved to be the most viable and affordable technique to purify salt water [37–39].

Solar energy can be utilized in solar desalination systems in both direct and indirect methods. In direct solar desalination systems, solar energy is directly used to produce freshwater in solar stills. Indirect desalination systems are composed of two subsystems, including the solar collector and desalination unit [28, 40, 41]. Solar energy can be directly converted into electricity using photovoltaic (PV) systems or indirectly through concentrating solar power (CSP) technology [32, 42]. Fig. 8.2 shows the most feasible combination of solar systems with desalination processes to supply heat or electricity. The cost of the freshwater produced by PV-RO plants is usually estimated to be in the range of 7.98–29 US\$ m^{-3} with the capacity between 120 and 12 m^3 day^{-1} to desalinate seawater and about 7.25 US\$ m^{-3} with the capacity of 250 m^3 day^{-1} to desalinate brackish water [44–46]. The final produced water cost for PV-ED desalination systems is between 16 and 5.8 US\$.m^{-3} [47–50].

8.5.1 PV-integrated solar desalination systems

Using PV power generation systems is most beneficiary in remote locations with low power demand applications where the distribution line costs are not reasonable [51]. PV cells are the semiconductor devices that generate direct current (DC) electricity. Cells are then grouped into modules, with a frontal transparent glass cover, a weatherproof cover for the back, and often a frame. The modules can then be combined to form strings and arrays [52, 53].

PV systems are commonly classified into two main configurations, grid-connected/on-grid and stand-alone/off-grid systems. The stand-alone PV

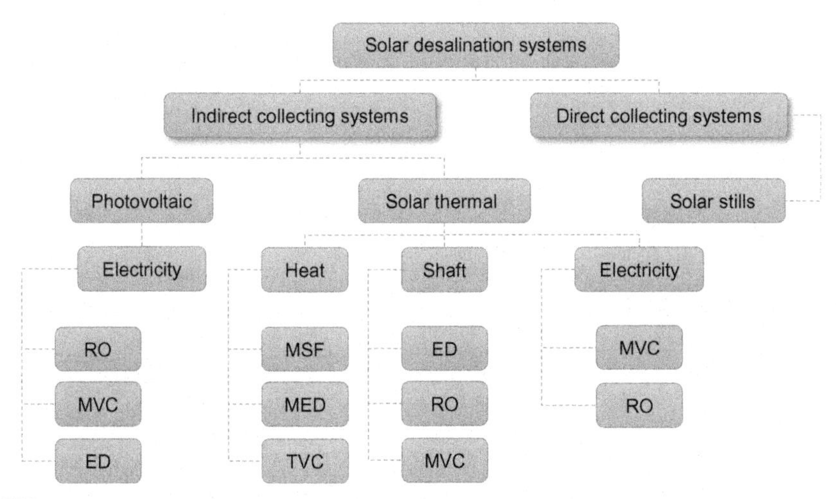

FIG. 8.2 Integration between solar power systems and conventional desalination technologies [43].

systems do not transfer power to the grid (Fig. 8.3) while on-grid (grid-connected) systems transmit the DC electricity generated by the solar panels to an inverter and then into the distribution system (Fig. 8.4). The use of PV-powered desalination technology is an excellent choice to provide water for small to medium communities located in remote and isolated areas with high availability of both solar radiation and saltwater [47].

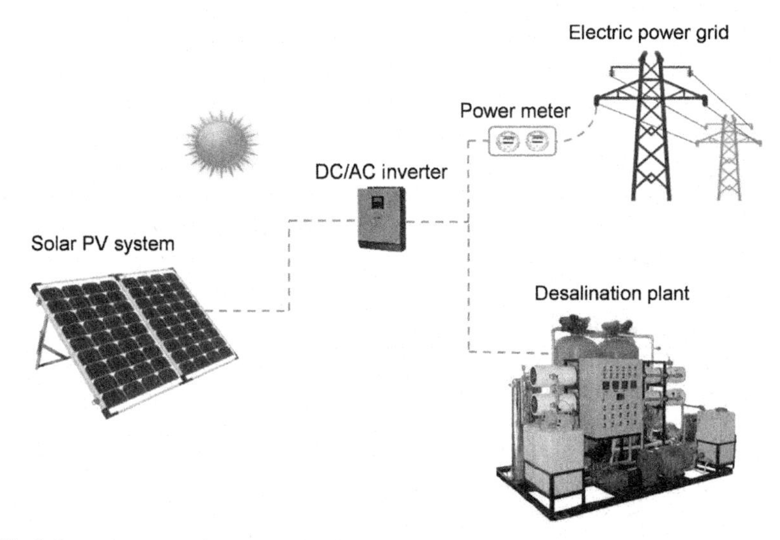

FIG. 8.3 Schematic of an on-grid solar PV-powered desalination system.

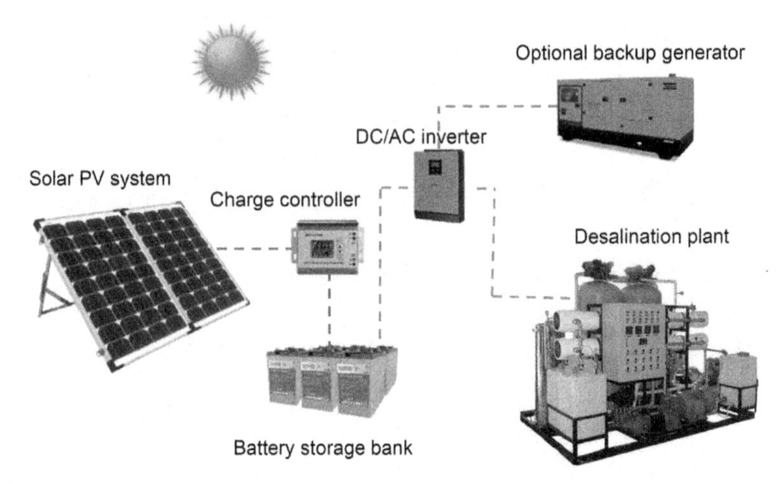

FIG. 8.4 Schematic of an off-grid solar PV-powered desalination system.

8.5.1.1 PV-powered RO desalination systems

Reverse osmosis (RO) systems are composed of a high-pressure pump, pretreatment and posttreatment units, and membranes. Considering the sensitivity of the membrane module to fouling, the feed water pretreatment section has a crucial role in the durability and lifetime of RO membranes. The operating pressure of RO systems for brackish water (BW, TDS: 1000–8000 mg/L) and seawater (SW, TDS: 20,000–45,000 mg/L) are around 17–27 and 55–82 bars, respectively [54].

In large plants, the energy of rejected pressurized brine is recovered through employing an energy recovery device (ERD), including Peloton-Wheel turbines that retrieve around 20–40% of the consumed energy [55, 56]. PV power systems could be a decent choice to supply the required power of RO desalination systems and make them more cost-effective [57].

In order to enhance efficiency and increase the amount of produced freshwater, various studies have been conducted on the integration of RO plants with solar PV systems. The schematic view of a typical PV-RO desalination unit is shown in Fig. 8.5.

Wu et al. [58] optimized the size of a hybrid diesel-PV-RO desalination plant to enhance the yield rate of freshwater and meet power demand using the Tabu Search method. For this purpose, some consequential parameters such as the battery bank size, the PV module area, and the diesel generator fuel consumption were optimized to achieve the minimum life cycle cost (LCC). The results indicated that the LCC of the overall system was calculated as 28,130 $ while the cost of the produced water was calculated between 1.59 and 2.39 $/m^3. The levelized cost of electricity (LCOE) also ranged from 0.3975 to 0.5975 $/kWh.

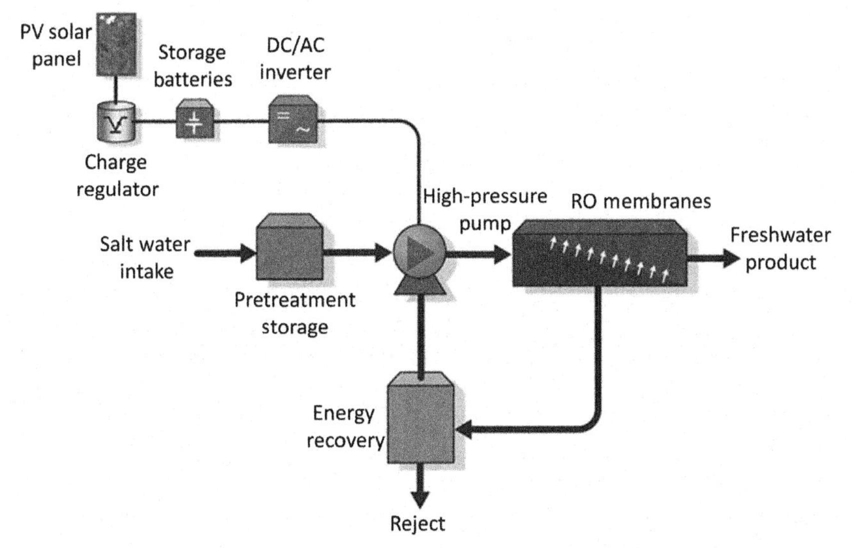

FIG. 8.5 Schematic view of a PV-RO desalination system [47].

Ahmad et al. [59] simulated the performance of a PV-RO desalination plant and verified the results with the experimental data. They investigated the effects of fixed and tracking panels on the amount of absorbed insolation by the PV modules. The membrane unit was also simulated to estimate the feed water pressure as well as the permeate flow rate. Based on the results, the annual permeate gain with the optimal monthly and annual tilt angle of PV modules was calculated as 10% and 19%, respectively. Moreover, the annual permeate gains of the PV modules with the single- and double-axis trackers were reported as 43% and 62%, respectively.

Alghoul et al. [60] evaluated the effects of climatic, design, and operational parameters on the performance of a PV-BWRO unit. They fabricated and tested a prototype with a 2kW PV system, and for five various membrane modules. The results demonstrated that the optimum load of the RO, the type of membrane and the configuration of the plant are 600W, 4″ × 40″ TW30-4040, and two-stage, respectively. The system could produce 5.1 m^3 of freshwater for 10h in a day with 1.1 kWh/m^3 specific energy consumption.

Gökçek [61] studied the performance of an RO desalination plant with a capacity of 1 m^3/h integrated with different off-grid power generation systems, including wind, PV, diesel, and battery, located in Bozcaada Island, Turkey. They used HOMER[b] software to do the techno-economic

[b]Hybrid optimization model for electric renewable.

evaluation, and ROSA[c] was used to specify the energy requirements. From the results, the electricity and water costs for the optimal state of the system consisting of 10 kW wind turbines, 20 kW PV panel, and a 8.9 kW diesel generator were obtained as 0.308 \$/kWh and 2.2 \$/m^3.

8.5.1.2 PV-powered ED desalination systems

Generally, an ED unit is comprised of some components including a pretreatment unit, a membrane module, a low-pressure circulation pump, a DC power supply, and a posttreatment section. The schematic view of a solar PV-ED system is presented in Fig. 8.6. In the ED process, the electrodes that are placed in the saltwater solution container are connected to an external DC power supply (battery or PV). The electrodes are usually fabricated from niobium and titanium with platinum coatings. Considering the tendency of ions to move toward the opposite charge electrode, the electrical current is conveyed through the solution. The passing of brackish water over a membrane located between two electrodes removes the salinity of the water [62]. A decent way to make the ED desalination method more cost-effective is integrating them with renewable energies as the source of power. Currently, several PV-driven ED systems have been installed throughout the world.

Ortiz et al. [63] conducted a feasibility study of using a PV-ED system for the desalination of brackish water. They developed a mathematical model that was able to estimate the system performance accurately. A good agreement was achieved between the experimental and numerical outputs. In another study, Ortiz et al. [48] investigated the feasibility of purifying the aquifer BW (with TDS range of 2300–5100 mg/L) using a PV-ED desalination system. Also, they deployed previous mathematical models and employed them at real conditions. The results revealed that the cost of the desalinated water is between 0.14 and 0.23 €m^{-3} for irrigation applications and around 0.17–0.32 €m^{-3} for potable water. Moreover, the electric consumption for the corresponding applications was around 0.7–1.4 kWhm^{-3} and 0.9–1.7 kWhm^{-3}, respectively.

Fernandez-Gonzalez et al. [64] focused on the sustainability of solar PV-ED desalination systems considering some substantial criteria such as energy consumption as well as economical and social issues for a case study of Canary Island. The energy consumption for the BWED system was obtained as 0.49–0.91 kWh/m^3. A summary of some selected studies on solar PV-RO and PV-ED desalination technologies is presented in Table 8.1.

[c]Reverse osmosis system analysis.

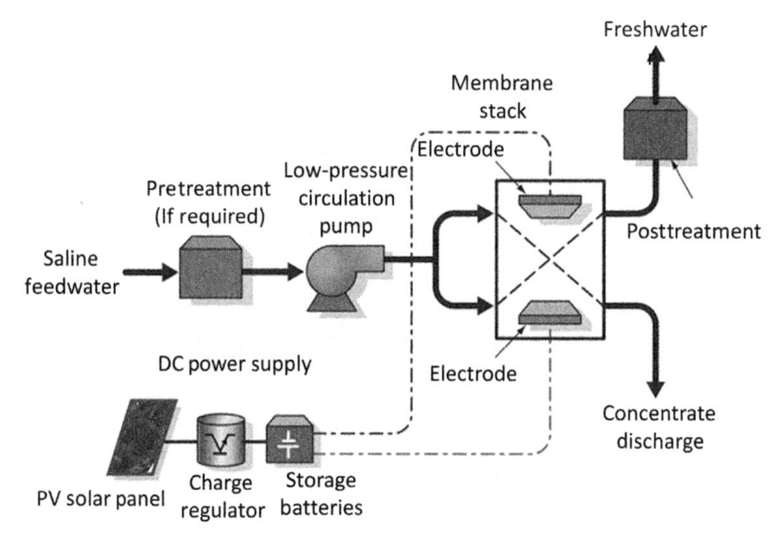

FIG. 8.6 Schematic of a solar PV-driven electrodialysis system [47].

8.5.1.3 *PV-integrated membrane distillation (MD) systems*

Membrane distillation (MD) technology operates based on a difference in the vapor pressure over a microporous membrane that is hydrophobic. The merits of this technology include the capability of using waste-grade heat, producing high-quality water, and compactness. Besides, this technology is also employed in the separation process of pharmaceutical and dairy compounds, juices, and purification of waters contaminated with oil. Fig. 8.7 shows a schematic view of the MD desalination technology [73].

Considering both the slow development and the high levelized cost of produced water, which are impediments for maturity and large-scale application of solar-powered MD systems [74], a plethora of solar-powered MD systems have been operated in various countries such as Spain, China, Saudi Arabia, Singapore, Australia, the United States, and Mexico [73, 75, 76]. Owing to the fact that thermal energy is the main source of energy to drive MD systems, most of the thermal solar collectors are integrated with this technology, and typically the solar PVs are employed to provide the required electricity of other accessories.

Moore et al. [77] developed a solar-powered MD system that supplied both thermal collectors and PV modules. Their results demonstrated that the water cost for an optimized state is 85 $\$/m^3$ while the cost of a 5.16 $\$/m^3$ air gap MD was reported for the other studies. Fig. 8.8 indicates the solar-powered MD system in two configurations, assisted and stand-alone.

Typically, various solar collectors, including flat-plate collectors (FPCs), evacuated-tube collectors (ETCs), parabolic concentrators, and solar ponds, are common solar thermal collecting systems used to supply

TABLE 8.1 Summary of selected PV-RO and PV-ED desalination studies.

System type	Location	Methodology		Main Results	Ref.
		Theoretical	Experimental		
PV-ED	Heverlee, Belgium		×	• The ED system desalinated saltwater and produced water with high quality; • The concentrate was recycled to the forward osmosis (FO) membranes and was reused as the feed water; • The water production cost was 3.32 to 4.92 €/m^3 for a small-scale system.	Zhang et al. [65]
PV-ED	Gran Canaria Island, Spain		×	• For a solar radiation range of 250–1100 W/m^2, the EDR plant produced high-quality water; • PV-ED is a very promising combination in terms of flexibility, energy performance, and produced water quality; • The off-grid operation was feasible and presented better working features in comparison with on-grid connections.	Penate et al. [66]
PV-RO	–		×	• Annual permeate gain of 10% and 19% with yearly and monthly optimal tilt angles of PV modules respectively; • Annual permeate gains were obtained as 43% and 62% when PV modules are connected with single- and double-axis sun-trackers, respectively.	Ahmad et al. [59]
PV-RO	–		×	• 43.22% increase in electrical energy output due to the use of V-trough concentrators on PV modules and a thermal energy recovery system; • 54.88% increase in the average mechanical power transferred to the feedwater; • Electrical power saving of 56% and 26% in the RO plant operation with the battery bank and the PV array, respectively.	Vyas et al. [67]

PV-RO	–		×	• Produced water with acceptable quality under constant solar irradiance of $400–1200\,W\,m^{-2}$; • Constant operation with a fluctuating energy source, especially when additional power is available.	Richards et al. [68]
PV-RO	Bozcaada Island, Turkey	×		• Levelized cost of water for the hybrid system was obtained $2.20/m^3; • For the system composed of wind turbines and a diesel generator, the electricity cost was obtained as $0.308/kWh; • RO plant with a hybrid power system is affordable for use in remote areas.	Gökçek [61]
PV-ED	Gaza Strip	×		• The system reduces a carbon footprint as well as environmental burdens of groundwater abstraction; • The specific energy consumption of 0.82 kWh/m^3 for desalination, 1.199 kWh/m^3 for pumping, a production rate of 0.917 m^3/h, and a recovery ratio of 0.91 were calculated.	Ramanujan et al. [69]
PV-RO	Delvar and Deylam ports, Iran	×		• The annual electricity production for Delvar and Deylam ports was calculated as 72,336 kW and 47,915 kWh, respectively; • The maximum and minimum amounts of produced potable water for Delvar and Deylam ports were obtained as 228 m^3 and 148 m^3 per day, respectively; • For Delvar and Deylam ports, the freshwater cost was calculated as 1.96 and 3.02 $/m^3, respectively.	Mostafaeipour et al. [70]
PV-RO	–	×		• For double-stage plants with production of more than 5 m^3/day, a 65% recovery rate was obtained; • The use of a battery reduced the costs because the extra cost is compensated by using fewer RO modules.	Monnot et al. [71]

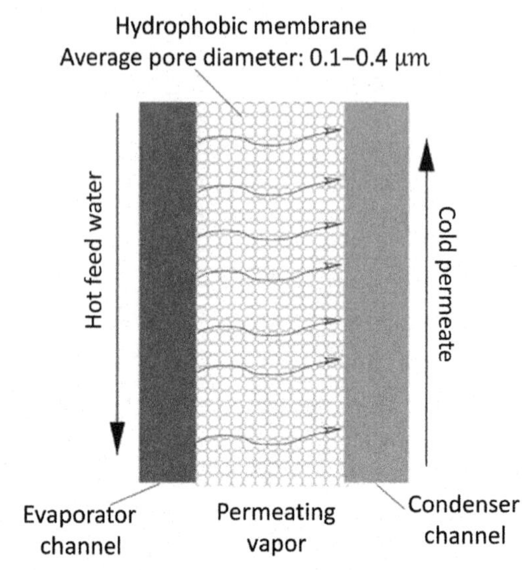

FIG. 8.7 Principle of membrane distillation technology [72].

the required thermal energy for MD desalination systems. The electricity demand for these systems is also provided by solar PV systems (stand-alone or grid-connected) [73].

Hughes et al. [78] investigated the integration of a concentrated PV (CPV) module with a thermally driven membrane distillation (MD) system. Achieving maximum distilled water of $3.4\,\mathrm{L/m^2\,h}$, they concluded that although the quality and quantity of the distilled water vary with power fluctuations supplied to the modules, it is not so significant to affect the integration of MD systems with solar energy. In another study, Baskaran [79] evaluated the integration of a PV-powered MD system in India (Fig. 8.9). Their experiments revealed that the solar PV-MD system

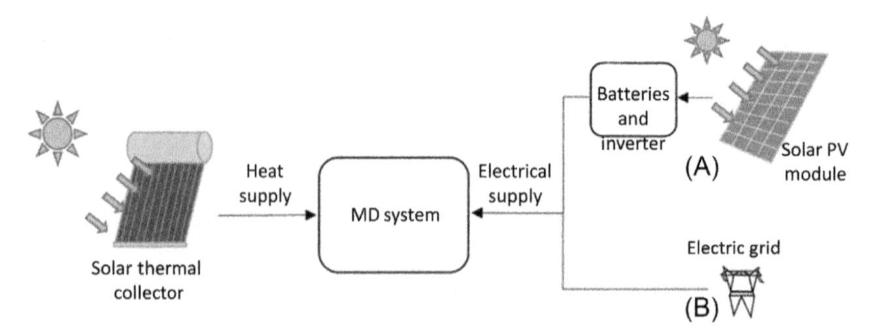

FIG. 8.8 A view of the solar-driven MD system in: (A) The Stand-alone, and (B) The assisted way [73].

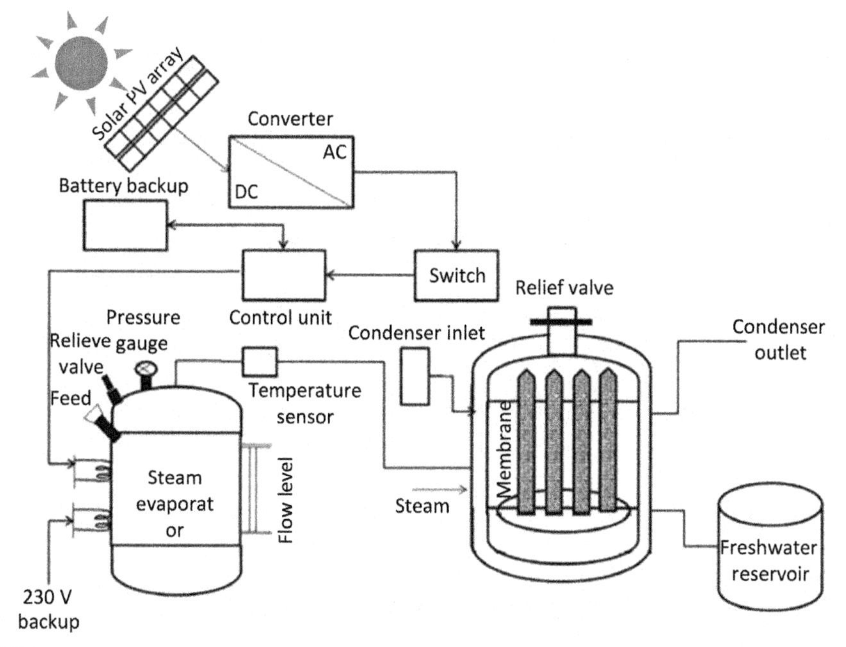

FIG. 8.9 Schematic of a PV-powered MD desalination unit [79].

could be a promising system to meet the freshwater demand in remote areas. Based on the results, the PV-MD system could achieve a thermal efficiency of up to 95% compared to 83% of conventional systems.

8.5.2 Solar distillation systems

Distillation is one of the most ancient methods used by humans to purify water around the world. Thermal distillation causes a physical transformation in the state of water from a heat source that can be powered by different sources of energy. Basically, in this method, seawater, brackish water, or any other impaired water is heated at the operating pressure to reach the boiling point and produce steam, where the steam is condensed to freshwater in a condenser.

In conventional passive solar stills, the heat form of solar energy is initially used to raise the water temperature and provide the required energy to change from the liquid to the vapor phase (Fig. 8.10). However, the integration of PV modules with solar stills is a novel method that has great potential to solve some of the problems of conventional passive systems. One of the essential needs in current advanced water production units is electric power. Therefore, integrated PV active solar stills are predicted to represent a better performance compared to non-PV-powered systems

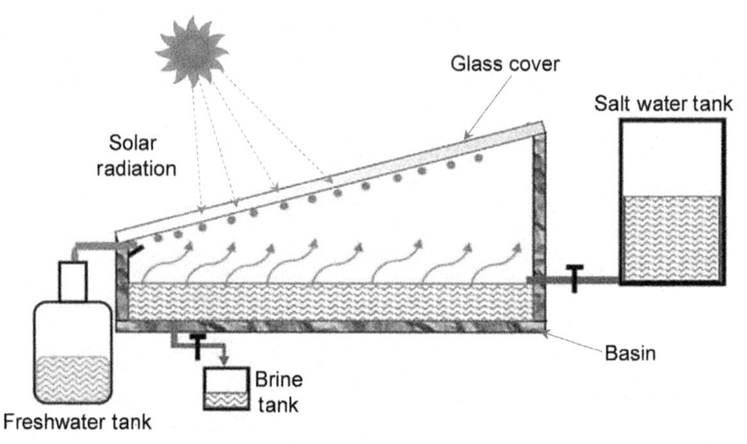

FIG. 8.10 Schematic view of a simple solar still [80].

due to additional features that different pieces of electric equipment such as the pump, fan, shaft, etc., can bring to the operation.

8.5.2.1 Conventional solar stills

Conventional passive solar stills utilize the available solar radiation to evaporate saline water and produce freshwater, with productivity ranging between 4 and 6 L/m^2, which is the lowest level among all desalination technologies. A simple solar still can purify saltwater with a concentration up to 10^4 ppm [81, 82]. For this reason, scientists have paid considerable attention to augmenting the freshwater yield, and the desired fields are using techniques to:

- **Improve water evaporation** by increasing basin temperature and heat transfer between water and basin, using capillary action and phase change materials (PCMs), exploiting recovered heat, breaking the surface tension, and making turbulent flow;
- **Improve water condensation** by decreasing glass temperature, using forced cooling effects, and increasing condensation area.

8.5.3 Integrated PV solar stills

Solar distillation processes mainly utilize heat for evaporation while PV modules provide electricity that can assist different distillation techniques. This section provides several techniques for PV integration with solar distillation systems.

In order to obtain the overall thermal efficiency (η_{th}) of a solar PV-powered distillation system, taking into account the solar energy used for powering a PV module, the efficiency of power generation has to be included in the efficiency equation as:

$$\eta_{th} = \frac{m_w h_{fg}}{G_s A_s + G_s A_{pv}} \quad (1)$$

where m_w is the rate of water condensation (kg/s), h_{fg} is the latent heat of vaporization (J/kg), A_s is the basin area (m^2), G_s is the solar energy incident on a horizontal surface (W/m^2), and A_{pv} is the area of the PV module (m^2).

8.5.3.1 PV-based mechanical power generation

Integrated PV-powered pump

Jijakli et al. [83] proposed a solar distillation unit for off-grid areas. The premise was to use PV modules as a power source for pumping water integrated into a solar still. In their study, the model included a $0.57\,m^2$ surface of the PV module, a 500 W inventor, and a pump to feed water from 72.5 m below ground level, all integrated with a conventional solar still by $1.25\,m^3$/day capacity to distillate saline water. The schematic view of the proposed design is shown in Fig. 8.11.

They assumed that primary water is fed from the groundwater table to a solar still using a PV-powered pump. The solar energy received on PV modules was converted to electric energy and the inventor supplied the generated power to the pump. The still also exploits solar energy to evaporate saline water without an external power source for the desalination process.

The hydraulic power needed for water pumping in a desalination system is a function of different characteristics that must be defined for system design. Therefore, hydraulic sizing can be estimated using the following equation [84];

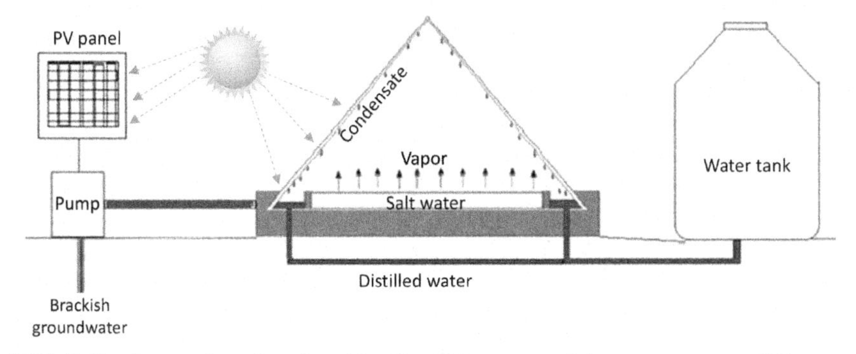

FIG. 8.11 Integration of a solar still with a PV system to drive a water pump [83].

$$P_{pump} = \frac{\rho g(h + \Delta H)Q}{\eta_p \eta_e} \tag{2}$$

where P_{pump} (W) is the required power to pump water with system requirements, ρ (Kg/m^3) is the water density, g (m/s^2) is the local acceleration due to gravity, h (m) is the total pumping head, ΔH (m) is hydraulic losses, Q (m^3/s) is the volume of the flow, η_p is the pump efficiency, and η_e is the electric motor efficiency.

Integrated PV-powered fan

In an experimental study carried out by Rahbar and Esfahani [85] on the utilization of heat pipes and thermoelectric modules for a portable solar still, a black Plexiglas surface was used in the bottom side as the basin for solar absorption where an inclined front side was used as a condenser in addition to a horizontal plate located at the top side to distillate freshwater, as shown in Fig. 8.12.

The thermoelectric module attached to an aluminum plate was incorporated to enhance water condensation with the aid of a fan. The results indicated that the PV module could be used to drive a 1.9W fan to cool down the heat pipe. The system was evaluated during five autumn days

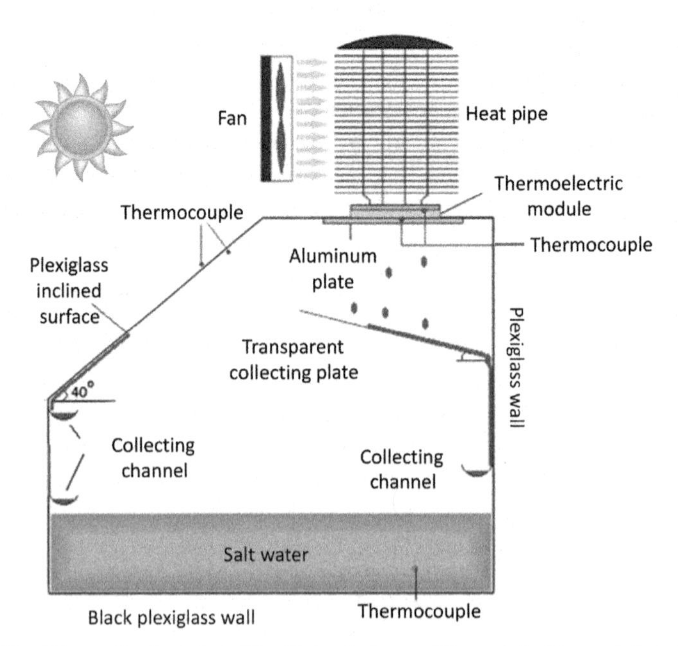

FIG. 8.12 Schematic of a solar still with heat pipe and thermoelectric module cooled by a PV-powered fan [85].

under outdoor climate conditions, and experimental results showed system performance enhancement.

In addition to cooling duties, fans can also be employed particularly to cause forced convection inside a solar still. In this case, the evaporation rate accelerates upon the higher airspeed and consequently the yield increases. Wang et al. [86] conducted research investigating the effect of forced convection inside a humidification-dehumidification (HDH) desalination unit using a PV-powered fan. They suggested a modification factor (F) to estimate evaporation mass flow rate as:

$$m_{evap,\ mod} = F \times m_{evap,theo} \tag{3}$$

where $m_{evap,\ mod}$ and $m_{evap,\ theo}$ are, respectively, modified and theoretical mass flow rate in kg/s. When the experimental condition is free convection, F as a function of water temperature (T_{wo}) can be obtained as:

$$F = 0.0043 \times \exp\left(\frac{T_{wo}}{11}\right) + 5.818 \tag{4}$$

A detailed analysis of the relation between modified and theoretical evaporation in an HDH system may be found in ref. [86].

Integrated PV-powered vacuum pump

Vacuum pumps are one of the most energy-consuming parts of the vacuumed distillation process. Hamawand et al. [87] presented a laboratory setting to introduce a new utilization of PV modules to drive a vacuum pump. In their study, a solution was suggested to drain the concentrated salty water from other desalination plants such as RO, where seawater is introduced inside a double-walled glass column to be sprayed and exposed to sunlight (Fig. 8.13). The vapor pressure inside the column was reduced by a PV-powered vacuum pump to decrease the boiling point and, thus, increase the evaporation rate. A chilled column with cold surfaces condenses vapor to recover the remaining freshwater.

In their design, a dark soluble dye was dissolved with salty water for light absorption enhancement. They suggested that the required heat for the evaporation process can be supplied from a number of options, including sunlight, the heat generated by the vacuum pump, the double glass effect, the heat generated by the PV modules, or latent heat recovered from the vapor. They concluded that the innovative design can produce 15 kg/m^2/day of freshwater in addition to the byproduct of salt.

Integrated PV-powered stirrer

Arun Kumar et al. [88] proposed a novel design of PV integration with conventional solar stills to improve productivity. They devised a system consisting of a still, agitators, a fan, and an external condenser. In their

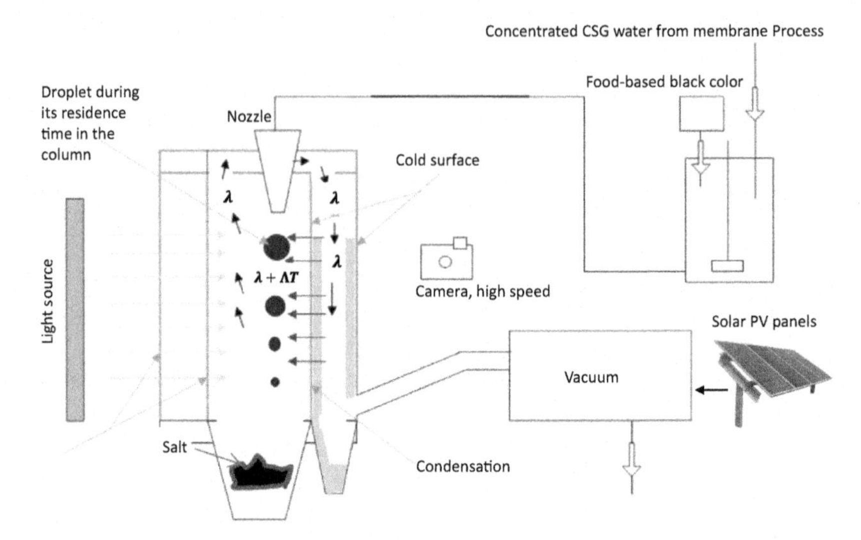

FIG. 8.13 An active vacuum solar distillation process [87].

study, a shaft was coupled with a 4W DC motor and a 0.45W exhaust fan, both driven by solar PV modules, for the agitation of water and extracting vapor to an external condenser, respectively. In the modified basin, five 15mm blades were attached to a shaft and placed above the saline water body to disturb the water surface and reduce its surface tension, and the exhaust fan was placed on the top corner where the exhaust duct was attached to a secondary condenser. In addition, cooling pipes made of copper tubes were placed in the condenser to distillate the water vapor (Fig. 8.14).

The agitation effect resulted in the disturbance of water in the basin and caused an increase in the contact area between the air and water, in which a higher evaporation rate was achieved. In addition, due to the existence of

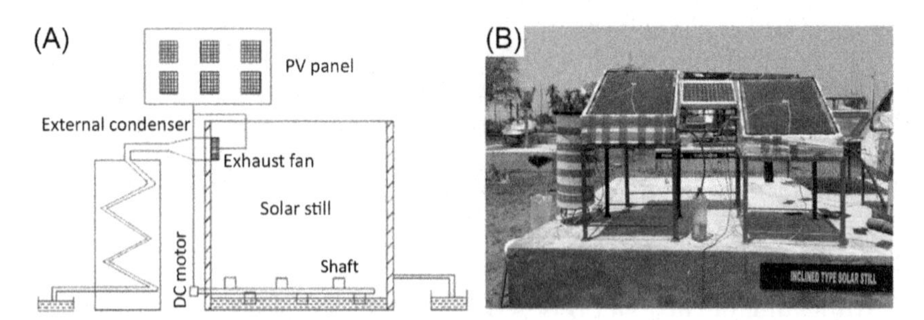

FIG. 8.14 (A) Schematic view, and (B) photo of a solar still with a PV-powered fan and agitator [88].

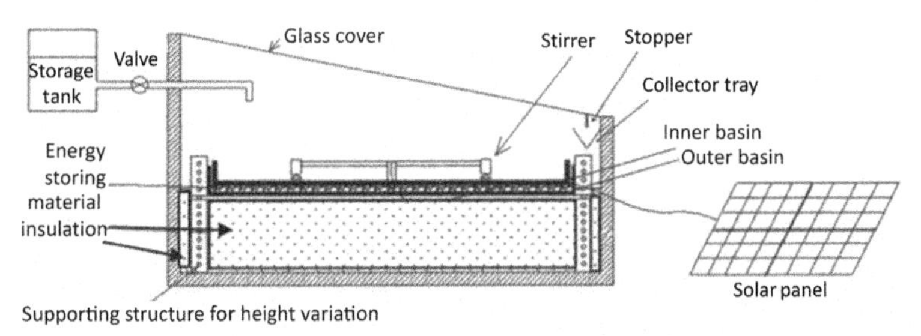

FIG. 8.15 The schematic view of the design using PV power to drive mechanical stirrers [89].

an extra condenser, a portion of the vapor was carried out and cooled the cover glass, resulting in more temperature differences with vapor. It was also proved that an efficiency enhancement of 30.57% can be achieved while the modified basin is more cost-effective compared to the conventional designs.

In a similar study, Rajaseenivasan et al. [89] investigated the accumulation effects of the use of mechanical stirrers and the energy storage materials on the productivity enhancement of the solar still. In their design, PV panels were employed as the power source for stirrers to accelerate the evaporation rate. Four stirrer DC motors were positioned over the basin and powered by a 40 W solar module. Charcoal and paraffin wax were placed below the basin and inside an insulated casing with the back of the still for heat storage and more thermal power in the absence of sunlight (Fig. 8.15). The experimental data showed a 30% increase in distillate compared to conventional stills as well as nocturnal freshwater output with a maximum daily yield of 5.23 kg/m^2.

Integrated PV-powered scraper

The implementation of a rubber scraper was introduced by Al-Sultani et al. [90] to obtain maximum freshwater during the operation of a double-slope solar still. In their design, two 12 V DC motors were used to move the rubber scrapers inside the basin and underneath the glass cover to facilitate drop collection using channels. This wiping effect resulted in higher solar transmittance through the glass and avoided water drops falling toward the basin, which together accelerated the productivity.

Two DC motors were driven using a 12 V-PV module with the power of 200 W/h incorporated with a charge controller and two 12 V batteries with the capacity of 150 Ah. The scrapers' movements were controlled by direction, time interval, and speed, which were governed by an electric control board (Fig. 8.16). Experimental results revealed that a solar still integrated with rubber scrapers indicates enhanced internal heat transferred plus higher productivity compared to conventional solar stills.

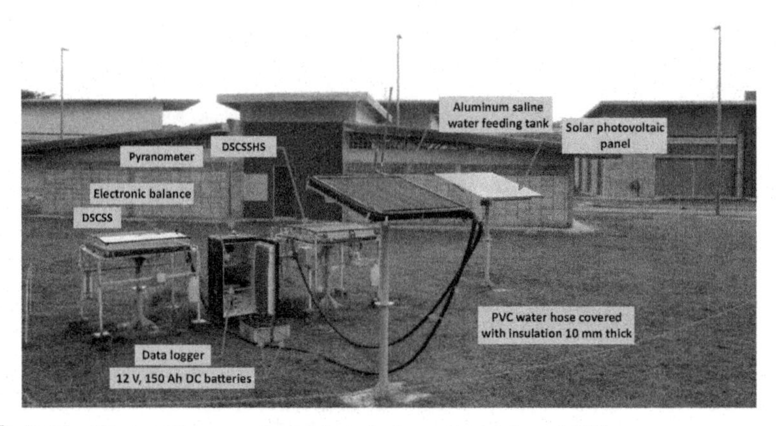

FIG. 8.16 Photo of the experimental study investigated in ref. [90].

Integrated PV-powered rotating bed

One of the novel designs is to drive a vertical rotating wick inside a still to act as an additional evaporator surface beside the basin. Haddad et al. [91] integrated a rotating wick belt against the rear wall of a single-slope solar still with the basin area of $0.36\,m^2$. In order to provide rotating motion, a DC motor, usually used as an auto window lifter, was used and controlled with a TE233 controller for variable speed adjustment; it was assumed that the required electric power was generated by a 25 W solar module. A motor drive and belt mechanism were implemented to rotate the wick belt through a small slit in the rear glass, where polyethylene foam was used to prevent vapor leakage (Fig. 8.17A). The results indicated that the design enabled an extra area for both absorber and

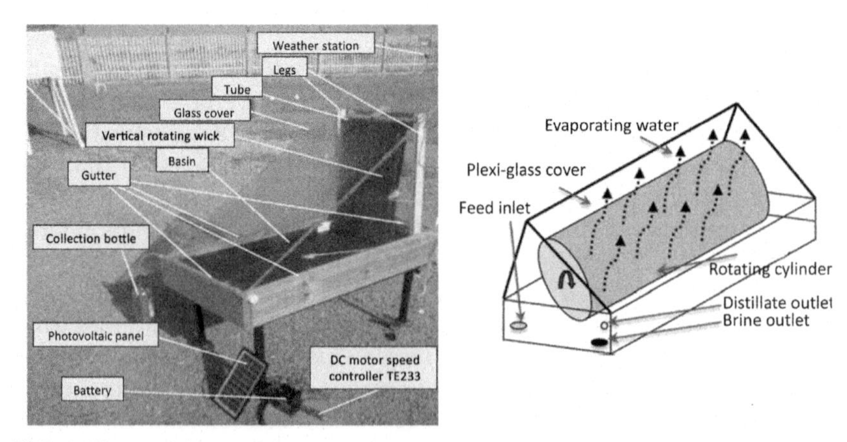

FIG. 8.17 (A) Photo of a solar still with PV-powered rotating wick [91], and (B) schematic of a solar still with a PV-powered rotating bed [92].

evaporator, which leads to higher vapor generation and consequently more productivity. The daily productivity was also improved by 14.72% and 51.1% for the summer and winter seasons, respectively.

In a design that was proposed by Malaeb et al. [93], a PV-powered DC motor was coupled with a rotating drum supported by bearings. The aluminum drum was coated black to maximize heat absorption from solar irradiance and was wrapped as a lightweight hollow cylinder with open ends. During operation, the drum rotated a thin water film formed around the drum circumference, both on the inner and outer faces. The formed water film was quickly evaporated by the high temperature of the drum and this increased the evaporation rate (Fig. 8.17B).

8.5.3.2 PV-based thermal power generation

Integrated PV-powered electric heater

An HDH process powered by a solar PV panel was proposed by Wang et al. [86]. In this design, a set of PV arrays was used to drive a pump for water displacement between the tank, heat exchanger, and humidifier. The PV array also powered an electric heater positioned inside the humidifier and a fan for forced convection between the humidifier and dehumidifier. It was premised that saline water was first preheated in the heat exchanger by absorbing the heat from water vapor. As it reached the humidifier, the heater increased its temperature to the desired evaporation temperature. In the humidifier, hot water and cool air exchanged heat, and humidified hot air was introduced to the dehumidifier through a fan-induced forced convection. During the dehumidification process, air exchanged its latent heat with the cold surface of the heat exchanger and produced freshwater.

As Fig. 8.18 shows, a 3500 W auxiliary heater was fixed at the bottom of the humidifier adjacent to a temperature sensor that monitored the water temperature. A fan with 5.9 m/s wind speed was fixed in the channel between the humidifier and dehumidifier to induce forced convention. Sponge heat insulation was used for wrapping the HDH unit to prevent any heat losses. Results indicated that 0.873 kg/m^2 day of freshwater was yielded at 64.3°C of evaporation temperature as the maximum productivity, which demonstrated the yield augmentation after forced convection was applied.

In an experimental study, Elbar and Hassan [94] incorporated a PV module with a modified solar still integrated with paraffin wax as the PCM. In their system, a 200 W PV-powered heater was fixed at the bottom of the basin to provide auxiliary heat. A 155 W PV module with $A_{PV} = 1\,m^2$ was installed above the condenser area to play the role of solar reflector for the still as well as a power generation unit (Fig. 8.19). The main purpose of using PCM was to store heat losses from the absorber beside the excessive heat produced by the electric heater and reuse them at night. Results demonstrated that the basin temperature increased with a new integration

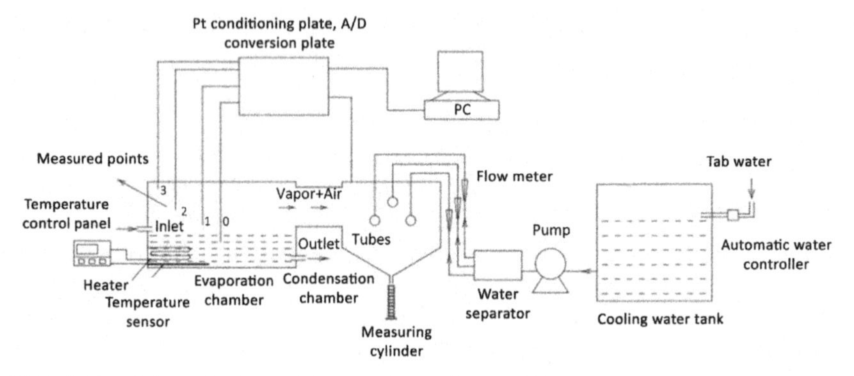

FIG. 8.18 Schematic view of the experimental setup [86].

setting and this led to higher production. Also, the thermal analysis showed that PV integration as the reflector can improve productivity by 3.5%, PV integration as the power supplier of the heater increases productivity by 9%, and the PCM incorporation with the PV module culminates in a 19.4% increase in the daily distillate.

8.5.4 PVT-integrated solar distillation systems

In the previous section, PV-integrated solar distillation systems were discussed and it was demonstrated that PV technology is beneficial and would result in the economic and technical augmentation of current

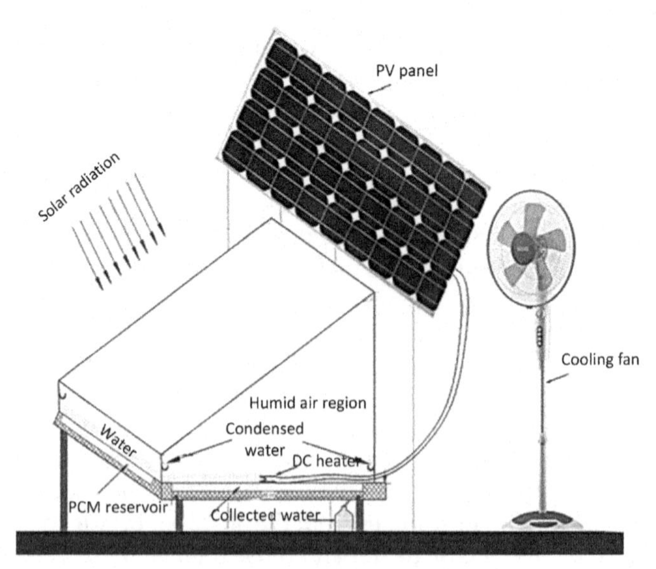

FIG. 8.19 A solar still incorporated with a PV panel to modify water distillation [94].

technologies. In this section, the incorporation of hybrid PVT collectors with different solar distillation processes will be presented and discussed.

8.5.4.1 Integrated flat-plate PVT solar still

Singh et al. [95] proposed a system consisting of a conventional double slope solar still and a flat-plate hybrid PVT collector to improve the productivity of the desalination system, as shown in Fig. 8.20.

In this work, two flat-plate collectors were used with a switchable connection to test different operational modes. A 40 W solar panel was also mounted on the absorber plate of one of the collectors in the way that converted energy from the back of the module was utilized for water heating. A 40 W DC motor powered by a solar PV panel was coupled to the water

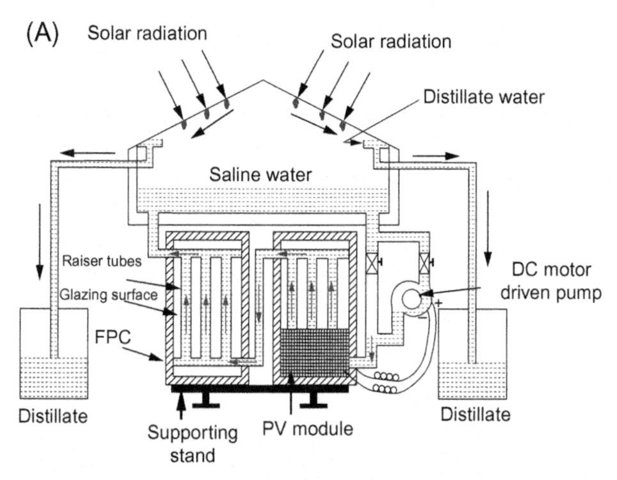

FIG. 8.20 (A) Schematic view, and (B) Photo of a solar still integrated with a flat-plate PVT collector [95].

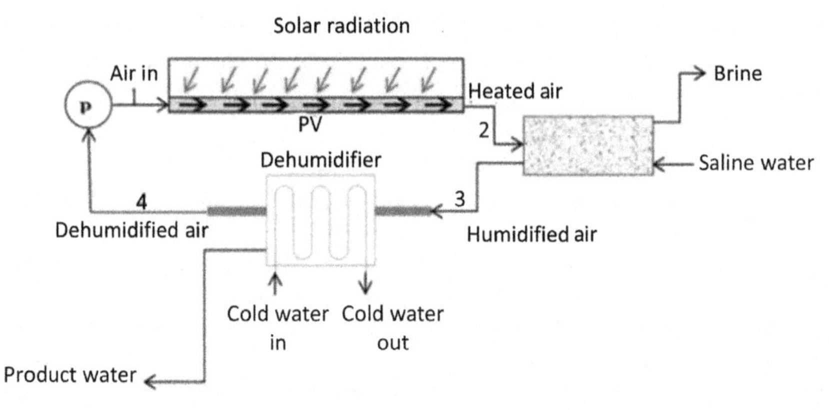

FIG. 8.21 Schematic of an HDH process powered by an air-type flat-plate PVT collector [96].

pump to circulate water from the first collector to the basin and through the second collector. Results indicated that the production rate would be accelerated 1.4 times more than a conventional solar still where the daily efficiency of the still reached 17.4%. In a study presented by Giwa et al. [96], the HDH distillation was powered by an air-type flat-plate PVT collector, and governing equations were developed. In their work, it was assumed that the generated electric power by the PV module was used to operate the electric components of the system, including the fan, pump, and inverter.

As can be seen in Fig. 8.21, a closed air cycle was considered in the system where the air is fed to the insulated air gap at the bottom layer of the PVT module using a fan. As the air passes through the channel, it cools down the PV module and becomes heated before the interface with sprayed saline water. In the humidifier, the air becomes saturated, and as the outlet air passes through the dehumidifier, the cooling pipes dehumidify saturated air and produce fresh water.

8.5.4.2 *Integrated concentrating PVT (CPVT) solar stills*

Gupta et al. [97] developed a single-slope solar still integrated with N identical fully covered CPC-PVT collectors. They arranged each collector to be connected to each other in a series configuration to provide higher temperatures in the basin (Fig. 8.22). In the proposed setup, the saltwater absorbs heat by passing through the CPC collectors, which increases the evaporation rate inside the still. Then, the vapor is condensed on the inclined glass cover while the remaining water is recirculated by a pump powered by the PV modules. The circulating water increases the rate of evaporation and therefore the system productivity.

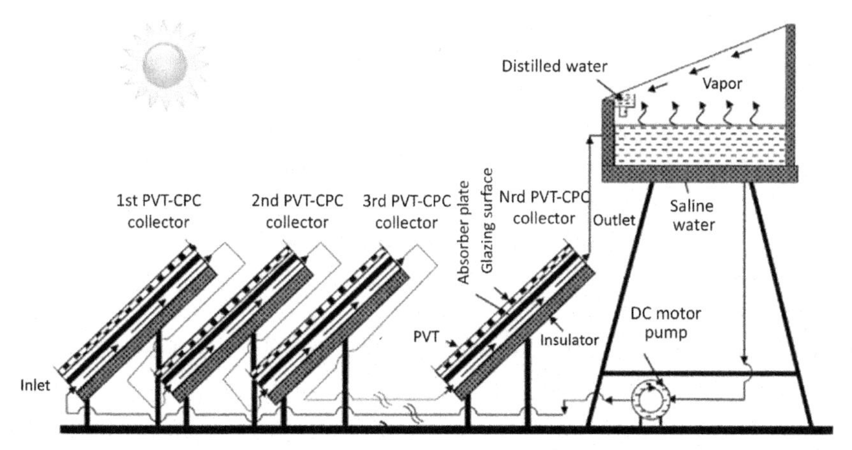

FIG. 8.22 Schematic diagram of N-identical CPC-PV/T collectors coupled to a solar still [97].

Kumar et al. [98] developed a hybrid PVT system to augment the productivity of the active solar still. Their study aimed to enhance the electrical efficiency by cooling down the PV module and simultaneously preheating the seawater before feeding to the basin area. A solar-powered NiCr spiral wire was also used and placed inside the basin to raise the water temperature and enhance the evaporation rate (Fig. 8.23). The novelty of their work is dedicated to the free flow of water using spreading pipes over the PV panel for both cooling and cleaning effects on the module.

The PV system used in this work consisted of a 0.65 m^2 PV panel, a battery, and a battery charge controller that supplied electric components

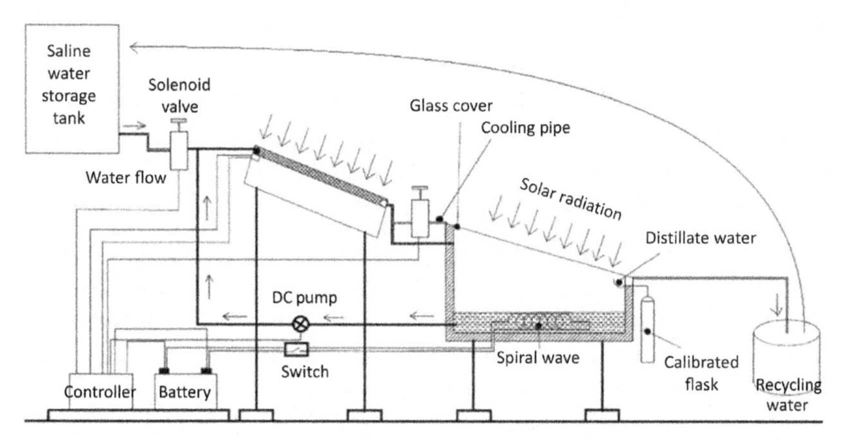

FIG. 8.23 Schematic view of a solar distillation unit incorporated with a cooling effect on the solar PV module [98].

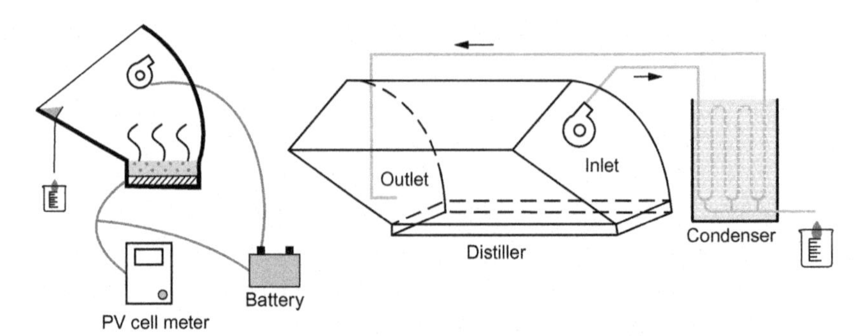

FIG. 8.24 Schematic view of the novel concentrating distiller proposed in ref. [99].

such as the pump, the heating wire, and two solenoid valves: one for a periodic cooling effect on the glass cover for condensation enhancement and one for water flow over the PV panel. Results demonstrated that the daily yield is six times higher than conventional passive stills.

An et al. [99] used a nanofluid-based spectral splitting PVT method to power a solar still with an external condenser. As shown in Fig. 8.24, the premise was to utilize the rejected heat from a PV panel immersed in the basin using gold and silver nanoparticles with selective spectral absorption. The PV module was used to operate a 3.2 W centrifugal fan to convey the steam to an external condenser for a higher distillation rate. The PV system consisted of 12 mono-Si cells, which were connected in a series arrangement. In addition, a concentrating distiller was designed with a glass cover at the aperture area while reflective sheets were placed inside the concentrator to focus the solar radiation at different incident angles. Results demonstrated that the implementation of nanoparticles enhances water productivity up to 79.9% with the major increase in the collector thermal efficiency.

8.6 PV-powered desalination plants installed around the world

In many areas around the world, the desalination of sea and brackish water is a common solution to alleviate increasing global freshwater shortage.

In 1986, a small unit of a PV-ED desalination plant was installed in Tanote, Rajasthan, India. This plant consisted of a PV system with a 450 W_p capacity including 42 cell pairs while the ED system was constructed from three stages with a produced freshwater capacity of 1 m^3/day with a TDS of 5000 ppm. The energy consumption of the system was calculated as 1 kWh/kg of the removed salt [100].

Another solar desalination plant was installed in Spencer Valley, New Mexico, United States. This unit included two tracking PV arrays (1000 Wp, 120 V) to drive pumps, and three fixed PV arrays with 2.3 kW and 50 V to power the ED system. The energy consumption and cost of the system were about 0.82 kWh/m^3 and 16 US$/$m^3$, respectively [62].

In 1996, a PV-RO desalination plant with a capacity of 1.2 m^3/day was installed in Gillen Bore, Australia, by SunTec Ltd.[d] The required electric power of this plant was provided by eight solar PV panels (520 W_p), which could be stored in batteries to be utilized when required [101].

In 2007, a PV-RO desalination plant within the Spanish-Tunisian cooperation framework was installed in the village of Ksar Ghilène, located in the Sahara Desert in the south of Tunisia. This plant can produce 15 m^3/day of freshwater from the brackish water wells located in the nearby oases. It is an autonomous solar-powered unit integrated with a PV system with a power capacity of 10.5 kWh, including batteries. The produced freshwater is distributed in the village through five public fountains [102].

In 2013, the renewable energy company *Conergy* installed the second-largest solar PV power plant in Tunisia with a capacity of 210 kW to drive a groundwater RO desalination plant in the Ben Guardane arid regions of the Medenine province in southeast Tunisia. The PV system covers 70% of the desalination plant's energy requirements. The desalination plant with a daily capacity of 1.8 million liters of freshwater can supply the water demand of the local population [103].

In 2017, another desalination plant sponsored by Mascara and Masdar was confirmed with the water utility Sonede in Tunisia. It is a hybrid project because the RO desalination plant is driven by solar panels during the daytime, but reverts to grid power during the night. This plant can produce 1000 m^3 of freshwater per day [104].

In 2015, a solar PV-powered desalination plant was officially inaugurated during UAE innovation week in Ghantoot, Abu Dhabi. The project was owned by Masdar, Abu Dhabi's company that helps in developing renewable energy technologies in the United Arab Emirates and the Gulf region. Also, the commercial partners were *Abengoa, Suez, Veolia, Trevi Systems*, and *Mascara Renewable Water*. Therefore, the plant was composed of five novel pilot projects with the latest using electricity produced by an off-grid solar PV system with a power capacity of 30 kW [105].

The *Elemental Water Makers Organization* is a leading company providing desalination plants powered by renewable energies. The company was founded in 2012 with clusters in more than seven countries involved in solar desalination projects for communities, private properties, municipalities, and industries [106]. In 2017, the company achieved the Mohammed bin Rashid Al Maktoum Global Water Award, selected as the first prize winner among 138 international companies that are active in solar

[d]Solar Energy Research Institute of Western Australia.

TABLE 8.2 Several projects in progress by the *Elemental Water Makers Organization.*

Country	Start date	Production (m^3/day)
Indonesia	December 2011	3
British Virgin Islands	November 2014	12.5
Mozambique	July 2016	9
Belize	December 2016	5
Spain (Canary Islands)	November 2017	9
Philippines	November 2017	11
Cape Verde	June 2018	50

desalination. Several projects are in progress by this company, and they are presented in Table 8.2.

In 2017, a PV-RO desalination plant with a produced freshwater capacity of 80 m^3/day came into operation in Bora Bora, an island in French Polynesia. In this project, the *Osmosun* technology, which is a seawater desalination solution based on the integration of PV technology with RO desalination without using batteries, was used. As a result, access to drinking water at lower costs would be possible [107]. Therefore, this project is completely designed to be off-grid to evaluate system reliability in very remote, challenging conditions [104, 108, 109].

In 2018, a PV-RO desalination plant was inaugurated in Caverne Bouteille in Rodrigues, an island located more than 580 km east of Mauritius. Although this region is surrounded by seawater, the residents were forced to travel long distances to access potable water. For this reason, a PV-RO desalination unit was built in this region by Quadran, which is a French start-up. This plant can supply 80 m^3 of water per day for 2400 people living in Rodrigues. However, the company claimed that the plant could produce 400 m^3 of freshwater by running in hybrid energy mode. The European Union, the Indian Ocean Commission (IOC), and the French Global Environment Facility funded this project, assisting this young company with 230 employees [107].

In 2019, a PV-RO desalination plant was officially launched in Witsand on the southern Cape coast. It produced freshwater with a cost of only $0.52–0.60 for 1000 L, which is less than a quarter of the water cost from the temporary desalination plant of Cape Town at Strandfontein, with a cost of $2.62–3 for every 1000 L [110, 111]. The lower costs are because of two reasons. One is that the diesel fuel used to drive the desalination plant is expensive while solar radiation is free, and the other is that the Strandfontein desalination plant was temporary, under a City of Cape Town contract for only two years. This scale economy made the produced freshwater expensive under this contract [110].

TABLE 8.3 Selected PV-RO desalination plants currently installed around the world.

Location of installation	Feedwater (ppm)	Capacity (m^3/day)	Online year
Jeddah, S. Arabia	42,800	3.2	1981
Lampedusa, Italy	SW	40	1990
Concepción del Oro, Mexico	3000	–	1993
Saint Lucie, Florida	32,000	0.64	1995
ITC Canaries Island, Spain	SW	3	1998
Lisbon, INETI, PRT	2549	0.02	2000
Cress, Laviro, Greece	36,000	<1	2001
Massawa, ERI	40,000	3.9	2002
Hassi-Kheba, Algeria	3200	<1	2003
Aqaba, JOR	4000	58	2005
Baja California Sur, MEX	4000	11.5	2005
Athens, GRC	30,000	0.35	2006
Abu Dhabi, UAE	45,000	20	2008
United States	BW	75	2012
Tunisia	BW	1800	2013
Qatar	SW	12,000	2013
Vanuatu	SW	96	2013
Vanuatu	SW	96	2013
Mexico	BW	840	2014
Mexico	BW	48	2014
Brazil	BW	3600	2014

Data from Dawoud MA. Economic feasibility of small scale solar powered RO desalination for brackish/saline groundwater in Arid regions. Int J Water Resour Arid Environ 2017;6:103–14, Ahmed FE, Hashaikeh R, Hilal N. Solar powered desalination—technology, energy and future outlook. Desalination 2019;453:54–76. doi:10.1016/j.desal.2018.12.002.[112, 113].

The *Osmosun* technology used in the Witsand desalination plant was developed by the French company of *Mascara Renewable Water* and implemented by *TWS-Turnkey Water Solutions* in South Africa. This new technology involves a specialized "intelligent" membrane that is able to continue delivering reverse osmosis, even when solar radiation is reduced. The "intelligent" membrane can "soften" the variability in the delivered solar energy, preserving the RO membranes [110]. Table 8.3 summarizes the PV-RO desalination plants installed around the world.

8.7 Challenges and prospects

Solar energy-driven desalination technology has significantly grown with steep progress in the last few years. Increasing the global desalination capacity and the urgent need for mitigating the adverse impacts of burning fossil fuels have resulted in huge efforts to drive desalination plants with renewable power sources. Using solar energy to power desalination plants is a practical way to produce freshwater in several locations around the globe. Solar desalination is especially a feasible option in remote regions where there is either freshwater scarcity and high potential of solar energy under a situation that there is no or too expensive accessibility to the power grid.

Solar PV-powered desalination systems are those working based on mechanical processes that are driven by electricity generated from the sun. Therefore, they are quite suitable to be integrated with desalination processes that entirely or partially consume electricity such as RO, ED, MD, and even solar distillation systems.

In general, a large part of the overall cost of PV-powered desalination plants is allocated to PV systems. For this reason, the selection of the proper-sized PV system is crucial. The installation site is one of the most important parameters affecting the final costs of the produced freshwater. A specified solar PV-powered desalination system will produce higher electricity and, consequently, higher amounts of water at the locations with higher availability of solar radiation. Fig. 8.25 shows the worldwide share of different solar-powered desalination technologies. As can be seen in this figure, a large share of the global solar-powered desalination capacity is allocated to PV-RO technology.

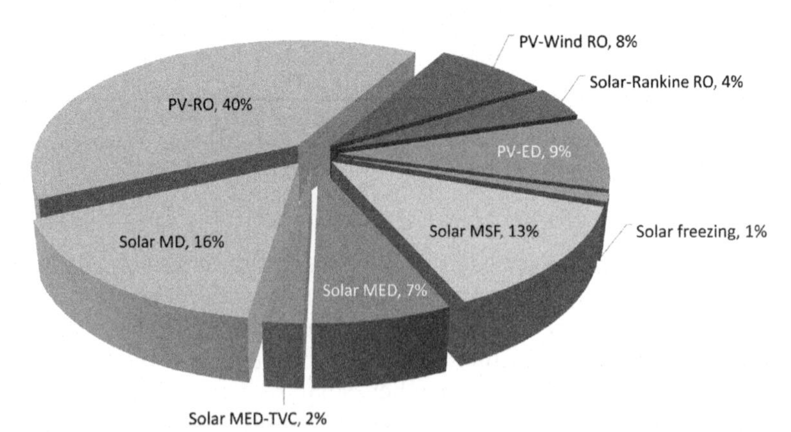

FIG. 8.25 Desalination technologies powered by solar energy installed worldwide [28].

The wide application of PV-RO desalination systems remains restricted on both the large and small scale due to high energy costs. In this case, the following solutions are suggested to improve the energy efficiency of these systems [114]:

- Use an algorithm that can evaluate the performance of the desalination plant according to the installation location and the amount of available solar radiation;
- Determine a standard variation range for some parameters, including feed pressure, temperature, water concentration, and capacity;
- Conduct more experiments to optimize the performance of the plants to make them more efficient and cost-effective.

Integrated PV systems: For integrated PV systems, the intermittency of solar radiation can be addressed by using more cost-effective energy storage units or batteries. This can assist in making solar energy compatible with conventional energy sources. Energy storage allows continuous operation regardless of insufficient solar radiation, which leads to water productivity increments [115, 116]. It should be noted that using batteries would significantly increase operating costs in spite of addressing insufficient solar radiation. However, several factors should be considered to size the battery charging system such as the storage capacity of the batteries, the desalination load, the PV rating requirements, the daily power output, and the estimated size of the PV array [117].

RO desalination unit: For RO desalination systems, energy efficiency can be improved by employing proper membrane configurations and using energy recovery devices (ERDs). The ERDs recover the pressure energy to pressurize the feed carried by the brine flow. This leads to a drastic decrease in the energy demand of the PV-RO systems. However, the use of ERDs is far from its potential for use in PV-RO systems because further improvements are still required [118, 119].

Pressure retarded osmosis (PRO) is another solution in which the water passes from a low- to a high-salinity solution through a membrane. As a result, the generated power by a hydroturbine claims the energy of the pressurized permeate. But high-performance and affordable PRO membranes are not available. Therefore, it can be concluded that before considering PRO as a viable energy recovery source, more research and development are required [120, 121].

PV-ED desalination technology is one of the most promising methods for the desalination of brackish water. Currently, the main barriers for the deployment of PV-ED systems are matching the intermittent output of solar energy with the water demand, the membrane lifetime, and the solar panel efficiency. In the case of solar-powered MD, most studies are allocated to modeling and simulation. Therefore, there is a requirement to close the gap between the theoretical studies and experimental works to progress the solar MD technology toward commercialization. However, using solar energy to supply the required power for the desalination processes has great potential for further performance improvements, energy savings, and cost reductions.

Acknowledgment

The authors would like to gratefully acknowledge the advisory support of the *Renewable Energy Research Institute*, an affiliation of Tarbiat Modares University (TMU), and the financial support [grant number IG/39705] of the *Renewable Energy Research Group* of the university.

References

[1] The United Nations. Water for people water for life. https://doi.org/10.1017/CBO9781107415324.004.

[2] Shiklomanov IA. World water resources. A new appraisal and assessment for the 21st century; 1998.

[3] Falkenmark M, Widstrand C. Population and water resources : a delicate balance. Vol. 47. Washington, DC, USA: Population Reference Bureau; 1992.

[4] Gorjian S, Ghobadian B. Solar desalination: a sustainable solution to water crisis in Iran. Renew Sustain Energy Rev 2015;48:571–84. https://doi.org/10.1016/j.rser.2015.04.009.

[5] Gorjian S, Ghobadian B, Tavakkoli Hashjin T, Banakar A. Experimental performance evaluation of a stand-alone point-focus parabolic solar still. Desalination 2014;352:1–17. https://doi.org/10.1016/j.desal.2014.08.005.

[6] Hanjra MA, Qureshi ME. Global water crisis and future food security in an era of climate change. Food Policy 2010;35:365–77. https://doi.org/10.1016/j.foodpol.2010.05.006.

[7] Nations U. Water; 2016.

[8] McEvoy J. Desalination and water security: the promise and perils of a technological fix to the water crisis in Baja California Sur, Mexico. Water Altern 2014;7:518–41.

[9] Kaur V, Mahajan R. Water crisis: towards a way to improve the situation. Int J Eng Technol Sci Res 2016;3. IJETSR WwwIjetsrCom ISSN 2394–3386.

[10] Rommerskirchen A, Gendel Y, Wessling M. Single module flow-electrode capacitive deionization for continuous water desalination. Electrochem Commun 2015;60:34–7. https://doi.org/10.1016/j.elecom.2015.07.018.

[11] Wilder MO, Aguilar-Barajas I, Pineda-Pablos N, Varady RG, Megdal SB, McEvoy J, et al. Desalination and water security in the US–Mexico border region: assessing the social, environmental and political impacts. Water Int 2016;8060:1–20. https://doi.org/10.1080/02508060.2016.1166416.

[12] Elimelech M, Phillip WA, Service RF, Shannon MA, Schiermeier Q, Fritzmann C, et al. The future of seawater desalination: energy, technology, and the environment. Science (80-) 2011;333:712–7. https://doi.org/10.1126/science.1200488.

[13] Tsiourtis NX. Desalination and the environment. Desalination 2001;141:223–36. https://doi.org/10.1016/S0011-9164(01)85001-3.

[14] Porter MG, Downie D, Scarborough H, Sahin O, Stewart RA. Drought and desalination: Melbourne water supply and development choices in the twenty-first century. Desalin Water Treat 2014;3994:1–18. https://doi.org/10.1080/19443994.2014.959743.

[15] Water Desalination Using Renewable Energy; 2012.

[16] Ghaffour N, Missimer TM, Amy GL. Combined desalination, water reuse, and aquifer storage and recovery to meet water supply demands in the GCC/MENA region. Desalin Water Treat 2013;51:38–43. https://doi.org/10.1080/19443994.2012.700034.

[17] Zotalis K, Dialynas E, Mamassis N, Angelakis A. Desalination technologies: Hellenic experience. Water 2014;6:1134–50. https://doi.org/10.3390/w6051134.
[18] Yu T-H, Shiu H-Y, Lee M, Chiueh P-T, Hou C-H. Life cycle assessment of environmental impacts and energy demand for capacitive deionization technology. Desalination 2016;399:53–60. https://doi.org/10.1016/j.desal.2016.08.007.
[19] Virgili F, Pankratz T, Gasson J. IDA desalination yearbook 2015–2016. Media Analytics Limited; 2016.
[20] Ziolkowska JR. Is desalination affordable? Regional cost and price analysis. Water Resour Manag 2015;29:1385–97. https://doi.org/10.1007/s11269-014-0901-y.
[21] This stock aims for big gains turning seawater into fresh water; n.d. https://www.tsinetwork.ca/daily-advice/wealth-management/stock-aims-big-gains-turning-seawater-fresh-water/ [Accessed 22 September 2019].
[22] Miller JE. Review of water resources and desalination techniques. Sand Rep 2003;1–54. doi: SAND 2003-0800.
[23] Bombar G, Dölgen D, Alpaslan MN. Environmental impacts and impact mitigation plans for desalination facilities. Desalin Water Treat 2015;3994:1–12. https://doi.org/10.1080/19443994.2015.1089198.
[24] Burn S, Hoang M, Zarzo D, Olewniak F, Campos E, Bolto B, et al. Desalination techniques—a review of the opportunities for desalination in agriculture. Desalination 2015;364:2–16. https://doi.org/10.1016/j.desal.2015.01.041.
[25] Subramani A, Badruzzaman M, Oppenheimer J, Jacangelo JG. Energy minimization strategies and renewable energy utilization for desalination: a review. Water Res 2011;45:1907–20. https://doi.org/10.1016/j.watres.2010.12.032.
[26] Elimelech M, Phillip WA. The future of seawater desalination: energy, technology, and the environment. Science (80-) 2011;333:712–7.
[27] Li C, Goswami Y, Stefanakos E. Solar assisted sea water desalination: a review. Renew Sustain Energy Rev 2013;19:136–63. https://doi.org/10.1016/j.rser.2012.04.059.
[28] Shatat M, Worall M, Riffat S. Opportunities for solar water desalination worldwide: review. Sustain Cities Soc 2013;9:67–80. https://doi.org/10.1016/j.scs.2013.03.004.
[29] Mezher T, Fath H, Abbas Z, Khaled A. Techno-economic assessment and environmental impacts of desalination technologies. Desalination 2011;266:263–73. https://doi.org/10.1016/j.desal.2010.08.035.
[30] Deng R, Xie L, Lin H, Liu J, Han W. Integration of thermal energy and seawater desalination. Energy 2010;35:4368–74. https://doi.org/10.1016/j.energy.2009.05.025.
[31] WaterDisaltingCommittee. Desalination of seawater. New York: American Water Works Association (AWWA); 2011.
[32] Goosen MFA, Mahmoudi H, Ghaffour N, Bundschuh J, Al Yousef Y. A critical evaluation of renewable energy technologies for desalination. Appl Mater Sci Environ Mater, World Scientific; 2016, p. 233–58. https://doi.org/10.1142/9789813141124_0032.
[33] Ghaffour N, Bundschuh J, Mahmoudi H, Goosen MFA. Renewable energy-driven desalination technologies: a comprehensive review on challenges and potential applications of integrated systems. Desalination 2015;356:94–114. https://doi.org/10.1016/j.desal.2014.10.024.
[34] Ghaffour N, Lattemann S, Missimer T, Ng KC, Sinha S, Amy G. Renewable energy-driven innovative energy-efficient desalination technologies. Appl Energy 2014;136: 1155–65. https://doi.org/10.1016/j.apenergy.2014.03.033.
[35] Hassanien RHE, Li M, Dong LW. Advanced applications of solar energy in agricultural greenhouses. Renew Sustain Energy Rev 2016;54:989–1001. https://doi.org/10.1016/j.rser.2015.10.095.
[36] Baljit SSS, Chan H-Y, Sopian K. Review of building integrated applications of photovoltaic and solar thermal systems. J Clean Prod 2016;137:677–89. https://doi.org/10.1016/j.jclepro.2016.07.150.

[37] Kalogirou S. Solar energy engineering: processes and systems. Academic Press; 2009.
[38] Vishwanath Kumar P, Kumar A, Prakash O, Kaviti AK. Solar stills system design: a review. Renew Sustain Energy Rev 2015;51:153–81. https://doi.org/10.1016/j.rser.2015.04.103.
[39] Gorjian S, Hashjin TT, Ghobadian B. Seawater desalination using solar thermal technologies: state of the art. In: 10th Int. Conf. Sustain. Energy Technol., Istanbul, Turkey; 2011. p. 4–7.
[40] Qiblawey HHM, Banat F. Solar thermal desalination technologies. Desalination 2008;220:633–44. https://doi.org/10.1016/j.desal.2007.01.059.
[41] Gorjian S, Zadeh BN, Eltrop L, Shamshiri RR, Amanlou Y. Solar photovoltaic power generation in Iran: development, policies, and barriers. Renew Sustain Energy Rev 2019;106:110–23. https://doi.org/10.1016/j.rser.2019.02.025.
[42] Al-Karaghouli A, Kazmerski LL. Energy consumption and water production cost of conventional and renewable-energy-powered desalination processes. Renew Sustain Energy Rev 2013;24:343–56. https://doi.org/10.1016/j.rser.2012.12.064.
[43] Cipollina A, Tzen E, Subiela V, Papapetrou M, Koschikowski J, Schwantes R. Renewable energy desalination : performance analysis and operating data of existing RES desalination plants. Desalin Water Treat 2015;55:3126–46. https://doi.org/10.1080/19443994.2014.959734.
[44] Helal AM, Al-Malek SA, Al-Katheeri ES. Economic feasibility of alternative designs of a PV-RO desalination unit for remote areas in the United Arab Emirates. Desalination 2008;221:1–16.
[45] De Carvalho PCM, Pontes RST, Oliveira DS, Riffel DB, De Oliveira RGV, Mesquita SB. Control method of a photovoltaic powered reverse osmosis plant without batteries based on maximum power point tracking. In: Transm Distrib Conf Expo Lat Am 2004 IEEE/PES, IEEE; 2004. p. 137–42.
[46] Mohamed ES, Papadakis G, Mathioulakis E, Belessiotis V. A direct coupled photovoltaic seawater reverse osmosis desalination system toward battery based systems—a technical and economical experimental comparative study. Desalination 2008;221:17–22.
[47] Al-Karaghouli A, Renne D, Kazmerski LL. Technical and economic assessment of photovoltaic-driven desalination systems. Renew Energy 2010;35:323–8. https://doi.org/10.1016/j.renene.2009.05.018.
[48] Ortiz JM, Expósito E, Gallud F, García-García V, Montiel V, Aldaz A. Desalination of underground brackish waters using an electrodialysis system powered directly by photovoltaic energy. Sol Energy Mater Sol Cells 2008;92:1677–88. https://doi.org/10.1016/j.solmat.2008.07.020.
[49] Al-Karaghouli AA, Kazmerski LL. Renewable energy opportunities in water desalination. In: Desalination, Trends Technol. InTech; 2011.
[50] Ortiz JM, Expósito E, Gallud F, García-García V, Montiel V, Aldaz A. Electrodialysis of brackish water powered by photovoltaic energy without batteries: direct connection behaviour. Desalination 2007;208:89–100.
[51] Singh GK. Solar power generation by PV (photovoltaic) technology: a review. Energy 2013;53:1–13. https://doi.org/10.1016/j.energy.2013.02.057.
[52] Http://Www.Iea.Org. Technology roadmap: solar photovoltaic energy; 2014.
[53] Solanki CS. Solar photovoltaic technology and systems: a manual for technicians; 2013.
[54] Richards BS, Schäfer AI. Design considerations for a solar-powered desalination system for remote communities in Australia. Desalination 2002;144:193–9. https://doi.org/10.1016/S0011-9164(02)00311-9.
[55] Avlonitis SA, Kouroumbas K, Vlachakis N. Energy consumption and membrane replacement cost for seawater RO desalination plants. Desalination 2003;157:151–8. https://doi.org/10.1016/S0011-9164(03)00395-3.

[56] Garg MC, Joshi H. A review on PV-RO process: solution to drinking water scarcity due to high salinity in non-electrified rural areas. Sep Sci Technol 2015;50:1270–83. https://doi.org/10.1080/01496395.2014.951725.

[57] Thomson M, Gwillim J, Rowbottom A, Draisey I, Miranda M. Batteryless photovoltaic reverse osmosis desalination system, S/P2/00305/REP, ETSU, DT1, UK n.d.

[58] Wu B, Maleki A, Pourfayaz F, Rosen MA. Optimal design of stand-alone reverse osmosis desalination driven by a photovoltaic and diesel generator hybrid system. Sol Energy 2018;163:91–103. https://doi.org/10.1016/j.solener.2018.01.016.

[59] Ahmad N, Sheikh AK, Gandhidasan P, Elshafie M. Modeling, simulation and performance evaluation of a community scale PVRO water desalination system operated by fixed and tracking PV panels: a case study for Dhahran city, Saudi Arabia. Renew Energy 2015;75:433–47. https://doi.org/10.1016/j.renene.2014.10.023.

[60] Alghoul MA, Poovanaesvaran P, Mohammed MH, Fadhil AM, Muftah AF, Alkilani MM, et al. Design and experimental performance of brackish water reverse osmosis desalination unit powered by 2 kW photovoltaic system. Renew Energy 2016;93:101–14. https://doi.org/10.1016/j.renene.2016.02.015.

[61] Gökçek M. Integration of hybrid power (wind-photovoltaic-diesel-battery) and seawater reverse osmosis systems for small-scale desalination applications. Desalination 2017;1–11. https://doi.org/10.1016/j.desal.2017.07.006.

[62] Kuroda O, Takahashi S, Kubota S, Kikuchi K, Eguchi Y, Ikenaga Y, et al. An electrodialysis sea water desalination system powered by photovoltaic cells. Desalination 1987;67:33–41. https://doi.org/10.1016/0011-9164(87)90229-3.

[63] Ortiz JM, Expósito E, Gallud F, García-García V, Montiel V, Aldaz A. Photovoltaic electrodialysis system for brackish water desalination: modeling of global process. J Membr Sci 2006;274:138–49. https://doi.org/10.1016/j.memsci.2005.08.006.

[64] Fernandez-gonzalez C, Ibañez R, Irabien A. Sustainability assessment of electrodialysis powered by photovoltaic solar energy for freshwater production. Renew Sustain Energy Rev 2015;47:604–15. https://doi.org/10.1016/j.rser.2015.03.018.

[65] Zhang Y, Pinoy L, Meesschaert B, Van Der Bruggen B. A natural driven membrane process for brackish and wastewater treatment: photovoltaic powered ED and FO hybrid system. Environ Sci Technol 2013;47:10548–55. https://doi.org/10.1021/es402534m.

[66] Peñate B, Círez F, Domínguez FJ, Subiela VJ, Vera L. Design and testing of an isolated commercial EDR plant driven by solar photovoltaic energy. Desalin Water Treat 2013;51:1254–64. https://doi.org/10.1080/19443994.2012.715071.

[67] Vyas H, Suthar K, Chauhan M, Jani R, Bapat P, Patel P, et al. Modus operandi for maximizing energy efficiency and increasing permeate flux of community scale solar powered reverse osmosis systems. Energ Conver Manage 2015;103:94–103. https://doi.org/10.1016/j.enconman.2015.05.076.

[68] Richards BS, Capão DPS, Früh WG, Schäfer AI. Renewable energy powered membrane technology: impact of solar irradiance fluctuations on performance of a brackish water reverse osmosis system. Sep Purif Technol 2015;156:379–90. https://doi.org/10.1016/j.seppur.2015.10.025.

[69] Ramanujan D, Wright N, Truffaut S, von Medeazza G, Winter A. Sustainability analysis of PV-powered electrodialysis desalination for safe drinking water in the Gaza Strip. Int Desalin Assoc World Congr; 2017. IDA17WC-58045_Ramanujan.

[70] Mostafaeipour A, Qolipour M, Rezaei M, Babaee-Tirkolaee E. Investigation of off-grid photovoltaic systems for a reverse osmosis desalination system: a case study. Desalination 2018. https://doi.org/10.1016/j.desal.2018.03.007.

[71] Monnot M, Carvajal GDM, Laborie S, Cabassud C, Lebrun R. Integrated approach in eco-design strategy for small RO desalination plants powered by photovoltaic energy. Desalination 2018;435:246–58. https://doi.org/10.1016/j.desal.2017.05.015.

[72] Koschikowski J, Wieghaus M, Rommel M. Solar thermal driven desalination plants based on membrane distillation. Water Sci Technol Water Supply 2003;3:49–55. https://doi.org/10.1016/S0011-9164(03)00360-6.
[73] González D, Amigo J, Suárez F. Membrane distillation: perspectives for sustainable and improved desalination. Renew Sustain Energy Rev 2017;80:238–59. https://doi.org/10.1016/J.RSER.2017.05.078.
[74] Zhang Y, Sivakumar M, Yang S, Enever K, Ramezanianpour M. Application of solar energy in water treatment processes: a review. Desalination 2018;428:116–45. https://doi.org/10.1016/J.DESAL.2017.11.020.
[75] Hogan PA, Sudjito FAG, Morrison GL. Desalination by solar heated membrane distillation. Desalination 1991;81:81–90. https://doi.org/10.1016/0011-9164(91)85047-X.
[76] Thomas N, Mavukkandy MO, Loutatidou S, Arafat HA. Membrane distillation research & implementation: lessons from the past five decades. Sep Purif Technol 2017;189:108–27. https://doi.org/10.1016/J.SEPPUR.2017.07.069.
[77] Moore SE, Mirchandani SD, Karanikola V, Nenoff TM, Arnold RG, Eduardo SA. Process modeling for economic optimization of a solar driven sweeping gas membrane distillation desalination system. Desalination 2018;437:108–20. https://doi.org/10.1016/J.DESAL.2018.03.005.
[78] Hughes AJ, O'Donovan TS, Mallick TK. Experimental evaluation of a membrane distillation system for integration with concentrated photovoltaic/thermal (CPV/T) energy. Energy Procedia 2014;54:725–33. https://doi.org/10.1016/j.egypro.2014.07.313.
[79] Selvi SR, Baskaran R. Solar photovoltaic-powered membrane distillation as sustainable clean energy technology in desalination. Curr Sci 2015;109:1247. https://doi.org/10.18520/v109/i7/1247-1254.
[80] Sharshir SW, Yang N, Peng G, Kabeel AE. Factors affecting solar stills productivity and improvement techniques: a detailed review. Appl Therm Eng 2016;100:267–84. https://doi.org/10.1016/j.applthermaleng.2015.11.041.
[81] Tiwari GN, Sahota L. Review on the energy and economic ef fi ciencies of passive and active solar distillation systems. DES 2016;1. https://doi.org/10.1016/j.desal.2016.08.023.
[82] Hosseini A, Banakar A, Gorjian S. Development and performance evaluation of an active solar distillation system integrated with a vacuum-type heat exchanger. Desalination 2018. https://doi.org/10.1016/j.desal.2017.12.031.
[83] Jijakli K, Arafat H, Kennedy S, Mande P, Theeyattuparampil VV. How green solar desalination really is? Environmental assessment using life-cycle analysis (LCA) approach. Desalination 2012;287:123–31. https://doi.org/10.1016/j.desal.2011.09.038.
[84] Al-Smairan M. Application of photovoltaic array for pumping water as an alternative to diesel engines in Jordan Badia, Tall Hassan station: Case study. Renew Sustain Energy Rev 2012;16:4500–7. https://doi.org/10.1016/j.rser.2012.04.033.
[85] Rahbar N, Esfahani JA. Experimental study of a novel portable solar still by utilizing the heatpipe and thermoelectric module. Desalination 2012;284:55–61. https://doi.org/10.1016/j.desal.2011.08.036.
[86] Wang J-h, Gao N-Y, Deng Y, Li Y-l. Solar power-driven humidification-dehumidification (HDH) process for desalination of brackish water. Desalination 2012;305:17–23. https://doi.org/10.1016/j.desal.2012.08.008.
[87] Hamawand I, Lewis L, Ghaffour N, Bundschuh J. Desalination of salty water using vacuum spray dryer driven by solar energy. Desalination 2017;404:182–91. https://doi.org/10.1016/j.desal.2016.11.015.
[88] Kumar RA, Esakkimuthu G, Murugavel KK. Performance enhancement of a single basin single slope solar still using agitation effect and external condenser. Desalination 2016;399:198–202. https://doi.org/10.1016/j.desal.2016.09.006.

[89] Rajaseenivasan T, Prakash R, Vijayakumar K, Srithar K. Mathematical and experimental investigation on the influence of basin height variation and stirring of water by solar PV panels in solar still. Desalination 2017;415:67–75. https://doi.org/10.1016/j.desal.2017.04.010.
[90] Al-Sulttani AO, Ahsan A, Rahman A, Nik Daud NN, Idrus S. Heat transfer coefficients and yield analysis of a double-slope solar still hybrid with rubber scrapers: an experimental and theoretical study. Desalination 2017;407:61–74. https://doi.org/10.1016/j.desal.2016.12.017.
[91] Haddad Z, Chaker A, Rahmani A. Improving the basin type solar still performances using a vertical rotating wick. Desalination 2017;418:71–8. https://doi.org/10.1016/j.desal.2017.05.030.
[92] Ayoub GM, Malaeb L. Economic feasibility of a solar still desalination system with enhanced productivity. Desalination 2014;335:27–32. https://doi.org/10.1016/j.desal.2013.12.010.
[93] Malaeb L, Aboughali K, Ayoub GM. Modeling of a modified solar still system with enhanced productivity. Sol Energy 2016;125:360–72. https://doi.org/10.1016/j.solener.2015.12.025.
[94] Elbar ARA, Hassan H. Experimental investigation on the impact of thermal energy storage on the solar still performance coupled with PV module via new integration. Sol Energy 2019;584–93. https://doi.org/10.1016/j.solener.2019.04.042.
[95] Singh G, Kumar S, Tiwari GN. Design, fabrication and performance evaluation of a hybrid photovoltaic thermal (PVT) double slope active solar still. Desalination 2011;277:399–406. https://doi.org/10.1016/j.desal.2011.04.064.
[96] Giwa A, Fath H, Hasan SW. Humidification-dehumidification desalination process driven by photovoltaic thermal energy recovery (PV-HDH) for small-scale sustainable water and power production. Desalination 2016;377:163–71. https://doi.org/10.1016/j.desal.2015.09.018.
[97] Sagar V, Bandhu D, Mishra RK, Kumar S, Tiwari GN. Development of characteristic equations for PVT-CPC active solar distillation system. Desalination 2018;445:266–79. https://doi.org/10.1016/j.desal.2018.08.009.
[98] Praveen Kumar B, Prince Winston D, Pounraj P, Muthu Manokar A, Sathyamurthy R, Kabeel AE. Experimental investigation on hybrid PV/T active solar still with effective heating and cover cooling method. Desalination 2018;435:140–51. https://doi.org/10.1016/j.desal.2017.11.007.
[99] An W, Chen L, Liu T, Qin Y. Enhanced solar distillation by nano fluid-based spectral splitting PV/T technique: preliminary experiment. Sol Energy 2018;176:146–56. https://doi.org/10.1016/j.solener.2018.10.029.
[100] Adiga MR, Adhikary SK, Narayanan PK, Harkare WP, Gomkale SD, Govindan KP. Performance analysis of photovoltaic electrodialysis desalination plant at Tanote in Thar desert. Desalination 1987;67:59–66. https://doi.org/10.1016/0011-9164(87)90232-3.
[101] Harrison DG, Ho GE, Mathew K. Desalination using renewable energy in Australia. Renew Energy 1996;8:509–13. https://doi.org/10.1016/0960-1481(96)88909-9.
[102] Bourouni K, Chaibi M. Solar energy for application to desalination in Tunisia: description of a demonstration project. Renew Energy Middle East, Springer; 2009, p. 125–49.
[103] Conergy builds first solar power plant in Tunisia for desalination plant; n.d.
[104] Learning from Ghantoot—Pushing Desalination Systems to the Limit; n.d.
[105] Solar PV powers desalination plant in Abu Dhabi; n.d.
[106] Elemental Water Makers—Member of the World Alliance, n.d. https://solarimpulse.com/companies/elemental-water-makers [Accessed 19 October 2019].
[107] MAURITIUS: Solar-powered desalination plant inaugurated in Rodrigues; n.d.

[108] OSMOSUN®: Solar Desalination; n.d.
[109] Masdar Desal Pilot: The Results So Far; n.d.
[110] Revolutionary solar-power desalination plant could provide cheaper option, but it needs space; n.d.
[111] The first solar powered desalination plant in South Africa was officially launched on 11 February 2019 at Witsand, Hessequa Municipality, in the Western Cape; n.d.
[112] Dawoud MA. Economic feasibility of small scale Solar powered RO desalination for brackish/saline groundwater in arid regions. Int J Water Resour Arid Environ 2017;6:103–14.
[113] Ahmed FE, Hashaikeh R, Hilal N. Solar powered desalination—technology, energy and future outlook. Desalination 2019;453:54–76. https://doi.org/10.1016/j.desal.2018.12.002.
[114] Kasaeian A, Rajaee F, Yan W-M. Osmotic desalination by solar energy: a critical review. Renew Energy 2019; 134:1473–90. https://doi.org/10.1016/j.renene.2018.09.038.
[115] Karavas C-S, Arvanitis KG, Kyriakarakos G, Piromalis DD, Papadakis G. A novel autonomous PV powered desalination system based on a DC microgrid concept incorporating short-term energy storage. Sol Energy 2018;159:947–61.
[116] Gálvez JB, Rodríguez SM, Delyannis E, Belessiotis VG, Bhattacharya SC, Kumar S. Solar energy conversion and photoenergy systems: thermal systems and desalination plants. vol. I. EOLSS Publications; 2010.
[117] Gude VG. Energy storage for desalination [chapter 10]. In: Gude VG, editor. Renew. Energy Powered Desalin. Handb. Elsevier; 2018. p. 377–414.
[118] Stover RL. Seawater reverse osmosis with isobaric energy recovery devices. Desalination 2007;203:168–75.
[119] MacHarg JP, Sessions B. Modular pumps bring efficiencies. IDA J Desalin Water Reuse 2013;5:106–13.
[120] Achilli A, Childress AE. Pressure retarded osmosis: from the vision of Sidney Loeb to the first prototype installation. Desalination 2010;261:205–11.
[121] Yang Z, Tang C. Novel membranes and membrane materials—membrane-based salinity gradient processes for water treatment and power generation; 2018 [chapter 7].

CHAPTER

9

Applications of solar PV systems in hydrogen production

Francesco Calise, Francesco Liberato Cappiello, and Maria Vicidomini

Department of Industrial Engineering, University Federico II, Naples, Italy

Nomenclature

E	energy (J)
G	Gibbs energy (kJ/mol)
F	Faraday constant (C/mol)
f_{H_2}	hydrogen production rate (Nm^3/h)
H	enthalpy (kJ/mol)
I	intensity of current (A)
S	entropy (kJ/mol K)
T	temperature (°K)
V	voltage (volt)
z	electron moles

Greek Symbols

Δ	difference (–)
η	efficiency (–)

Subscripts

act	activation
AC	alternate current
AEM	anion exchange membrane
cell	refers to cell
con	concentration
DC	direct current
DNI	direct normal irradiance
H_2	refers to hydrogen
MPP	maximum power point
ohm	ohmic

Photovoltaic Solar Energy Conversion
https://doi.org/10.1016/B978-0-12-819610-6.00009-0

PEM polymer exchange membrane electrolysis
PV photovoltaic
SOE solid oxide electrolysis
rev reversible

9.1 Introduction

In modern society, environmental issues related to global warming have pushed public opinion as well as scientific and industrial research to find new energy vectors that have low or null environmental impact [1, 2]. Hydrogen represents a promising energy carrier that matches these expectations [3, 4].

Hydrogen is not available in the Earth, and it may be obtained by the conversion of other substances by many different techniques. In fact, it may be produced by means of electrochemical processes [5–7], fossil fuel reforming [8, 9], coal gasification [10, 11], and biomass and biological methods [12–14]. In particular, electrochemical processes are interesting options for coupling renewable energy sources and hydrogen production [15–17].

The hydrogen higher heating value (HHV) is significantly higher than that of crude oil or natural gas. The hydrogen HHV is equal to 140 MJ/kg, whereas crude oil and natural gas HHVs are equal to 45 and 50 MJ/kg, respectively [3]. Moreover, hydrogen is a good energy vector for many applications. In fact, it exhibits high performance when used as: (i) electric energy storage (through reversible fuel cells or fuel cells combined with electrolyzers [18–22]); and (ii) fuel for vehicles [23, 24].

Another useful aspect of hydrogen is related to the fact that it may be produced by exploiting several different sources: (i) nonfood crops and biomass [25]; (ii) fossil fuels (mainly natural gas) [26]; (iii) nuclear energy [27], and (iv) renewable energy sources [4, 15]. This plurality of potential supply sources represents why hydrogen is considered a promising energy vector for the future. Moreover, hydrogen production is scalable; in fact, hydrogen may be produced by large centralized plants, middle-scale plants, and small distributed plants located close to the point of use. The use of photovoltaic (PV) energy for driving electrolysis is a promising strategy for reaching fully renewable and cheap hydrogen production [28, 29]. This is the most important advantage of this technology. Conversely, some issues affect this kind of technology related to electrical matching between the electrolysis stack and the PV module, and the high cost of the electrolyzer [30–33].

This chapter specifically focuses on the available techniques that can be used to employ solar energy to obtain hydrogen from water.

9.2 Electrochemical hydrogen production

The production of hydrogen from water, exploiting renewable energy sources, is a fully renewable process [3]. It is well known that hydrogen combustion is obtained by a chemical reaction of hydrogen and oxygen. This reaction is extremely exothermic and produces a significant amount of energy, which can be used for different purposes [34]. In addition, this chemical reaction also produces water [3]. Therefore, the water cycle of hydrogen is a closed renewable loop: (i) a renewable energy source is employed for producing hydrogen from water, and (ii) the energy of the hydrogen is exploited by combining hydrogen and oxygen, producing water and energy.

Hydrogen may be produced by water and by electrolysis [3], also using several renewable energy sources [28, 35]. This process allows separating hydrogen from water using direct current.

Electrolysis is based on an electrolytic cell that consists of two electrodes [36] immersed in a conductive liquid, called electrolyte. The electrolyte consists of a solution of water and a certain salt, the latter of which is dissociated in positive and negative ions. An electric potential difference is applied between the two electrodes; therefore, the direct current flows from the positive electrode (anode) to the negative one (cathode) [36]. Then, the water is split into oxygen at the anode and hydrogen at the cathodic side; see Fig. 9.1. The electrolysis reaction is well explained by the following chemical equation:

$$\text{Total reaction}: H_2O + \text{heat} + \text{electricity} \rightarrow \frac{1}{2}O_2 + H_2 \tag{9.1}$$

The total energy required for this reaction is equal to the enthalpy of water decomposition (ΔH_{water}), calculated as follows [38]:

$$\Delta H_{water} = \Delta G_{water} + T\Delta S \tag{9.2}$$

where ΔG_{water} is the Gibbs energy of water decomposition, and $T\Delta S$ is the product of temperature with the entropy of the reaction of the water splitting. In particular, ΔG_{water} represents the minimum electric energy required for the reaction [38], and $T\Delta S$ represents the minimum heat required for the reaction [38]. Therefore, as shown in Fig. 9.2, the electrolysis reaction may occur at different thermodynamic conditions.

The efficiency of the electrolysis method (η_{elc}) is evaluated as the ratio of the output energy per unit of time of the produced hydrogen on the input energy per unit of time ($\dot{E}_{input}$). $\dot{E}_{input}$ is calculated as follows:

$$\dot{E}_{input} = I_{DC} V_{cell} \tag{9.3}$$

where I_{DC} is the direct current that flows between the electrodes, and V_{cell} is the dissociation voltage cell voltage. V_{cell} may be evaluated as the sum of

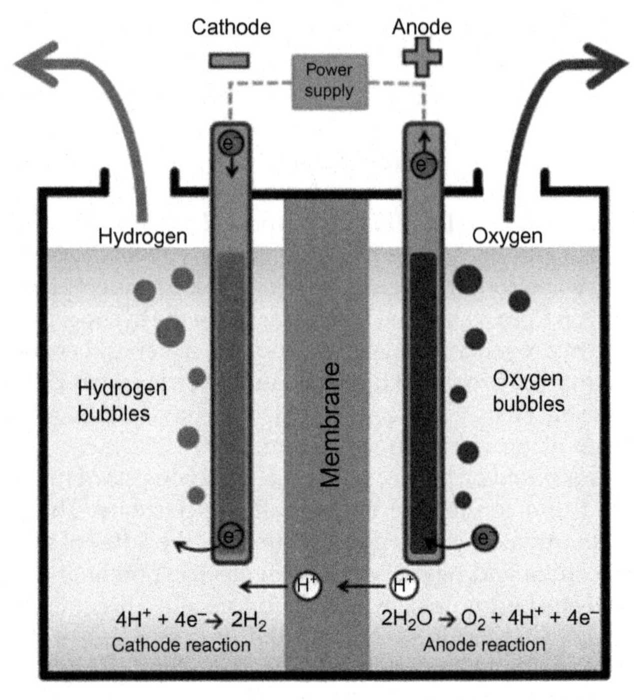

FIG. 9.1 Electrolyzer cell [37]. *Credit: US Office of Energy Efficiency and Renewable Energy.*

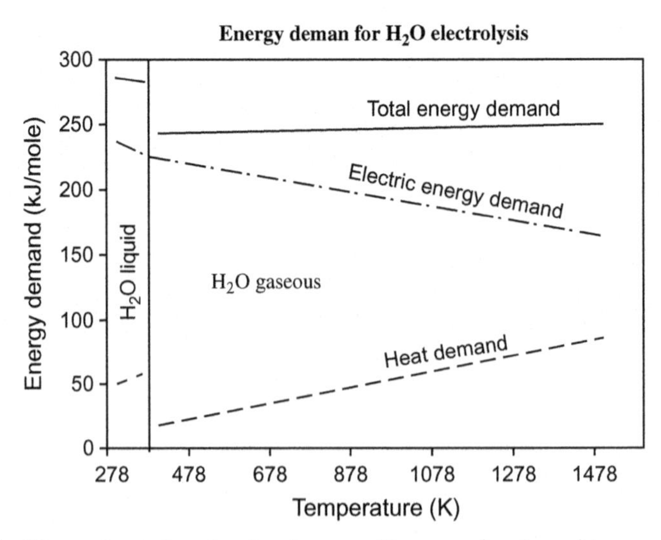

FIG. 9.2 Thermodynamics of water decomposition as a function of temperature [39]. *Credit: Hydrogen and synthetic fuel production from renewable energy sources.*

the reversible voltage and the additional overvoltage that appears in the cell [38, 40]:

$$V_{cell} = V_{rev} + V_{act} + V_{con} + V_{ohm} \tag{9.4}$$

where V_{act} (called activation overvoltage) depends on the electrode kinetics. It is needed to supply energy to allow the charge to move between the electrodes and chemical species [38]. This energy barrier leads to an overvoltage across the electrodes [38]. V_{con} (called concentration overvoltage) represents the mass transport processes (diffusion and convection) that affect the reaction [38]. This term describes the nonhomogeneous distribution of reactants and products due to transport limitations [38]. The reactants exhibit a low concentration while the products are concentrated at the interface between the electrolyte and electrode. V_{ohm} represents the overvoltage due to the ohmic losses [38].

V_{cell} should be reduced, for a given current, in order to increase the efficiency of the electrolysis reaction. Then, the surface of the electrodes may be modified in order to accelerate the reaction and to reduce the voltage losses.

Usually, the surface of the electrodes is coated with highly active catalysts. Platinum is the best catalyst, but it is very expensive. Therefore, the commercial electrolytic cells used in practice consist of nickel-plated steel electrodes. This solution exhibits limited cost and good chemical performance.

In conclusion, the efficiency of hydrogen production through electrolysis varies from 60% to 100% [3, 41, 42] for every single cell. Obviously, the global efficiency of the system is lower, ranging from 50% to 80% [3, 41, 42].

Therefore, the analyzed process has limited efficiency. However, it allows one to store photovoltaic electric energy (and renewable electric energy in general) without environmental impact [3, 34].

Fig. 9.3 shows the polarization curve (*I-V*) for a typical electrolysis cell. The polarization curve well describes cell performance. Note that a high-performance electrolyzer should exhibit a relatively low voltage at high current densities [43]. Considering the V_{cell} equation (Eq. 9.4), the polarization curve of a typical electrolysis cell has three specific regions (Fig. 9.3): (i) activation polarization region; (ii) ohmic polarization region, and (iii) concentration polarization region [44]. The activation polarization region regards the low current density, and it is dominated by the activation phenomena (V_{act}). The ohmic polarization region regards the intermediate current densities, and it is dominated by the ohmic losses (V_{ohm}) [44]. The concentration polarization region regards the high current densities, and is dominated by the mass transfer effect (V_{con}) [43].

Different kinds of electrolysis techniques have been developed in the last few years: alkali electrolysis, polymer membrane electrolysis, solid

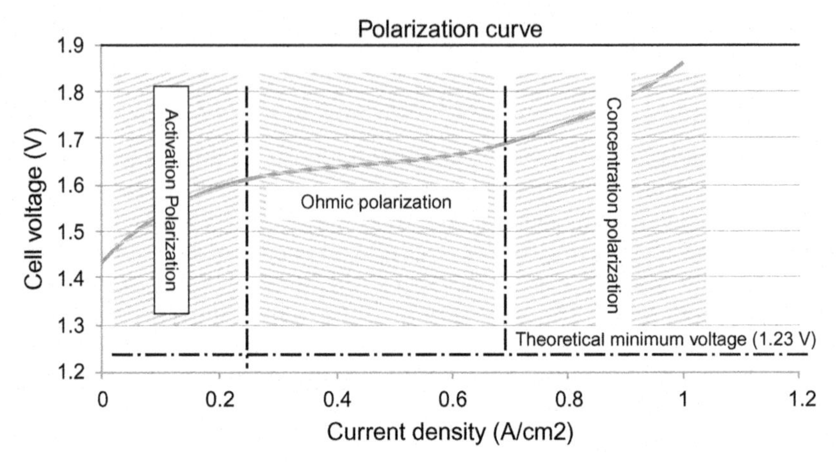

FIG. 9.3 *I-V* curve of an electrolysis cell [43]. *Credit: Characterization tool development for PEM electrolyzers.*

oxide exchange electrolysis, and anion exchange membrane electrolysis. These methods are analyzed in the following sections.

9.2.1 Alkali electrolysis

The most common and mature technology for hydrogen production is alkali electrolysis [38]. The alkali electrolysis cell consists of two electrodes immersed in a liquid alkaline electrolyte made of a caustic potash solution at a level of 20–30% KOH [18]. As shown in Fig. 9.4, a diaphragm separates the two electrodes to avoid contact between the product gases for the sake of safety and efficiency. Note that the diaphragm has to be permeable to the water molecules and hydroxide ions (OH^-) [18]. When the direct current flows between the two electrodes, the hydrogen (H_2) and oxygen (O_2) are separated from the water (Fig. 9.4). During this reaction, OH^- is an ionic charge carrier between the two electrodes. The following equations summarize the electrochemical reaction of alkali electrolysis:

$$\begin{aligned} \text{Cathode} &\rightarrow 2H_2O + 2e^- \rightarrow H_2 + 2OH^- \\ \text{Anode} &\rightarrow 2OH^- \rightarrow H_2O + {}^1/{}_2O_2 + 2e^- \\ \text{Total} &\rightarrow H_2O \rightarrow H_2 + {}^1/{}_2O_2 \end{aligned} \tag{9.5}$$

In the past, the diaphragm of alkali cells was made of porous white asbestos, $Mg_3Si_2O_5(OH)_4$. However, the use of this material has led to remarkable and well-known health issues. In fact, it is proved that asbestos causes cancer [45–47]. Therefore, other materials are used for the diaphragm such as polyphenylene sulfide (PTFE) and a composite of polytetrafluoroethylene (PTFE) and potassium titanate (K_2TiO_3) fibers [48]. In addition, during the last decades, much research was performed

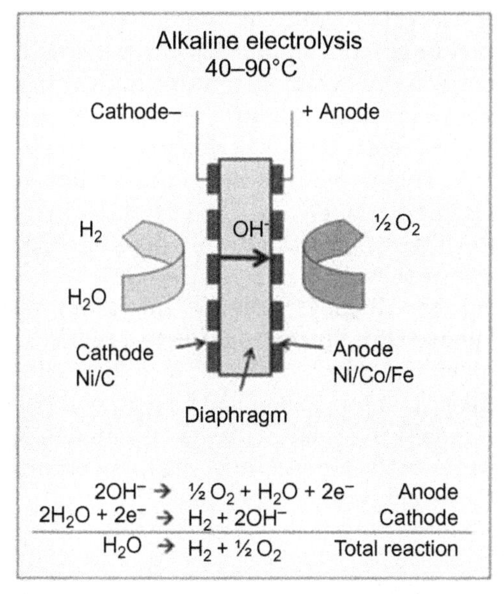

FIG. 9.4 Alkali electrolyzer scheme [18]. *Credit: A comprehensive review of PEM water electrolysis.*

with the aim of developing membranes made of new materials, such as hydroxide-conducting polymers [49], a porous composite composed of ZrO_2 and a polysulfone matrix (Zirfon) [50] and on polyphenyl sulfide (Ryton) [51].

The electrodes are made by steel plates treated with nickel (Ni) for enhancing the resistance of the plate to corrosion. In fact, nickel is a corrosion-resistant material and is largely employed because of its low cost. In addition, other metals are used for increasing the performance of the electrode. Vanadium and iron are added to the cathode while the anode is treated with cobalt.

The hydrogen produced using this process has a purity of 99% [41, 42]. Further purification processes may be performed in order to match the purity level required for hydrogen fuel cells. The global efficiency of alkaline electrolysis is about 70% [41, 42].

One of the major problems related to alkaline electrolyzers is the low maximum achievable current density [18]. This problem is caused by the ohmic losses across the diaphragm and the electrolyte [18]. Secondly, the alkali electrolyzer has a limited load range [18].

System efficiency may be increased by performing the reaction at low current densities. Usually, the commercial alkaline electrolyzers work at a current density ranging from 240 to 450 mA/cm^2. In addition, the zero-gap configuration is developed to reduce the ohmic losses of the alkali cell [52]. This configuration aims at reducing the distance between the electrodes.

The specific energy cost of hydrogen production with alkaline electrolyzers ranges from 4.6 to 6.8 kWh/Nm^3.

The alkali electrolysis reaction is well described by the *I-V* characteristic curve; see Fig. 9.5A. This curve represents the relationship between the cell current (I_{cell}) and voltage (V_{cell}). This curve (*I-V*) is heavily affected by the temperature of the cell. In fact, the temperature dramatically affects the alkali electrolysis reaction.

In particular, Fig. 9.5A displays three *I-V* curves for three temperatures, showing that the temperature increase leads to a reduction of the voltage at a fixed current density. Moreover, Fig. 9.5 also shows the activation voltage, ohmic voltage (i.e., ohmic losses), and Faraday efficiency. Note that activation voltage hugely increases with the current density. The Faraday efficiency decreases with the temperature: the higher the cell temperature, the lower the Faraday efficiency of the cell [53].

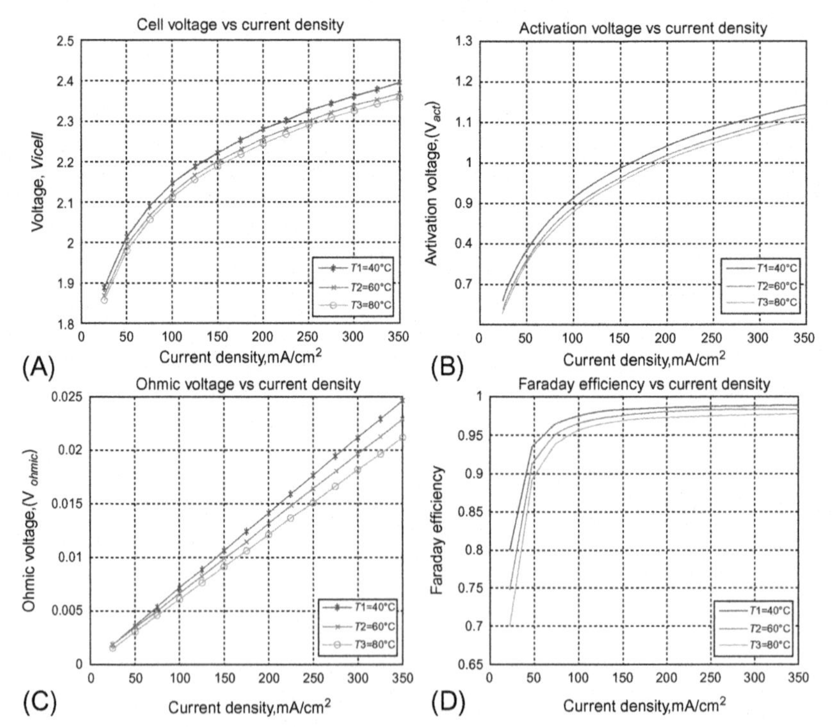

FIG. 9.5 Polarization curve, (A) *I-V* curve; (B) activation overpotentials contribution; (C) ohmic losses contribution; and (D) Faraday efficiency [53]. *Credit: Mathematical modeling and simulation analysis of advanced alkaline electrolyzer systems for hydrogen production.*

9.2.2 Polymer exchange membrane electrolysis

Polymer exchange membrane (PEM) electrolysis was developed to overcome the drawbacks of alkaline electrolysis [54, 55]. In PEM electrolyzers, a solid polysulfonated membrane is employed as the electrolyte. As shown in Fig. 9.6, the water is pumped to the anode, where water splitting in protons (H^+), oxygen (O_2), and electrons (e^-) occurs. The formed protons pass through the membrane, which allows only the protons to permeate. The electrons are forced to move from the positive electrode (anode) to the negative one (cathode) through an external circuit. A potential difference is applied to the extreme of the cell for providing the energy in order to allow the electrons to move to the negative pole. Finally, on the cathode side, the electrons combine with protons to form hydrogen (Fig. 9.6). The reaction that occurs in the PEM cells may be summarized as follows:

$$\begin{gathered}\text{Cathode} \rightarrow 2H^+ + 2e^- \rightarrow H_2 \\ \text{Anode} \rightarrow H_2O \rightarrow 2H^+ + {}^1/{}_2O_2 + 2e^- \\ \text{Total} \rightarrow H_2O \rightarrow H_2 + {}^1/{}_2O_2\end{gathered} \tag{9.6}$$

The membrane represents the backbone of the PME electrolyzer. The material used for the PEM membrane is the perfluorosulfonic acid polymer. In particular, the following commercial membranes are usually employed:

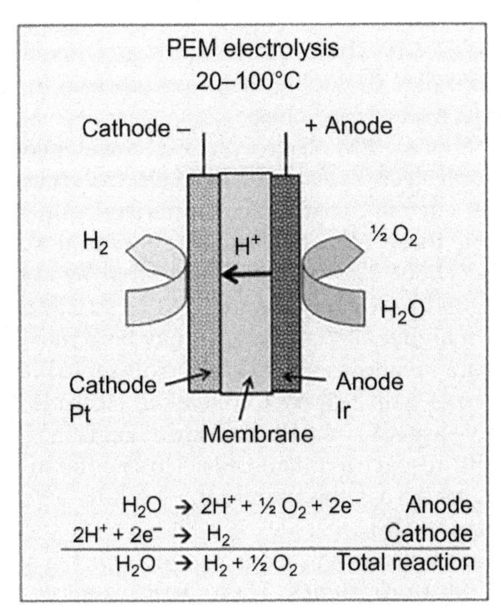

FIG. 9.6 PEM electrolyzer scheme [18]. *Credit: A comprehensive review of PEM water electrolysis.*

Fumapem, Nafion, Aciplex, and Flemion [56, 57]. The main features of these membranes are high efficiency, high strength, high oxidative stability, good durability, dimensional stability with a change of temperatures, and high proton conductivity [58]. In particular, currently, Nafion membranes are largely employed because this kind of membrane (Nafion) has remarkable advantages in comparison with other kinds of membranes. In fact, Nafion membranes can operate at higher current density (i.e., $2\,A/cm^2$) [18]. Moreover, Nafion membranes can achieve higher durability, higher proton conductivity, and higher mechanical stability [18].

PEM electrodes operate at harsh corrosive conditions (pH lower than 2). Therefore, a suitable material has to be employed to prevent corrosion of the electrodes and membranes [18]. In addition, the PEM electrolyzer must tolerate very high voltage conditions at high current densities [18].

Only a few materials can work at these extreme conditions. Therefore, noble catalysts and expensive and rare materials are employed for PEM electrolyzers, such as platinum (Pt), iridium (Ir), and ruthenium (Ru) [18]. The use of these materials is limited by their rarity and cost. In fact, iridium has an average mass fraction of 0.001 ppm in the crust of the Earth, whereas platinum and gold are 10 times and 40 times more abundant, respectively. Therefore, the high demand for these materials may dramatically increase their costs [59–61]. This trend may negatively affect the penetration of PEM electrolysis technology [18].

In the last decades, many studies were performed to face this issue. These works aim to reduce the content of noble metals by using transition metal oxides [58, 62–64]. Then, these oxides are mixed with the noble metal, both to minimize the noble metal presence in the electrodes and to preserve the electrode proprieties.

As mentioned above, PEM electrolysis was developed for overcoming the problem of alkali electrolysis. First, PEM electrolyzers are able to operate at much higher current densities in comparison with the alkali electrolyzer [18]. In fact, these electrolyzers may work at values of current densities equal to about $2\,A/cm^2$ [18]. Note that by considering a thin membrane with good proton conductivity ($0.1 \pm 0.02\,S/cm$) [65], limiting the ohmic losses, a higher current density may be achieved [18]. This feature is crucial because it allows the PEM electrolyzer to reduce the electrolysis operational costs with respect to the alkali electrolyzer.

Another positive aspect of the PEM electrolyzer is that it shows a lower crossover rate with respect to alkali cells. This is due to the fact that the transport of protons across the membrane exhibits a rapid response to power input variation [18].

The use of a solid electrolyte leads to a compact system layout with remarkable structural proprieties. These features allow the PEM cell to achieve high operative pressures [66]. An electrolyzer that is able to operate at high pressure is extremely useful because it provides high-pressure hydrogen. This feature reduces the energy for compressing and storing the produced hydrogen [18]. In addition, the high operating pressure reduces the amount of the gaseous phase at the electrodes, remarkably enhancing the product gas removal according to Ficks diffusion law [67].

On the one hand, the PEM electrolyzer may operate at a higher pressure with respect to the alkali one. On the other hand, PEM cells exhibit some problems working at significant high pressure (i.e., higher than 100 bar) because cross-permeation phenomena may occur [68, 69]

Usually, the membranes are filled with some materials for limiting cross-permeation [70], but this solution dramatically reduces the conductivity of the membrane.

Finally, Fig. 9.7 displays the *I-V* curve variation as a function of the type of membrane and the temperature of the cell. In particular, Fig. 9.7A

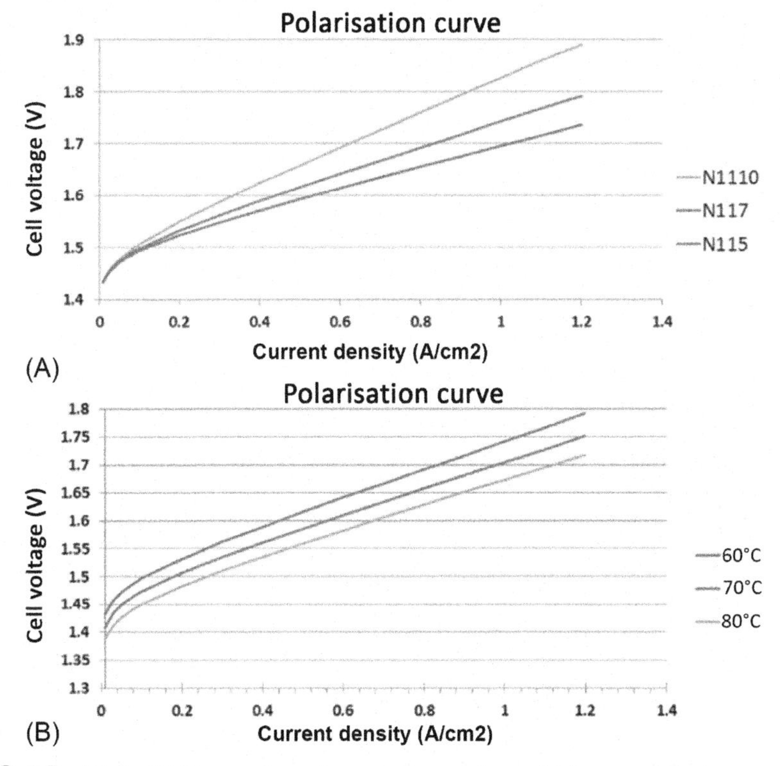

FIG. 9.7 PEM cell *I-V* curve dependence on the membrane thickness and the temperature of the cell [43]. *Credit: Characterization tool development for PEM electrolyzers.*

displays the *I-V* curve by varying the thickness of the membrane, whereas Fig. 9.7B plots the *I-V* curve varying the temperature of the cell. The increase of the thickness of the membrane leads to a worsening of cell performance due to the increase of the ohmic losses. In fact, the *I-V* curve exhibits an increase of the cell voltage at a fixed density current [43]; see Fig. 9.7. Conversely, the increase of the temperature leads to a slight improvement in electrolyzer performance [43]; see Fig. 9.7. In conclusion, 1.48 V is the minimum voltage needed for activating the PEM reaction [58] while for industrial applications, the accepted reference is $2\,A/cm^2$ at 1.9 V at 80°C [71].

9.2.3 Solid oxide electrolysis

Solid oxide electrolysis (SOE) has received much attention from the industrial and scientific communities because this technology is able to convert electrical energy in chemical energy, producing ultrapure hydrogen with high efficiency [72, 73].

SOE employs water in the form of steam and works at high pressures and temperatures, ranging between 500°C and 850°C [58].

Fig. 9.8 shows the scheme of a solid oxide electrolyzer. The electrolyte of an SOE cell consists of a thin ion-conducting membrane; usually the width ranges from 5 to 200 μm. The electrodes are porous and are made of composite or metallic materials.

The water steam is pumped to the negative electrode side (cathode), where the H_2O is reduced to hydrogen. Oxide ions (O^{2-}), after passing across the membrane, recombine to oxygen near the positive electrode (anode), releasing electrons; see Fig. 9.8.

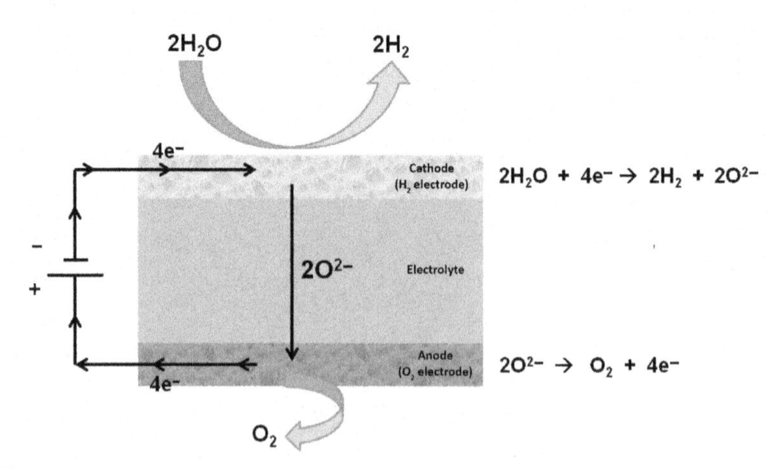

FIG. 9.8 SOE electrolyzer scheme [74]. *Credit: Solid oxide electrolysis cell analysis by means of electrochemical impedance spectroscopy: a review.*

The following equations describe the chemical reactions that occur in the SOE cell:

$$\begin{aligned} \text{Cathode} &\rightarrow H_2O + 2e^- \rightarrow H_2 + O^{2-} \\ \text{Anode} &\rightarrow O^{2-} \rightarrow {}^1/{}_2O_2 + 2e^- \\ \text{Total} &\rightarrow H_2O \rightarrow H_2 + {}^1/{}_2O_2 \end{aligned} \tag{9.7}$$

Many different layouts of solid oxide cells have been developed in past years, such as: (i) SOE coupled with syngas production [75]; (ii) SOE used as an energy storage system [76], and (iii) SOE coupled with methane production [77]. Nevertheless, the SOE electrolyzer is still at a research and development stage [52, 78].

The material for SOE cells must be able to operate at considerable high temperatures (i.e., lower than 1000°C). Usually, for the electrolytes, gadolinium-doped ceria (GDC), scandia-stabilized zirconia (ScSZ, 9 mol % Sc_2O_3—ScSZ), and yttria-stabilized zirconia (YSZ) (8 mol% YSZ) [79, 80] are employed. For the electrodes, lanthanum-strontium-cobalt-ferrite (LSCF), lanthanum-strontium-manganite (LSM), and NiO-YSZ [79] are used. Then, solid oxide electrolyzers employ no precious or expensive elements. However, because the short-term and medium-term availability of yttrium may be critical [79], scandium has been studied to replace yttrium in SOE electrolyzers [81].

SOE electrolyzers are able to operate at very high temperatures. This feature makes SOE technology very attractive compared to low-temperature electrolysis [58]. Fig. 9.2 clearly explains this trend: by increasing the temperature, the electric input needed for driving the reaction decreases. Moreover, the high operative temperature leads to an improvement of the reaction kinetics and a reduction of the losses in the electrode reaction.

Anyway, SOE electrolysis also suffers from premature degradation and a lack of stability [82–84]. These problems must be solved in order to promote large-scale commercialization of SOE electrolysis.

9.2.4 Anion exchange membrane electrolysis

In recent years, a new water electrolysis process has been developed: anion exchange membrane (AEM) electrolysis. Note that AEM electrolysis is still at the development stage, meaning it is at the laboratory stage [85]. The AEM cell operates in a similar way with respect to the PEM cells [52, 86]. The only difference between PEM and AEM cells is that in the AEM cell, the charge carrier is the anion OH^- instead of the proton H^+, as occurs in PEM cells. Then, the anion OH^- permeates across the membrane; see Fig. 9.9. Conversely, the reaction that occurs near the electrodes is the same as for the alkali electrolyzer.

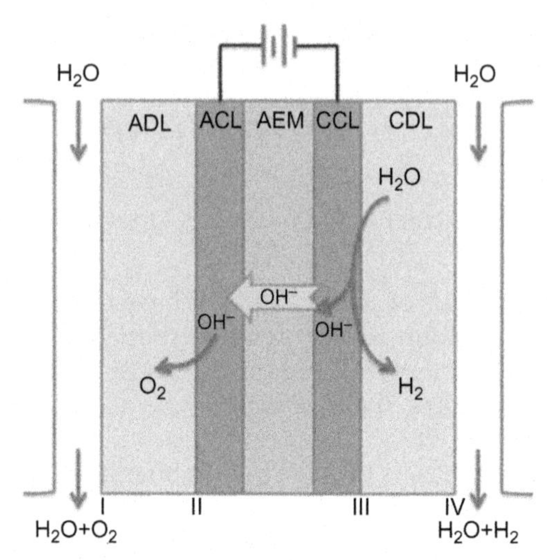

FIG. 9.9 AEM electrolyzer scheme [86]. *Credit: Mathematical modeling of an anion-exchange membrane water electrolyzer for hydrogen production.*

The following equations summarize the anion exchange membrane electrolysis reaction:

$$\begin{aligned} \text{Cathode} &\rightarrow 2H_2O + 2e^- \rightarrow H_2 + 2OH^- \\ \text{Anode} &\rightarrow 2OH^- \rightarrow {}^1/{}_2O_2 + 2e^- + 2H_2O \\ \text{Total} &\rightarrow H_2O \rightarrow H_2 + {}^1/{}_2O_2 \end{aligned} \tag{9.8}$$

The backbone of the AEM electrolyzer is the membrane. This membrane allows the hydroxide ions (OH^-) to permeate, but at the same time, the membrane has to act as a barrier for gases and electrons that are the product of the chemical reaction [85]. Usually, the core of the membrane consists of polysulfone (PSF) or polystyrene cross-linked with divinylbenzene (DVB) [85]. On the core of the membrane, some ion-exchange groups are linked, such as -RNH_2^+, =RNH_1^+ -NH_3^+, and -R_3P^+ or quaternary ammonium salts [87]. The membrane should exhibit the following proprieties: (i) high thermal, mechanical, and chemical stability; (ii) barrier action against gases and electrons, and (iii) good ionic conductivity. Both the ion exchange groups (functional groups) and polymer matrix cause the chemical stability of the membrane [87].

A positive aspect of AEM is related to the fact that the AEM cell operates in a basic condition, whereas the PEM cell works in acid conditions. Then, for the AEM electrodes, platinum-group-metal (PGM) use is not needed. Cheaper materials may be employed for this purpose. Some papers ([88–91]) pointed out that transition material may be used without worsening the AEM electrolyzer performance. However, their use makes AEM cheaper and competitive from an economic point of view.

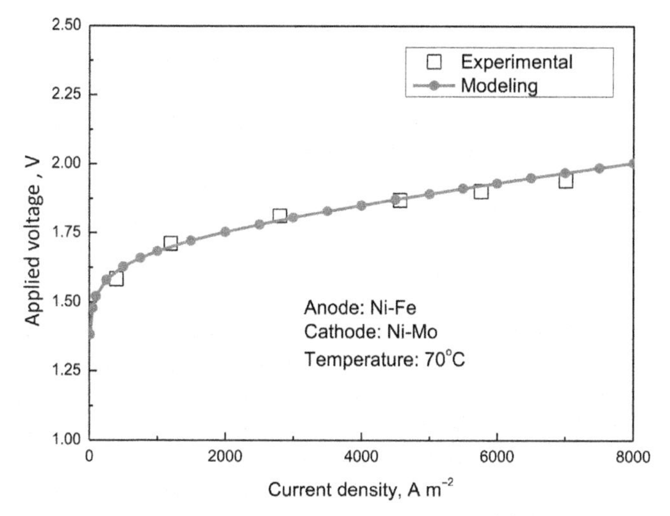

FIG. 9.10 AEM electrolyzer *I-V* curve [86, 93]. *Credit: Mathematical modeling of an anion-exchange membrane water electrolyzer for hydrogen production.*

AEM electrolyzers show important improvements, compared with traditional strategies. First, AEM electrolyzers exhibit ohmic losses, due to the fact that the AEM cell employs thinner membranes with respect to the traditional method. In addition, the precipitation of carbonates does not occur [52]. Another positive aspect is that a concentrated KOH solution is not mandatory, making the operation and installation conditions less critical and easier.

Anyway, the AEM cell has shown low durability with respect to the other water electrolysis methods [92]. In addition, the AEM cell operating temperature is not that high. In fact, high operating temperatures may determine severe chemical damages of the AEM cell, causing lower ionic conductivity [92].

Fig. 9.10 displays the comparison between the *I-V* curves obtained by (i) experiments in ref. [93] and (ii) the mathematical model presented in Ref. [86]. This plot is extremely useful for studying AEM cell performance and comparing its performance with the other electrolysis methods. Fig. 9.11 displays the efficiency and *I-V* curve as a function of the thickness of the membrane; note that these figures are numerically carried out [86]. The increase of the membrane width leads to a decrease in the AEM cell performance. In fact, by increasing the membrane thickness, the voltage increases at a fixed density current. Obviously, the efficiency dramatically decreases.

9.2.5 Summary of the available technology

Table 9.1 reports the main pros and cons of alkaline, PEM, and AEM electrolysis. Table 9.2 summarizes the main data and features of the alkali,

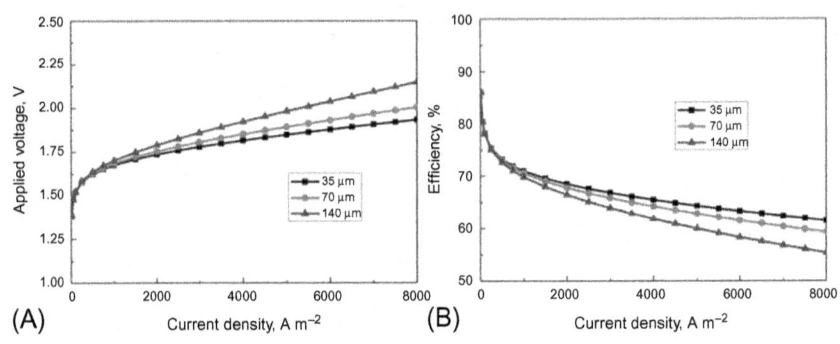

FIG. 9.11 Effect of the membrane thickness on the performance. (A) Applied voltage, and (B) efficiency [86]. *Credit: Mathematical modeling of an anion-exchange membrane water electrolyzer for hydrogen production.*

TABLE 9.1 Pros and cons of alkaline, PEM, SOE, and AEM electrolysis [18, 85].

Alkaline	PEM	SOE	AEM
		Advantages	
Mature technology	Higher performance	Efficiency up 100%	Nonnoble metal catalyst
Non-PGM catalyst	Higher voltage efficiencies		Noncorrosive electrolyte
Long term stability	Good partial load	Nonnoble catalysts	Compact cell design
Low cost	Rapid system response		Low cost
Megawatt range	Compact cell design	High-pressure operation	Absence of leaking
Cost-effective	Dynamic operation		High operating pressure
		Disadvantages	
Low current densities	High cost of components	Laboratory stage	Laboratory stage
Crossover of gas	Acidic corrosive components	Bulky system design	Low current densities
Low dynamic	Possible low durability	Durability (brittle ceramics)	Durability
Low operating pressure	Noble metal catalyst		Membrane degradation
Corrosive liquid electrolyte	Stack below megawatt range	No dependable cost information	Excessive catalyst loading

Credit: (i) A comprehensive review of PEM water electrolysis and (ii) low-cost hydrogen production by anion exchange membrane electrolysis: a review.

TABLE 9.2 Main parameters of the three electrolysis technologies [41, 42].

Parameters	Unit	Alkali	PEM	SOE
Cell temperature	°C	60–80	50–80	700–100
Cell pressure	bar	<30	20–50	1–15
Current density	A/cm^2	<0.45	1.0–2.0	0.3–1.0
Load flexibility	%	20–100	0–100	−100/+100
Cold start-up time	min	15	<15	>60
Warm start-up time	s	60–300	< 10	900
Nominal stack efficiency	%	63–71	60–68	100
Specific energy consumption	kWh/Nm^3	4.2–4.8	4.4–5.0	2.5–3.5
Nominal system efficiency	%	51–60	46–60	76–81
Specific energy consumption	kWh/Nm^3	5.0–5.9	5.0–6.5	3.7–3.9
Max. nominal power per stack	MW	6	2	<0.01
H_2 production per stack	Nm^3/h	1400	400	<10
H_2 purity	%	>99.8	99.999	
Cell area	m^2	<3.6	<0.13	<0.06
Life time	kh	55–120	60–100	8–20
Efficiency degradation	%/year	0.25–1.5	0.5–2.5	3–50
Investment costs	€/kW	790–1500	1330–2040	>2000
Maintenance costs	%/investment	2–3	3–5	n.a.

Credit: (i) Current status of water electrolysis for energy storage, grid balancing, and sector coupling via power to gas and power to liquids: a review and (ii) review and evaluation of hydrogen production methods for better sustainability.

PEM, and SOE electrolyzers. Alkali electrolysis is the most mature and used electrolysis method. PEM cells are competing with alkali cells for higher current density, higher hydrogen purity, and greater operating ranges. However, the biggest problem for PEM cells is the durability of the components [94] and the higher cost of the rare materials employed for PEM electrodes. As explained before, many works are investigating the possible reduction of the use of rare materials in PEM electrodes. Solid oxide exchange electrolysis exhibits a higher value of efficiency, but it is far from reaching commercialization status.

Finally, Fig. 9.12 shows how the cost of the alkali electrolyzer is competitive with respect to PEM technology. Note that Fig. 9.12 displays the investment, operation, and maintenance costs per unit of produced power of hydrogen (i.e., the lower heating value of the produced hydrogen).

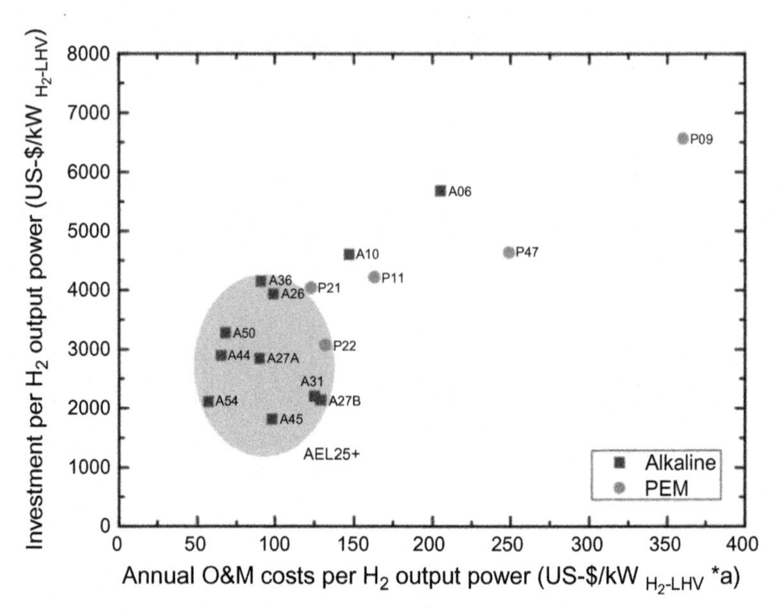

FIG. 9.12 Investment cost and cost of operation and maintenance by produced power of hydrogen (H_2) [95]. *Credit: State of the art of commercial electrolyzers and onsite hydrogen generation for logistic vehicles in South Carolina.*

9.3 Solar syngas production

Hydrogen production may be combined with CO_2 sequestration to produce synthesis gas, also known as syngas. Syngas is a mixture of hydrogen (H_2) and carbon monoxide (CO). Syngas may be produced by capturing the existing CO_2 and using solar energy for reducing the CO_2 in CO. At the same time, solar energy is employed for separating hydrogen from water. Therefore, syngas represents a fuel/feedstock material with no environmental impact. In addition, it may be employed as an energy vector or raw material for producing fuels or other chemical products (Fig. 9.13) [96–103].

In particular, the ratio of CO/H_2 plays a key role in the use of syngas. Varying this ratio, many different products can be achieved.

Note that CO_2 is a very stable molecule. Therefore, a considerable amount of energy is needed to reduce it in CO.

The use of solar energy represents one of the main methods to produce syngas in a renewable way. One of the main strategies for coupling solar energy and syngas production is the photoelectrochemical process.

The photoelectrochemical process is based on exploiting the photoelectric effect. The photons are absorbed by a semiconductor photoelectrode.

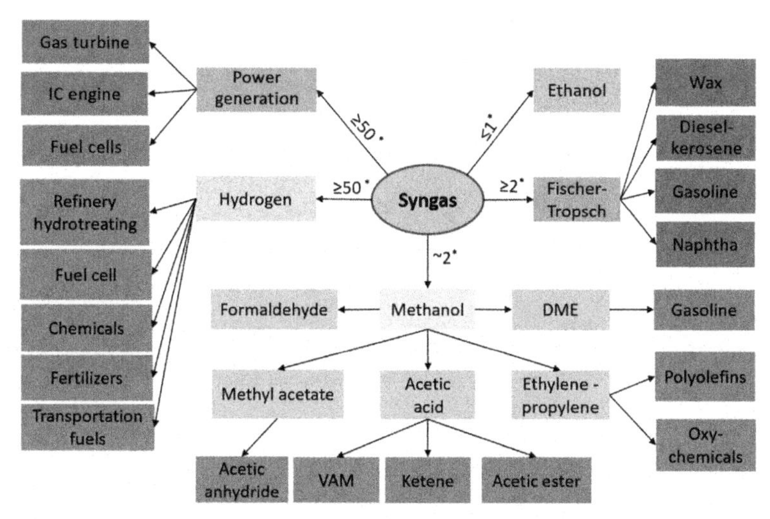

FIG. 9.13 Syngas applications [96]. *Credit: Syngas production from the electrochemical reduction of CO_2: current status and prospective implementation.*

Then, an electron-hole pair is generated. This pair contributes to the surface reactions at the anode and cathode side.

The photochemical cell consists of a chamber separated by a proton-exchange membrane; see Fig. 9.14. Three different configurations are usually employed: (i) anode-photocathode, (ii) photoanode-cathode, and (iii) - photocathode-photoanode.

In the first layout, anode-photocathode (Fig. 9.14A), the photocathode absorbs photons and photogenerates electrons. These electrons are employed by the reaction that reduces CO_2 in CO and that separates hydrogen from water, producing the CO-H_2 mixture.

In the photoanode-cathode layout (Fig. 9.14B), the photoanode absorbs the photons and generates hole-electron pairs. These pairs join the reduction reaction that produces CO and H_2.

In the photocathode-photoanode system (Fig. 9.14C), both electrodes are able to absorb photons. The photoanode generates holes, and these holes are able to oxidize H_2O in O_2 and H_2 while the photocathode generates electrons that reduce CO_2 in CO.

Another way to employ solar electric energy to produce syngas is to directly couple a photovoltaic field with electrochemical cells for producing H_2 and CO [105]. Note that this strategy may show some problems related to the matching of the devices involved in the system. Therefore, in order to not reduce the electric performance of the system, the impedance of the electrochemical cells has to match the impedance of the photovoltaic field.

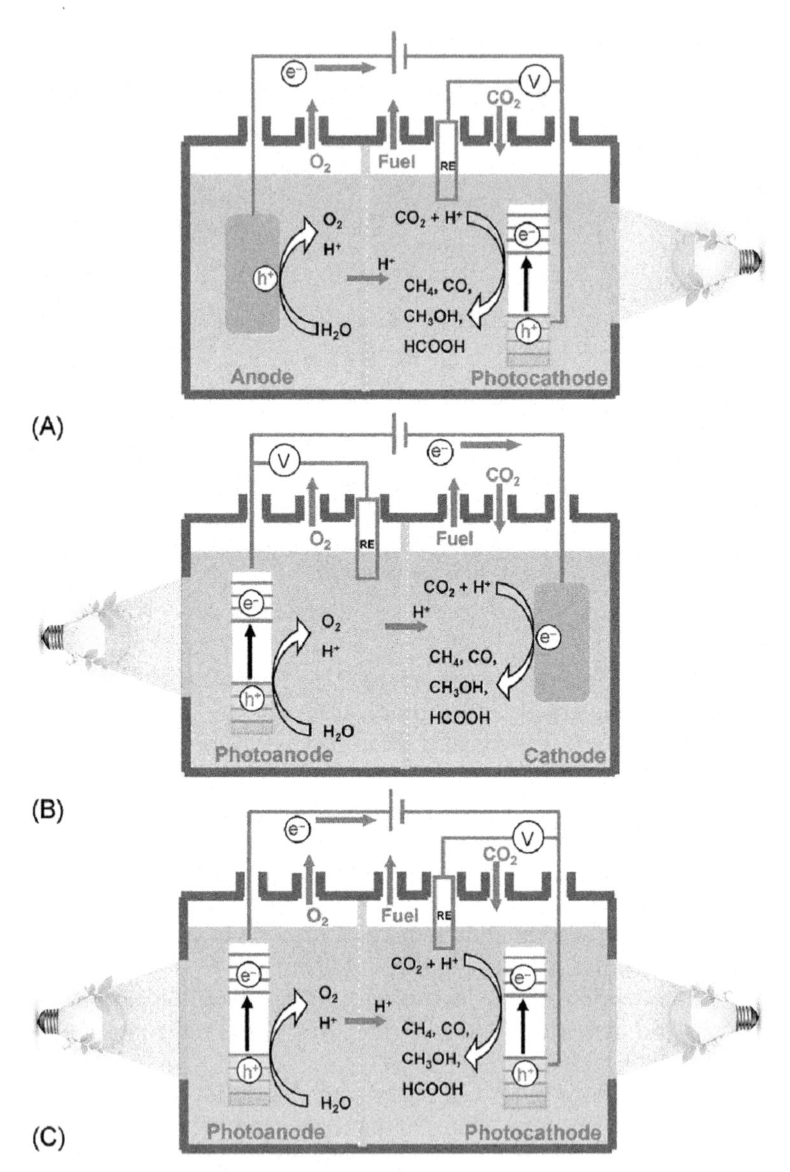

FIG. 9.14 Photoelectrochemical cell operation [104]. *Credit: Photocatalytic and photoelectrocatalytic reduction of CO_2 using heterogeneous catalysts with controlled nanostructures.*

9.4 PV-based hydrogen production using electrolysis

As mentioned before, a number of different techniques are commercially available for the production of hydrogen by electrolysis. Similarly, several research papers investigated such techniques from numerical, theoretical, and experimental points of view. Nevertheless, the use of renewable energy to drive electrolysis is still far from any commercial stage and only a limited number of papers have investigated the hybrid combination of PV panels and electrolyzers. In this framework, electrolysis water splitting driven by PV energy is the most straightforward strategy for achieving commercial large-scale hydrogen production [28, 29, 106]. It is based on the coupling of the PV solar field with a commercial electrolysis cell/stack [106, 107], as shown in Fig. 9.15.

Usually, solar hydrogen production through the alkaline cell is achieved by coupling the alkali cell with the PV field using an inverter (Fig. 9.16) [32, 108]. In fact, the voltage of the PV field has to be modified to match the

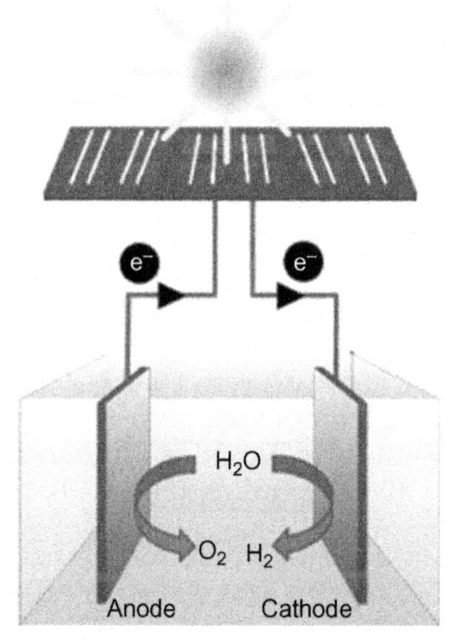

FIG. 9.15 PV-based production of hydrogen [106]. *Credit: Research advances toward large-scale solar hydrogen production from water.*

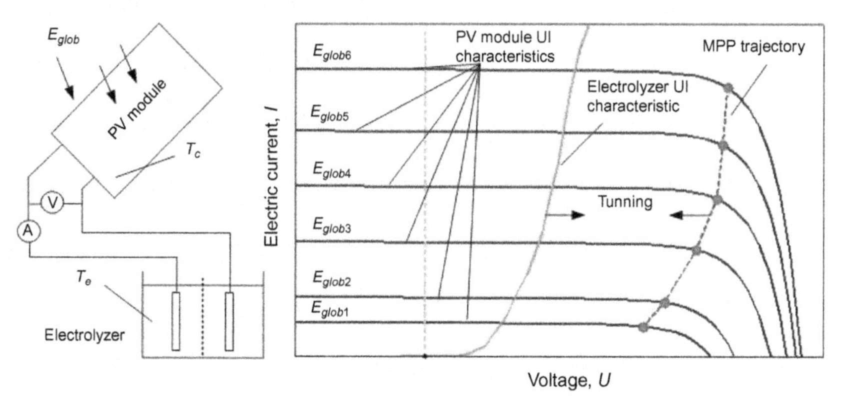

FIG. 9.16 PV field *I-V* curve and alkali cell *I-V* curve [32, 108]. *Credit: Autonomous hydrogen production system.*

I-V curve of the alkali cell, reaching the optimum operative condition (Fig. 9.16) [32, 108]. Any difference between the maximum power point (*MPP*) curve of the PV and the *I-V* curve of the alkali cell represents the electric losses of the system [32]. In an ideal configuration, the *MPP* curve and the *I-V* curve of the alkali cell should overlap [32, 33].

The performance of the system PV-electrolysis cell is deeply affected by the fluctuations of PV power production.

Analyzing the performances of the PV-alkali cell system, some studies point out that the temperature of the electrolyte increases during the central part of the day [109, 110]. This trend is caused by the increase of power production of the panels during the central part of the day [109, 110]. Fig. 9.17 clearly shows this trend. Note that the increase of the temperature of the electrolyte is useful: the higher the electrolyte temperature, the lower the power needed for electrolysis; see Fig. 9.2.

The fact that the NaOH electrolyte reaches slightly higher temperatures with respect to the KOH electrolyte is caused by the difference between the ionic conductivity of Na^+ and K^+ [110]. The maximum temperature is detected at 13:00, and it is equal to about 85°C [110], Fig. 9.17.

Concerning the electric performance of the alkali cell, the intensity of the current achieves the maximum value at about 13:00 (Fig. 9.18) [109, 110], following the power production of the PV field. Note that the voltage measured for the KOH electrolyte is higher than the NaOH electrolyte voltage [110]. This trend is due to the fact that NaOH presents higher electrical resistance compared with KOH [110].

The maximum flow rate of hydrogen production is achieved in the central part of the day, about 0.8 mL/s (Fig. 9.19), following PV power production [109, 110].

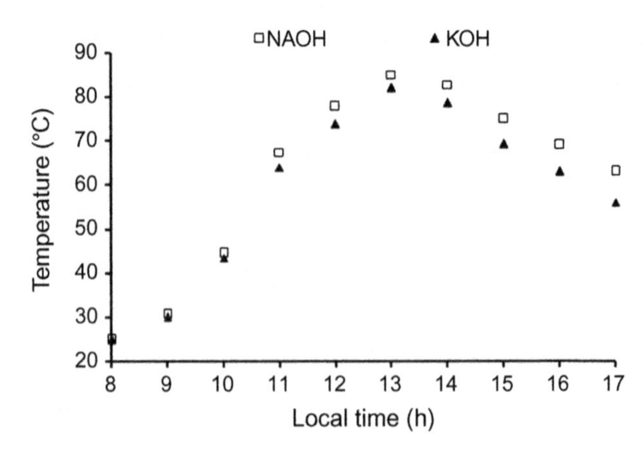

FIG. 9.17 Electrolyte temperature versus local time [110]. *Credit: Electrolyte behavior during hydrogen production by solar energy.*

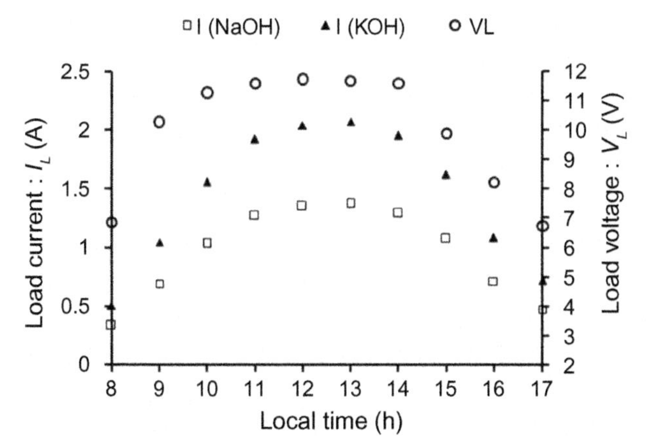

FIG. 9.18 Load voltage and load current for electrolyzers versus local time [110]. *Credit: Electrolyte behavior during hydrogen production by solar energy.*

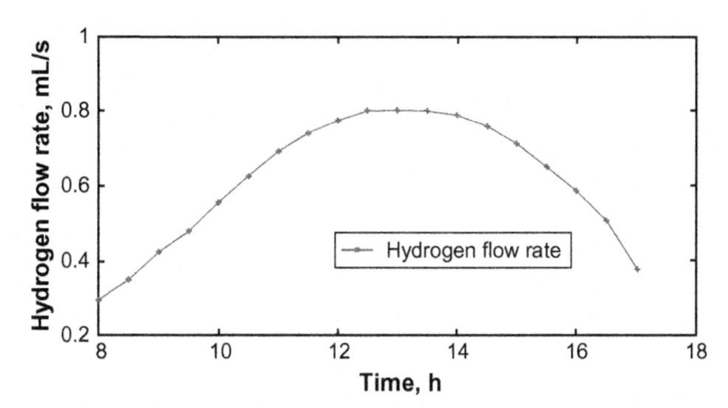

FIG. 9.19 Hydrogen flow rate [109]. *Credit: Photovoltaic-assisted alkaline water electrolysis: basic principles.*

Finally, Fig. 9.20 displays the relationship between the amount of hydrogen produced and solar irradiation. It is clear from this figure that the higher the solar radiation, the higher the amount of hydrogen produced [110].

Note that according to Fig. 9.2, it is recommended to make the electrolyte pass through the PV module [32]. This trick preheats the electrolyte and reduces the PV module temperature (the efficiency of the PV module decreases with the growth of the temperature of the module).

Ref. [111] presents a new electrolytic cylindrical cell coupled with a PV panel. The cell is made of 304 stainless steel, acrylic and sodium hydroxide (2–5 $mol\,L^{-1}$) is employed as the electrolyte. A PV panel of 0.176 m^2 provides the direct current for driving the water electrolysis process. This study proves that the main problem regarding the coupling of PV technology and alkali electrolysis is the mismatching between the *MMP* curve and the alkali cell *I-V* curve. In particular, the cell *I-V* curve forces the PV module to work outside the maximum power point [111]. Thus the system achieves

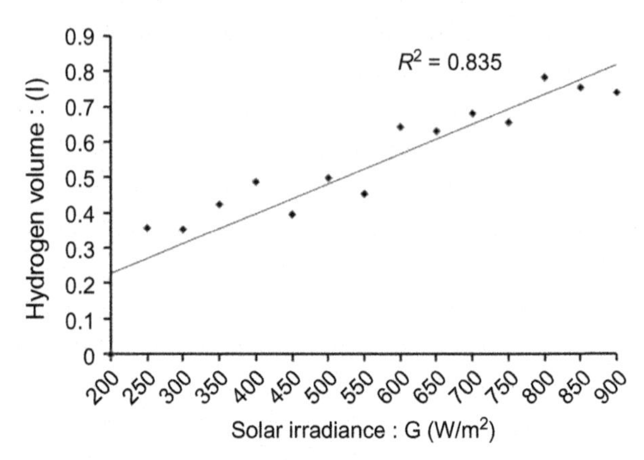

FIG. 9.20 Hydrogen volume produced by a KOH electrolyte versus solar irradiance [110]. *Credit: Electrolyte behavior during hydrogen production by solar energy.*

an overall efficiency ranging from 1.13% to 1.15% [111]. These values are in accordance with results in the literature [112–114].

The exergy efficiency of the PV-alkaline hydrogen production ranges from 8% to 16% [115].

Also, for coupling the PEM cell and PV field, the first problem is evaluation of the optimal configuration. In order to achieve this configuration, the gap between the *MPP* curve of the PV field and the *I-V* curve of the PEM cell must be reduced. As explained before, this gap causes electric losses for mismatching between the cell and PV field [116, 117].

An inverter equipped with a DC-DC converter with a maximum power point tracking algorithm may be used to modify the operating condition of the panels in order to match the *I-V* curve of the PEM cell [116, 118–121]. Note that the modern inverters equipped with maximum power point tracking algorithms are able to operate at high efficiency across a wide range of solar irradiance [116, 122].

Another strategy for achieving the optimal layout is to evaluate the number of cells in parallel and in series, including in the electrolyzer stack, in order to minimize the gap between the *MMP* curve of PV and the *I-V* curve of the cell [117, 123, 124]. This choice allows the system to achieve the maximum hydrogen production rate and the minimum energy losses between the PV module and PEM stack [117, 123, 124], due to the lack of cables, inverters, and transformers. Usually, this selection may be performed by a graphical approach, intersecting the *I-V* curve of the PV module and the load curve of the PEM stack; see Fig. 9.21. Note that by varying the incident radiation on the PV module, the *I-V* curve of the module curve changes [28, 125]. Thus, this approach leads to intermittent operation and highly variable hydrogen production [31].

Fig. 9.22 displays the performance of the layout in which the PV field and the PEM cell are linked by means of an inverter equipped with an

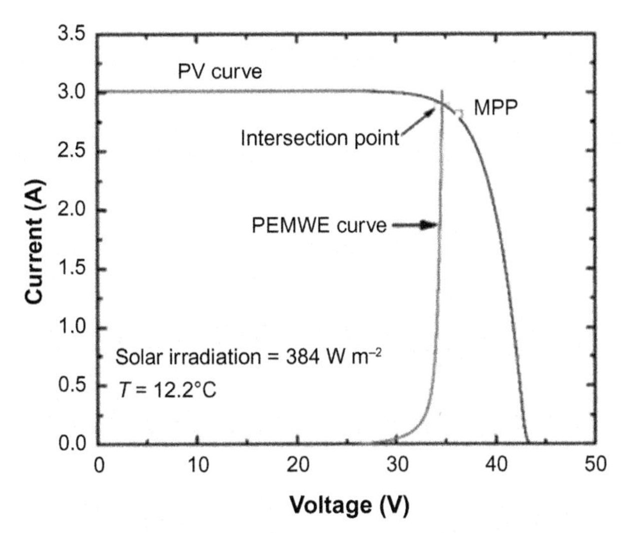

FIG. 9.21 Intersection between the *I-V* curve of the PV panel and the *I-V* curve of the PEM stacks [123]. *Credit: Modeling of solar photovoltaic-polymer electrolyte membrane electrolyzer direct coupling for hydrogen generation.*

MMP tracking algorithm. The maximum rate of hydrogen production is achieved at midday, following solar irradiance (Fig. 9.22) [116]. In fact, the higher the solar irradiance, the higher the power supplied to the PEM stack. For a sunny day in March, the studied system achieved a peak of hydrogen production equal to 155 L/h (Fig. 9.22) [116]. A clear inverse correlation between stack efficiency and stack power is detected (Figs. 9.22 and 9.23) [116, 123].

Note that the thermal insulation of the PEM stack has to be considered to prevent heat losses during nonoperating time: night hours and hours of low solar irradiance [116]. With the aim of better exploiting the direct coupling between the PV field and the PEM stack, many works [126–130] were developed in which the PV field was directly integrated with the anode of the PEM stack; see Fig. 9.25. Note that a Fresnel lens is also employed for concentrating solar radiation on the PV module [126–130]. This direct coupling is useful for reducing electrical losses due to cables, inverters, and transformers [130]. In addition, the water that flows near the anode is able to cool the anode and the PV module, increasing the PV module efficiency [130]. The increases in the temperature of the water are useful for enhancing electrolyzer performance (Fig. 9.2).

At low operating pressures, the lowest electrolysis voltage at high current density levels (0.16 V at 1.5 A cm^{-2}) may be achieved [130]. In conclusion, this layout is mainly attractive for the higher temperature achieved by the PEM stack due to solar heating [130].

The use of PV-thermal (PVT) collectors to preheat the water that will be split in the PEM cell increases system performance [131] (see Fig. 9.2).

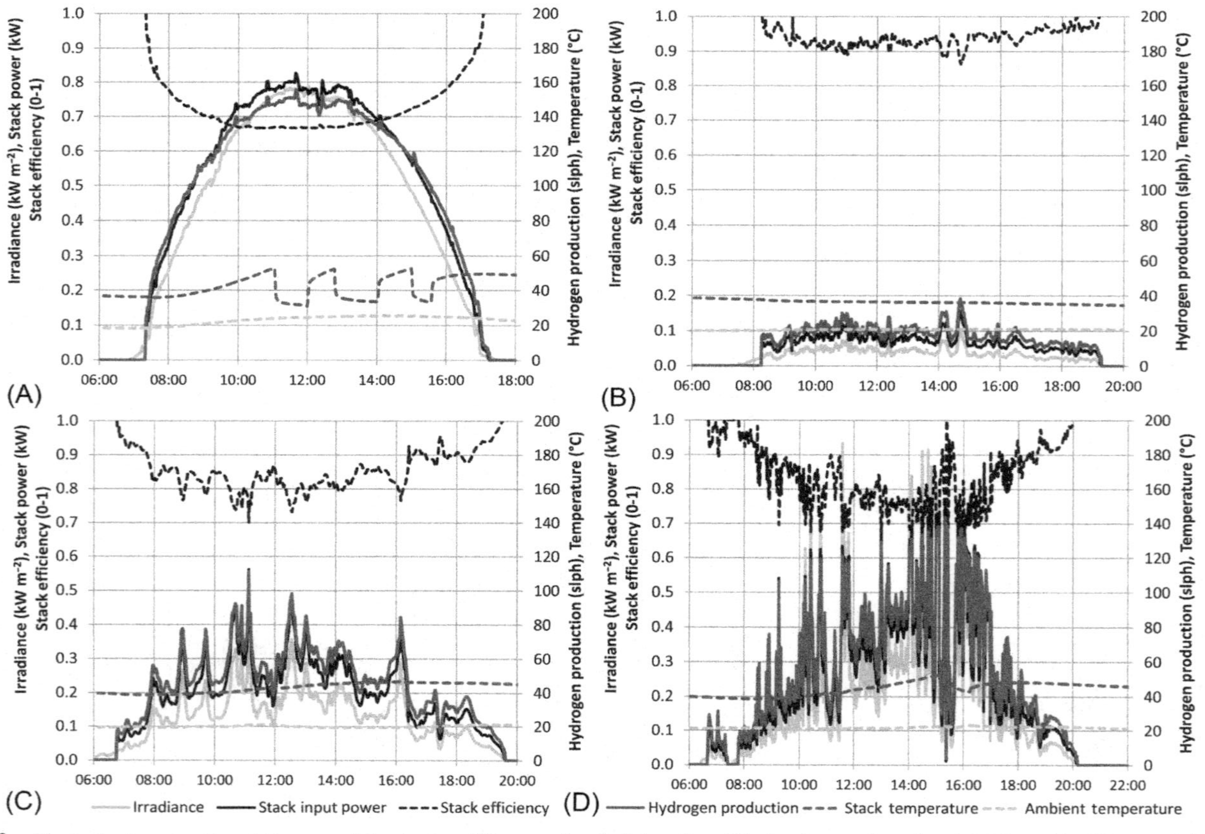

FIG. 9.22 Electrolysis operation: (A) sunny March day; (B) very cloudy May day; (C) slowly varying cloud thickness in May; and (D) rapidly varying cloud cover in May [116]. *Credit: Off-grid solar-hydrogen generation by passive electrolysis.*

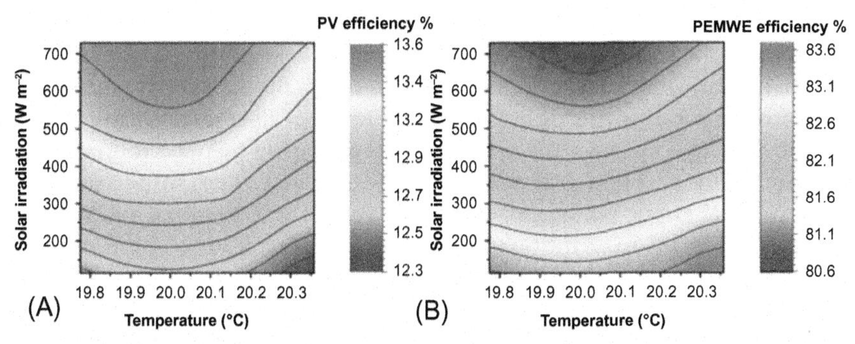

FIG. 9.23 Distribution of (A) PV efficiency and (B) PEMWE efficiency versus solar irradiation and temperature [123]. *Credit: Modeling of solar photovoltaic-polymer electrolyte membrane electrolyzer direct coupling for hydrogen generation.*

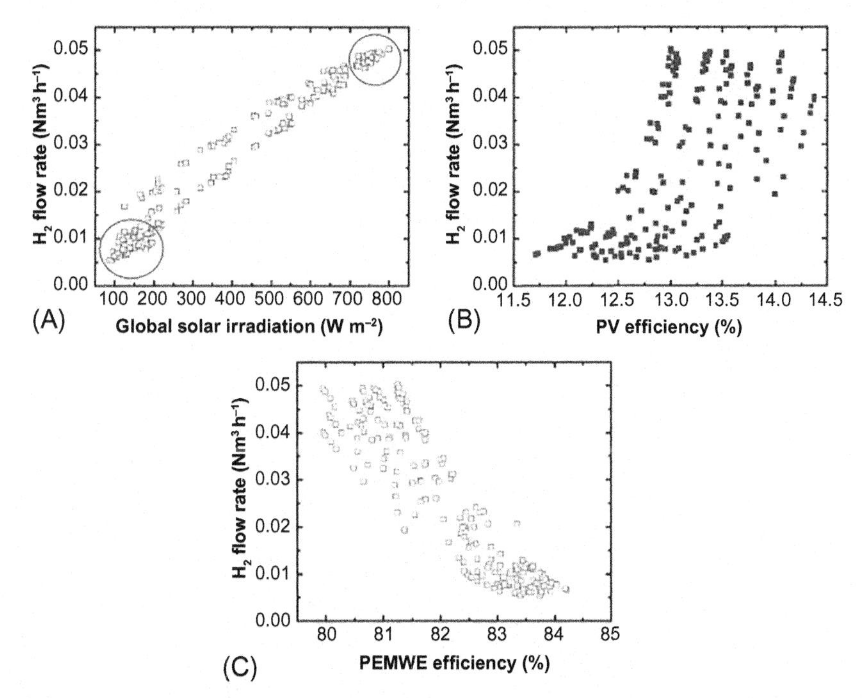

FIG. 9.24 Hydrogen production rate (A) versus solar irradiation (B) versus PV efficiency (C) versus PEMWE efficiency [123]. *Credit: Modeling of solar photovoltaic-polymer electrolyte membrane electrolyzer direct coupling for hydrogen generation.*

In conclusion, the higher the solar radiation, the higher the hydrogen rate production (Fig. 9.24) [123]. Moreover, PEM efficiency has a negative effect on the rate of hydrogen production (Fig. 9.24) [123].

Due to the fact that SOE is in precommercial status [41] and AEM is at the laboratory stage [85], there is a lack of knowledge in the literature about works in which these kinds of electrolysis are coupled with a PV field.

Ref. [132] suggests a novel plant based on the separation of the solar spectrum to simultaneously generate heat and electricity; see Fig. 9.26. In particular, the radiation not involved in the photoconversion is exploited to provide heat that is supplied to the SOE stack. A share of this heat is employed to heat the water that flows inside the electrolyzer, and the remaining part of this heat is used to keep the temperature of the SOE stack high, at about 1100 K. As explained above, SOE requires a high temperature to occur. In addition, the PV cells are cooled down by the water flow inside the electrolyzer, thus a twofold advantage is achieved: (i) the PV efficiency is increased, and (ii) the water that will be electrolyzed is preheated. Note that a heat exchanger to recover heat from the produced gas (H_2 and O_2) is also considered, with the aim of increasing the overall efficiency of the plant [132].

With a direct normal irradiance (DNI) of 899 W/m^2, the plant achieved a conversion efficiency of 36.5% and a hydrogen production rate of 850 g/h ([132], Fig. 9.27). Obviously, the system is dominated by solar irradiation, and the higher the solar irradiation the higher the hydrogen production rate; see Fig. 9.27. This proves that the use of the entire solar spectrum may be useful for the development of SOE power plants [132].

Note that this plant is at the laboratory stage [132], and therefore the work does not include a suitable economic analysis to prove the real feasibility of the plant.

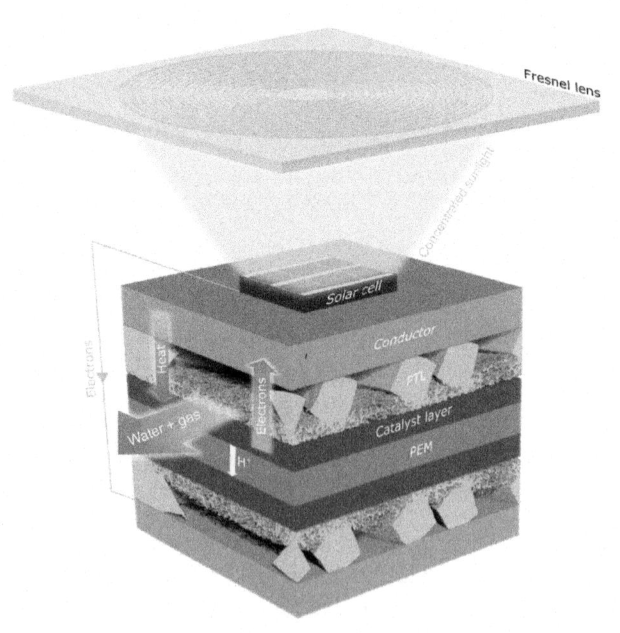

FIG. 9.25 Direct coupling of a PV module and a PEM electrolyzer, with the integration of the module on the anode of the PEM cell [129]. *Credit: Investigation of a novel concept for hydrogen production by PEM water electrolysis integrated with multijunction solar cells.*

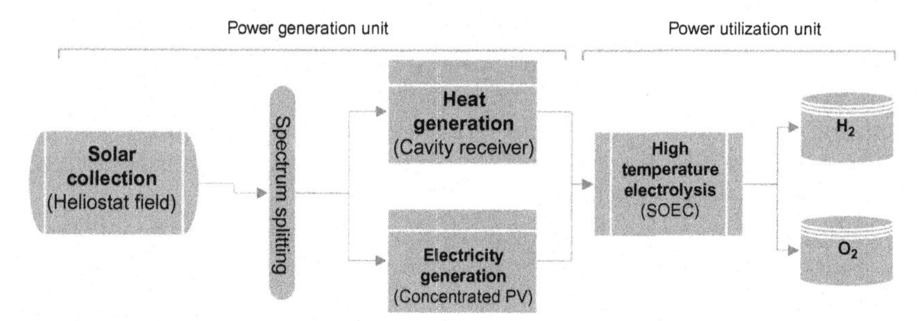

FIG. 9.26 Novel PV-SOE power plant scheme [132]. *Credit: Solar-driven high-temperature hydrogen production via integrated spectrally split concentrated photovoltaics (SSCPV) and solar power tower.*

SOE is still in the precommercial stage [41], thus there is a lack of literature regarding the photovoltaic-driven SO process.

9.4.1 Summary of the available PV-based hydrogen production plants

In conclusion, the major issue related to PV electrolysis is the electrical matching between the PV field and the electrolier stack. Two strategies are described: a direct connection between the photovoltaic field and the electrolyzer stack, and coupling by means of inverters.

The direct coupling determines a significant reduction of the system electrical losses due to the absence of electrical devices such as inverters, cables, and transformers. But this strategy may force, in some conditions, the photovoltaic module to work far from the maximum power point,

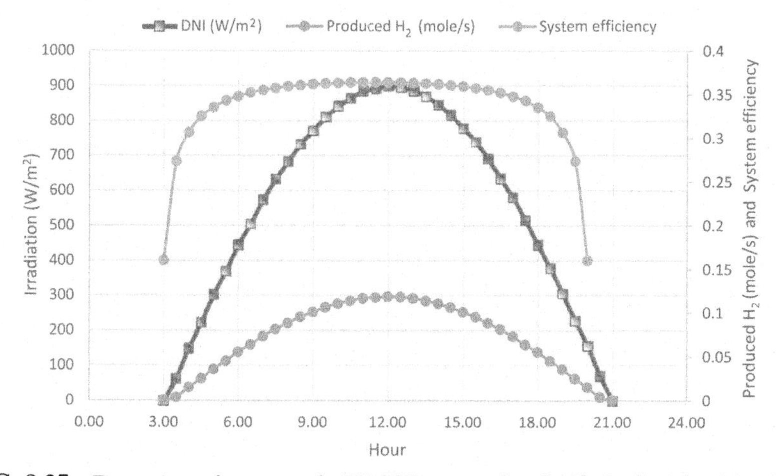

FIG. 9.27 Dynamic performance of a PV-SOE power plant [132]. *Credit: Solar-driven high temperature hydrogen production via integrated spectrally split concentrated photovoltaics (SSCPV) and solar power tower.*

reducing the overall efficiency of the process. Therefore, this approach exhibits higher nominal efficiency but leads to intermittent operation with a fluctuating rate of hydrogen production [31–33, 108, 111]. Conversely, coupling by means of inverters optimizes the dynamic operation but reduces the overall efficiency of the system, increasing the number of electrical devices installed [31, 111, 116, 118–121].

Table 9.3 summarizes the main advantages and disadvantages of alkali-PV hydrogen production and proton membrane exchange-PV hydrogen production.

The main problem related to alkaline electrolysis is the delay in the response of the cell stack varying the input current. This feature reduces the operative condition of the alkaline electrolyzer, which exhibits a considerable problem during partial load operation [18, 85]. Conversely, the proton exchange membrane electrolyzer exhibits a faster response to the current density variation [18, 85, 123]. Therefore, proton exchange stacks can suitably follow the fluctuations of PV energy production. The main issue of photovoltaic hydrogen production based on the proton membrane exchange stack is the high cost of this electrolyzer [18, 85, 108, 123].

Other significant advantages of alkaline electrolysis compared to the proton exchange membrane are simplicity and reliability [108]. These aspects make photovoltaic-driven alkaline electrolysis suitable for use in transportation, industries, and small households [108].

Finally, Table 9.4 reports the main PV hydrogen systems presently in operation and their mean features. Note that the majority of these systems were installed using funds provided by institutions, governments, and

TABLE 9.3 Pros and cons of alkaline and PEM PV hydrogen production.

PV-Alkaline	PV-PEM
Advantages	
Mature technology	Higher voltage efficiencies
Long-term stability	Good partial load
Low cost	Rapid system response
Simple and scalable	Dynamic operation
Disadvantages	
Partial load	High cost of components
	Acidic corrosive components
Low dynamic	Possible low durability
	Noble metal catalyst

TABLE 9.4 Existing PV hydrogen production power plants [30].

Project name	Time	Source		Electrolyzer		Hydrogen storage		
		Type	Installed power (kWp)	Type	Power (kW)	Type	Volume capacity (Nm^3 H_2)	Energy capacity (kWh)
FIRST	2000–04	PV	1.4	PEM	1	Metal hydrides, 30 bar	70	248
HARI	2002–	PV–wind–micro hydro	13–50–3.2	Alkaline	36	Pressurized tanks, 137 bar	2856	10,127
HRI	2001–	PV–wind	1–10	Alkaline	5	Pressurized tanks, 10 bar	40	142
INTA	1989–97	PV	8.5	Alkaline	5	Metal hydrides–pressurized tanks, 200 bar	24–9	85–32
PHOEBUS	1993–2003	PV	43	Alkaline	26	Pressurized tanks, 120 bar	3000	10,638
SAPHYS	1994–97	PV	5.6	Alkaline	5	Pressurized tanks, 200 bar	120	426
SCHATZ	1989–96	PV	9.2	Alkaline	6	Pressurized tanks, 8 bar	60	213
Solar house	1992–95	PV	4.2	PEM	2	Pressurized tanks, 28 bar	400	1418
Solar hydrogen pilot plant	1990–92	PV	1.3	Alkaline	0.8	Pressurized tanks, 25 bar	200	709
SWB	1989–96	PV	370	Alkaline	100	Pressurized tanks, 30 bar	5000	17,730
CEC	2007–	PV	5	PEM	3.35	Metal hydrides, 14 bar	5.4	19

Credit: A review of solar-hydrogen/fuel cell hybrid energy systems for stationary applications.

universities. In fact, this kind of plant is very expensive due to the high cost of the electrolysis stack (both proton exchange membrane and alkaline) [30].

Nowadays a 3.1$ project that couples PV panels and hydrogen production was announced by the Australian Renewable Energy Agency (AREA). This pilot project would consist of a 100 kW PV field coupled with 220 kW electrolyzer stacks. This plant would be able to produce about 2400 kg/day dihydrogen [133].

9.5 Conclusions

This chapter describes the most common electrolysis technologies: (i) alkaline electrolysis; (ii) proton exchange membrane electrolysis; (iii) solid oxide electrolysis; and (iv) anion exchange membrane electrolysis.

Through an in-depth literature analysis, it was revealed that alkali electrolysis is the most studied, mature, and used electrolysis method because the alkali cell is easy to design and it employs inexpensive materials. Proton exchange membrane electrolysis was developed to overcome the alkaline limits. Thus, proton exchange membrane technology is competing with alkali for higher current density, higher hydrogen purity, and a greater operating range. But this technology is affected by some problems that limit its commercial use, such as the durability of the components and the high cost of the rare materials employed for its electrodes. To overcome the problems of the proton exchange membrane and alkali electrolysis, solid oxide electrolysis was developed. Despite its higher efficiency, it is still in precommercial status. Finally, anion exchange membrane electrolysis is in prelaboratory status.

The integration between the photovoltaic and electrolysis cell is very useful for producing renewable hydrogen. The main problem of this integration is the electrical matching between the photovoltaic field and the electrolysis stack.

Two strategies were described to solve this problem: direct coupling and coupling by means of an inverter.

Direct coupling is able to significantly reduce system electrical losses due to the absence of electrical equipment such as cables, inverters, and transformers. But this strategy may force, in some conditions, the photovoltaic module to work far from the maximum power point, leading to intermittent operation.

The coupling of photovoltaic field and an electrolysis stack thought inverters is useful in order to achieve a wider range of operating conditions. But this strategy increases the number of electrical devices employed, significantly reducing the overall efficiency of the plant.

Photovoltaic-based alkaline hydrogen production plants are characterized by high reliability and nonexcessive costs because alkaline electrolysis is a mature technology. But these plants exhibit poor performance at partial load operation because the alkali stack is not able to follow the fast variation of the input current density.

Conversely, the system photovoltaic-proton exchange stack exhibits remarkably better performance at partial load operation and higher efficiency. But the high cost of the rare material employed in the electrodes of the proton exchange stack limits its commercialization.

Note that some studies are ongoing to replace rare materials with cheaper ones in order to reduce the proton exchange membrane cell cost, allowing this strategy to be easier commercialized.

References

[1] http://www.ren21.net/gsr-2019/pages/foreword/foreword/, R., *http://www.ren21.net/gsr-2019/pages/foreword/foreword/*; 2019.

[2] Ogbonnaya C, et al. The current and emerging renewable energy technologies for power generation in Nigeria: a review. Thermal Sci Eng Progr 2019;13:100390.

[3] Chi J, Yu H. Water electrolysis based on renewable energy for hydrogen production. Chin J Catal 2018;39(3):390–4.

[4] Hanley ES, Deane JP, Gallachóir BPÓ. The role of hydrogen in low carbon energy futures—a review of existing perspectives. Renew Sust Energ Rev 2018;82:3027–45.

[5] Barbir F. PEM electrolysis for production of hydrogen from renewable energy sources. Sol Energy 2005;78(5):661–9.

[6] Atlam O, Kolhe M. Equivalent electrical model for a proton exchange membrane (PEM) electrolyser. Energy Convers Manag 2011;52(8):2952–7.

[7] Siracusano S, et al. An electrochemical study of a PEM stack for water electrolysis. Int J Hydrogen Energy 2012;37(2):1939–46.

[8] Rahimpour MR, Jafari M, Iranshahi D. Progress in catalytic naphtha reforming process: a review. Appl Energy 2013;109:79–93.

[9] Boyano A, et al. Exergoenvironmental analysis of a steam methane reforming process for hydrogen production. Energy 2011;36(4):2202–14.

[10] Burmistrz P, et al. Carbon footprint of the hydrogen production process utilizing subbituminous coal and lignite gasification. J Clean Prod 2016;139:858–65.

[11] Seyitoglu SS, Dincer I, Kilicarslan A. Energy and exergy analyses of hydrogen production by coal gasification. Int J Hydrogen Energy 2017;42(4):2592–600.

[12] Elsharnouby O, et al. A critical literature review on biohydrogen production by pure cultures. Int J Hydrogen Energy 2013;38(12):4945–66.

[13] Sivagurunathan P, et al. Enhancement strategies for hydrogen production from wastewater: a review. Curr Org Chem 2016;20(26):2744–52.

[14] Mujeebu MA. Hydrogen and syngas production by superadiabatic combustion—a review. Appl Energy 2016;173:210–24.

[15] Levene JI, et al. An analysis of hydrogen production from renewable electricity sources. Sol Energy 2007;81(6):773–80.

[16] Gahleitner G. Hydrogen from renewable electricity: an international review of power-to-gas pilot plants for stationary applications. Int J Hydrogen Energy 2013;38(5):2039–61.

[17] Sgobbi A, et al. How far away is hydrogen? Its role in the medium and long-term decarbonisation of the European energy system. Int J Hydrogen Energy 2016;41(1):19–35.

[18] Carmo M, et al. A comprehensive review on PEM water electrolysis. Int J Hydrogen Energy 2013;38(12):4901–34.

[19] Bartolucci L, et al. Fuel cell based hybrid renewable energy systems for off-grid telecom stations: data analysis and system optimization. Appl Energy 2019;252:113386.
[20] Serincan MF. Validation of hybridization methodologies of fuel cell backup power systems in real-world telecom applications. Int J Hydrogen Energy 2016;41(42):19129–40.
[21] Guney MS, Tepe Y. Classification and assessment of energy storage systems. Renew Sust Energ Rev 2017;75:1187–97.
[22] Das V, et al. Recent advances and challenges of fuel cell based power system architectures and control—a review. Renew Sust Energ Rev 2017;73:10–8.
[23] Li M, et al. Review on the research of hydrogen storage system fast refueling in fuel cell vehicle. Int J Hydrogen Energy 2019;44(21):10677–93.
[24] Murugan A, et al. Measurement challenges for hydrogen vehicles. Int J Hydrogen Energy 2019;44(35):19326–33.
[25] Bundhoo ZMA. Potential of bio-hydrogen production from dark fermentation of crop residues: a review. Int J Hydrogen Energy 2019;44(32):17346–62.
[26] Muradov N. Low to near-zero CO2 production of hydrogen from fossil fuels: status and perspectives. Int J Hydrogen Energy 2017;42(20):14058–88.
[27] El-Emam RS, Khamis I. Advances in nuclear hydrogen production: results from an IAEA international collaborative research project. Int J Hydrogen Energy 2019; 44(35):19080–8.
[28] Ismail TM, et al. Using MATLAB to model and simulate a photovoltaic system to produce hydrogen. Energy Convers Manag 2019;185:101–29.
[29] Lodhi MAK. Photovoltaics and hydrogen: future energy options. Energy Convers Manag 1997;38(18):1881–93.
[30] Yilanci A, Dincer I, Ozturk HK. A review on solar-hydrogen/fuel cell hybrid energy systems for stationary applications. Prog Energy Combust Sci 2009;35(3):231–44.
[31] Mraoui A, Benyoucef B, Hassaine L. Experiment and simulation of electrolytic hydrogen production: case study of photovoltaic-electrolyzer direct connection. Int J Hydrogen Energy 2018;43(6):3441–50.
[32] Đukić A. Autonomous hydrogen production system. Int J Hydrogen Energy 2015; 40(24):7465–74.
[33] Firak M, Đukić A. An investigation into the effect of photovoltaic module electric properties on maximum power point trajectory with the aim of its alignment with electrolyzer UI characteristic. Therm Sci 2010;14(3):729.
[34] Acar C, Dincer I. Review and evaluation of hydrogen production options for better environment. J Clean Prod 2019;218:835–49.
[35] Abdallah MAH, Asfour SS, Veziroglu TN. Solar–hydrogen energy system for Egypt. Int J Hydrogen Energy 1999;24(6):505–17.
[36] dos Santos KG, et al. Hydrogen production in the electrolysis of water in Brazil, a review. Renew Sust Energ Rev 2017;68:563–71.
[37] https://www.energy.gov/eere/fuelcells/hydrogen-production-electrolysis; 2019. Available from: https://www.energy.gov/eere/fuelcells/hydrogen-production-electrolysis.
[38] Ursua A, Gandia LM, Sanchis P. Hydrogen production from water electrolysis: current status and future trends. Proc IEEE 2012;100(2):410–26.
[39] Jensen SH, Larsen PH, Mogensen M. Hydrogen and synthetic fuel production from renewable energy sources. Int J Hydrogen Energy 2007;32(15):3253–7.
[40] Hansen JB. Solid oxide electrolysis—a key enabling technology for sustainable energy scenarios. Faraday Discuss 2015;182:9–48.
[41] Buttler A, Spliethoff H. Current status of water electrolysis for energy storage, grid balancing and sector coupling via power-to-gas and power-to-liquids: a review. Renew Sust Energ Rev 2018;82:2440–54.
[42] Dincer I, Acar C. Review and evaluation of hydrogen production methods for better sustainability. Int J Hydrogen Energy 2015;40.

[43] van der Merwe J, et al. Characterisation tools development for PEM electrolysers. Int J Hydrogen Energy 2014;39(26):14212–21.
[44] Wu J, et al. Diagnostic tools in PEM fuel cell research: part I electrochemical techniques. Int J Hydrogen Energy 2008;33(6):1735–46.
[45] Phanprasit W, et al. Asbestos exposure among mitering workers. Saf Health Work 2012;3 (3):235–40.
[46] Jöckel K-H, et al. Occupational risk factors for lung cancer: a case-control study in West Germany. Int J Epidemiol 1998;27(4):549–60.
[47] Brüske-Hohlfeld I, et al. Occupational lung cancer risk for men in Germany: results from a pooled case-control study. Am J Epidemiol 2000;151(4):384–95.
[48] Rosa VM, Santos MBF, da Silva EP. New materials for water electrolysis diaphragms. Int J Hydrogen Energy 1995;20(9):697–700.
[49] Pletcher D, Li X. Prospects for alkaline zero gap water electrolysers for hydrogen production. Int J Hydrogen Energy 2011;36(23):15089–104.
[50] Vermeiren P, et al. Evaluation of the Zirfon® separator for use in alkaline water electrolysis and Ni-H2 batteries. Int J Hydrogen Energy 1998;23(5):321–4.
[51] Rajeshwar, K., Solar hydrogen generation [electronic resource]: toward a renewable energy future. Springer, 2008.
[52] David M, Ocampo-Martínez C, Sánchez-Peña R. Advances in alkaline water electrolyzers: a review. J Energy Storage 2019;23:392–403.
[53] Tijani AS, Yusup NAB, Rahim AHA. Mathematical modelling and simulation analysis of advanced alkaline electrolyzer system for hydrogen production. Procedia Technol 2014;15:798–806.
[54] Khan MA, et al. Recent progresses in electrocatalysts for water electrolysis. Electrochem Energy Rev 2018;1(4):483–530.
[55] Baykara SZ. Hydrogen: a brief overview on its sources, production and environmental impact. Int J Hydrogen Energy 2018;43(23):10605–14.
[56] Mališ J, et al. Nafion 117 stability under conditions of PEM water electrolysis at elevated temperature and pressure. Int J Hydrogen Energy 2016;41(4):2177–88.
[57] Millet P, et al. Electrochemical performances of PEM water electrolysis cells and perspectives. Int J Hydrogen Energy 2011;36(6):4134–42.
[58] Shiva Kumar S, Himabindu V. Hydrogen production by PEM water electrolysis—a review. Mater Sci Energy Technol 2019;2(3):442–54.
[59] Parry SJ. Abundance and distribution of palladium, platinum, iridium and gold in some oxide minerals. Chem Geol 1984;43(1):115–25.
[60] Mitchell RH, Keays RR. Abundance and distribution of gold, palladium and iridium in some spinel and garnet lherzolites: implications for the nature and origin of precious metal-rich intergranular components in the upper mantle. Geochim Cosmochim Acta 1981;45(12):2425–42.
[61] Hunt L. A history of iridium. Platin Met Rev 1987;31(1):32–41.
[62] Ma R, et al. Oxygen evolution and corrosion behavior of low-MnO2-content Pb-MnO2 composite anodes for metal electrowinning. Hydrometallurgy 2016;159:6–11.
[63] Kumar SS, et al. Preparation of Ru x Pd 1-x O 2 electrocatalysts for the oxygen evolution reaction (OER) in PEM water electrolysis. Ionics 2018;24(8):2411–9.
[64] Ju H, Giddey S, Badwal SPS. The role of nanosized SnO2 in Pt-based electrocatalysts for hydrogen production in methanol assisted water electrolysis. Electrochim Acta 2017;229:39–47.
[65] Slade S, et al. Ionic conductivity of an extruded Nafion 1100 EW series of membranes. J Electrochem Soc 2002;149(12):A1556–64.
[66] Medina P, Santarelli M. Analysis of water transport in a high pressure PEM electrolyzer. Int J Hydrogen Energy 2010;35(11):5173–86.

[67] Grigor'ev S, et al. Electrolysis of water in a system with a solid polymer electrolyte at elevated pressure. Russ J Electrochem 2001;37(8):819–22.
[68] Millet P, et al. PEM water electrolyzers: from electrocatalysis to stack development. Int J Hydrogen Energy 2010;35(10):5043–52.
[69] Millet P, et al. Scientific and engineering issues related to PEM technology: water electrolysers, fuel cells and unitized regenerative systems. Int J Hydrogen Energy 2011;36 (6):4156–63.
[70] Baglio V, et al. Solid polymer electrolyte water electrolyser based on Nafion-TiO2 composite membrane for high temperature operation. Fuel Cells 2009;9(3):247–52.
[71] Debe M, et al. Initial performance and durability of ultra-low loaded NSTF electrodes for PEM electrolyzers. J Electrochem Soc 2012;159(6):K165–76.
[72] Xu W, Scott K. The effects of ionomer content on PEM water electrolyser membrane electrode assembly performance. Int J Hydrogen Energy 2010;35(21):12029–37.
[73] Brisse A, Schefold J, Zahid M. High temperature water electrolysis in solid oxide cells. Int J Hydrogen Energy 2008;33(20):5375–82.
[74] Nechache A, Cassir M, Ringuedé A. Solid oxide electrolysis cell analysis by means of electrochemical impedance spectroscopy: a review. J Power Sources 2014;258: 164–81.
[75] Zhang H, et al. Solid-oxide electrolyzer coupled biomass-to-methanol systems. Energy Procedia 2019;158:4548–53.
[76] Wendel CH, et al. Modeling and experimental performance of an intermediate temperature reversible solid oxide cell for high-efficiency, distributed-scale electrical energy storage. J Power Sources 2015;283:329–42.
[77] Jeanmonod G, et al. Trade-off designs of power-to-methane systems via solid-oxide electrolyzer and the application to biogas upgrading. Appl Energy 2019;247:572–81.
[78] Badwal SP, Giddey S, Munnings C. Emerging technologies, markets and commercialization of solid-electrolytic hydrogen production. Wiley Interdiscip Rev Energy Environ 2018;7(3).
[79] Venkataraman V, et al. Reversible solid oxide systems for energy and chemical applications—review & perspectives. J Energy Storage 2019;24:100782.
[80] Liang M, et al. Preparation of LSM–YSZ composite powder for anode of solid oxide electrolysis cell and its activation mechanism. J Power Sources 2009;190(2):341–5.
[81] Thijssen J. Solid oxide fuel cells and critical materials: a review of implications. National Energy Technology Laboratory: US Department of Energy; 2011.
[82] Moçoteguy P, Brisse A. A review and comprehensive analysis of degradation mechanisms of solid oxide electrolysis cells. Int J Hydrogen Energy 2013;38(36):15887–902.
[83] Knibbe R, et al. Solid oxide electrolysis cells: degradation at high current densities. J Electrochem Soc 2010;157(8):B1209–17.
[84] Laguna-Bercero MA. Recent advances in high temperature electrolysis using solid oxide fuel cells: a review. J Power Sources 2012;203:4–16.
[85] Vincent I, Bessarabov D. Low cost hydrogen production by anion exchange membrane electrolysis: a review. Renew Sust Energ Rev 2018;81:1690–704.
[86] An L, et al. Mathematical modeling of an anion-exchange membrane water electrolyzer for hydrogen production. Int J Hydrogen Energy 2014;39(35):19869–76.
[87] Merle G, Wessling M, Nijmeijer K. Anion exchange membranes for alkaline fuel cells: a review. J Membr Sci 2011;377(1):1–35.
[88] Pavel CC, et al. Highly efficient platinum group metal free based membrane-electrode assembly for anion exchange membrane water electrolysis. Angew Chem Int Ed 2014; 53(5):1378–81.
[89] Faraj M, et al. New LDPE based anion-exchange membranes for alkaline solid polymeric electrolyte water electrolysis. Int J Hydrogen Energy 2012;37(20):14992–5002.

[90] Tang X, et al. Noble fabrication of Ni–Mo cathode for alkaline water electrolysis and alkaline polymer electrolyte water electrolysis. Int J Hydrogen Energy 2014;39(7): 3055–60.
[91] Leng Y, et al. Solid-state water electrolysis with an alkaline membrane. J Am Chem Soc 2012;134(22):9054–7.
[92] Park JE, et al. High-performance anion-exchange membrane water electrolysis. Electrochim Acta 2019;295:99–106.
[93] Xiao L, et al. First implementation of alkaline polymer electrolyte water electrolysis working only with pure water. Energy Environ Sci 2012;5(7):7869–71.
[94] Saul Gago A, et al. Degradation of proton exchange membrane (PEM) electrolysis: the influence of current density. ECS Trans 2018;86:695–700.
[95] Felgenhauer M, Hamacher T. State-of-the-art of commercial electrolyzers and on-site hydrogen generation for logistic vehicles in South Carolina. Int J Hydrogen Energy 2015;40(5):2084–90.
[96] Hernández S, et al. Syngas production from electrochemical reduction of CO2: current status and prospective implementation. Green Chem 2017;19(10):2326–46.
[97] Leonzio G. State of art and perspectives about the production of methanol, dimethyl ether and syngas by carbon dioxide hydrogenation. J CO_2 Util 2018;27:326–54.
[98] Akhtar A, Krepl V, Ivanova T. A combined overview of combustion, pyrolysis, and gasification of biomass. Energy Fuel 2018;32(7):7294–318.
[99] Griffin DW, Schultz MA. Fuel and chemical products from biomass syngas: a comparison of gas fermentation to thermochemical conversion routes. Environ Prog Sustain Energy 2012;31(2):219–24.
[100] Munasinghe PC, Khanal SK. Biomass-derived syngas fermentation into biofuels: opportunities and challenges. Bioresour Technol 2010;101(13):5013–22.
[101] Köpke M, et al. 2, 3-Butanediol production by acetogenic bacteria, an alternative route to chemical synthesis, using industrial waste gas. Appl Environ Microbiol 2011; 77(15):5467–75.
[102] Phillips J, Huhnke R, Atiyeh H. Syngas fermentation: a microbial conversion process of gaseous substrates to various products. Fermentation 2017;3(2):28.
[103] de Souza Pinto Lemgruber R, et al. Systems-level engineering and characterisation of Clostridium autoethanogenum through heterologous production of poly-3-hydroxybutyrate (PHB). Metab Eng 2019;53:14–23.
[104] Xie S, et al. Photocatalytic and photoelectrocatalytic reduction of CO2 using heterogeneous catalysts with controlled nanostructures. Chem Commun 2016;52(1):35–59.
[105] White JL, et al. Light-driven heterogeneous reduction of carbon dioxide: photocatalysts and photoelectrodes. Chem Rev 2015;115(23):12888–935.
[106] Liu G, et al. Research advances towards large-scale solar hydrogen production from water. EnergyChem 2019;100014.
[107] Acar C, et al. Transition to a new era with light-based hydrogen production for a carbon-free society: an overview. Int J Hydrogen Energy 2019.
[108] Kovač A, Marciuš D, Budin L. Solar hydrogen production via alkaline water electrolysis. Int J Hydrogen Energy 2019;44(20):9841–8.
[109] Djafour A, et al. Photovoltaic-assisted alkaline water electrolysis: basic principles. Int J Hydrogen Energy 2011;36(6):4117–24.
[110] Sellami MH, Loudiyi K. Electrolytes behavior during hydrogen production by solar energy. Renew Sust Energ Rev 2017;70:1331–5.
[111] de Fátima Palhares DDA, Vieira LGM, Damasceno JJR. Hydrogen production by a low-cost electrolyzer developed through the combination of alkaline water electrolysis and solar energy use. Int J Hydrogen Energy 2018;43(9):4265–75.
[112] Knob D. Geração de hidrogênio por eletrólise da água utilizando energia solar fotovoltaica. Universidade de São Paulo; 2017.

[113] Shedid MH, Elshokary S. Hydrogen production from an alkali electrolyzer operating with Egypt natural resources. Smart Grid Renew Energy 2015;6(01):14.
[114] Ahmad GE, El Shenawy ET. Optimized photovoltiac system for hydrogen production. Renew Energy 2006;31(7):1043–54.
[115] Bhattacharyya R, Misra A, Sandeep KC. Photovoltaic solar energy conversion for hydrogen production by alkaline water electrolysis: conceptual design and analysis. Energy Convers Manag 2017;133:1–13.
[116] Scamman D, et al. Off-grid solar-hydrogen generation by passive electrolysis. Int J Hydrogen Energy 2014;39(35):19855–68.
[117] Maroufmashat A, Sayedin F, Khavas SS. An imperialist competitive algorithm approach for multi-objective optimization of direct coupling photovoltaic-electrolyzer systems. Int J Hydrogen Energy 2014;39(33):18743–57.
[118] Nafeh AESA. *Hydrogen production from a PV/PEM electrolyzer system using a neural-network-based MPPT algorithm.* Int J Numer Modell Electron Netw Devices Fields 2011; 24(3):282–97.
[119] Harrison K, et al. Renewable electrolysis integrated system development and testing. National Renewable Energy Laboratory; 2007.
[120] Garcia-Valverde R, et al. Optimized photovoltaic generator–water electrolyser coupling through a controlled DC–DC converter. Int J Hydrogen Energy 2008;33(20): 5352–62.
[121] Gibson TL, Kelly NA. Optimization of solar powered hydrogen production using photovoltaic electrolysis devices. Int J Hydrogen Energy 2008;33(21):5931–40.
[122] Enrique JM, et al. Theoretical assessment of the maximum power point tracking efficiency of photovoltaic facilities with different converter topologies. Sol Energy 2007;81(1):31–8.
[123] Laoun B, et al. Modeling of solar photovoltaic-polymer electrolyte membrane electrolyzer direct coupling for hydrogen generation. Int J Hydrogen Energy 2016;41(24):10120–35.
[124] Atlam O, Barbir F, Bezmalinovic D. A method for optimal sizing of an electrolyzer directly connected to a PV module. Int J Hydrogen Energy 2011;36(12):7012–8.
[125] Ibrahim H, Anani N. Variations of PV module parameters with irradiance and temperature. Energy Procedia 2017;134:276–85.
[126] Rau S, et al. Highly efficient solar hydrogen generation—an integrated concept joining III–V solar cells with PEM electrolysis cells. Energy Technol 2014;2(1):43–53.
[127] Dimroth F, et al. Hydrogen production in a PV concentrator using III-V multi-junction solar cells, In: 2006 *IEEE 4th world conference on photovoltaic energy conference*; 2006.
[128] Peharz G, Dimroth F, Wittstadt U. Solar hydrogen production by water splitting with a conversion efficiency of 18%. Int J Hydrogen Energy 2007;32(15):3248–52.
[129] Fallisch A, et al. Investigation on PEM water electrolysis cell design and components for a HyCon solar hydrogen generator. Int J Hydrogen Energy 2017;42(19):13544–53.
[130] Ferrero D, Santarelli M. Investigation of a novel concept for hydrogen production by PEM water electrolysis integrated with multi-junction solar cells. Energy Convers Manag 2017;148:16–29.
[131] Stansberry J, et al. Experimental analysis of photovoltaic integration with a proton exchange membrane electrolysis system for power-to-gas. Int J Hydrogen Energy 2017;42(52):30569–83.
[132] Shafiei Kaleibari S, Yanping Z, Abanades S. Solar-driven high temperature hydrogen production via integrated spectrally split concentrated photovoltaics (SSCPV) and solar power tower. Int J Hydrogen Energy 2019;44(5):2519–32.
[133] ARENA, https://arena.gov.au; 2019.

CHAPTER

10

Solar PV power plants

Mohammadreza Aghaei[a,b], *Aref Eskandari*[c], *Shima Vaezi*[d], *and Shauhrat S. Chopra*[e]

[a]Eindhoven University of Technology (TU/e), Design of Sustainable Energy Systems, Energy Technology, Department of Mechanical Engineering, Eindhoven, The Netherlands, [b]Department of Microsystems Engineering (IMTEK), University of Freiburg, Solar Energy Engineering, Freiburg im Breisgau, Germany, [c]Electrical Engineering Department, Amirkabir University of Technology (Tehran Polytechnic), Tehran, Iran, [d]Chemical Engineering Department, Islamic Azad University South Tehran Branch, Tehran, Iran, [e]School of Energy and Environment, City University of Hong Kong, Kowloon, Hong Kong

10.1 Introduction

In recent years, the photovoltaic (PV) industry has emerged as one of the key drivers of the electric power sector. The contribution of PVs in energy utility also has increased. The main reasons for this are the cost decline and technological advancement seen in the PV module industry, where conversion efficiencies have greatly increased. The growth seems to be much faster when compared to the early stages. This spectacular growth has boosted the market for its aligned groups (for example, inverter manufacturers, battery manufacturers, etc.). At present, the global market for PV has exceeded 100GW [1, 2]. The growth trend for PV varies in different countries, as the PV growth rate in China is a bit lower due to its policy change. It was expected that through the end of 2019, at least 32 countries "representing every region had a cumulative capacity of 1GW or more." PV installation size would range from small to large scale. But, at present, most nations are very interested in the development of large-scale PV projects. In some countries, they even call these ultramegawatt solar projects. The main driving force for this growth is

Photovoltaic Solar Energy Conversion
https://doi.org/10.1016/B978-0-12-819610-6.00010-7

global policy change and new regulatory schemes. The worldwide growth for the PV market in various countries is shown in the Table 10.1. Fig. 10.1 depicts solar PV market scenarios between 2018 and 2022 worldwide.

TABLE 10.1 PV market and policy in various countries [3].

Country	Highlights
China	– China, being one of the giants in solar energy harvesting, created a record by installing around 53.3% of solar installation in one particular year. This percentage is believed to be more than half of global solar PV installations. – The Chinese solar market has seen tremendous growth of around 128% when compared to 2015. In 2017 alone, the growth of solar PV in China stunned all the global players due to the feed-in-tariff policy and other benefits given in terms of market creation and subsidies.
United States	– In the United States, the installed capacity of grid-connected PV systems has captured the PV market when compared to the other types of PV systems. – Globally, the United States secured the second position in 2017s PV market. – Due to various uncertainties related to tax credit schemes and tariff rates, the growth of PV has declined and the trend is not as expected.
India	– India is one of the pioneers in solar PV installation, and the dream of India in renewable harvesting was favored by the Ministry of New and Renewable Energy (MNRE). – At present, the total PV installation capacity exceeds 19 GW. – The manufacturing of the PV module in India is quite limited when compared to countries such as China, hence, India has to import modules, where they experience difficulties in terms of import taxes, module price hikes, etc. These issues hinder PV growth in India. – Even though India has experienced many issues, it is at the forefront in terms of PV installation with 45% of the new additions.
Japan	– Japan's solar PV market seems to be slowing. The statistics reveal that in 2017, Japan installed around 7.2 GW of PV capacity, but in the present its installations seem to be 9% less compared to 2017. – The feed-in-tariff rates and various policies seem to be favoring the PV market. – The experts from the Japanese PV Energy Association (JPEA) highlight that the solar downturn would continue until 2024 as the country is trying to capture the market and infrastructure needs for progress.
Europe	– Europe is one of the pioneers in solar PV installation. In 2017 alone, a 30% increase in PV installation was seen, and this was in comparison with the previous year's installation. – Turkey's contribution is crucial to PV market growth in Europe. Recently, Turkey added around 2.6 GW of PV installation.

TABLE 10.1 PV market and policy in various countries [3]—cont'd

Country	Highlights
	– The total PV market in Europe is the collective combination of 28 European nations. When these 28 countries are considered, the growth seems to be not much.
Latin America	– The PV market in Latin America is considerably lower when compared to other countries. – Among the many Latin American countries, Brazil and Chile are progressing well in the PV market. – As per 2017s statistics, Brazil (with an installed capacity of 1 GW) and Chile (with an installed capacity of 788 MW) secured the first and second positions, respectively. – Countries such as Mexico and Colombia are in the PV race and are trying to capture the market with attractive energy policies and auctions.
Australia	– The market for PV installation is mostly based on residential PV systems. – At present, the total installed capacity in Australia is around 1.3 GW, which had a great downfall of 15% in 2016. – The new policies related to the levelized cost of electricity are increasing the attention of the PV market toward the development of large-scale ground-mounted PV systems.
Middle East	– In the Middle East, the United Arab Emirates (UAE) are at the forefront in the PV market. – UAE recently added a PV capacity of 262 MW to its utility. – Apart from countries such as Abu Dhabi and Dubai, countries such as Saudi Arabia and Oman are encouraging solar installation and have called for tenders recently. – Middle East regions have set the records on achieving the lowest feed-in-tariff rates.
Africa	– Africa has a very low installation capacity for PV of less than 100 MW. – Even though the current installation capacities are very low, countries such as South Africa, Egypt, and Nigeria are trying to contribute secure financial assistance in building new solar power plants. – South Africa is one of the largest markets for PV on the entire African continent. – Countries such as Zambia, Madagascar, Senegal, and Ethiopia have called for tenders in building megawatt-level power plants recently. – The major drawbacks seem to be the existing power infrastructure and financial assistance toward capital investment.

10.2 Development

While developing a solar PV power plant, the following six steps are necessary. The first is the solar resource assessment, a step that will allow the installer to have a detailed understanding of the available solar potential. The second step is site selection and analysis. Here, apart from the solar

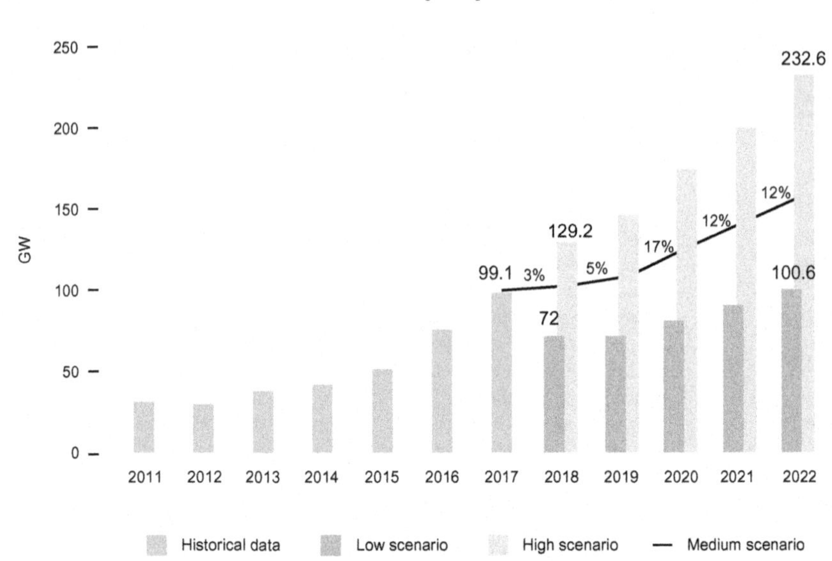

FIG. 10.1 World annual solar PV market scenarios 2018–22 [3].

irradiance potential, the other parameters related to environmental conditions should be considered at the site location and also during the construction phase. The third step is the environmental stress assessment on the PV plant components. The fourth step is the predesign and optimization of the PV system, where the sizing of the PV plant is done. Once the sizing is done, in the fifth step, the feasibility study is carried out to assess the energy potential and other economic and environmental emission parameters. In the fifth step, mostly multicriteria decision making is carried out to see whether the PV plant is performing better in terms of the considered analysis (for example energy, emission, exergy, economic). The last step is the social consideration where the interests of individuals, groups, communities, and society are considered in relation to the PV plant installation at that site.

10.2.1 Solar resource assessment

Solar resource assessment is a necessary step in PV plant design that allows understanding the feasibility of a plant in a given location. One of the ultimate objectives of the assessment is to find out the amount of solar potential that is available and how much energy from a PV power plant with typical PV technology can be annually produced [4]. There are certain factors that vary from place to place and with time, hence it is important to gain knowledge of these factors before establishing. These factors include the solar irradiance at a horizontal plane, the irradiance at a tilted position of the PV module, and a sun path diagram. Solar resource assessment generally involves collecting meteorological data from the site such as weather data, the amount of sunlight received in the location, wind speed, air temperature, etc. There are two methods in which the assessment is done:

Onsite ground measurements and collecting satellite data:

Over a lengthy period of time, lasting one to several years, a variety of sensors are used such as pyranometers to measure solar irradiance at the location [4, 5].

Secondary or primary data from satellites:

The second method involves studying obtained secondary or primary data from satellites, such as statistically aggregated solar data, to conduct the SRA. The SRA is merely the feasibility study. There are a variety of other factors to consider in order to optimize the yield of a solar power plant [4, 5].

The solar resource assessment potential would vary from location to location. Here, in Fig. 10.2 we show how global horizontal solar irradiance varies across the globe. The solar resource map below reveals that the long-term average of global horizontal irradiation varies from 2.2 to 7.4 kWh/m^2/day worldwide.

Also, the related PV potential is shown in Fig. 10.3. With the use of specific PV technology, the maximum possible PV energy varies from 2 to 6.4 kWh/KWp on a daily basis. The yearly PV potential would vary from 730 to 2337 kWh/kWp.

10.2.2 Site selection and analysis

Irrespective of PV plant size, site selection is a vital part of the PV plant design, especially in a large-scale PV system. It directly affects PV plant performance. For site selection, the preferred tool is the geographical information system (GIS), which allows the installer to analyze the location parameters and see the impacts on the overall performance. The GIS also allows editing the maps and the aim in energy planning with a typical layout arrangement. Based on the initial understanding, one can frame the decision support model with various parameters that influence the overall performance. Consequently, effective site selection allows improvements in performance. In addition, it maximizes output capacity, helps to minimize project costs, and helps in planning future infrastructure projects [4, 5].

10.2.3 Environmental stress assessment

Stress assessment is one of the crucial activities that should be considered in the PV plant design and installation stage. The installation site would experience adverse weather conditions and even risks (natural hazards). These risks would affect the PV plant's overall performance and can

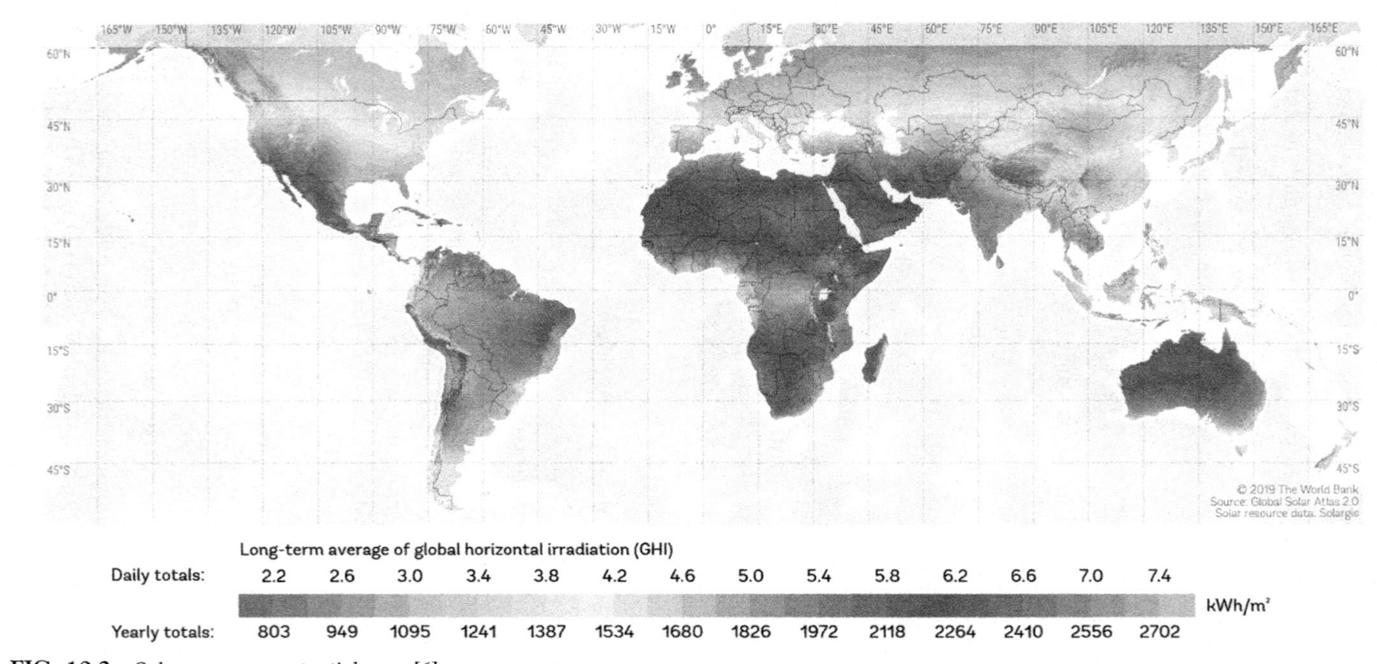

FIG. 10.2 Solar resource potential map [6].

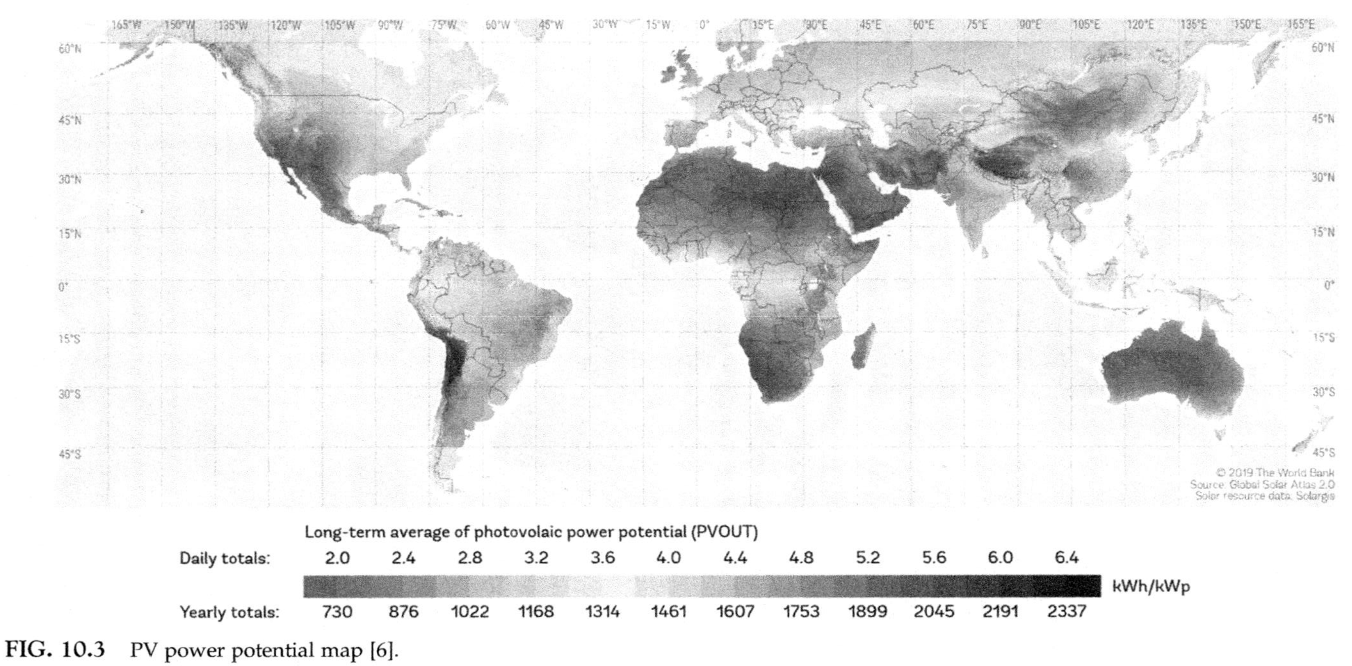

FIG. 10.3 PV power potential map [6].

TABLE 10.2 Environmental stress analysis.

Classification	Activity to be considered
Extreme weather events	Sandstorms
	Heavy snowfall
	Flooding
Continuous climatic loads	UV exposure
	Humidity and temperature
	Salt content in the air
	Atmospheric corrosivity
	Soiling
	Abrasion from sand
Natural hazards	Global risks: earthquakes, volcanic eruptions, etc.
	Local risks: moving sand dunes, avalanches, etc.

even cause reliability issues. In Table 10.2, a list of parameters that should be considered for the environmental stress assessment is given [7, 8].

10.2.4 Predesign, optimization, and feasibility studies

The feasibility study of PV plants is not an easy task. There are different ways one can do feasibility studies. One is by simulation models. The other is based on experimental data. However, the most familiar way is simulation models. Here, we have reviewed various simulations and listed them in Table 10.3 [9].

10.2.5 Social consideration

In the process of maximizing profits, it is the company's responsibility to ensure that third parties are not affected. They must first obtain public acceptance to construct the power plant. For example, the project would likely not be well accepted in locations where there is a shortage of land, thus the projects have to maintain a distance from residential and urban areas [10]. Moreover, the site for the power plant should be able to generate a source of employment. The site chosen should not be fertile, which would impact the livelihoods of farmers, achieving the opposite of the desired effect [11]. Therefore, it is important to support those disadvantaged parts of the population such as poor people and women by generating employment opportunities and contributing to the economic

TABLE 10.3 Various software tools used for feasibility studies of PV power plants [9].

Name of software	Organization behind development	Operating system	Assessments to be carried out	Access option (subscription/open access)
iHOGA	University of Zaragoza, Spain	– Windows XP, – Vista,7 or 8C++	– Techoeconomic assessment – Single and multiobjective function-based optimizations – Genetic algorithm is used	– The PRO version of the tool is under subscription option – EDU version is free for education purposes only
iGRHYSO	University of Zaragoza, Spain	– Windows C++	– Technical assessment – Economic assessment	Subscription
INSEL	German University of Oldenburg	– Windows – Fortran and C/C++	– Assessment of planning – Monitoring of electrical energy system – Monitoring of thermal energy systems	Subscription
HOMER	National Renewable Energy Laboratory, United States	– Windows Visual C++	– Technical assessment – Economic assessment – Emission assessment – Single objective function-based optimization	Subscription
HYBRID2	University of Massachusetts, United States, and NREL (Hybrid1 in 1994, Hybrid 2 in 1996)	– Windows XP Visual BASIC	– Technical assessment – Economic assessment	Subscription

Continued

TABLE 10.3 Various software tools used for feasibility studies of PV power plants [9]—cont'd

Name of software	Organization behind development	Operating system	Assessments to be carried out	Access option (subscription/open access)
PVsyst	PVsyst SA, Switzerland	– Windows version, – LINUX through VirtualBox	– 3D and 2D design of PV system – Shading analysis – Energy performance assessment – Economic assessment – Environmental assessment	Subscription
RETScreen	Ministry of Natural Resources, Canada	– Windows 2000, – XP, – Vista Excel, – Visual Basic, C	– Financial assessment – Environmental analysis – Portfolio assessment – Energy planning – Statistical assessment of energy networks	Open access and subscription
RAPSIM	University Energy Research Institute, Australia	– Windows	– Hybrid power system – Technical performance assessment – Power capacity range evaluation based on energy source and load	Open access
TRNSYS	University of Wisconsin and University of Colorado (1975)	– Windows	Simulate transient system behavior	Subscription

development of the region. To do so, companies ought to provide training to the local population in impoverished areas instead of importing labor from elsewhere.

10.3 Engineering, procurement, and construction

10.3.1 PV plant design

PV systems can be designed with varying complexities. The performance of PV system components can change due to environmental conditions and climate changes. These large changes in system performance may alter the primary design significantly. Therefore, these changes should be taken into account in system design. Before designing and selecting components, customer expectations and the needs of the system must be determined. The money invested by the customer affects the system size. Note that designers must familiarize themselves with national standards due to key points in the system design that are described by these standards in most countries. According to the aforementioned, many compromises must be made to achieve a desirable balance between cost and performance. This section explains some of the major issues in designing a PV power plant [12].

10.3.1.1 System layout

The site conditions can affect the general layout and the distance selected between rows in a PV plant. The purpose of the plant layout is to minimize expenses while gaining the maximum income from the plant [4_new]. Generally, this means:

- Reducing the effect of shading by standard designing and sufficient spacing between the PV strings.
- Electrical loss reduction by minimizing cable runs in designing the plant layout.
- Maintenance objectives by creating availability paths and enough space between rows.
- Optimization of the annual energy yield by selecting an accurate tilt angle according to the latitude of the location.
- Receiving maximum annual income from power generation with correct module orientation.

10.3.1.2 Choice of system components

Component selection plays an important role in system design. In this section, the main factors for selecting components are listed, and then the design process is discussed in the next section.

10.3.1.3 Modules

Selecting the proper module is very important because it is the heart of a PV plant. Also, it has the highest cost of equipment in the PV plant. Many factors affect that selection such as performance, warranty, aesthetics, cost, and environmental conditions. The important factors are discussed in the following [13].

The identified environmental conditions during site evaluation are summarized below:

- Temperature range: The temperature range of the installation location is important in the selection of PV modules because the modules operate at a specified temperature range in the datasheet. For instance, a low-temperature coefficient in modules is required for locations with hot weather.
- Coastal environments: Salt stain examination of the PV system should be carried out according to IEC 61701 for modules at a 1 km distance from shore.
- Snow environments: Module installation with a load capacity of 5400 Pa is very important in snow areas.

The warranty for PV modules is divided into two categories: materials and output power. Modules are covered by a 10-year warranty on materials and 25 years on output power. Module warranty for output power means that the PV module can generate 80–85% of its nominal power during the warranty duration. For instance, if the output power of a module is 270 W under STC conditions, it must be able to generate at least 216 W after 25 years. Also, the nameplate of each module should have UL and IEC certifications. Certifications used in modules are located on the datasheet or the manufacturer's website [14].

10.3.1.4 Mounting structure

After selecting the system modules, designers must specify the type of mounting structure. The mounting structure holds PV modules and also anchors modules to the ground or roof in a flat plane. The choice of the type of structure depends on factors such as module type, installation location, and environmental conditions. Environmental conditions such as the amount of heavy rain, closeness to the coast, snow loading, and wind loading are very important for choosing the type of mounting structure. Also, the aesthetics of the system can be considered according to customer expectations [13, 14].

10.3.1.5 Inverters

There are four types of inverters: central, multistring, string, and modular. The following considerations are important for selecting an inverter [13, 15]:

- Inverter efficiency: The efficiency of the newest inverters is similar. Transformer-less inverters have more efficiency and are widely applied.
- Direction and tilt angle of PV modules: If modules are at various tilt angles and directions, the system should be divided into different strings. That's better to apply a string inverter for each string or a multistring inverter. Typically, using a few small inverters instead of a large inverter is more expensive, although the benefit of more energy efficiency may outweigh the extra expenses.
- Inverter location: The inverters must be impermeable to dust and water when installed outdoors. In this case, the protection rating is specified in the inverter datasheet that includes Ingress Protection 65 or National Electrical Manufacturers Association 3R.
- Cost of various inverters.
- The average annual energy production.

The designers should specify the pros and cons of each solution depending on the system cost and performance and need to consult with the customer for the final decision.

10.3.1.6 Transformers

Grid and distribution transformers are the essential components in PV plants. Distribution transformers step up the output voltage of the inverter at the distribution level. Electricity can be injected into the grid directly when the plant is connected to the distribution grid. The grid transformer is required to further increase the voltage when the plant is connected to a transmission network. The efficiency and cost of ownership are key criteria in the selection of transformers and they directly affect the annual income from the plant. The selection criteria consist of power rating, system voltage, duty cycle, guarantee, site conditions, vector group, sound power, and efficiency. The warranty provided for transformers is different among manufacturers and 18 months is the minimum warranty for typical transformers. Mean time to failure (MTTF) for distribution transformers is 30 years, according to academic studies and manufacturer data. This depends on the duty cycle and load profile [16]. The transformers must be manufactured to the following standards:

- IEC 60076
- BS EN 50464-1

Choosing and sizing transformers: Typically, the inverters provide power at voltages between 300 and 450 V. However, PV power plants are connected to the grid with a voltage level of 11 kV and above. Therefore, one or more transformers is needed to increase the voltage. The primary and secondary side voltages of the transformer determine its position in the electrical system.

Several main issues should be considered when choosing a transformer for PV plants:

- Position in the electrical system
- Transformer capacity
- Environmental conditions
- Physical location

The maximum power extracted from the PV array will define the transformer capacity. The technical requirements of the electricity grid should be considered in the selection of the transformer because the transformers will organize a main component of the substation design. Therefore, they should comply with local or international characteristics, including IEC 60076. The following points should also be considered [16]:

Losses: Transformers can lose energy in two ways, through magnetic current in the core (called iron losses), and through copper losses in the windings. The reduction of losses is an essential requirement in transformers because this will increase the delivered energy to the grid and hence bring more revenue to the PV power plant.
Test: A number of routine tests must be carried out on each transformer manufactured based on the IEC 60076 standard.
Delivery: Delivery and construction of transformers require a period of time. Especially, most large transformers are manufactured as ordered, and will therefore need a long time between the start and end of the production process.

10.3.1.7 Balance of system equipment

A system needs other components to operate safely and properly. These components are called the balance of the system equipment (BOS). The BOS is used to connect and protect the PV array and inverter. The key BOS are listed below [17].

PV combiner box

It is located between the PV array and the inverter. Generally, it is applied to interconnect the output cables from the PV array. Connections are usually built with screw terminals and should be of high quality to prevent overheating. In addition, it must be allowed for use in outdoor locations, for instance, ingress protection (IP) [18].

DC and AC switching

The task of these devices is to isolate and protect the equipment of the PV power plant. These devices include the DC and AC main disconnect/isolator, and DC and AC fuses/circuit breakers, which are briefly discussed below [19].

DC and AC main disconnect/isolator

Disconnects/isolators are manual devices that provide electrical isolation for the equipment of a PV power plant. Therefore, these devices are needed during maintenance and installation.

Circuit breakers and fuses

Circuit breakers and fuses are usually applied in order to protect overcurrent in a PV plant. These devices are installed on both the array side and the grid side. DC fuses usually are set by local codes; also ensure that the fuses used are based on the DC rating. Moreover, breaking a direct current is more difficult, which causes differences with AC fuses applied in normal applications. Typically, DC fuses are often costly because most of them can be applied in AC applications. It should be noted that due to PV is the current-limited source, DC circuit breakers/fuses operate differently to AC circuit breakers/fuses. The circuit breakers/fuses operate very fast in the AC circuits because the generated fault current is very large, whereas the produced fault current in the DC side is not higher than the normal operating current [18].

Cabling

The proper choice of cables affects the overall performance of the PV plant. This means that cables should be suitably UV-resistant for outdoor use and also the size of the cables should be designed according to the system current and voltage. However, there are local codes that cover cable sizing, and also they vary widely between countries. In addition, cable sizing varies according to the location, for example, the size of cable used in a string is thoroughly different from the cable size used in an array. The cables for installation should be properly sized so that the power losses between the inverter and PV array do not exceed the specified value. These cables must be able to carry the system current because if the cable crossing current is greater than the capacity of the installed cable, then the cable is heated and may damage other elements of the circuit [13, 17].

Voltage and current sizing

Cables have a maximum current and voltage, and these values should never be exceeded. Codes for cable sizing vary between various countries.

For instance, according to the United Kingdom and Australian codes, DC cables must be able to carry at least $1.25 \times I_{SC}$. This value is different in the United States and is equal to $1.5625 \times I_{SC}$.

System protection

In order to operate the PV system safely, the structure of the protection system is of great importance because the PV plant can fail for many reasons. The protective structures and devices used in the PV plant will be described in the safety issues section. In addition, the details of these topics are set in national standards.

10.3.1.8 AC switchgear

The proper type of switchgear depends on the operating voltage. The switchgear for voltages below 33kV usually includes an internal metal-clad cubicle type with air- or gas-insulated busbars and SF6 or vacuum breakers. The suitable choice for high voltage switchgear in outdoor is air-insulated whereas, gas-insulated is typically used for indoor applications. All switchgear should [16]:

- Have proper labels in the ON and OFF positions.
- Be built based on IEC standards and national codes
- Be used for suitable grounding
- Be ranked for operating voltage and current

10.3.1.9 Plant substation

The plant substation comprises equipment such as supervisory control and data acquisition (SCADA), switchgear, LV/MV transformers, a metering system, and protection. The substation layout should optimize the space used in accordance with building codes. A secure working environment should be provided for staff in the plant. The electrical equipment generates heat in the plant environment, hence suitable air conditioning should be installed. Note that a large substation must be built and designed based on the requirements of the grid.

Segregation between converter rooms, MV switch rooms, storerooms, control rooms, and offices is an essential requirement. In addition, lighting and secure access should be provided. Lightning protection must also be installed to diminish the lightning effect on equipment [16].

10.3.1.10 Grounding and surge protection

In order to protect against electric shock and fire hazards, grounding should be provided in a PV plant that encompasses lightning and surge protection, the noncurrent carrying parts of a PV array, and inverter grounding. In designing the grounding, the national codes and special features of the site location should be considered. The appropriate designing diminishes the electric shock risk to users onsite and fire hazards caused by a fault or lightning [20].

10.3.2 Design process

There are three main components for designing a PV plant: the modules, the mounting structure, and the inverters. The design process consists of a series of successive steps [14]:

- Goals: The goal is to determine the size of the DC system based on AC annual energy generation.
- Module selection: In order to achieve this goal and also the considered budget, a proper module is chosen based on module dimensions, output, and adaptability with mounting structure.
- Array layout: In this step, the number of modules is determined and the array layout on the mounting structure is designed.
- Electrical calculations: String and array output values (voltage, current, and power) are computed using the module specification.
- Inverter selection: According to the inverter specification, its compatibility is checked with the calculated array layout.
- Mechanical calculations: In this step, the quantities of mounting structure elements used are determined.
- Meters, disconnect, etc.: The rest of the PV plant components are easily completed.

Here, a realistic example will be explained to better learn the PV plant design procedure. The sample design will be carried out step-by-step according to the above descriptions.

10.3.2.1 Sample design

This design is a sample of a 4.5 kw grid-connected PV plant with a string inverter. The PV array includes 18 modules that are located in two rows with nine modules per row, as shown in Fig. 10.4. These modules are mounted on a rooftop, which has an area equal to 32 × 14 feet. The following steps are presented to complete the design of the related grid-connected PV plant according to the above template.

Step 1. Site assessment: In this step, a complete map of the roof is essential to identify the existing details in the roof such as penetrations,

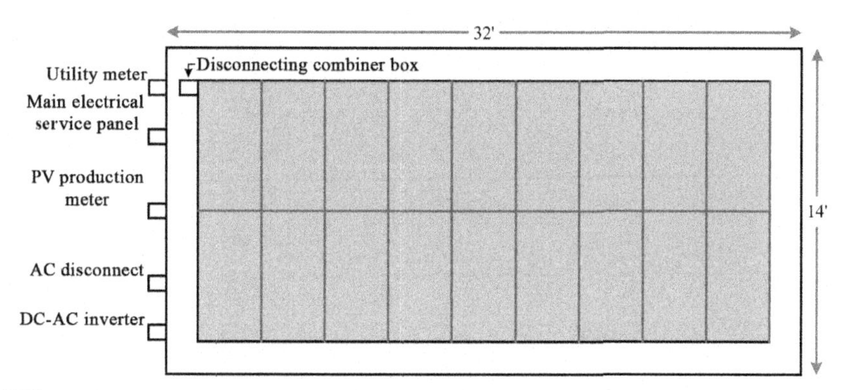

FIG. 10.4 PV array layout.

obstructions, etc., that are needed during installation. Also, the roof tilt and azimuth are 23 and 205 degrees respectively, and the module orientation is portrait.

Step 2. Module specifications: The selected module is the Helios 6T-250. Table 10.4 lists the detailed parameters of the PV module under STC.

Step 3. Number of modules: The number of modules is obtained by dividing the total DC system power (4.5 kW) into the individual module power:

$$\text{Number of modules} = 4500\,\text{W} \div 250\,\text{W} = 18 \tag{10.1}$$

Note that this PV plant can generate 6.294 kWh of AC energy per year according to the location specification.

Step 4. Array layout: In this step, we determine the placement of the modules on the rooftop. First, the roof dimension should be converted to inches:

$$32\,\text{ft} \times 12 = 384\ \text{in.} \tag{10.2}$$

Next, the number of modules in a row is determined according to module orientation:

$$384\,\text{in.} \div 39\,\text{in.}(\text{module width}) = 9.85 \tag{10.3}$$

Therefore, the number of modules in a row is nine and also the remaining space on each side of the row is 16.5 in.:

$$9\,\text{modules} \times 39\ \text{in.} = 351\ \text{in.} \tag{10.4}$$

$$384 - 351 = 33\,\text{in.}$$

$$33 \div 2 = 16.5\,\text{in.}$$

TABLE 10.4 PV module specifications.

Parameters	Values
P_{mpp}	250 W
V_{mpp}	30.3 V
I_{mpp}	8.22 A
V_{oc}	37.4 V
I_{sc}	8.72 A
Length	66.1 in.
Width	39.0 in.

Now, the number of rows on the roof is determined:

$$14\,\text{ft} \times 12 = 168\,\text{in.} \tag{10.5}$$

$$168\,\text{in.} \div 66.1\,\text{in.}\,(\text{module length}) = 2.54 \tag{10.6}$$

Therefore, the number of rows is two and also the remaining space is 36 in.:

$$\begin{gathered} 2\,\text{rows} \times 66.1\,\text{in.} = 132.2\,\text{in.} \\ 168 - 132 = 36\,\text{in.} \end{gathered} \tag{10.7}$$

Step 5. String size: The string size is determined based on the number of modules and the installation location. For the best performance, the strings must have the same number of modules per string. Therefore, this system contains two strings, and each string has nine modules.

Step 6. String output power/inverter input power: The string voltage is computed by multiplying the number of modules per string by the module operating voltage at the maximum power point. The output power of the string is determined by having the module operating current at the maximum power point as follows:

$$\text{String voltage} = 9\,\text{modules} \times 30.3\,V\,(V_{mpp}) = 272.7\,\text{V} \tag{10.8}$$

$$\text{String power} = 272.7\,\text{V} \times 8.22\,\text{A}\,(I_{mpp}) = 2.242\,\text{kW} \tag{10.9}$$

$$\text{Inverter input power (array power)} = 2\,\text{strings} \times 2.242\,\text{kW} = 4.484\,\text{kW} \tag{10.10}$$

Step 7. Maximum string voltage and current: The maximum string voltage is computed by multiplying the number of modules per string by open-circuit voltage, and then multiplying by 1.25. The National Electrical Code (NEC) determines this value due to the temperature limit. Also, the maximum string current is computed by multiplying the short-circuit current by 1.25 and then multiplying by 1.25. These values are used due to the maximum radiation value and NEC, respectively.

$$\text{Max voltage} = 9\,\text{modules} \times 37.4\,\text{V}\,(V_{oc}) \times 1.25 = 421\,\text{V} \tag{10.11}$$

$$\text{Max current} = 8.72\,\text{A}\,(I_{sc}) \times 1.25 \times 1.25 = 13.63\,\text{A} \tag{10.12}$$

Step 8. Inverter: An inverter is chosen based on the DC system size. Also, the inverter rating should exceed the array power. Therefore, the Fronius IG Plus 5.0 is the selected inverter that has 5 kW power. Table 10.5 lists the inverter specifications.

Now, the inverter specifications should be checked in order to match with the string and array layout. The string operating voltage is 272.7 V, which is within the MPPT voltage range (230–500 V) of the inverter. Also, the maximum string voltage (421 V) is lower than the maximum system voltage (600 V).

Because this system has two strings, the array output current is twice the string current (8.22 A). Therefore, the array current (16.44 A) is lower than the maximum DC input current (23.4 A).

10.3.3 Yield prediction

Calculating the expected energy of the PV power plant is one of the important phases in the project feasibility process. The goal is to forecast the average annual energy produced by the PV power plant during the operation of the system (25–30 years). In order to accurately predict the energy yield, the full detailed PV plant should be simulated by software. To this end, in addition to the detailed specs of the plant equipment, the data such as temperature situations and solar resources of the site are also required. Generally, the process of system simulation through software includes the following steps [17]:

1. Using satellite sources and meteorological stations for the measurement of environmental information. The time series of irradiance on the horizontal plane is one of the most important pieces of information needed at the site location.
2. Computing the irradiance on the tilted plane.
3. Modeling the PV plant according to the temperature and irradiance variations in order to compute the energy yield prediction.
4. Applying losses with accurate information on equipment and environmental conditions. This information includes DC and AC

TABLE 10.5 Inverter specifications.

Parameters	Values
MPPT voltage range	230–500 V
Max. system voltage	600 V
Max. DC input current	23.4 A
AC output voltage	208/240/277 V

wiring, module configuration, transformers, inverters, PV modules, soiling characteristics, and auxiliary equipment.

5. Using the evaluation of uncertainty in values and the statistical analysis of information to achieve the proper level of uncertainty.

10.3.3.1 Losses in PV systems

There are several factors that hinder the performance of PV systems at maximum efficiency. These factors depend on the plant design and site characteristics. A procedure called derating accounts for these factors. This section describes these factors and the calculation of derating.

1. PV module temperature:

As previously mentioned, the output power of the PV array is affected by the operating temperature. This effect on the PV module can vary depending on the type of module. The amount of impact can be computed through the temperature coefficient presented in the datasheet of the modules. Remarkably, if the temperature of the module is more than the standard temperature condition (25 °C), it can lead to power losses [21].

2. Soiling:

Losses caused by soiling (bird dropping and dust) depend on the following factors:

- The site location, as the loss in deserts can be large compared to other locations. Also, local pollution caused by industry and agriculture can significantly reduce the solar resource.
- Rainfall frequency
- The cleaning strategy outlined in the O&M.
- Module tilt angles; the modules with a high tilt angle are expected to have fewer power losses because they can better benefit from rainwater.

Derating caused by soiling can be estimated. Losses in very dirty sites are considered to have a 10% reduction in the output power. For clean sites that have regular rainfall, there is a 5% reduction in the output power [13, 22].

3. Shading:

Shading can play an important role in the loss of system efficiency for two reasons [15, 23]:

- If the size of the shadow on the PV modules is large enough, it will cause a large voltage drop in the system. Then the inverter will turn off if this voltage is outside the inverter's operating voltage.
- The shadow of one string can affect other strings as well because the irradiance received affects the current and voltage directly. In this case, unshaded strings produce the same voltage as the shaded strings.

4. Module quality:

The specifications presented in the module datasheet do not precisely match the actual PV modules used in a system. There is a tolerance in the nominal peak power of the PV modules. Deviation in the characteristics of the real module affects the energy yield. Generally, the output power of PV modules is larger than the power presented in the datasheet. Also, some manufacturers provide positive tolerance in power that can be used for the energy yield [17].

5. Module mismatch:

The performance losses caused by mismatch impact, are related to the dissimilar values of the current and voltage of the modules in a PV string. This variation leads to an increase in power loss. The tolerance in the module power causes this loss [24].

6. Voltage drop:

Voltage drop refers to the losses in the cables. These losses due to electrical resistance in cables are called ohmic losses. There are two types of cable in the PV power plants: DC cables and AC cables. DC cables are located between the modules and the inverter input and AC cables are located between the inverter output and the metering point. Typically, losses in DC cables are larger than losses in AC cables [17]. Generally, there are two main ways to minimize the voltage drop:

- Using larger cables
- Reducing the length of cables

National codes allow the minimum voltage drop in the system to be between 2% and 5% and the voltage drop in DC cables to be 1%.

7. Inverter efficiency:

Heat dissipation generated from the transformer and electronics devices leads to losses of power in the inverter. Also, transformer-less inverters have higher efficiency. The inverters' efficiency is provided in the datasheet. The inverters should be installed in areas with good air conditioning.

10.4 Operation and maintenance

This section describes the maintenance needed for PV power plants. Following the user manual and the maker's recommendations for all the maintenance of all components are very important. The owner can undertake some maintenance functions, but most of them need an eligible technician. The task of the maintenance company is to inspect and examine the system in order to assure proper functioning of the equipment and system, which is performed once or twice a year. Inspection results should be documented in a maintenance schedule and component booklet that must be presented to the owner before the commissioning of the system. Documents delivered to the user can be useful for the following reasons [25]:

- Information extracted from inspections can demonstrate variations in important system parameters over time and deviations from normal conditions.
- Recorded information about system performance over a year can help to compare performance in the future.

Performance monitoring is one of the most important features of maintenance because it can detect and modify system problems before serious hazards threaten the system.

10.4.1 PV array maintenance

PV arrays can operate without problems for many years due to the lack of moving parts. However, to ensure the efficient and safe operation of PV arrays, regular maintenance is required. Electrical and mechanical surveys and cleaning the modules are common maintenance functions. The maintenance of PV modules should be done by trained personnel. Although rain cleans any dirt on PV modules, the technicians need to deal with bird droppings, any soil created, or falling leaves. In this case, two points are recommended: (1) use freshwater with soap for cleaning the modules, (2) solvents are never used. In general, the presence of snow on PV arrays is not a problem until the PV modules conform to standards. However, if there is a lot of snow for a long time, the owner can consider cleaning them because they will overshadow the PV array and affect the efficiency. Also, frequent pruning of trees that cast shade on PV modules is very important [26].

The tasks of a qualified technician consist of the inspection of connection points to assure that they are impermeable and checking the mounting structure to find symptoms of corrosion. In addition, wiring and cabling must be checked in order to ensure that they are mechanically healthy and weatherproof.

10.4.2 Inverter maintenance

Generally, inverter maintenance includes:

- Keeping the inverter installation place clean
- Cleaning and checking the electrical and mechanical connections
- Checking the inverter performance

Each inverter manufacturer lists requirements and instructions for maintenance in the manual. Inverter maintenance must be performed through a qualified technician. Also, any maintenance on the inverter should be documented in the system booklet [27].

10.4.3 Troubleshooting

Although a properly designed and well-installed PV power plant can operate without flaws for many years, defects may occur in the system. One of the following two reasons can lead to malfunctions in the system:

- Fault in the equipment
- Problem in the grid

Similar to the maintenance section, a qualified technician should undertake system troubleshooting because the technician is aware of the dangers of working with high voltage and at height [13, 28].

10.4.3.1 Problem identification

The monitoring system alerts the owner of the PV power plant when the plant is not operating or underperforming. Visual inspection is the first step for identifying any problems such as shading of modules, and the owner should contact a technician if the problem is not identified.

Troubleshooting should be done in daylight to ensure that the PV modules have sufficient power for testing and that there are sufficient hours to conduct a full inspection. In the first troubleshooting step, the DC and AC main disconnect/isolator should be checked. If the disconnects/isolators are on, the next inverter is checked. The inverter failure is usually caused by electronic component failure. In this case, the inverter may display an error message. The problem may not be identified immediately if the inverter is off because no messages are displayed; it should be determined whether the power of the DC and AC is present at the inverter. If the AC power is not available, the ways back to the supply point should be checked in order to detect faults in the system. If the inverter does not contain a fault and the AC power is present in the inverter, the PV array should be inspected [13,28].

10.4.3.2 Troubleshooting of the PV array

Depending on the system configuration, this troubleshooting will be different. It is assumed that the system configuration includes only a PV combiner box, which is located between the PV array and the DC main disconnect/isolator. If there is a fault on the DC side of the inverter, it can be caused by receiving inadequate power from the array or the system may not function properly because the PV array does not produce the expected power. The open-circuit voltage of the PV array should be measured in the inverter input terminal to ensure that the DC power reaches the inverter. If DC power was not available at the inverter, each piece of equipment and connection from the inverter to the PV array should be checked. If DC voltage is not observed in the inverter, then the first common function should check the status of the DC main disconnection/isolation (it may be operated). The next possibility is that DC voltage is not available at the combiner box. Therefore, there is a fault in the PV array. This fault can be caused by the following reasons:

- Cable disconnection
- Module failure
- Weak connection

Measurement of the voltage and current and a physical examination of the modules and cables can be used to detect the actual fault [26, 27].

10.4.4 Autonomous monitoring for PV plants

Autonomous monitoring is a novel concept to integrate various monitoring methods to enhance the reliability and service life of PV plants. Autonomous monitoring aims to identify the failures and faults of the different components of PV systems in a short time and facilitate decision making for remedial actions.

In this concept, the control system is integrated with inspection, recognition of the problem, preprocessing/postprocessing of the data, and decision making. Therefore, the entire monitoring services that are required for the operation and maintenance (O&M) of PV plants can be provided by this system. Fig. 10.5 depicts the schematic of the first attempt to

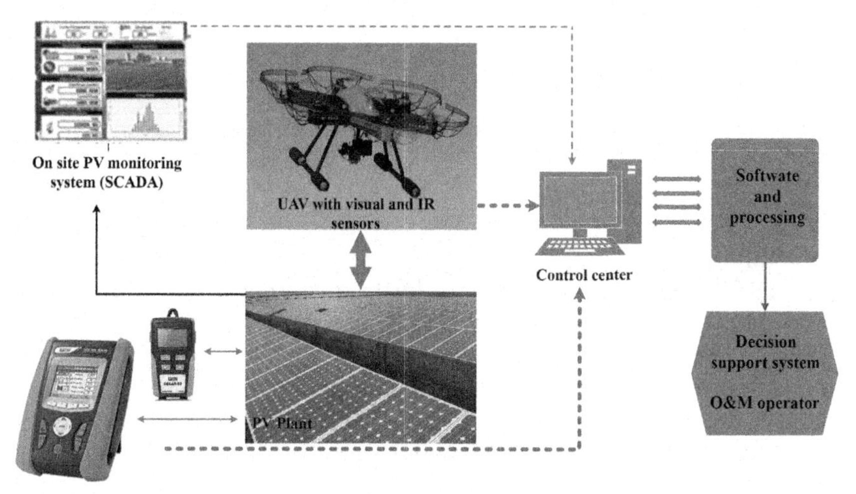

FIG. 10.5 Schematic of the concept of autonomous monitoring for PV plants [29].

develop an autonomous monitoring system. This system can perform the visual and thermography assessments by an infrared (IR) imaging sensor and a high-resolution visual camera mounted on the unmanned aerial vehicle (UAV). The aerial images captured are automatically transferred to the ground control station (GCS) by the RF channel. Later on, the classified data are sent by the GCS to the database for processing. Subsequently, the signal characteristic data (e.g., current, voltage, power, temperature, irradiation, energy production, etc.) are acquired by SCADA. After preprocessing and postprocessing of the aerial images and data, the entire processed information is transferred to the decision support system and the O&M operator for future actions. In this phase, the system evaluates the information in order to identify and analyze the specific failure or fault of the PV systems.

Autonomous monitoring aims to increase the reliability and durability of PV plants, as it can contribute to reducing the time for monitoring procedures and decision-making for appropriate solutions, which are very important tasks, especially for large-scale PV plants. In addition, autonomous monitoring provides reliable information for the stakeholders of PV plants in a short time. The accuracy of the system is very high in comparison with traditional monitoring methods. Autonomous monitoring systems can identify not only defects or failures but also the location of the specifically degraded components in the PV plants.

Autonomous monitoring systems can be an integration of separate systems, platforms, and devices, namely UAV, sensors (IR/EL/UV-FL), a meteorological station, a monitoring cabinet, a ground control station, SCADA, a database, the Internet of Things (IoT) platform, the Big Data Analytics (BDA) platform, and the decision support system [2, 30–41].

10.5 Quality assurance services and certification

The primary stages of the PV plant project are crucial to achieve favorable outcomes. Quality assurance services are offered by many audit institutions and companies (e.g., Fraunhofer Institute for Solar Energy Systems and VDE Renewables in Germany) for each step of a PV plant project (from development to long-term operation) in order to achieve high and optimal efficiency. The quality assurance service begins with the site assessment, an investigation of environmental conditions, and an energy yield assessment. The proper and accurate design of the PV plant and component selection are also essential to achieve a high energy yield [42, 43].

10.5.1 Development

Climate data is collected from different geographic information systems (GIS) and then meteorological and solar irradiance data are used to assess the special sites. Moreover, an evaluation of the site's environmental conditions is very important for the system lifespan and the produced output energy. The following factors are investigated for the special site:

- Ultraviolet irradiance
- Humidity
- High temperature
- Natural hazards
- Soiling
- Salting
- Corrosion

These factors are needed for:

- Detailed reports of solar resource evaluation
- Assessment of annual and long-term changes in radiation
- The first analysis and optimization of plant design
- Reliability test for devices at the special site
- Feasibility studies

10.5.2 Engineering

Performance prediction reports are highly credited for ensuring project bankability. To this end, a complete and accurate model of the PV plant is generated according to design documents and all components to simulate the system performance. The resulting reports contain detailed information on the factors affecting system performance. These results include the following:

- Prediction of performance and annual output energy over 25 years
- The trend of possible variations in radiation
- Degradation factors
- Detailed uncertainty evaluation

In order to optimize customer cost and time, the component benchmarking method was created to comparatively evaluate efficiency, workmanship, and reliability. Considering the most important aspects of quality, the various modules selected are comparatively characterized. With a specific test process, module performance is verified according to datasheet specs. The following tests are required for component benchmarking:

- Power confirmation
- Electroluminescence imaging to detect hidden flaws

- Power rating in accordance with IEC 61853-1, including low light performance
- Reliability testing (e.g., corrosion, climate loads, cell cracks)
- Material testing

10.5.3 Procurement

In order to guarantee the quality of components in the procurement step, an examination of an independent sample of components delivered according to reliability and performance is required. A representative sample is chosen based on a sampling technique and the required measurement and test are performed with maximum precision.

10.5.4 Commissioning

A comprehensive test plan is presented by the aforementioned institutions to ensure that the PV plant is made to the highest standards, and also that the expected power rate can be achieved. These services include:

- Visual inspections
- Power, current, and voltage measurements
- *I-V* characteristic measurements to specify the installed capacity
- Thermography
- Infrared camera to identify defective modules

The inspections assist in detecting faults and flaws in the PV installation. If the fault is detected early, operators can react quickly for required repairs.

10.5.5 Operation

The verification of quality and the efficiency of PV components is equally valuable for EPCs (engineering, procurement, and construction), manufacturers, and financiers. Ongoing monitoring of the PV plant ensures the detection of faults early and long-term yield analysis. Also, reliable and accurate performance reports are offered over periods from a couple of weeks to many years. The report contains benchmarking and also analysis of the measured indicators against the expected performance ratios. These reports and monitoring also can be applied for the next developments in design and component selection for PV plants. Moreover, the monitoring system investigates the long-term reliability of PV components.

10.5.6 Certifications process

For the return on capital cost, quality in the design, safety, and operation of PV plants is very important. International standards for the certification of a component provide the first degree of quality assurance. However, a standardized evaluation of PV plants from design to operation is needed for the industry. The VDE Institute is a certification entity with deep knowledge in the electrical industry, and Fraunhofer ISE is an outstanding applied research institution with abundant experience in solar energy field. These two organizations have developed comprehensive testing quality assurance and certifications procedure for PV plants. Its duty is that it documents the advance operation and electrical safety of the PV plant at the commissioning time. Therefore, quality assurance aims to reduce the technical and economical risks for investors, financiers, operators, and other shareholders of the PV plants. There are two main reasons from a financial aspect to perform quality tested certification. First, a PV plant that is quality tested (QT) will be recognized as a higher quality plant with better performance and availability. Second, the cost of certification will be compensated by a reduction in financial cost. An overview of the certification procedure is explained in [42–44].

In the first phase, documents related to PV plant design and engineering are confirmed and evaluated. In the next phase, all main selected components are checked in terms of dimension, conformity with international standards, and correct rating. Then, documents related to the installation equipment package (foundation, mounting structure, PV module, inverter, other electrical equipment, and monitoring system) are investigated. The competence of the chosen installer is confirmed. A sample of the selected PV module is taken to Fraunhofer ISE for testing. Irradiance dependence, temperature coefficient, and power at STC are checked. During the onsite inspection of the PV power plant, the installation status and the components are checked visually. This inspection consists of checking the safety and performance of the PV plant with the latest methods and standards. In order to assure the forecasting and expected performance, the output power of the PV module and the performance ratio of the whole PV plant are computed [42–44].

10.6 Economics of PV power plants

In the development of solar PV plants, the financial aspect is the most crucial one that needs to be given special attention. The assessment related to finance can be broadly understood with the economic assessment, which highlights the internal rate of return (IRR) and the net present value (NPV) of the solar PV project. To understand the economics, financial modeling of the solar power plant can be carried out. Any financial model depends on the pricing model of the PV system. The most commonly used pricing model is the feed-in-tariff (FiT). The financial model of the PV plant would require some key data related to the project. These data include the cost parameters, the performance parameters, and sometimes the parameters related to incentives in any form that can be used. Each of these parameters is briefly explained in Table 10.6.

Based on the above-given data, the financial assessment of the PV plant can be understood. A few key financial parameters used are the internal rate of return (IRR), the net present value (NPV), and payback periods (PBP).

The internal rate of return (IRR): This parameter generally accounts for the cash flow in the total PV system lifetime. In an evaluation, IRR generally considers the discounted cash flow. It measures based percentages rather than a specific currency. This is quite different from other economic parameters.
Net present value (NPV): This is the most commonly used cost metric in the economic assessment of solar PV system. NPV evaluation generally

TABLE 10.6 Key parameters required for financial modeling of PV power plants.

Parameter	Description	Remarks
Peak installation capacity	Represents the overall capacity of the power plant in kW or MW.	– Cost of the PV capacity will vary based on the PV technology used – The available area and location-specific conditions would decide capacity
Lifetime	Total life span of the project along with the individual component's life span. Generally, the PV system life span is expected to be between 20 and 25 years.	– The life of PV modules may not be the same as other PV system components such as the battery and power converters
Degradation rate	It represents the percentage of energy production that decreases per year. It has a long-term impact on the overall economic outcomes from the PV plant.	– The degradation rate is PV technology-specific – It also varies with respect to climatic conditions
Pricing model	Pricing model represents an agreement between power producers and the utility. The tariff rate is considered one of the best pricing models and many of the national follow the same.	– The tariff rate is generally decided on many technoeconomic parameters – Generally, it depends on the size of the PV plant – In some situations, the type of PV installation configuration could influence the tariff rate

gives a detailed idea of the possible revenue from the PV system. It is given based on the following formula [45].

$$\text{NPV} = \sum_{y=1}^{n} \frac{\text{Cash flow}_y}{(1 + \text{discount rate})^y} \tag{13}$$

where n is the PV system lifetime and y is the operating year in its total lifetime.

Payback periods (PBP): These are indicators to determine the entire period for a PV system to recover its capital investment. It is formulated as [46]:

$$\text{PBP (years)} = \frac{\text{Initial cost of the system (INR)}}{\text{Annual cash flow (INR)}} \tag{14}$$

As highlighted in the above table, these financial parameters would depend on the type of PV technology as well as the PV plant type and capacity. Table 10.7 represents the economic assessment of building-integrated PV (BiPV) and a ground-mounted PV as per the feed-in-tariff pricing model for France's weather conditions.

10.7 Safety issues

Considering the distributed characteristics of PV systems that can be installed in all public places, the safety issues and fault protection in PV systems are most important. Although PV systems need low maintenance due to not having moving components, they are subject to different faults that lead to generating hazardous currents and voltages in the system. Technology and codes have made great strides in developing fault

TABLE 10.7 Economic assessment of different PV systems in France [47].

Peak capacity (kW)	Type of PV installation	PBP (years)	IRR (%)	NPV (k€)
3	BIPV	16	2.26	−0.77
20	Simplified BIPV	19	0.63	−8.54
100	Simplified BIPV	17	1.21	−29.42
500	Ground-mounted	>25	−3.80	−525.43
1000	Ground-mounted	>25	−3.22	−925,235

protection techniques to prevent or reduce unwanted voltage and current components [48].

10.7.1 Overcurrent protection

Overcurrent protection devices such as fuses, breakers, and others limit the current to the proper rate and disconnect short circuits. The type of overcurrent protection system used and the maximum current passing through the circuit components determine the size of the overcurrent protection system. The design and size of the overcurrent protection system are outlined by national codes. In many countries, the short-circuit current of the array determines the size of the overcurrent protection devices. The main reason for selecting the short-circuit current of the array is that the PV is a limited current source, that is, the short-circuit current is the highest current produced by this source. Also, the maximum array current is equal to the sum of the short-circuit currents of the strings. An important point is the dependence of the short-circuit current on the solar irradiance and temperature of the PV module, both of which should be considered in the design and selection of the overcurrent protection device in PV systems [13, 24].

10.7.2 Grounding and ground fault protection

A ground fault is an electrical short-circuit between the ground and current-carrying conductors. Factors such as environmental conditions, fault impedance, and fault location affect the amplitude of the fault current. If the ground fault protection device (GFPD) cannot clear the ground-fault current, the fault connection leads to hazardous DC arcs and fire. The most common type of fault in conductors is a ground fault because just one cable is all that needs to be damaged. The most important reason for the ground fault occurrence is the insulation failure of the cables [49]. The type of DC system grounding affects the protection devices and techniques. Grounding is an accidental or deliberate conducting connection of the DC circuit to the ground. The basic goals of grounding are as follows:

- Fire hazard minimization
- Electrical shock minimization
- Electromagnetic interference reduction
- Equipment protection from faults

Equipment grounding is needed in all countries, that is, the conductive parts that do not carry current must be grounded. However, system grounding is optional among countries. Therefore, local codes specify the type of grounding system.

10.7.3 Arc fault protection

A PV array has many connections all over the array. An arc is created when current is transferred through the air-informed gap between conductive parts due to discontinuity and insulation failure. An arc fault is dangerous and harmful for the PV array because conductors may be melted from the generated intense heat and also arc fault leads to fire hazards. Compared to an AC system, an arc in the DC system can lead to a sustained arc because the current through the DC arc is not periodic and also does not have zero-crossing [50].

In PV systems, connections are made by connectors, soldering, and screw terminals. Typically, arc faults occur in these connections. Series arcs happen due to discontinuity in the current-carrying conductors caused by the corrosion of connectors, destruction in solder joints, damage by rodents, cell damage, etc. The NEC-2014 requires a protection device to detect series arc faults in all PV systems with operating voltages above 80 V; this device is known as an arc-fault circuit interrupter (AFCI) [51].

10.7.4 Safety of the people

Ensuring the safety of the equipment exposed to touch provides for the safety of the people. Efficient grounding and safety testing of equipment ensures the desired safety. For example, IEC or UL standards certify module safety, which means that construction requirements are dictated in order to diminish the voltage on subjected parts. The efficient grounding of mounting structures is truly comprehended, but there is a considerable novelty in this field to create a ground path by using mechanical connections instead of ground conductors [52].

Also, building fires with the presence of the PV systems need specific procedures via emergency personnel. The first step in the event of a fire is to turn off building electricity. The PV systems turn off automatically when the utility voltage is interrupted, so the system current tends to zero. Because the PV modules and live equipment are always energized in daylight, there is a potential risk if contact is created. This problem has received much attention in recent studies as well as in the industry [24].

10.8 Conclusion

The increasing demand for PV plants in the energy market indicates the high potential of photovoltaic technology to supply energy of the highest quality. PV power plants are classified into small-scale PV systems (e.g., 1–100 kW) that are used for commercial and residential rooftops. There are also utility-scale PV systems (e.g., >100 kW), namely ground-mounted systems that supply electric power for urban and industrial applications.

The efficient implementation of a solar PV plant guarantees the reliability of the project. The development phase consists of six steps, namely, solar resource assessment, site selection, environmental stress assessment, predesign and optimization, feasibility studies, and social considerations. Engineering, procurement, and construction so-called EPC, are the main phases of a PV plant project. EPC consists of designing, system layout, choice of system components (PV module technology, inverter, transformers, the balance of system equipment. BoS also includes combiner box, DC and AC switching, circuit breakers, fuses, cabling, AC switchgear, substation, grounding, and surge protection.

The design process and yield prediction are the most significant initial actions to prevent failures and other problems such as soiling, module mismatching, and voltage drops, which can cause serious loss during the PV plant lifetime. Moreover, operation and maintenance are significant keys to continue the solar PV plant performance over the long term. It should be taken into account that the quality assurance service begins with the site assessment, an investigation of environmental conditions, and an energy yield assessment.

For the return on capital cost, quality in the design, safety, and operation of PV plants is very important. International standards for the certification of a component provide the first degree of quality assurance. Typically, PV plants can be installed in all public places, hence safety issues and fault protection in PV systems are most important.

This chapter presented detailed aspects of the implementation phases of solar PV plants, namely development, engineering, procurement, construction, operation, and maintenance. Also, this chapter provided detailed information about quality assurance services, the norms, and the certification procedure. Furthermore, the quality assurance services and economic issues, namely financial models, feed-in-tariff (FiT), internal rate of return (IRR), net present value (NPV), and payback periods (PBP), were also discussed. Moreover, autonomous monitoring was introduced as a novel concept for integrating various monitoring methods to enhance the reliability and service life of solar PV plants.

References

[1] Renewable energy policy network for the 21st century (REN21). Renewables 2018 global status report, *Paris Renew. energy policy Netw. 21st Century*; 2018.

[2] Aghaei M, Dolara A, Leva S, Grimaccia F. Image resolution and defects detection in PV inspection by unmanned technologies. In: *IEEE power and energy society general meeting*, vol. 2016-November; 2016.

[3] SolarPower Europe. Global market outlook for solar power 2018–2022; 2018

[4] Renne D, George R, Wilcox S, Stoffel T, Myers D, Heimiller D. Solar resource assessment. In: Renewable energy grid integration: building and assessment; 2010.

[5] Al Garni HZ, Awasthi A. Solar PV power plants site selection: a review. In: Advances in renewable energies and power technologies; 2018.

[6] "Global Solar Atlas website," [Online]. Available: https://globalsolaratlas.info., 2019.
[7] Critto A, Suter GW. Environmental risk assessment. In: Decision support systems for risk-based management of contaminated sites; 2009.
[8] Chiabrando R, Fabrizio E, Garnero G. The territorial and landscape impacts of photovoltaic systems: definition of impacts and assessment of the glare risk. Renew Sustain Energy Rev 2009;13(9):2441–51.
[9] Sinha S, Chandel SS. Review of software tools for hybrid renewable energy systems. Renew Sustain Energy Rev 2014.
[10] Sindhu S, Nehra V, Luthra S. Investigation of feasibility study of solar farms deployment using hybrid AHP-TOPSIS analysis: case study of India. Renew Sustain Energy Rev 2017.
[11] Yunna W, Geng S. Multi-criteria decision making on selection of solar-wind hybrid power station location: a case of China. Energy Convers Manag 2014.
[12] Reinders A, Verlinden P, van Sark W, Freundlich A. Photovoltaic solar energy: from fundamentals to applications. Wiley; 2017.
[13] Stapleton G, Neill S. Grid-connected solar electric systems: the Earthscan expert handbook for planning, design and installation. Taylor & Francis; 2012.
[14] Burdick J, Schmidt P. Install your own solar panels: designing and installing a photovoltaic system to power your home. Storey Publishing, LLC; 2017.
[15] Smets AHM, Jäger K, Isabella O, van Swaaij R, Zeman M. Solar energy: The physics and engineering of photovoltaic conversion technologies and systems. UIT; 2016.
[16] Cabrera-Tobar A, Bullich-Massagué E, Aragüés-Peñalba M, Gomis-Bellmunt O. Topologies for large scale photovoltaic power plants. Renew Sustain Energy Rev 2016;59:309–19.
[17] J. Payeras and International Finance Corporation (IFC) World Bank. Utility-scale solar photovoltaic power plants: a projects developer's guide. *World Bank Gr.*2015. p. 216.
[18] Berwal AK, Kumar S, Kumari N, Kumar V, Haleem A. Design and analysis of rooftop grid tied 50 kW capacity solar photovoltaic (SPV) power plant. Renew Sustain Energy Rev 2017;77(February):1288–99.
[19] Eltawil MA, Zhao Z. Grid-connected photovoltaic power systems: technical and potential problems-a review. Renew Sustain Energy Rev 2010;14(1):112–29.
[20] Rosa-Clot M, Tina GM. Introduction to PV plants; 2018.
[21] Faiman D. Assessing the outdoor operating temperature of photovoltaic modules. Prog Photovoltaics Res Appl 2008;16(4):307–15.
[22] Pavan AM, Mellit A, De Pieri D. The effect of soiling on energy production for large-scale photovoltaic plants. Sol energy 2011;85(5):1128–36.
[23] Deline C, Dobos A, Janzou S, Meydbray J, Donovan M. A simplified model of uniform shading in large photovoltaic arrays. Sol Energy 2013;96:274–82.
[24] Reinders A, Verlinden P, van Sark W, Freundlich A. Photovoltaic solar energy: from fundamentals to applications. Wiley; 2017.
[25] Hernández-Callejo L, Gallardo-Saavedra S, Alonso-Gómez V. A review of photovoltaic systems: design, operation and maintenance. Sol Energy 2019;188(March):426–40.
[26] Haney J, Burstein A. PV system operations and maintenance fundamentals solar america board for codes and standards; 2013 August.
[27] Whaley C. Best practices in photovoltaic system operations and maintenance 2nd edition NREL/Sandia/Sunspec alliance SuNLaMP. December; 2016.
[28] Global Sustainable Energy Solutions [GSES]. Installation, Operation & Maintenance of Solar PV Microgrid Systems.
[29] Aghaei M. Novel methods in control and monitoring of photovoltaic systems. Politecnico di Milano; 2016.
[30] Kirsten A, et al. In: Aerial infrared thermography of a utility—scale PV power plant a fter a meteorological tsunami in Brazil. *WCPEC*, no. 20; 2016. p. 1–3.
[31] Vidal de Oliveira AK, Amstad D, Madukanya UE, Rafael L, Aghaei M, Rüther R. Aerial infrared thermography of a CdTe utility-scale PV power plant. In: 46th IEEE PVSC; 2019.

[32] Sizkouhi AMM, Esmailifar SM, Aghaei M, Vidal de Oliveira AK, Rüther R. Autonomous path planning by unmanned aerial vehicle (UAV) for precise monitoring of large-scale PV plants. In: 46th IEEE PVSC; 2019.
[33] Aghaei M, Madukanya EU, de Oliveira AKV, Rüther R. Fault inspection by aerial infrared thermography in a PV plant after a meteorological Tsunami. VII Congr Bras Energ Sol—CBENS 2018;2018.
[34] Grimaccia F, Aghaei M, Mussetta M, Leva S, Quater PB. Planning for PV plant performance monitoring by means of unmanned aerial systems (UAS). Int J Energy Environ Eng 2015;6(1).
[35] Aghaei M, Grimaccia F, Gonano CA, Leva S. Innovative automated control system for PV fields inspection and remote control. IEEE Trans Ind Electron 2015;62(11).
[36] Grimaccia F, Leva S, Dolara A, Aghaei M. Survey on PV modules' common faults after an O&M flight extensive campaign over different plants in Italy. IEEE J Photovolt 2017;7(3).
[37] Leva S, Aghaei M. Failures and defects in PV systems review and methods of analysis; 2018. p. 56–84.
[38] Quater PB, Grimaccia F, Leva S, Mussetta M, Aghaei M. Light unmanned aerial vehicles (UAVs) for cooperative inspection of PV plants. IEEE J Photovolt 2014;4(4).
[39] Leva S, Aghaei M, Grimaccia F. PV power plant inspection by UAS: correlation between altitude and detection of defects on PV modules. In: 2015 *IEEE 15th international conference on environment and electrical engineering, EEEIC 2015—conference proceedings*; 2015.
[40] Aghaei M, Gandelli A, Grimaccia F, Leva S, Zich RE. IR real-Time analyses for PV system monitoring by digital image processing techniques. In: *Proceedings of 1st international conference on event-based control,* communication and signal processing, EBCCSP 2015; 2015.
[41] Aghaei M, Leva S, Grimaccia F. PV power plant inspection by image mosaicing techniques for IR real-time images. In: 2017 IEEE 44th photovoltaic specialist conference, PVSC 2017; 2017.
[42] Quality Assurance and Yield Optimization for PV Power Plants, *Fraunhofer Inst. Sol. Energy Syst. ISE*; 2015.
[43] Farnung B, Bostock P, Bruckner J, Kiefer K. Complete PV power plant certification: new standards for quality assurance of large scale PV power plants. In: 2014 IEEE 40th photovolt. spec. conf. PVSC 2014; 2014. p. 1917–20.
[44] Kiefer K, Reich NH, Dirnberger D, Reise C. Quality assurance of large scale PV power plants. In: 2011 37th IEEE photovoltaic specialists conference; 2011. p. 1987–92.
[45] Kumar NM, Vishnupriyan J, Sundaramoorthi P. Techno-economic optimization and real-time comparison of sun tracking photovoltaic system for rural healthcare building. J Renew Sustain Energy 2019;11(1).
[46] Manoj Kumar N, Sudhakar K, Samykano M. Techno-economic analysis of 1 MWp grid connected solar PV plant in Malaysia. Int J Ambient Energy 2019.
[47] Dusonchet L, Telaretti E. Comparative economic analysis of support policies for solar PV in the most representative EU countries. Renew Sustain Energy Rev 2015;42:986–98.
[48] Falvo MC, Capparella S. Safety issues in PV systems: design choices for a secure fault detection and for preventing fire risk. Case Stud Fire Saf 2015;3:1–16.
[49] Albers MJ, Ball G. Comparative evaluation of DC fault-mitigation techniques in large PV systems. IEEE J Photovolt 2015;5(4):1169–74.
[50] Johnson J, Armijo K. Parametric study of PV arc-fault generation methods and analysis of conducted DC spectrum. In: *2014 IEEE 40th Photovoltaic Specialist Conference (PVSC)*; 2014. p. 3543–8.
[51] Alam MK, Khan F, Johnson J, Flicker J. A comprehensive review of catastrophic faults in mitigation techniques. IEEE J Photovolt 2015;5(3):982–97.
[52] Reil F, et al. Determination of fire safety risks at PV systems and development of risk minimization measures. In: *26th European Photovoltaic Solar Energy Conference and Exhibition, Hamburg (Ger)*; 2011.

CHAPTER

11

New concepts and applications of solar PV systems

Mohammadreza Aghaei[a,b], *Hossein Ebadi*[c], *Aline Kirsten Vidal de Oliveira*[d], *Shima Vaezi*[e], *Aref Eskandari*[f], *and Juan M. Castañón*[b,g]

[a]Eindhoven University of Technology (TU/e), Design of Sustainable Energy Systems, Energy Technology, Department of Mechanical Engineering, Eindhoven, The Netherlands, [b]Department of Microsystems Engineering (IMTEK), University of Freiburg, Solar Energy Engineering, Freiburg im Breisgau, Germany, [c]Biosystems Engineering Department, Faculty of Agriculture, Shiraz University, Shiraz, Iran, [d]Federal University of Santa Catarina, Florianopolis, Brazil, [e]Chemical Engineering Department, Islamic Azad University South Tehran Branch, Tehran, Iran, [f]Electrical Engineering Department, Amirkabir University of Technology (Tehran Polytechnic), Tehran, Iran, [g]Symtech Solar Group, Lewes, DE, United States

11.1 Introduction

Photovoltaic (PV) energy generation as a versatile technology has numerous applications and integrations with significant impacts on world development and the amenities of life. PV deployment has been much faster than the anticipated value, where it is likely to be twice the expected level in 2020 [1]. What has driven this rapid deployment are the higher rising trend of electrification and the relative energy demands coupled with the falling costs of manufacturing. By the end of 2017, the total installed worldwide PV power was measured at 403GW, mostly grid-connected [2–4]. Having a notable diversity in the niche market, PV technology needs an in-depth look through penetration in various energy sectors to meet growing energy demands and facilitate human life. With electricity demand, particularly

Photovoltaic Solar Energy Conversion
https://doi.org/10.1016/B978-0-12-819610-6.00011-9

in buildings and road transportation, expected to double by 2050 due to the rise in electric vehicles and electrified heating and cooling as well as different appliances, it is expected to make up half of the requisite energy by renewable energies [5].

The key feature that is reflected by PV systems is the flexibility of end-use applications, ranging from large-scale power plants to on-device portable gadgets. Therefore, this wide variety of applications can be categorized under four main groups, (i) utility-interactive systems, (ii) stand-alone systems, (iii) hybrid systems, and (iv) building-integrated systems [6]. In the first type, the electricity generated from PV modules is connected to the grid, and the DC is converted to AC via an inverter before the distribution network. The power can be used by electric appliances in the house (self-consumption); otherwise, it can be supplied to the central power network (smart grids). As the main source (solar radiation) is intermittent, a backup system is usually integrated to keep power in the system irrespective of weather disturbances. Today, modern designs employ this technique to offset the energy costs in residential buildings.

The second type of PV system is a stand-alone system, which does not rely on grid/utility connections. In this case, a battery storage unit is always needed when nocturnal operation or constant energy supply is concerned, and solar insulation is not sufficient. However, the third type of design presents a hybrid mode of power generation in which PV systems are coupled with diesel or wind power generators. In an experimental study conducted by Khatib et al. [7] to optimize a BIPV-diesel hybrid system in Malaysia, it was seen that the hybrid mode operation cost relatively 30% less than the stand-alone PV or diesel types. Therefore, a technoeconomical assessment is necessary to determine the viability of PV technology integration with an application. The fourth type of PV system is the integration of PV modules into the building envelope, which is known as a building-integrated PV (BIPV) system. This new technique not only provides power generation through PV materials, but also reduces material consumption for construction as the PV materials are used as envelope materials for the roof or façade. In recent studies, the potential of BIPV systems has been highlighted as the annual energy collected by the four vertical façades facing four perpendicular directions exceeds by two times the total energy absorbed on a horizontal plate at the same location. Data also state that in winter, the energy increases by threefold [8].

So far, PV power generation has become a plausible technology for remote areas where distribution lines are not affordable, or the running costs are not justifiable. Because the cost of PV technology is decreasing, PV use for grid-connected applications is on the rise [9]. BIPV systems are also gaining attention as they bring benefits by replacing conventional materials with new construction, at which monetary advantages emerge after the electricity consumption cost decreases. BIPV is also architecturally more interesting compared to roof-mounted PV structures. Stand-alone systems are in practice for different purposes in remote areas, such as spatial communications, terrestrial communications sites, electrifying villages, etc. Hybrid systems are also viable when a renewable or nonrenewable source of energy is implemented by means of generators to compensate for the required net energy that occurs at insufficient solar radiation levels.

11.1.1 PV impacts on disaster relief

One of the outstanding outcomes of solar-powered facilities is the integration of natural disaster relief efforts. According to the Centre for Research on the Epidemiology of Disasters (CRED), 281 climate-related and geographical events were recorded in 2018, with 62 million people affected by these disasters resulting in 10,733 casualties [10]. Power outages are likely at the very beginning of every catastrophic event due to major damage to infrastructure and power stations. This local shutdown may last for days and leaves homes, schools, hospitals, food stores, and other vital services without electric power. Hence, mobile stand-alone PV systems are a boon to residents, supplying power to needed areas in the aftermath of the disasters. The mobility of these solar-powered automatic devices facilitates their displacement, which has shown considerable advantage. Being lightweight is another positive feature of portable PV systems, particularly when local roads and bridges are damaged and tools have to be carried over rough ways. In addition to providing electricity, some compact PV systems can purify contaminated water plus offer lighting features [11].

11.1.2 PV effects on urban development

According to one of the sustainable development goal (SDG) subsets, energy access and sanitation are the cornerstones in improving people's living standards [12]. PV integration is booming and has increased electricity access by reducing energy costs and raising energy availability. Furthermore, the growing trends of digital applications in daily life plus new decarburization policies are becoming tipping points for achieving sustainable development in urban areas. Electric mobility is one of the outcomes of this rapid change toward sustainability and has shown a significant reduction in urban pollution. Employing PV systems to empower electric applications has introduced new, innovative, and ecofriendly devices and machines into the energy niche market. Alongside the prementioned applications, some new, innovative, and less-investigated areas are emerging as PV-related technologies become more advanced. Solar unmanned aerial vehicles (UAVs) are light flying objects that are used for surveillance missions that focus on agricultural, geographical, military, and other purposes [13–17]. Utilizing a PV system extends their flying time, which provides longer times for photography in mapping operations, enhancing the final product [18]. In a mature technique, the effort to power flying objects dates backs to 2015 when the first zero-emission plane took a 5 month journey around the globe. This successful attempt allowed the integration of PV systems, which was before merely used for remote communications via satellites, in aviation. Road, water, and rail transportation systems also benefited from PV technologies while electric boats, buses, passenger cars, and trains have penetrated into the current means of transportation. Equipping bus stations with solar panels has brought the opportunity to recharge electric buses while in traffic. The integration may also be applied for lightning or public network supply. PV-integrated carports are also beneficial to consumers as they offset their energy demands from the grid and dependency on the grid [19]. One of the subsets of PV technology is BIPV, which offers a promising solution for cleaner electricity by introducing new integrations with different building appliances. BIPV is advantageous to society because it offers both environmental and health benefits in addition to the reduction in carbon emissions and added viability to the economy [20].

11.1.3 Effects of PV on rural development

Access to electricity improves people's living standards; however, being deprived of energy access has resulted in various drastic health problems throughout history. Solar electrification employing PV technology is starting what looks like a boom. Delivering numerous social impacts, electrified villages are on the rise and have evolved the residents' lives and economies. Increasing their working hours, bringing information, promoting small enterprises, and reducing transportation costs are some of the social impacts in rural areas. PV applications are in extensive use for solar home systems, vaccine refrigeration, telecommunications, radio transceivers, rural telephone and security systems, and agricultural purposes [21]. An educational impact is also expected as the experimental data prove that the literacy rate in solar-electrified households is higher than in nonelectrified households [22]. Eliminating the frenzied use of kerosene lamps by introducing solar-powered lights gives new impetus to mitigating health-related problems [23]. Interestingly, the global trend of the integration of PV applications into health facilities is growing rapidly. A recent

WHO-led review of sub-Saharan African health facilities reported that 15% and 43% of hospitals in Uganda and Sierra Leone, respectively, use PV systems in combination with other electric sources [24].

11.2 Building-integrated photovoltaics (BIPVs)

As PV installations grow globally combined with decreasing prices in silicon solar cells, building-integrated photovoltaic (BIPV) systems—although still considered a niche product for the market—will play a significant role in the near future for the solar industry. BIPV systems today are viewed as state-of-the-art systems that offer innovative solutions for PV integration into buildings. Different from building-attached PV systems (BAPV) that require additional components and materials to be mounted onto existing surfaces, BIPV systems are aesthetically pleasing, following architectural flows within their design. Although BIPV is still in the early stages for stable mass production, in the future BIPV will represent a noteworthy share of total global solar installation. This chapter focuses on BIPV systems and offers a general overview and description of this type of system.

BIPVs are the result of integrating or combining building construction materials with photovoltaic solar cells in order to make urban edification structures (buildings) power generators, therefore sustainable [25, 26]. The building construction materials typically used in BIPV systems can be seen in either the façade, namely walls, windows, and curtains, or in roofs as skylights, shingles, and tiles [27]. The main objective of BIPV installation is to achieve multifunctionality in buildings by offering energy optimization to all users and at the same time serve as a protective, sustainable, and aesthetically pleasing construction [28].

11.2.1 BIPV modules

A key component of a BIPV system is its module. These are available in a wide range of semiconductor technologies, such as crystalline silicon (c-Si) in both mono and poly, as well as thin films, including amorphous silicon (a-Si), copper indium selenide (CIS), copper indium gallium selenide (CIGS), cadmium telluride (CdTe), and dye solar cells (DSC). For standard c-Si BIPV modules, their cells need to be encapsulated or adhered into the cover material, typically glass, while in thin-film and dye-sensitized solar modules, the semiconductor can be inserted directly into the substrate, typically glass, metal, or plastic. All commercially available cell technologies will eventually influence the appearance, efficiency, and overall cost of the finished BIPV module [25, 29].

11.2.2 Comparison of BIPV and BAPV

In urban locations, PV systems are mounted on roofs or building surfaces to avoid shading nearby objects. These types of systems are typically known as building-added or attached BAPV. Both BAPV and BIPV systems share similarities. However, it is important to differentiate between them. In BAPV systems, the use of mechanical fixtures known as mounting systems allows the installer to clamp and connect PV modules on an existing roof or surface. BIPV systems, on the other hand, are considered materials of the building itself and for this reason, they do not require additional materials to be fixed in a roof or building surface [30–32]. When comparing BIPV to BAPV systems, although knowing beforehand that their main difference relates to aesthetics, the following differences are listed in Table 11.1.

TABLE 11.1 Brief comparison between BIPV versus BAPV.

BIPV	BAPV
- Few suppliers available for BIPV modules	- Commercially available standard manufactured modules
- Low market demand, which results in a high waiting time for product acquisition	- High market demand for standard manufactured modules, which results in fast product acquisition
- Sizes need to be matched to the architectural design of the building	- Semistandard sizes for all components
- Higher costs due to the complex manufacturing process	- Lower cost on modules due to mass production capacity
- Risk of increasing cell temperature due to material selection and inside building environments	- Less risk of increasing cell temperature due to exterior installation, especially in fixed-tilt mounted systems
- Design and installation based on communication between multidisciplinary groups developing the project	- Design and installation standards available
- Superior and elegant aesthetics	- Not very pleasing appearance

11.2.3 BIPV design

There have been multiple discussions for the design of BIPV systems in the architectural and construction communities that aim to set parameters that can define the commonalities of a BIPV project. Most of these have a goal to balance energy efficiency, aesthetics, and environmental and economic contemplations [30]. Typically, designs form part of the roof materials, for example, a skylight, a roof tile, a shingle, or the façade of the building. The main purpose of BIPV is to integrate solar materials into existing building materials, which can then generate power while achieving aesthetically pleasing views and architectural sense [33]. Selecting the type of PV module to use for the BIPV project can help improve the aesthetics of the design because some materials are offered in transparent or opaque glazing covers. These will eventually need to be considered for the visual appearance of the project.

11.2.4 BIPV building codes and regulations

BIPV components require certification because they are used for electrical and construction purposes and need to be reliable. In the United States, for example, BIPV modules comply with basic electrical safety standards related to the IEC 61730 and UL norms [34]. For other standards, because BIPV modules are still to some extent novel and because their level of installation requires a more complex process, agreeing on a generic standard for these modules might be challenging at present, and the building regulations may vary from country to country. In the European Union, thanks to the adoption of the Energy Performance of Buildings Directive (EPBD), more institutions and countries are setting compliant regulations for BIPV [29, 35].

11.3 Luminescent solar concentrator photovoltaic (LSC PV)

The integration of PV modules has been beset with difficulties through adaptation in several applications. PV cells are still not the most cost-effective alternative to conventional energy providers, and they also lack flexibility in shape, color, and weight. On the other hand, solar energy intensity is generally low and solar facilities are recommended to be incorporated with concentrators to augment their optical efficiency. These acute challenges have abated the growing interest in PV in the market. However, several techniques have been employed to break down technical barriers toward PV flexibility in terms of integration with other applications. A luminescent solar concentrator (LSC) is a new device that brings a number of benefits over other alternative solutions such as color reflecting films or organic cells, which result in low efficiency and a short lifetime, respectively [36, 37]. In addition, LSCs are able to concentrate solar insulation, avoiding sun-tracking devices or dish-type reflectors, particularly for utilization in an urban environment. This technology has shown significant improvement in PV integration with buildings and the architectural sector, offering a wider range of design choices [38].

11.3.1 LSC description

Typically, the LSC is a planar transparent plate comprised of plastic or glass waveguides as the host matrix (with refractive index n) embedded with dispersed organic dyes, inorganic phosphors, or quantum dots (guests). As Fig. 11.1A illustrates, when the light penetrates the transparent material, the guests (luminescent molecules or luminophores) absorb photons and isotropically reemit them at longer wavelengths as the result of the difference between the absorption and emission wavelengths, known as the Stokes shift (Fig. 11.1B). The emitted lights are trapped within the enclosure and concentrated toward the edges, where PV modules are located to receive the solar incident. According to Snell's law, when lights reach the interface between air and material at the angle within the cone angle (critical angle), the fraction will be lost from the

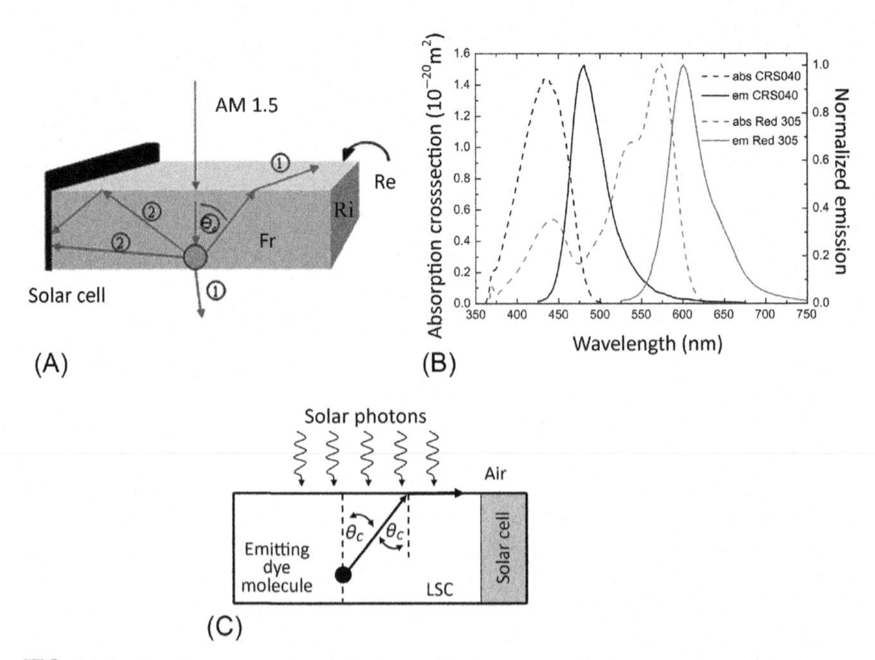

FIG. 11.1 (A) A conceptualized diagram of the operation of a luminescent solar collector, (B) Stokes shift, and (C) the principle of escape cone loss [39, 40].

surfaces of the luminescent concentrator (Fig. 11.1C) [38–40]. The critical angle (θ_C) and the fraction f of photons trapped in the LSC are given as:

$$\theta_C = \sin^{-1}\left(\frac{1}{n}\right) \tag{11.1}$$

$$f = \cos\theta_C = \left(1 - \frac{1}{n^2}\right)^{1/2} \tag{11.2}$$

Solar cells absorb photons that are the results of internal reflections, therefore, the photon flux gain or geometry gain G_p of the LSC can be defined as [40]:

$$G_P = \frac{A_{face}}{A_{edge}} Q_A \eta f \tag{11.3}$$

where A_{face} is the aperture area to the solar incident, A_{edge} is the area of the LSC edge that is exposed to the PV, Q_A is the fraction of absorbed photons, and η is the quantum efficiency.

11.3.1.1 LSC performance

LSC performance is dependent on several loss mechanisms, as shown in Fig. 11.2. As the light reaches the top surface of the waveguide, a portion is reflected due to the Fresnel reflection. Photons that penetrate the host material result in light emission by luminophores and the lights lie under the cone angle and refract out of the waveguide (escape cone loss). The other loss refers to the reabsorption of emitted photons by other luminophores. In this case, the overlaps between the absorption and emission bands of luminophores (insufficient Stokes shifts) lead to an incomplete reemission by the adjunct luminophore. The other absorption degradation stems from the luminophore's inability to absorb all incidents due to the limited absorption band range, which causes a portion of the penetrated lights to leave the waveguide medium from the bottom surface and without being absorbed by the luminescent materials.

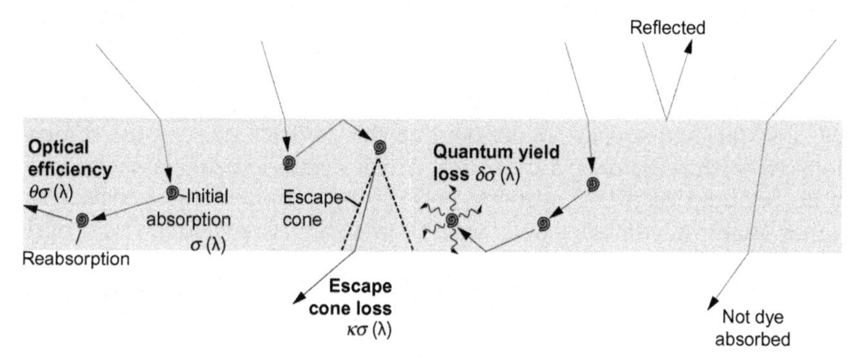

FIG. 11.2 Configuration of a waveguide and doped luminescent dyes with different loss mechanisms associated with dye functions [41].

The other loss comes from the strong UV photons that degrade the luminophore's molecules and result in malfunctioning in a certain part of the LSC. In this case, high energy photons break down the molecules, resulting in energy loss in forms of heat or vibration instead of photon reemission (quantum yield loss). There are still some other losses such as self-absorption, light scattering, and surface scattering that indicate the transmission efficiency of the light through the waveguide material. Therefore, the efficiency of an LSC (η_{LSC}) not only relies on the optical efficiency (η_{opt}), but also is a function of PV efficiency (η_{PV}) and can be expressed as:

$$\eta_{LSC} = \eta_{opt} \times \eta_{PV} \tag{11.4}$$

where

$$\eta_{opt} = f(1-R)\eta_{abs}\eta_{LQE}\eta_{s}\eta_{mat}\left(1-\eta_{self}\right)\eta_{TIR} \tag{11.5}$$

where R denotes surface reflection coefficient, η_{abs} is the fraction of light that is absorbed by luminophores, η_{LQE} is the quantum efficiency by the luminescent species, η_s is the Stokes' efficiency, η_{mat} is the light transmission efficiency in the host matrix, η_{self} is the self-absorption of the luminescent species, and η_{TIR} is the efficiency of the total internal reflections.

The photon concentration ratio (C_P) can also be defined as:

$$C_P = \eta_{opt} \times G_P \tag{11.6}$$

The power conversion efficiency of the LSC is another important factor that can be determined as [42]:

$$PCE = \frac{V_{OC} I_{SC} FF}{A_{edge} P_{in}} \tag{11.7}$$

where V_{OC} is the open-circuit voltage (V), I_{SC} is the short-circuit current (A), FF is the fill factor, and P_{in} is the total flux density of the solar radiation (W/m^2 nm).

11.3.1.2 LSC developments

Scientists have conducted numerous studies to develop LSC technology and augment the optical efficiency. It has been proven that the growth of geometric gain raises the probability of reabsorption by luminescent material [43]. Distinctive research findings also demonstrated that the LSC loss mechanisms are dependent on the device's size. As the dimensions grow, the efficiency diminishes, which results in being less favorable to be adopted by BIPV systems [44]. Using luminescent species with higher quantum efficiency and larger Stokes shifts has yielded fewer optical losses [42].

In 2007, there was an attempt by Gallagher et al. [45] to introduce several tandem LSC devices, being covered by PV cells from one edge, in which polymethylmethacrylate (PMMA) was used as the host matrix doped with cadmium selenide/cadmium sulfide (CdSe/CdS) quantum dots (QDs) and Lumogen F Red-30 (BASF) organic dye. The results showed that the high fluorescence quantum yield of the dye material increases the PCE to 3.3% compared to 2.1% of the QD-based LSC while both materials suffered from high self-absorption due to the short Stokes shifts. One year later, Slooff et al. [46] developed tandem LSCs equipped with backside mirrors and used a PMMA material with a mixture of Lumogen Red-305 and Yellow-CRS040 organic dyes. They also investigated the device performance for different PV cells with the same size and configuration. The results suggested that the higher the number of solar cells, the better the power conversion efficiency.

In 2009, Goldschmidt et al. [47] proposed an LSC device incorporated with a photonic structure as a reflective filter to prevent the escape cone loss, which accounts for a 26% reduction in the total solar radiation absorption. In this work, they applied two approaches to increase the LSC efficiency. First, they investigated the effects of the combination of different dyes, which is expected to broaden the used spectral range. As a result, the system enjoying two different dyes had a higher efficiency at 6.7%. The second approach referred to the use of a commercially available filter, as shown in Fig. 11.3, where the technique increased the optical efficiency by 20% with a 3.1% concentration ratio. Incorporating a white bottom reflector was also beneficial in the performance augmentation and is well analyzed in this study.

Another technique to prevent surface loss is to control the alignment of organic luminophores. This effort enables control over emitted light distribution by managing the physical ordering of dyes in a macroscopic scale. Several techniques have been proven for aligning luminophores, such as employing self-aligning fluorescent nanorods [48] or using an alignment layer and additives to align guest molecules with liquid crystals [49].

Later in 2012, Chandra et al. [50] introduced an enhanced quantum dot solar concentrator (QDSC) using plasmonic interaction between QDs and metal nanoparticles (MNPs). In this work, it was premised that the plasmonic interaction causes a growth in the excitation and emission rates of QDs, which presumably increases the QDSC efficiency. Moreover, using

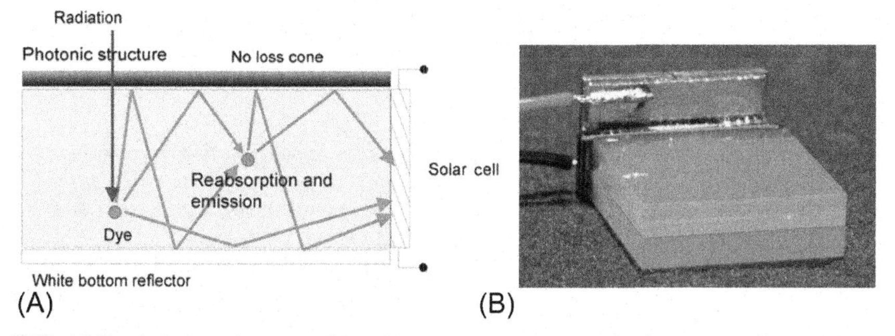

FIG. 11.3 (A) The schematic of the photonic structure used as a bandstop reflection filter, and (B) the photograph on the stack system used in this study [47].

QDs as the luminescent species brings a significant advantage over organic dyes. Using QDs offers the ability to tune the absorption threshold as a function of different dot diameters in addition to their stability and lesser vulnerability than organic dyes [51]. In results, scientists concluded that using the plasmon-induced electromagnetic field of Au NPs can boost the fluorescence properties of CdSe/ZnS quantum dots. Experiments also revealed that an increase in Au NP concentration enhances the optical absorption. Moreover, since the fluorescence emission is a function of Au NP concentration, the maximum emission efficiency was achieved in a certain NP concentration where a further increase in concentration reduced the emission efficiency. The reason was attributed to the nonradiative transfer of the excited carriers from the QDs to adjunct Au NPs [50]. In another study, El-Bashir et al. [52] developed an LSC with a plasmonic thin-film coated by polycarbonate substrates in addition to fluorescent PMMA film doped with coumarin dyes, nanogold, and nanosilver molecules, as shown in Fig. 11.4. Experimental data proved that the use of Au NPs extends the LSC lifetime and the maximum PCE was measured as 53.2% for amorphous silicon (a-Si) cells. This study puts forth the high potential of plasmonic applications in LSC technology.

Recently, Aghaei et al. [53] presented a novel configuration for luminescent solar concentrator photovoltaic (LSC PV) devices with vertically placed bifacial PV solar cells made of monocrystalline silicon (mono c-Si). This LSC PV device comprised multiple rectangular cuboid lightguides made of poly (methyl methacrylate) PMMA containing Lumogen dyes, in particular, either Lumogen red 305 or orange 240. The bifacial solar cells are located between these lightguide cubes and can, therefore, receive irradiance at both their surfaces. Ray tracing simulations by the LightTools software resulted in a maximum efficiency of 16.9% under standard test conditions (STC) for a $15 \times 15\,cm^2$ LSC PV device consisting of nine rectangular cuboid $5 \times 5 \times 1\,cm^3$ PMMA lightguides with 5ppm orange 240 dye with 12 $5 \times 1\,cm^2$ vertically positioned bifacial cells between the lightguides and nine 5x5 cm^2 PV cells attached to the back of the device. If not applying cells to the back of this LSC PV device configuration, the maximum PCE

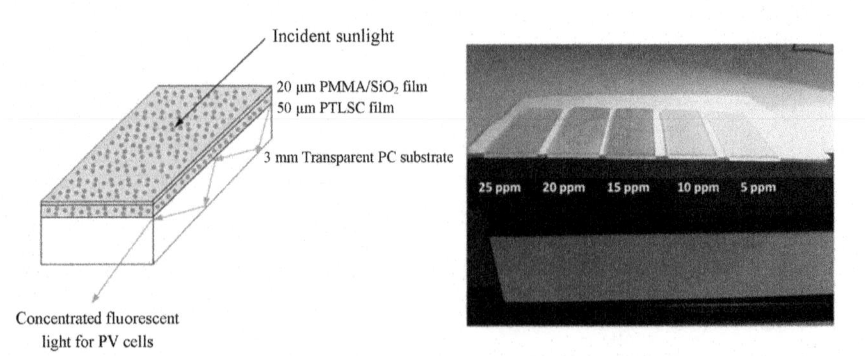

FIG. 11.4 The schematic and the photograph of this film double-layer plasmonic-LSC reported in [52].

will be 2.9% (at STC) where the LSC PV device consists of 25 cubical $1 \times 1 \times 1\,cm^3$ PMMA lightguides with 110 ppm red 305 dye with 40 vertically oriented bifacial PV cells of $1 \times 1\,cm^2$ between the lightguides. This study shows the future potential for LSC PV technologies with higher performance and efficiency than the common threshold PCE of 10%.

11.3.2 LSC applications

The current state-of-the-art LSC offers a wide and diverse set of applications by employing numerous shapes, colors, and materials. In this section, some of the latest advances are presented.

11.3.2.1 LSC integrated with houses

A. Colored windows

Having brightness and a fluorescent coloration makes the LSC devices a unique and adaptable technology to be employed in window areas of a house due to transparency and the value it adds by power generation. Fig. 11.5 represents different statues of a room affected by LSC integration,

FIG. 11.5 Photographs of an office room integrated with different percentages of the colored window: (A) no LSC covering, (B) 25% LSC covering, and (C) 75% LSC covering [54].

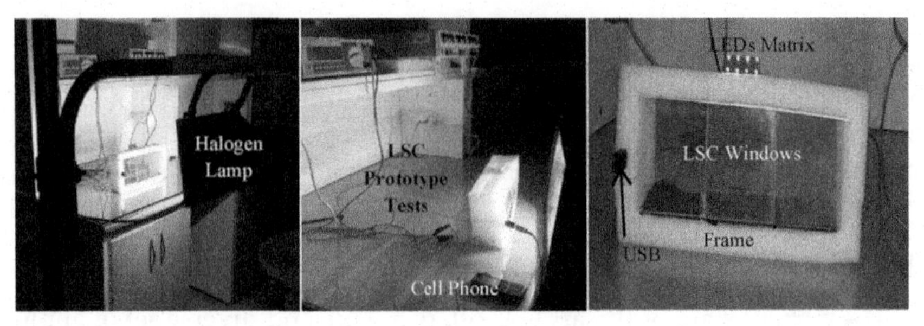

FIG. 11.6 Photographs of experimental test set-up and final product of LSC-integrated colored glass [55].

proposed by the investigation conducted by Vossen et al. [54]. Despite the benefits, there are still some legitimate concerns with visual comfort for both indoor and external users. Experimental results have indicated that the application of LSC to at least the top 25% of an office window brings electricity generation with minimum impact on the user's comfort level.

Incorporating small colored glass will also provide amenities for some on-the-board applications such as small-scale power provision prototypes. In a concept proposed by Fathi et al. [55], a luminescent window for supplying LED lighting and a USB port was tested; Fig. 11.6 depicts the photograph of this device.

B. Leaf roof

The "leaf roof" concept proposed by a collaboration between the Eindhoven University of Technology and the University of Twente in the framework of the Dutch Lighthouse program of 3TU [56] integrates LSC roof tiles with the roof areas of residential houses where the design is inspired by the leaf of a European birch. Fig. 11.7 illustrates the new roof tile including solar cells and the LSC in addition to an example of a leaf roof configuration. Considering this novel design, in contrast to conventional solar collecting roofs, the roof has full freedom to be faced toward any direction and inclined with any angle, leading to better urban planning and a more natural environment [58].

C. Smart windows

The term "smart window" is defined as an advanced window that participates in the energy performance of a building, offering a set of functions from energy saving to energy production by exploiting renewable energies [59]. LSC is one of the main components of smart windows (SWs) and provides electric energy to power the moving components, making it a stand-alone device. Fig. 11.8 depicts the main parts of an SW, which has some features such as solar electricity production, indoor daylight control, and energy control scenarios [60]. This window can be

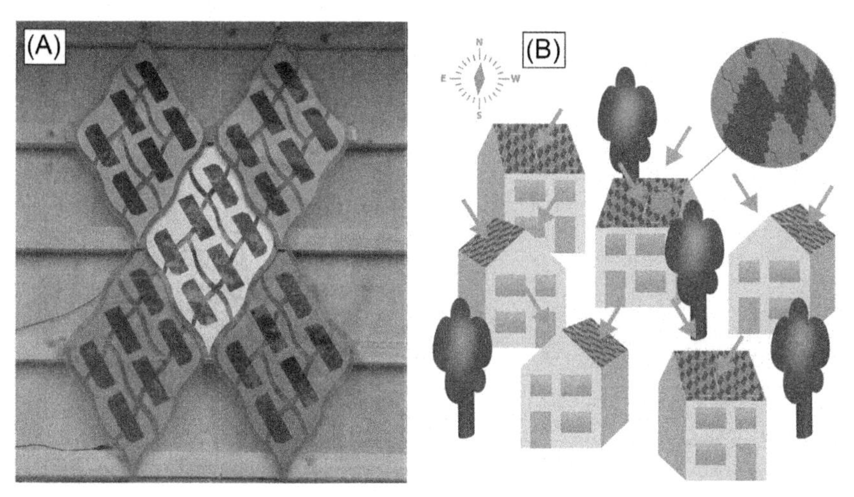

FIG. 11.7 (A) The photograph of an LSC roof tile, and (B) the concept of leaf roof utilization in urban planning [57].

mounted in new or retrofitted buildings where there are several considerations for controlling artificial lights, solar lights, thermal energy, and visual comforts. In the upper part, a fanlight with LSC devices embedded inside double glazing is located to provide energy supply and to improve the indoor daylight by its coloring. In the middle, the window is divided by an aluminum light shelf that prevents unwanted direct sunlight and reflects it to the ceiling, minimizing glare effects and boosting indoor light conditions in addition to diminishing the heat that enters the space. In the lower double glazing section, several metal reflective Venetian blinds are employed to operate by preprogrammed conditions (i.e., achieving different levels of transparency with regard to desired solar absorption, climate, and seasonal condition, which is done by displacing inside the section). These blinds are driven by electric motors that are powered by LSC PVs. A battery package embedded inside the frame is prepared for energy storage while a light measurement sensor is used to monitor dynamic solar changes and send the appropriate response.

11.3.2.2 LSC integrated with urban facilities

A. Noise barrier LSC

The solar noise barrier (SONOB) project was done by Kanellis et al. to study the feasibility of the integration of two large-scale LSCs as noise barriers; one facing north/south and the other east/west (shown in Fig. 11.9). Experimental data showed that the east/west facing panel culminates in more varied performance where structural elements interfere with the

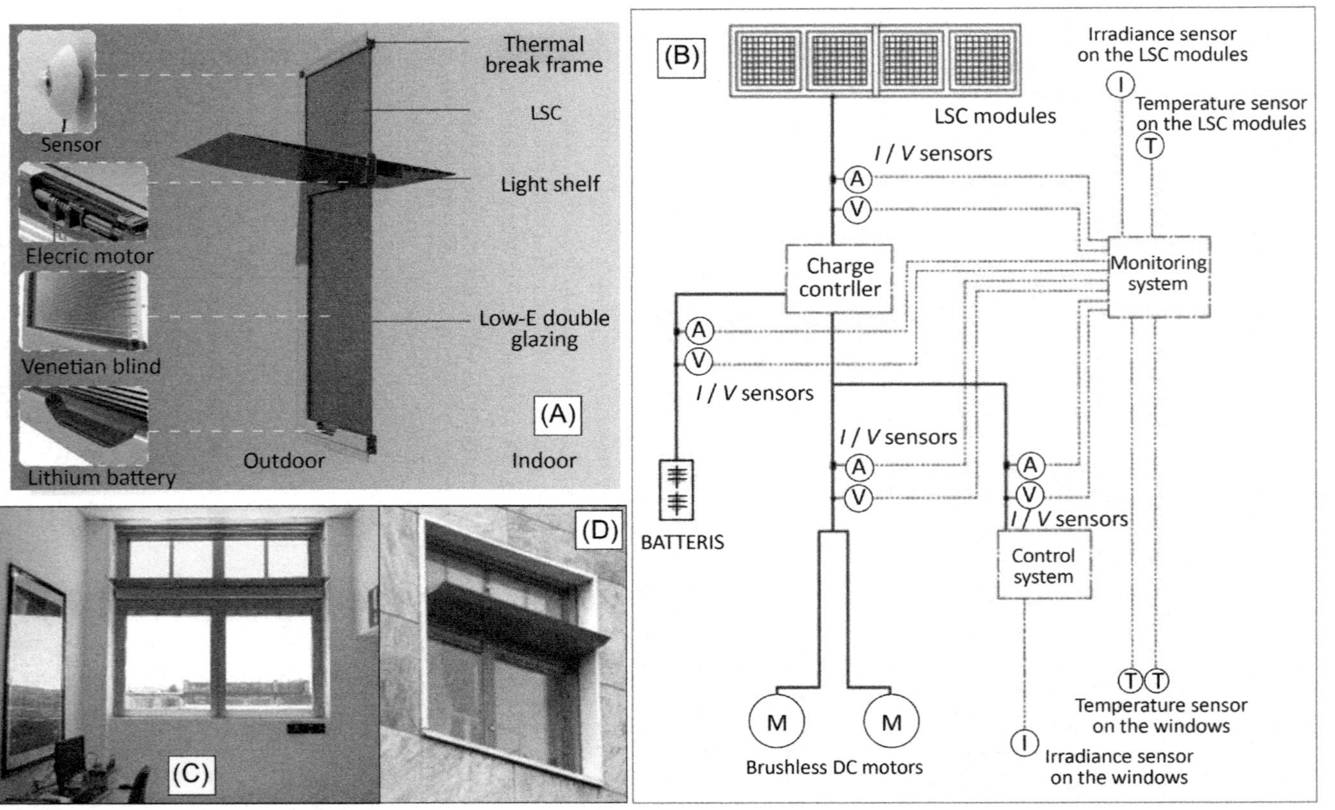

FIG. 11.8 (A) The main components of a smart window, (B) general wiring of the device, (C) inside, and (D) outside views of the installed prototyped with yellow LSC [60].

FIG. 11.9 Photographs of the noise barrier prototypes [61].

absorption efficiency by shading effects [61]. This work demonstrated the concept of LSC utilization in more and larger noise barriers.

B. Parking shed with integrated LSCs

In this concept, colorful LSCs are placed in the roof area to provide power for lighting facilities during the night. It is also assumed that e-bikes can benefit from the station by receiving electric charges produced by LSCs [62].

C. Outdoor LSC-integrated amenities

There are still some other concepts that arise from LSC integration in urban facilities to promote green development. In suggestions proposed by the University of Twente, outdoor street furniture equipped with LSC devices, a self-powered table, and a beach shelter can be introduced as examples of LSC-integrated facilities [62]. Furthermore, several other designs have been conceptualized so far such as an LSC-sustainable tent, LSC safety strips, an LSC garden fence, an LSC labyrinth, etc. [63], where the main aim is to generate electricity for self-consumption or to supply the gird.

11.4 Solar PV in mobility

Due to the current global concerns about CO_2 emissions and dependency on fossil fuels, electric vehicles are seen as an alternative to traditional vehicles. However, they increase the load on the power grid, creating a need for changes in the power system, which is a major source of greenhouse gas. Thus, the benefits of electromobility to the environment are strictly dependent on the use of renewable sources of energy [64].

At the same time, because of the cost reduction of PV systems and their flexibility, the integration of PV systems with electromobility is seen as a perfect solution for achieving CO_2 reduction goals. The next subsections will explore the integration of PV with different types of mobility.

11.4.1 Vehicle-integrated photovoltaics (VIPVs)

The integration of PV with vehicles has many advantages such as the improvement of PV power generation and the demand for charging, the reduction of peak loads on the electric grid, and optimization of the grid energy balance. Another potential benefit is the reduction of risks associated with fuel shortages and price spikes [65]. However, obtaining viable commercial ViPV solutions still faces many challenges such as the costs of the technology, standards and safety regulations to be developed, and the correct estimation of travel autonomy. The balance between design and PV energy production is a complex optimization problem that depends on the development of highly efficient, colored, and flexible PV cells; the vehicle roof area; and techniques for reducing cell mismatch during shadowing.

The first vehicles with solar panels on the roof appeared at the end of the last century, mostly in two-, three-, and four-wheeled vehicles. They use rigid solar modules to power small vehicles without the necessity of high velocity or propulsion. These are used to transport people and materials at universities, golf courses, parks, and companies. An example is shown in Fig. 11.10, which shows an electric vehicle developed in 2003 by Universidade Federal de Santa Catarina, located in Florianópolis, Brazil. The quadricycle is used to transport people and materials inside the university campus and has 192 Wp of PV installed capacity of a-Si.

For passenger vehicles, most of the development in the area began decades ago with solar car challenges, such as the American Solar Challenge.

FIG. 11.10 Electric solar quadricycle developed by the Universidade Federal de Santa Catarina in 2003 [66].

The challenges gathered many university engineering teams that have been building and designing solar-powered cars that are not intended for broad commercial applications, but ended up preparing the market that is now emerging [67].

In recent years, using the knowledge obtained from the solar car challenges, some passenger commercial solutions have been developed. One example is the Prius Plug-in Hybrid Solar that was brought to the market by Toyota in 2017, with an installed solar capacity of 180 W. In 2021, two companies are expected to deliver their first units of solar-powered electric cars, using solar cells of silicon-type IBC (interdigitate back contact): Sono Motors (1204 W of PV capacity and around 34 km of autonomy per day) and Lightyear (1000 W of PV capacity and around 50 km of autonomy per day). The last one, Lightyear, in spite of having a smaller PV installed capacity, has invested in better aerodynamics to obtain a larger travel range.

For larger vehicles, a wider roof area and the design of trucks and buses facilitate the integration of PV modules. Kronthaler et al. [68] demonstrated that up to 17,000 L of diesel over a 10 year vehicle lifetime could be saved with the deployment of VIPV in trucks and buses.

A. Solar cars

Solar cars convert sunlight into power by PV cells in order to drive the electric motor. The amount of energy imported into the car severely limits the solar car design. Solar cars are mainly constructed for public use and car races. Solar cars have to use ultralight bodies in order to reduce weight because they only obtain limited power. Unlike conventional vehicles, solar cars suffer from the lack of convenient charging facilities and safety issues [69, 70].

B. Solar buses

Solar energy moves solar buses, and this energy is completely or significantly obtained by stationary solar modules. In order to decrease the energy consumption and extend the researchable battery's life, bus services use electric buses that are supplied by solar modules installed on the bus roof [71].

C. Solar-powered spacecraft

Satellites and spacecraft often use solar energy to supply power for operating within the solar system because solar energy is able to supply energy for a long time. It should be noted that satellites consist of several radio transmitters that work constantly throughout their lives. Using primary batteries or fuel cells for such vehicles is not economical because they may orbit for years. Also, satellites do not use solar energy to adjust their position [72].

FIG. 11.11 Example of a solar boat with social purposes, the "Barco Amazônia" [66].

D. Solar boats

Until 2007, solar boats were used in rivers, but the first solar boats sailed the Atlantic from Seville to Miami in 2007. To date, different systems of solar boats have been built. None have been able to benefit from the cooling power of water. Using solar vessels is limited due to the low power density in the solar modules. Also, boats that have sails and do not use combustion engines to generate electricity depend on battery power for general uses such as lighting, communications, and refrigeration. Therefore, solar modules are very popular for recharging batteries because solar modules don't need fuel, don't make noise, and can be easily added to the deck space [73]. Fig. 11.11 shows an example of a solar boat that also serves social purposes. The "Barco Amazônia" is a catamaran with 4.4 kWp of PV installed capacity and 36 kWh of storage; it can transport 20 passengers (plus two tripulants). It is used for school transport of kids in Belem, Brazil.

E. PV-charged electric vehicles

Electric vehicles can be charged at PV stations. These vehicles can maintain their batteries' energy by using PV sources so that solar cells are added to the roofs of the electric vehicles. The triple hybrid vehicles are an interesting type of electric vehicle. Also, other vehicles can be charged by using PV stations [74].

PV applications in transport can include auxiliary units or motive power, in particular, when using combustion engines is prevented due to maintenance, fuel, or noise. However, speed can be limited when applied for motive power due to the limited space on vehicles [75].

F. Solar-powered unmanned aerial vehicles (UAVs)

UAVs have received considerable military attention. UAVs can stay overhead for months by applying solar energy. They are much cheaper

FIG. 11.12 Example of a solar-powered unmanned aerial vehicle [77].

devices for doing some functions that are performed by satellites today. The first solar-powered UAV flight took place for 48 hours under fixed power in September 2007 [76]. Fig. 11.12 depicts an example of a lightweight solar-powered unmanned aerial vehicle [77].

G. Solar-powered train

In addition to all the above-mentioned examples, railroads have an enormous potential to be integrated with solar PVs. In an experiment conducted in India, an electrified rail coach was implemented with PV modules mounted on the rooftop to introduce an alternative for diesel-powered trains (Fig. 11.13). Results indicated that the coach is able to run at speeds up to 120 km/h and it was estimated that one solar coach can generate 18 kWh of electricity in a day [78].

H. Solar-powered hybrid airship

Unlike conventional airships, hybrid airships, which are viable as high-altitude detection platforms, can be fabricated in unconventional shapes to improve aerodynamic characteristics, as shown in Fig. 11.14. Possessing a photovoltaic module array can provide the required energy for all airborne systems and make long-distance transportation achievable [79].

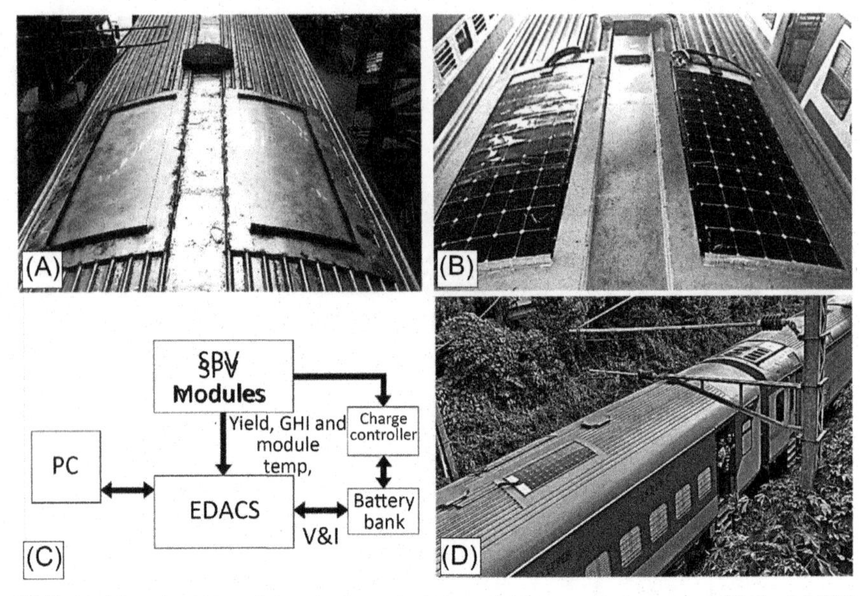

FIG. 11.13 The PV roof-mounted coach: (A) module mounting structure, (B) flexible PV modules mounted on a rooftop, (C) the diagram of the electric powering, and (D) the coach during the operation [78].

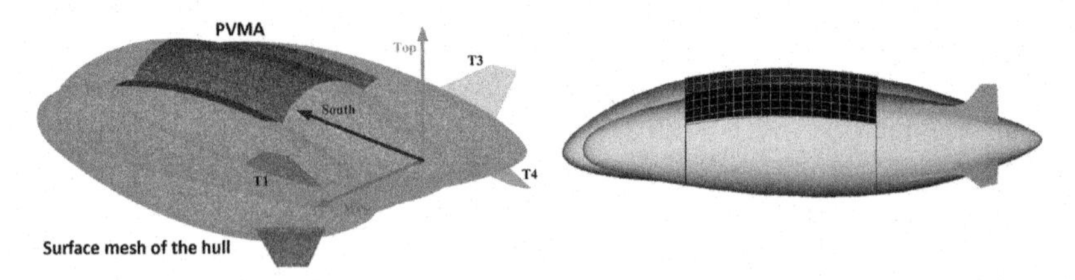

FIG. 11.14 A model of the high-altitude hybrid airship [79].

I. Solar-powered autonomous undersea vehicle (AUV)

Oceanography is highly dependent on undersampling. Despite the promising achievements in sampling technology, there are still some hardships such as energy, navigation over long distances, and communications with remote platforms that must be addressed in this way. Solar-powered autonomous undersea vehicles (SAUVs) can be a boon to scientists in conducting long-endurance sampling of oceans by solving the three mentioned problems. This device must surface to recharge the onboard energy system. A real example that was tested for long endurance missions (depicted in Fig. 11.15) has shown the ability to submerge

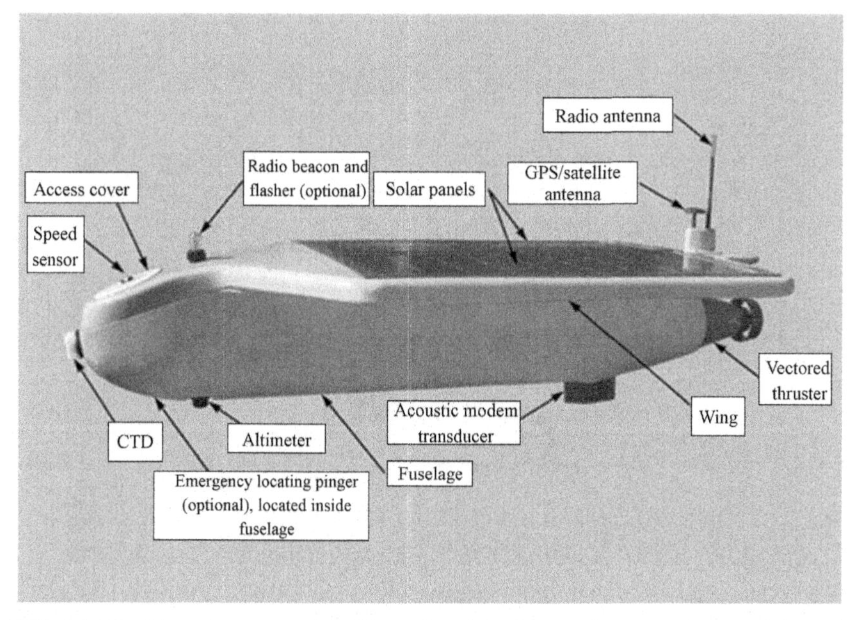

FIG. 11.15 The solar-powered AUV employed for long-endurance missions, Falmouth Scientific Inc [80].

to 500m, to transit to designated waypoints, and to surface for power charging and storing via a battery unit [81].

J. Solar-powered bikes

PV integration with electrified bikes (e-bikes) is usually defined in two ways: solar e-bikes and solar-charged e-bikes. The first term deals with electric bikes that carry PV solar cells mounted on wheels or other parts of the bike's structure. This design brings on-the-go charging opportunity as well as parked mode charging. However, in the second model, the electric power generated by PV is supplied via the special charging systems that might be found in charging stations; it merely works when e-bikes are parked [82]. In this respect, the value added to the previous e-bike designs is the leapfrog from conventional designs to more sustainable technologies in urban transportation systems.

11.4.2 Transportation facility integrated PVs

11.4.2.1 Solar PV lights for streets and roadways

Another mobility application of PV systems is their use to power urban illumination and roads. Solar lights that are mostly combined with LED lighting have great advantages, such as their easy installation in remote

FIG. 11.16 Example of solar light system installed in Fotovoltaica/UFSC solar lab [66,71].

areas and their low maintenance. Besides their higher initial investment, they are not dependent on the utility grid so they still work in the case of a grid shortage, which is a great benefit also for public security reasons [83].

The system is normally composed of a PV module that charges a self-contained battery during the day and the charger battery powers the LED light during the night. The battery is normally oversized in order to give autonomy to the system up to 3–5 days, even in low-light conditions [84]. An example of such a system is shown in Fig. 11.16. The system consists of a 30W LED light, a 60-cell PV module, charging controllers, and three second-life batteries taken from old electric cars. The system has autonomy of more than 3 days.

The social impact of PV lights is also very important. An example is the project Litro de Luz (Liter of Light), which uses a PV module and a battery that powers small LED lamps that are located inside PET bottles and are attached to PVC poles. The systems are installed in public areas that are remote and have been highly impactful, improving the quality of life and bringing safety to the communities [85].

11.4.2.2 PV charging stations

Given the need for high-efficiency cells and complex power electronic devices for ViPV, a solution to fuel electromobility with renewable energy is the use of charging stations powered by PV energy. An example of this

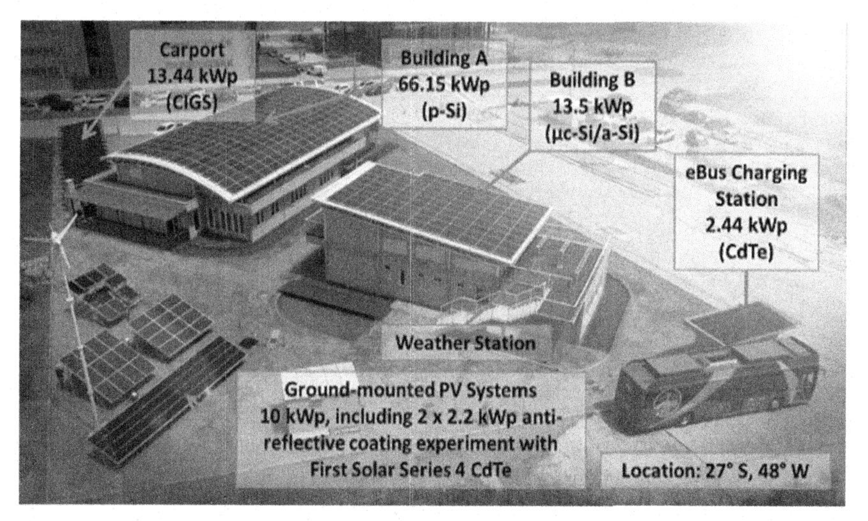

FIG. 11.17 The Fotovoltaica/UFSC solar lab, the 110kWp PV systems, and the eBus [66].

is the project eBus [86, 87], which travels 250km per day and has its energy needs totally fulfilled by the PV systems installed on the roofs of the Fotovoltaica/UFSC solar energy laboratory at Universidade Federal de Santa Catarina (see Fig. 11.17).

The integration of PV to a parking/shade structure equipped with a charging connection has many advantages, such as PV systems using existing structures to generate energy and provide shade and shelter to vehicles. They tend to be cheaper than ViPV integration because the weight and area occupied by the equipment is not a problem and the maintenance of fixed installed PV systems is simpler. The benefits are even higher when charging is combined with load management strategies, favoring self-consumption and increasing energy use during mid-day [88].

11.4.2.3 *Integrated PV bike path*

Harvesting abundant solar PV energy shed on the roadways can bring a considerable amount of energy, which reduces the grid dependency of road energy providers. Fig. 11.18 displays a practical example of a solar road (SR) in urban areas such as a bike path. A model tested successfully in the Netherlands used a transparent antiskid layer to improve traction force by increasing friction between the surface and the moving object [90]. This concept puts forth the idea of maximized land utilization in parallel with power generation.

FIG. 11.18 Solar bike pathway [89].

11.5 Solar PVs in life

The new arising benefits for life from solar photovoltaic systems are becoming incentives for more technology deployment. The dissemination of solar PV applications is rooted in the development of human living conditions, both in rural and urban areas. Although the implications are different, both developing and developed communities are benefiting from PV implementation. Economic profits and health-related promotions are two important outcomes of PV applications for developed communities. However, developing nations are reaching higher standards of living plus getting new amenities within their living places, in addition to new sanitation and health treatment levels. Being highly influential in both communities has led to new concepts for PV system development with wider applications throughout more aspects of life. PV systems not only offer new and alternative ways for energy generation, but also have become the only source of power in emergency cases where other conventional fuels are out of reach. Thus, in this section a wide variety of applications that have emerged from creative concepts are presented.

11.5.1 Solar PV in buildings

11.5.1.1 Solar PV in smart buildings

With the rise of global awareness about energy efficiency in buildings, there are still some concerns with comfort and safety for the building occupants that should be considered together, despite their conflicting relations. The only panacea for achieving these three criteria is the utilization of automation systems. A smart building uses modern information and communication technology to manage energy consumption automatically. PV integration into the energy management program of a smart building offers a reliable source of energy with minimum detrimental impacts on the environment. Because most smart buildings incorporate a building energy management system (BEMS), the balance between energy consumption and comfort levels is maintained [91]. Therefore, including PV systems in the BEMSs will result in a huge reduction in energy costs without affecting overall user comfort [92].

In the literature, smart buildings can be managed in two scales: individual or cluster types. In the second mode, several neighboring smart buildings are electrically interconnected via the same microgrid, which is the predominant consideration in renewable energy integration. This is due to the ability to reduce the uncertainties associated with renewable energy power generation on the power grid [92]. Fig. 11.19 illustrates the diagram of a smart building cluster (SBC) in which the PV source of each building and the load are connected to an intelligent electric meter to form one

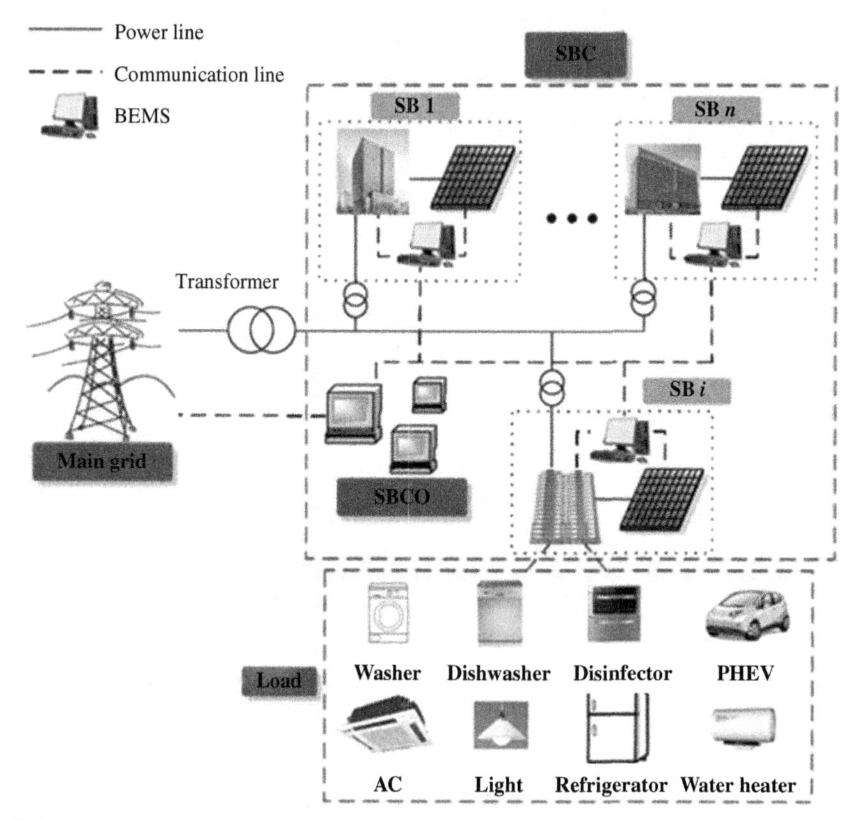

FIG. 11.19 The schematic of a cluster smart building [92].

individual smart building (SB). Connecting these individual SBs to the neighbors results in a cluster formation that is controlled with a smart building cluster operator (SBCO) to share the power generated via PV systems; it is also connected to the power grid networks. In this design, the primary source of energy is the PV source of each SB; however, when the load exceeds the generated amount of electricity via PV, the system purchases electricity through the SBCO. On the contrary, if the generated power runs over the load demand, the redundant power will be sold to the other SBs or transmitted to the grid power network at a different price. The SBCO is the main implement to control these power transitions, which ensures the maximum utilization of PVs in smart buildings [92]. Moreover, this datalogger also speculates on the future behaviors of PVs based on the models developed in past years to respond appropriately to weather disturbances [93].

11.5.1.2 *Solar PV in zero/low-energy buildings*

Bioclimatic architecture offers innovative designs that provide both visual and thermal comfort for users. This concept is globally accepted as a new approach to build energy-efficient buildings, where the remaining energy demand can be met by the utilization of renewable energy sources [94]. So far, several zero energy buildings (ZEB) have been developed by passive and active solar thermal applications to be the most efficient design while PV integration is necessary for electric energy demand. Connecting ZEBs to the utility power grid has been demonstrated as a boon for delivering storage energy to the grid. In this concept, although the first aim is to apply the passive strategies to diminish the entire load for cooling and heating, the PV power is used for the remaining energy demand [95]. Fig. 11.20 shows a diagram of the interconnections between different parts of a ZEB and a photograph of a proposed model.

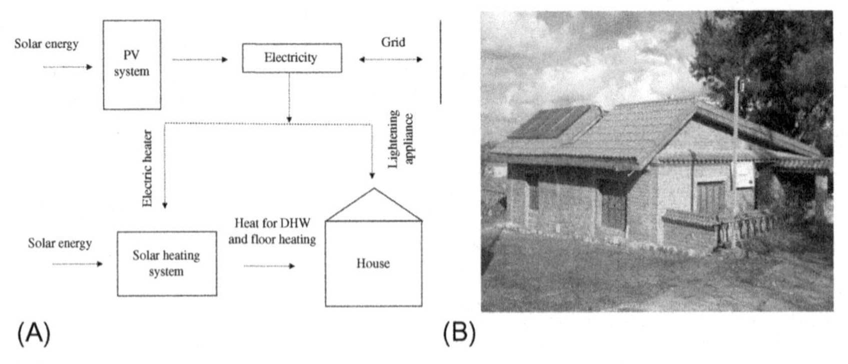

FIG. 11.20 (A) The interconnections between different parts of a ZEB, and (B) a photograph of a ZEB [95].

11.5.1.3 PV-powered air-conditioning systems

Air-conditioning devices consume the most energy in offices and residential houses in hot climate regions [96]. Experimental analyses state that air-conditioning equipment has the capacity to consume up to 80% of the total energy needed for residential, public, and commercial buildings [97]. Regarding the increasing trend of energy prices, coupling PV technology as an alternative source of electricity creates a drastic momentum toward both a noncarbonic lifestyle and economic benefits. Scientists have demonstrated the viability of PV-powered air-conditioning systems for large commercial buildings. In an attempt conducted by Al-Ugla et al. [98], three different cooling systems were tests for a 41,016 m^2 building in Saudi Arabia. Results showed that the feasibility of solar-powered systems improves as the size of the commercial building and the electricity rate increase. Additionally, it has been stated in the literature that a PV-powered air conditioner can conserve grid electricity by more than 67% and 77% during summer daytime and summer nighttime, respectively [99]. Fig. 11.21 shows a typical PV-powered air-conditioning system with its connections to other utilities.

11.5.1.4 PV windrail technology

One of the distinctive advantages of PV technology is its flexibility in combination with other power conversion systems. PV windrail technology is a hybrid energy harvesting system that exploits both solar PV and

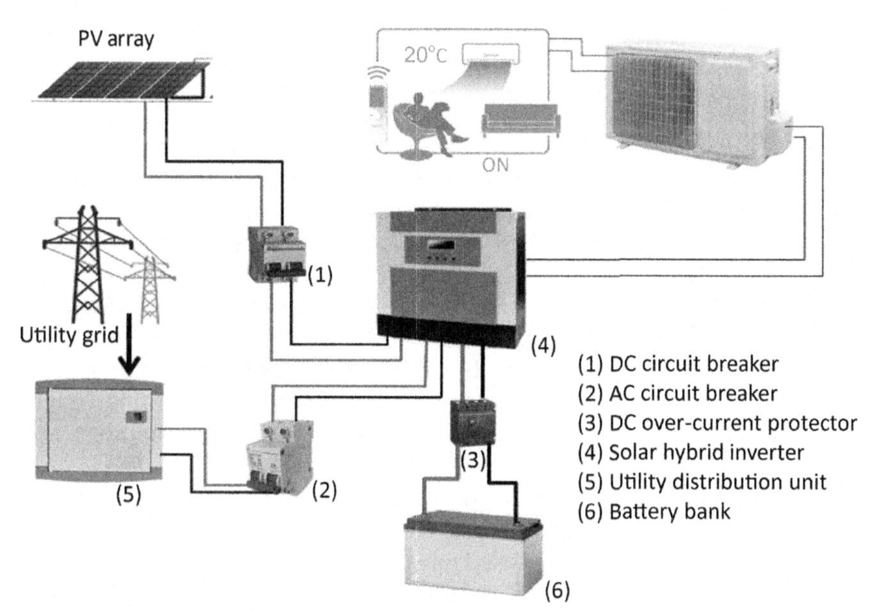

FIG. 11.21 The diagram of a PV-powered air-conditioning system [100].

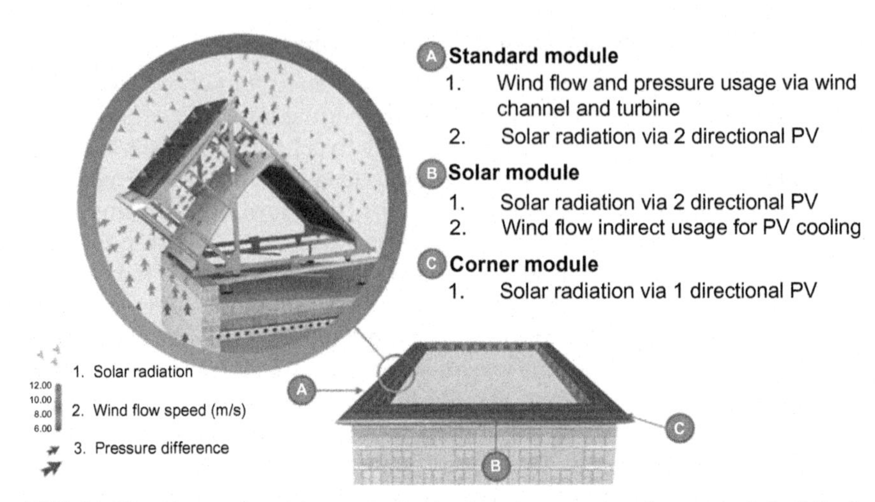

FIG. 11.22 The concept of the windrail technology integrated on the corner of a building's rooftop [101]. *Courtesy Anerdgy SA.*

wind energy simultaneously to meet the electric demand of a large building. This device is usually located on the side or corner of the rooftop of a building in order to capture the high-speed wind. As Fig. 11.22 illustrates, the pressure difference between the façade, the flow surface, and the rooftop out of the flow area is employed by the windrail channel to overspeed the wind velocity and generate electricity via turbine generators inside the channel and driven by the wind [101]. Using PV panels as the shelter is inclined toward the solar incident may contribute to a considerable amount of the total energy generation with improvements in the aesthetics of the building structure.

11.5.2 Solar PV applications in society

11.5.2.1 PV trees

The installation of PV modules is hindered by the difficulties of land requirements in land restricted areas. Furthermore, most of the installed PV systems suffer from poor public perception originating from a lack of aesthetics, which is a crucial factor when PV is an element in the environment. A solar PV tree is a concept in which art and technology are amalgamated together to form a solar PV sculpture [102]. In this decorative design, solar panels are arranged as leaves stretched in different directions to utilize solar energy despite the hourly changes in the incident angle. The structure is composed of a steel stem to support the solar panels, which generate electric energy to charge mobile phones, laptops,

FIG. 11.23 Two lift solar tree project, Spotlight Solar Inc. [105].

and other small gadgets [103]. Also, there are still more applications such as street lighting, household supply, industrial supply, charging of electric vehicles, and supplying the grid that can be met by this device [104]. Fig. 11.23 represents two lift solar trees installed near the Frost Museum of Science in Miami, Florida.

11.5.2.2 PV pavement

Respecting sustainable developments in urban areas, PV pavements have emerged as a promising solution for both green power generation and urban heat island mitigation. As pavement has covered 30–40% of the urban surface [106], there is a notable potential behind PV pavement integration while it is also able to provide electricity for peak energy demands. In a model proposed in the literature [107], the impact of PV pavement on mitigating heat islands was investigated where experimental data revealed that this technique is, directly and indirectly, influential on the local microclimate. It was also found out that PV pavement has 8 K lower surface temperatures than conventional pavement, which decreases the ambient temperature by 0.8 K. Developing walkable PV floor tiles puts forth the promotion of PV pavement integration into urban areas. Fig. 11.24 shows the layout of a PV floor tile in which the solar cell is encompassed by two PVB/EVA foils surrounded with two thick antislip tempered glasses. Experimental tests have reflected the features of efficient energy conversion, antislip, heat resistance, durability, and compressive strength,

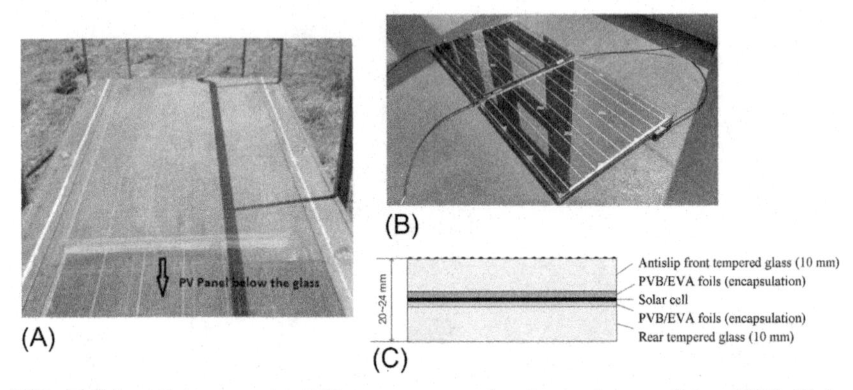

FIG. 11.24 (A) An example of PV pavement tested under outdoor conditions [107], (B) the photograph of a developed walkable PV floor tile, and (C) the layout of the proposed floor tile components [108].

demonstrating that such PV floor tiles can be used as replacements for pavement and cycling tracks [108].

11.5.2.3 Landscape PV integration

PV integration with natural landscapes is becoming an innovative path in urban architecture and is removing barriers for PV installation in green areas, specifically green roofs. A vertical bifacial PV system plays a significant role in this technology adaptation, which outperforms the typical mono facial installations on flat grounds. In an experiment conducted in Switzerland, vertical bifacial PV panels were installed on a flat green roof (Fig. 11.25) [109]. Scientists reported that the total output was highly dependent on the albedo, where plants with higher silvery leaves improve power generation due to the higher and more stable albedo at the unfavorable conditions. The other significant achievements of vertical orientation can be fewer soiling effects and in winter, minimum snow covering and the higher albedo factor of snow, which leads to power augmentation. The remarkable advantage of this concept is to lower the burden of cumulative covering with imperious material by artificial structures.

11.5.3 Product-integrated PVs (PIPVs)

PIPV has the advantage of offering users and designers the possibility to integrate PV into products that can satisfy consumer expectations, reduce primary battery consumption and waste, and serve as portable devices that are grid free. PIPV can be seen in indoor and outdoor applications such as watches, lamps, scales, phone chargers, keyboards, and many more. PIPV also utilizes a wide variety of solar cell technologies,

(A)

(B)

FIG. 11.25 (A) The vertically bifacial PV modules installed on a green roof in Winterthur, Switzerland, and (B) an individual module of a vertical bifacial PV [109].

which will eventually serve as the main power source of the device [110, 111].

According to a recent consumer-related solar-powered product study, PIPV seems attractive to lead users due to the design, aesthetics, usability, and performance [112]. These basic principles will determine the level of customer satisfaction and eventually will act as the decision factors in acquiring these products. Because PIPV seems to be at an early stage, even though calculators and watches have been using this technology since 1960, the design parameters and guidelines are still under development. However, optimized design methods for these types of products are proposed by offering basic guidelines, which associate the industrial design features of the product to their optimal cell efficiency in order to gain higher power yields to reduce their environmental impact [113].

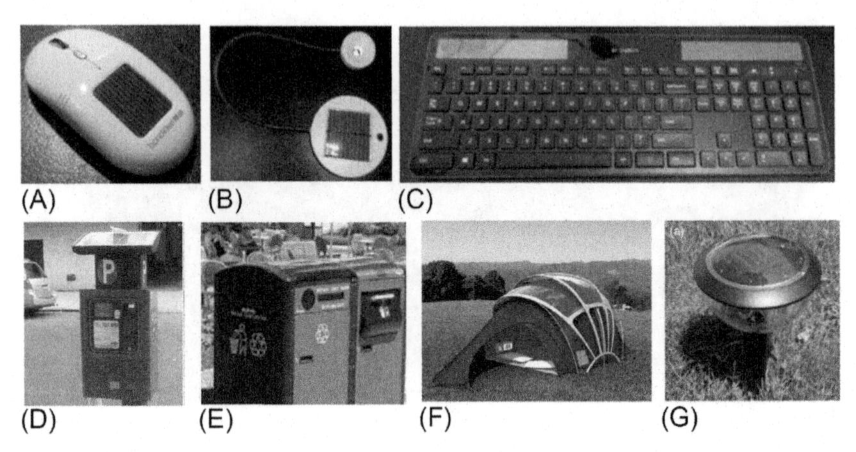

FIG. 11.26 Some of the instances of PV-integrated gadgets: (A) PV-powered computer mouse [114], (B) PV-powered luminaire [114], (C) PV-powered keyboard [114], (D) solar parking meter in New York [115], (E) automated solar trash bin Big Bell [115], (F) solar-powered tent [115], and (G) garden light [115].

11.5.3.1 PV-integrated gadgets

PV solar cells have been utilized in different types of products and gadgets for years, beginning vigorously in the 1970s (some examples are shown in Fig. 11.26). PV utilization with consumer products has culminated in several benefits such as environmental advantages [116] resulting from the elimination of primary batteries and efficiency augmentation by employing rechargeable batteries. Based on the applications, PV-integrated gadgets can be categorized into three main groups as follows.

11.5.3.2 PV-powered products

There are still other versatile products that have the ability to exploit solar PV energy in both indoor and outdoor conditions. Having the same structure as the other PV integrated products, they require an optimal design to work on low irradiance conditions. Some of the explicit designs are solar-powered toys, watches, calculators, mobile phones, PV chargers used for electric gadgets, etc.

PV-integrated products can also be classified by other factors such as the size of the product, where it gives small and thin or large and thick categories [111]. However, the other classification is aimed at the functionality and application of these products, which are given as consumer products, lighting products, business-to-business products, recreational products, vehicle and transportation products, and artistic products [115].

11.5.4 New horizons in PV utilization

11.5.4.1 Floating PV systems

Floating PV systems (shown in Fig. 11.27) bring new opportunities to increase the power generation capacity, particularly in countries with severe restrictions on limited land use and highly populated regions. One of the predominant abilities that emerged from this technology is the utilization of existing electricity transmission infrastructure at hydropower sites. Moreover, several advantages come after floating PV integration, such as the following [118]

- Providing shade on the below water body, leading to a reduction in water evaporation from water reservoirs
- Diminishing algae growth and improving water quality
- Avoiding the installation foundation and leveling structure, which are necessary for land-based installations
- Easy and fast installation with flexibility in deployment at different sites

This technology has accelerated during recent years and is nearly mature, beginning in 2007 where the first plant was commercialized in California in 2008 with a 175 kWp capacity. Analogous to ground-mounted PV plants, the floating PV systems enjoy the same components, other than the fact that the inverters are usually floated. Basically floated PV plants are mounted on a platform that is made of floats designed as self-buoyant bodies to which PV panels can be directly affixed [118].

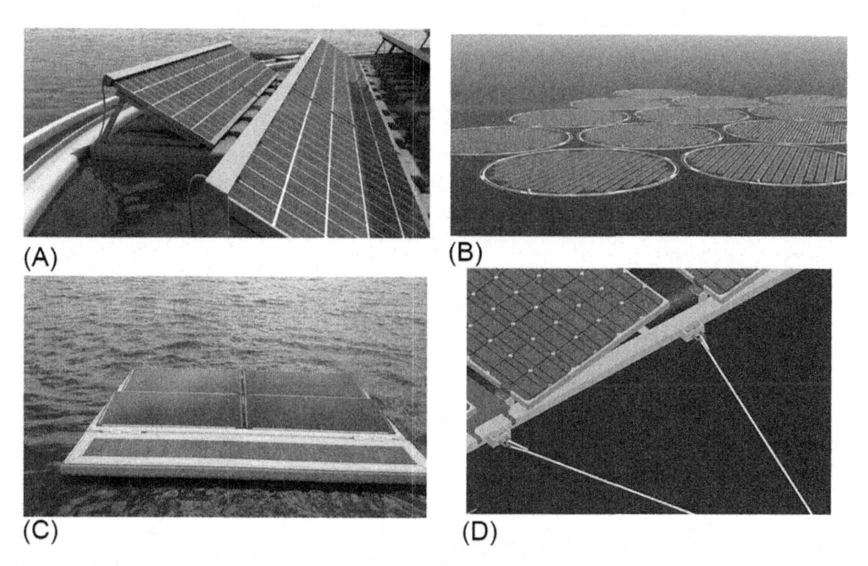

FIG. 11.27 Floating PVs; (A) incorporating a cooling mechanism (patent 2008), (B) an artist's view of a full plant, (C) oselic system in PVC (France), and (D) mooring cables [117].

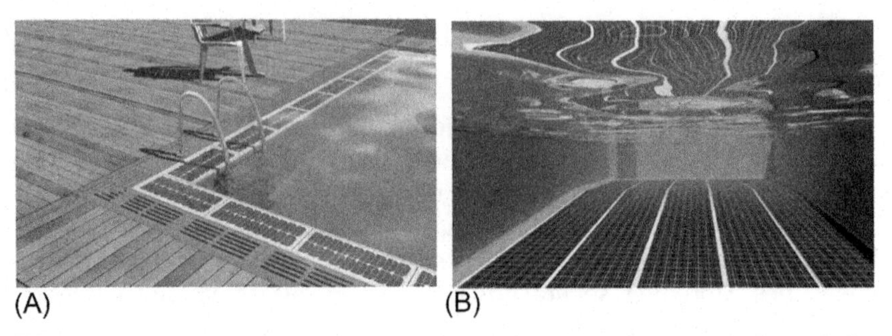

FIG. 11.28 The concept of PV integration with swimming pools: (A) PV on the pool edge, and (B) PV on the pool floor [119].

Mooring cables are used for both fixed and tracking floating PVs [117]; however, the floating platform can be anchored to the bottom or banks within shallow ponds [118].

11.5.4.2 Submerged PVs

The novelty behind the utilization of water as the operating ambient for PV arrays may result in some beneficial outcomes that are influential on power generation efficiency. In a concept proposed by scientists [119], the integration of PV panels with a swimming pool was introduced with the least environmental impact. Avoiding cleaning and overheating problems, the solar panels perform efficiently compared to conventional PV arrays. Water also behaves as a filter that blocks the long-wavelength photons and transmits lights within the visible spectrum, the optimum condition for PV cells. In this concept, a part of the PV panels was positioned on the pool edge and the others were placed on the pool floor, as shown in Fig. 11.28. In the first configuration, PV panels were protected by the glass and cooled by the water recirculating in the skimmer. A feasibility study suggested the viability of the design for integration with existing swimming pools while recreational and aesthetic provisions are maintained.

11.6 Conclusion

Renewable electricity provided by photovoltaic conversion technology is driving the world's development and improving public health, tackling air and water pollution and consequently premature death. With the advent of new integrations by different sorts of life sectors, this inexhaustible energy is becoming more reachable and is expanding energy access. Rural and urban areas, developing and underdeveloping countries, oil-dependent and nonoil-dependent economies, all embrace the increasing deployment of PV applications for local users to enhance the quality of life. Scientists are also developing new concepts to augment the efficiency of PV-integrated systems and expand their lifespan as well as preserve architectural aesthesis and prevent environmental impacts. New advances in PV cell technology have yielded pragmatic results for the flexibility of PV technology adaptation in daily applications.

Incorporating BIPV technology into a building façade combines energy strategy with modern architecture as an ecofriendly solution that enumerates several functions as well as energy generation. In the case of BIPV design, the performance is affected by some factors, including the installation angle and the surrounding objects that must be considered in the preliminary stages. LSC technology is also pacing with advances in material science, where several designs, shapes, and configurations have been developed to deploy this technology to numerous applications. The fabrication process is quite straightforward and displaying high quantum efficiency and low reabsorption loss are necessary to advance the current technology for higher mass production, commercialization, and more cost reductions. Introducing PV technology to mobile apparatus has brought significant opportunities for more sustainable means of transport and a less-polluted environment. This technology not only affects manned and unmanned vehicles, but also has considerable influence on the infrastructure involved in transportation systems such as parking lots and stations. More than roads and railways, there are buildings, gadgets, and life commodities that have absorbed PV applications, enabling a cleaner living environment with higher accessibility to electric energy and consequently more facilitated life. Therefore, the current trend will lead us to a severe reliance on PV technology integrated with smart, portable, and handy devices.

References

[1] I. E. Agency. Technology Roadmap. Encycl Prod Manuf Manag 2014;781–2.

[2] Masson G, Kaizuka I. The 23rd international survey report on trends in photovoltaic (PV) applications; 2018.

[3] Leva S, Mohammadreza A. Failures and defects in PV systems: review and methods of analysis; 2018.

[4] Aghaei M. Novel methods in control and monitoring of photovoltaic systems. Politecnico di Milano; 2016.

[5] McKinsey Global Institute. Global energy perspective 2019: reference case. In: Energy Insights; 2019 [January].

[6] Shankarappa N, Ahmed M, Shashikiran N, Naganagouda H. Solar photovoltaic systems–applications & configurations. Int Res J Eng Technol 2017;4(8):1851–5.

[7] Khatib T, Mohamed A, Sopian K, Mahmoud M. Optimal sizing of building integrated hybrid PV/diesel generator system for zero load rejection for Malaysia. Energy Build 2011;43(12):3430–5.

[8] Díez-Mediavilla M, Rodríguez-Amigo MC, Dieste-Velasco MI, García-Calderón T, Alonso-Tristán C. The PV potential of vertical façades: a classic approach using experimental data from Burgos, Spain. Sol Energy 2019;177(June 2018):192–9.

[9] Singh GK. Solar power generation by PV (photovoltaic) technology: a review. Energy 2013;53:1–13.

[10] CRED. Executive summary of natural disasters 2018; 2018.
[11] Qazi S. Portable standalone PV systems for disaster relief and remote areas. In: Standalone photovoltaic (PV) systems for disaster relief and remote areas. Elsevier; 2017. p. 113–38.
[12] Rao ND, Pachauri S. Energy access and living standards: some observations on recent trends. Environ Res Lett 2017;12(2).
[13] Aghaei M, Leva S, Grimaccia F. PV power plant inspection by image mosaicing techniques for IR real-time images. In: *Conference record of the IEEE photovoltaic specialists conference*, vol. 2016-November; 2016.
[14] Grimaccia F, Leva S, Dolara A, Aghaei M. Survey on PV modules' common faults after an O&M flight extensive campaign over different plants in Italy. IEEE J Photovolt 2017;7(3).
[15] Quater PB, Grimaccia F, Leva S, Mussetta M, Aghaei M. Light unmanned aerial vehicles (UAVs) for cooperative inspection of PV plants. IEEE J Photovolt 2014;4(4).
[16] Vidal De Oliveira AK, Aghaei M, Rüther R, Aghaei M. Automatic fault detection of photovoltaic arrays by convolutional neural networks during aerial infrared thermography. In: 36th EU PVSEC—European PV Solar Energy Conference and Exhibition; 2019.
[17] Aghaei M, Grimaccia F, Gonano CA, Leva S. Innovative automated control system for PV fields inspection and remote control. IEEE Trans Ind Electron 2015;62(11).
[18] Jung S, Jo Y, Kim YJ. Flight time estimation for continuous surveillance missions using a multirotor UAV. Energies 2019;12(5).
[19] Umer F, Aslam MS, Rabbani MS, Hanif MJ, Naeem N, Abbas MT. Design and optimization of solar carport canopies for maximum power generation and efficiency at Bahawalpur. Int J Photoenergy 2019;2019.
[20] Yang RJ, Zou PXW. Building integrated photovoltaics (BIPV): costs, benefits, risks, barriers and improvement strategy. Int J Constr Manag Jan. 2016;16(1):39–53.
[21] Obeng GY, Evers H. Solar PV rural electrification and energy-poverty: a review and conceptual framework with reference to Ghana. ZEF Work Pap Ser 2009;36 (17136):1–20.
[22] Buragohain T. Impact of solar energy in rural development in India. Int J Environ Sci Dev 2012;3(4):1–4.
[23] Mikul B, et al. Access to modern energy services for health facilities in resource-constrained settings; 2018.
[24] Adair-Rohani H, et al. Limited electricity access in health facilities of sub-Saharan Africa: a systematic review of data on electricity access, sources, and reliability. Glob Heal Sci Pract Aug. 2013;1(2):249–61.
[25] Jelle BP, Breivik C, Drolsum Røkenes H. Building integrated photovoltaic products: a state-of-the-art review and future research opportunities. In: Solar energy materials and solar cells; 2012.
[26] Henemann A. BIPV: built-in solar energy. Renew Energy Focus 2008;9(6, suppl.):14–9.
[27] Yoon JH, Song J, Lee SJ. Practical application of building integrated photovoltaic (BIPV) system using transparent amorphous silicon thin-film PV module. Sol Energy 2011;85 (5):723–33.
[28] A. Reinders, P. Verlinden, W. van Sark, and A. Freundlich, PV systems and applications (Chapter 11). In: Photovoltaic Solar Energy: From Fundamentals to Applications. Wiley, 2017.
[29] Tabakovic M, et al. Status and outlook for building integrated photovoltaics (BIPV) in relation to educational needs in the BIPV sector. Energy Procedia 2017;111:993–9.
[30] Cronemberger J, Corpas MA, Cerón I, Caamaño-Martín E, Sánchez SV. BIPV technology application: highlighting advances, tendencies and solutions through Solar Decathlon Europe houses. Energy Build 2014.

[31] Imenes AG. Performance of BIPV and BAPV installations in Norway. In: *2016 IEEE 43rd Photovoltaic Specialists Conference (PVSC)*; 2016. p. 3147–52.
[32] Zomer CD, Costa MR, Nobre A, Rüther R. Performance compromises of building-integrated and building-applied photovoltaics (BIPV and BAPV) in Brazilian airports. Energy Build 2013;66:607–15.
[33] Peng C, Huang Y, Wu Z. Building-integrated photovoltaics (BIPV) in architectural design in China. Energy Build Dec. 2011;43(12):3592–8.
[34] Scott K, Zielnik A. Developing codes and standards for BIPV and integrated systems international photovoltaic reliability workshop (IPRW) II removing barriers to photovoltaic technology adoption: reliability, codes/standards, and market acceptance-Tempe Mission Palms Hotel; 2009.
[35] Zero Energy Buildings in France: Overview and Feedback.: EBSCOhost.
[36] Yang HJ, et al. Adjusted colorful amorphous slicon thin film solar cells by a multilayer film design. J Electrochem Soc 2011;158(9).
[37] Zhan X, Zhu D. Conjugated polymers for high-efficiency organic photovoltaics. Polym Chem 2010;1(4):409–19.
[38] Debije MG, Verbunt PPC. Thirty years of luminescent solar concentrator research: solar energy for the built environment. Adv Energy Mater 2012;2(1):12–35.
[39] van Sark WGJHM. Luminescent solar concentrators—a low cost photovoltaics alternative. Renew Energy 2013;49:207–10.
[40] Hermann AM. Luminescent solar concentrators—a review. Sol Energy 1982;29(4): 323–9.
[41] Tummeltshammer C, Taylor A, Kenyon AJ, Papakonstantinou I. Losses in luminescent solar concentrators unveiled. Sol Energy Mater Sol Cells 2016;144:40–7.
[42] Rafiee M, Chandra S, Ahmed H, McCormack SJ. An overview of various configurations of luminescent solar concentrators for photovoltaic applications. Opt Mater (Amst) 2019.
[43] Correia SFH, et al. Scale up the collection area of luminescent solar concentrators towards metre-length flexible waveguiding photovoltaics. Prog Photovolt Res Appl Sep. 2016;24(9):1178–93.
[44] Norton B, et al. Enhancing the performance of building integrated photovoltaics. Sol Energy Aug. 2011;85(8):1629–64.
[45] Gallagher SJ, Norton B, Eames PC. Quantum dot solar concentrators: electrical conversion efficiencies and comparative concentrating factors of fabricated devices. Sol Energy Jun. 2007;81(6):813–21.
[46] Slooff LH, et al. A luminescent solar concentrator with 7.1% power conversion efficiency. Phys Status Solidi Rapid Res Lett 2008;2(6):257–9.
[47] Goldschmidt JC, et al. Increasing the efficiency of fluorescent concentrator systems. Sol Energy Mater Sol Cells 2009;93(2):176–82.
[48] Nobile C, Carbone L, Fiore A, Cingolani R, Manna L, Krahne R. Self-assembly of highly fluorescent semiconductor nanorods into large scale smectic liquid crystal structures by coffee stain evaporation dynamics. J Phys Condens Matter 2009;21 (26):264013.
[49] Verbunt PPC, Kaiser A, Hermans K, Bastiaansen CWM, Broer DJ, Debije MG. Controlling light emission in luminescent solar concentrators through use of dye molecules aligned in a planar manner by liquid crystals. Adv Funct Mater 2009;19(17): 2714–9.
[50] Chandra S, Doran J, McCormack SJ, Kennedy M, Chatten AJ. Enhanced quantum dot emission for luminescent solar concentrators using plasmonic interaction. Sol Energy Mater Sol Cells 2012;98:385–90.
[51] Alivisatos AP. Perspectives on the physical chemistry of semiconductor nanocrystals. J Phys Chem Jan. 1996;100(31):13226–39.

[52] El-Bashir SM, Barakat FM, AlSalhi MS. Double layered plasmonic thin-film luminescent solar concentrators based on polycarbonate supports. Renew Energy 2014;63: 642–9.
[53] Aghaei M, Nitti M, Ekins-Daukes NJ, Reinders AHME. Simulation of a novel configuration for luminescent solar concentrator photovoltaic devices using bifacial silicon solar cells. Appl Sci 2020;10(3):871.
[54] Vossen FM, Aarts MPJ, Debije MG. Visual performance of red luminescent solar concentrating windows in an office environment. Energy Build 2016;113:123–32.
[55] Fathi M, Abderrezek M, Djahli F. Experimentations on luminescent glazing for solar electricity generation in buildings. Optik (Stuttg) 2017;148:14–27.
[56] Reinders AHME, et al. Leaf roof—designing luminescent solar concentrating PV roof tiles, In: 2017 IEEE 44th photovolt. spec conf PVSC 2017; 2017. p. 1–5.
[57] Leaf Roof—4TU.Federation.
[58] Rosemann A, Doudart G, Grée D, Papadopoulos A. Leafroof. Spool 2015;2(2):21–3.
[59] Baetens R, Jelle BP, Gustavsen A. Properties, requirements and possibilities of smart windows for dynamic daylight and solar energy control in buildings: a state-of-the-art review. Sol Energy Mater Sol Cells 2010;94(2):87–105.
[60] Aste N, Buzzetti M, Del Pero C, Fusco R, Leonforte F, Testa D. Triggering a large scale luminescent solar concentrators market: the smart window project. J Clean Prod 2019;219:35–45.
[61] Kanellis M, de Jong MM, Slooff L, Debije MG. The solar noise barrier project: 1. Effect of incident light orientation on the performance of a large-scale luminescent solar concentrator noise barrier. Renew Energy 2017;103:647–52.
[62] Eggink W, Reinders A. Design it with LSCs; an exploration of applications for luminescent solar concentrator PV technologies. In: 2017 IEEE 44th photovoltaic specialist conference (PVSC); 2017. p. 2109–13.
[63] Reinders A, Kishore R, Slooff L, Eggink W. Luminescent solar concentrator photovoltaic designs. Jpn J Appl Phys 2018;57(8).
[64] Saber AY, Venayagamoorthy GK. Plug-in vehicles and renewable energy sources for cost and emission reductions. IEEE Trans Ind Electron Apr. 2011;58(4):1229–38.
[65] Letendre S, Perez R, Herig C. Vehicle integrated PV : a clean and secure fuel for hybrid electric vehicles. In: Annual Meeting of the American Solar Energy Society, Austin; 2003.
[66] Universidade Federal de Santa Catarina. Fotovoltaica UFSC; 2019
[67] Letendre SE. Vehicle integrated photovoltaics: exploring the potential.
[68] Kronthaler L, Maturi L, Moser D, Alberti L. Vehicle-integrated photovoltaic (ViPV) systems : energy production, diesel equivalent, payback time; an assessment screening for trucks and busses. In: 9th international conference on ecological vehicles and renewable energies (EVER); 2014. p. 1–8.
[69] Chevillard N. Solar mobility chosing solar for the Driver' s seat; 2018
[70] Reichel R. Solar mobility. no. 0049, 1985. p. 1–18.
[71] NAFTC eNews | West Virginia University.
[72] Helios Solar-Powered Aircraft Crashes—NetComposites.
[73] BBC NEWS | Europe | Solar boat makes Atlantic history.
[74] Solar Electrical Vehicles.
[75] Using Solar Roofs To Power Hybrids | TreeHugger.
[76] BBC NEWS | Science/Nature | Solar plane flies into the night.
[77] Malaver A, Motta N, Corke P, Gonzalez F. Development and integration of a solar powered unmanned aerial vehicle and a wireless sensor network to monitor greenhouse gases. Sensors (Switzerland) 2015;15(2):4072–96.
[78] Shravanth Vasisht M, Vashista GA, Srinivasan J, Ramasesha SK. Rail coaches with rooftop solar photovoltaic systems: a feasibility study. Energy 2017;118:684–91.

[79] Zhang L, Li J, Meng J, Du H, Lv M, Zhu W. Thermal performance analysis of a high-altitude solar-powered hybrid airship. Renew Energy 2018;125:890–906.
[80] Falmouth Scientific, Inc—AUVAC.
[81] Blidberg DR, Chappell S, Jalbert JC. Long endurance sampling of the ocean with solar powered AUV's. IFAC Proc Vol 2004;37(8):561–6.
[82] Apostolou G, Reinders A, Geurs K. An overview of existing experiences with solar-powered e-bikes. Energies 2018;11(8):1–19.
[83] Khan IJ, Khan YH, Rashid S, Khan JA, Campus I, Bannu T. Installation of solar power system used for street lights and schools in Khyber; 2015. p. 13–7.
[84] António J, Vieira B, Mota AM. Implementation of a Stand-Alone Photovoltaic Lighting System with MPPT Battery Charging and LED Current Control. In: 2010 IEEE International Conference on Control Applications; 2010.
[85] Liter of Light Brazil; 2019.
[86] de Oliveira AKV, Gomes AMF, Goulart VV, Montenegro AdA, Rüther R. Analysis of the integration of an electric bus and an electric vehicle with grid-connected PV systems and a storage system. In: *Solar World Congress*; 2019.
[87] Mattes P, de Oliveira AKV, Montenegro ADA, Rüther R. Performance of an electric bus, power by solar energy. In: *VII Congresso Brasileiro de Energia Solar*; 2018.
[88] Denholm P, Kuss M, Margolis RM. Co-benefits of large scale plug-in hybrid electric vehicle and solar PV deployment. J Power Sources 2013;236:350–6.
[89] Netherlands solar bike path - ABC News (Australian Broadcasting Corporation).
[90] Shekhar A, et al. Harvesting roadway solar energy-performance of the installed infrastructure integrated pv bike path. IEEE J Photovolt 2018;8(4):1066–73.
[91] Missaoui R, Joumaa H, Ploix S, Bacha S. Managing energy smart homes according to energy prices: analysis of a building energy management system. Energy Build 2014;71:155–67.
[92] Ma L, et al. Multi-party energy management for smart building cluster with PV systems using automatic demand response. Energy Build 2016;121:11–21.
[93] Wang Z, Wang L, Dounis AI, Yang R. Multi-agent control system with information fusion based comfort model for smart buildings. Appl Energy 2012;99:247–54.
[94] Badescu V. Case study for active solar space heating and domestic hot water preparation in a passive house. J Renew Sustain Energy 2011;3(2):23102.
[95] Missoum M, Hamidat A, Imessad K, Bensalem S, Khoudja A. Impact of a grid-connected PV system application in a bioclimatic house toward the zero energy status in the north of Algeria. Energy Build 2016;128:370–83.
[96] Aguilar FJ, Aledo S, Quiles PV. Experimental analysis of an air conditioner powered by photovoltaic energy and supported by the grid. Appl Therm Eng 2017;123:486–97.
[97] Pérez-Lombard L, Ortiz J, Pout C. A review on buildings energy consumption information. Energy Build 2008;40(3):394–8.
[98] Al-Ugla AA, El-Shaarawi MAI, Said SAM, Al-Qutub AM. Techno-economic analysis of solar-assisted air-conditioning systems for commercial buildings in Saudi Arabia. Renew Sust Energ Rev 2016;54:1301–10.
[99] Liu Z, Li A, Wang Q, Chi Y, Zhang L. Performance study of a quasi grid-connected photovoltaic powered DC air conditioner in a hot summer zone. Appl Therm Eng 2017;121:1102–10.
[100] Opoku R, Mensah-Darkwa K, Samed Muntaka A. Techno-economic analysis of a hybrid solar PV-grid powered air-conditioner for daytime office use in hot humid climates—a case study in Kumasi city, Ghana. Sol Energy 2018;165(February): 65–74.
[101] Kolokotsa D. Smart cooling systems for the urban environment. Using renewable technologies to face the urban climate change. Sol Energy 2017;154:101–11.
[102] Cao W, et al. 'Solar tree': Exploring new form factors of organic solar cells. Renew Energy 2014;72:134–9.

[103] Verma NN, Mazumder S. An investigation of solar trees for effective sunlight capture using Monte Carlo simulations of solar radiation transport. In: *ASME International Mechanical Engineering Congress and Exposition, Proceedings (IMECE)*, vol. 8A; 2014.
[104] Hyder F, Sudhakar K, Mamat R. Solar PV tree design: a review. Renew Sustain Energy Rev 2018;82(September 2017):1079–96.
[105] Solar Energy Trees | Project Gallery | Spotlight Solar—Solar Trees | Spotlight Solar.
[106] Akbari H, Rose LS. Characterizing the fabric of the urban environment: a case study of metropolitan Chicago, Illinois; 2001.
[107] Efthymiou C, Santamouris M, Kolokotsa D, Koras A. Development and testing of photovoltaic pavement for heat island mitigation. Sol Energy 2016;130:148–60.
[108] Ma T, Yang H, Gu W, Li Z, Yan S. Development of walkable photovoltaic floor tiles used for pavement. Energy Convers Manag 2019;183(February):764–71.
[109] Baumann T, Nussbaumer H, Klenk M, Dreisiebner A, Carigiet F, Baumgartner F. Photovoltaic systems with vertically mounted bifacial PV modules in combination with green roofs. Sol Energy 2019;190(March):139–46.
[110] Apostolou G, Reinders AHME. Overview of design issues in product-integrated photovoltaics. Energy Technol 2014;2(3):229–42. Wiley-VCH Verlag.
[111] Apostolou G, Reinders A, Verwaal M. Comparison of the indoor performance of 12 commercial PV products by a simple model. Energy Sci Eng 2016;4(1):69–85.
[112] Apostolou G, Reinders A. How do users interact with photovoltaic-powered products? Investigating 100 'lead-users' and 6 PV products. How do users interact with photovoltaic-powered products? 2016.
[113] Alsema EA, et al. Towards an optimized design methodfor PV-powered consumer and professional applications-the Syn-energy project Luminescent Solar Concentrator View project MeDoS. In: Parametric optimization of double-skin façades in the mediterranean climate to improve energy; 2005.
[114] Apostolou G. Investigating the use of indoor photovoltaic products towards the sustainability of a building environment. Procedia Environ Sci 2017;38:905–12.
[115] Reinders AHME, van Sark WGJHM. Product-integrated photovoltaics. In: Comprehensive renewable energy. vol. 1. Elsevier Ltd; 2012. p. 709–32.
[116] Durlinger B, Reinders A, Toxopeus M. Environmental benefits of pv powered lighting products for rural areas in South East Asia: a life cycle analysis with geographic allocation. In: *Conference Record of the IEEE Photovoltaic Specialists Conference*; 2010. p. 2353–7.
[117] Rosa-Clot M, Tina GM. The floating PV plant; 2018.
[118] "Where Sun Meets Water," *Where Sun Meets Water*, 2019.
[119] Clot MR, Rosa-Clot P, Tina GM. Submerged PV solar panel for swimming pools: SP3. Energy Procedia 2017;134:567–76.

CHAPTER

12

Life cycle assessment and environmental impacts of solar PV systems

Nallapaneni Manoj Kumar[a], *Shauhrat S. Chopra*[a], and *Pramod Rajput*[b]

[a]School of Energy and Environment, City University of Hong Kong, Kowloon, Hong Kong, [b]Department of Physics, Indian Institute of Technology Jodhpur, Jodhpur, Rajasthan, India

12.1 Introduction

In the new energy sector, solar photovoltaic (PV)-based electricity generation is increasing, due to which the PV industry has also seen tremendous growth over the years. The commercial use of solar PV systems began in early 2000 with an installed capacity of 1.3 GW. From 2001 to 2007, the growth rate was considerably high, and the cumulative installation reached a value of 9.2 GW by the end of 2007. By the end of 2008, the installations saw exponential growth to a value of 16 GW. In 2009, PV installation rose to 23 GW. In 2010 alone, PV installation saw tremendous growth, and this was due to the various benefits offered by national and local organizations. These benefits include subsidies, viable tariff policies for energy trade, and affordable PV technology prices [1–3]. All these benefits collectively contributed to PV growth, which finally resulted in a cumulative PV installation capacity of approximately 40 GW by the end of 2010. In subsequent years, there have been ups and downs in the growth rate, but the overall PV installation capacities are higher when compared to the preceding years. By the end of 2011, the installation rose to approximately 70.5 GW and in later years, PV installations reached a cumulative value of 101 GW, 138 GW, 179 GW, 230 GW, 308 GW, and 407 GW for the

Photovoltaic Solar Energy Conversion
https://doi.org/10.1016/B978-0-12-819610-6.00012-0

years 2012, 2013, 2014, 2015, 2016, and 2017, respectively. By the end of 2018, solar-powered solutions had ramped up, and the cumulative installation reached approximately 509 GW [1, 2]. The roles played by many government organizations that enforce renewable portfolio standards, policies, regulations, and governance related to solar and other renewables favored this growth. Consequently, large-scale solar PV plant deployment is seen all over the world. On the other hand, solar PV technology has matured as well as created novel installation methods that harness solar power so effectively. As a result, the growth rate of solar PV installation seems to continue to increase. Looking at PV installations by region, it is observed that Asia is the most happening place for solar PV, with China, India, and Japan the leaders [1, 2].

Fig. 12.1 shows that the European Union is no longer a solar-centric region worldwide. Countries such as China, the United States, India, and China have already surpassed the annual installation capacities of Germany, Italy, and France. Even in the case of cumulative installation capacities in 2018, Japan, the United States, and China had already surpassed the European Union PV market [2, 3]. From this, it is understood that solar is a key player in the modern-day electric utility sector. But recently, there have been many concerns raised over the environmental impacts associated with solar PV systems. The energy produced from solar PV plants seems to be clean and comparatively free from carbon emissions concerning conventional fossil fuel-based power plants [4]. Even though solar panels do not produce any harmful chemicals or noise in the operational stage, they seem to have potential environmental impacts while in the production process [5]. Also, in most situations, the expected life for PV panels appears to be 25 years, and afterward

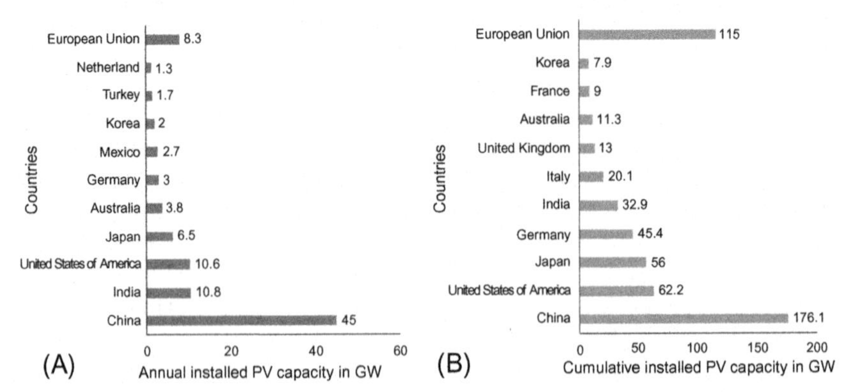

FIG. 12.1 Top 10 countries leading in the PV market in comparison with the European Union: (A) Annual installed PV capacity in GW for the year 2018; (B) cumulative installed PV capacity in GW in 2018 [2]. *Data Source: IEA PVPS 2019 snapshot of global PV market.*

their end-of-life management is not guaranteed with current practices [5, 6]. This also raises concerns over the environmental impacts due to improper management of solar power plant waste. Finally, this has resulted in many potential environmental, health, and safety hazards [6]. One of the best ways to understand these environmental impacts is by diving deep into the solar PV system life cycle. This allows us to understand the exact environment impacts possible with solar PV technologies.

In this context, this chapter highlights these topics:

- Methods to assess the environmental impacts both in a conventional way and by using advanced life cycle assessment (LCA) simulation tools.
- Summarizes the environmental impacts of PV systems in manufacturing, operation, and the end-of-life stage of the PV plant life cycle.
- Discusses how business models would have a role in mitigating environmental impacts.

12.2 Methods to assess environmental impacts

The environmental impacts associated with PV systems can be estimated in two different ways. The first is by using conventional methods that deal with energy balance and carbon footprint calculation. The second is the use of advanced simulation tools that have the entire life cycle data inventory support. These two methods are discussed in the below sections. In addition to these methods, various indicators used for environmental impact assessment are listed. A few advanced concepts such as the Internet of Things (IoT) and blockchain that support the LCA are also discussed.

12.2.1 Conventional method

In the traditional method, an energy balance-related indicator is used for LCA studies. This is used to estimate the total amount of greenhouse gas (GHG) emissions possible with the product [7].

$$\text{Annual } CO_2 \text{ emission} = (\text{Embodied energy} \times \text{Emission factor}) / \text{PV plant lifetime}$$

The emission factor can be of any source that is used for power generation, in the absence of solar power plants.

The conventional method does not accurately account for life cycle emissions. Hence, the use of advanced simulation tools is advised.

12.2.2 Life cycle assessments using industrial tools

This section deals with the brief introduction of a few commercial LCA simulation tools and the challenges associated with them as well as the scope of new technologies that support LCA.

In general, a product follows the life cycle stages, as shown in Fig. 12.2A. These life cycle stages include raw materials, manufacturing, distribution, use, and end-of-life management. For carrying out the LCA studies on such products, a framework is needed. Fig. 12.2B depicts the general framework for LCA. The standard LCA system has four main stages: goal and scope definition, inventory analysis, impact assessment, and interpretation [8].

LCA studies mainly need the system level of quantitative data. In LCA studies, more emphasis is put on quantitative data for analyzing environmental releases from the particular system, considering its overall life cycle. The required data include the raw materials used for manufacturing the product, the required embodied energy, and the process outputs [7]. As mentioned earlier, the LCA studies have three main stages, which are clearly described in Table 12.1.

In the LCA study, the interpretation stage, which is considered in each of the LCA stages and mainly helps in understanding data sensitivity, results in variation. Results communication will become much more comfortable with the help of the interpretation stage. The consistency in LCA results mainly depends on the type of data as well as the quality of data used in the analysis. The LCA study covers a wide range of the product cycle; it always becomes difficult to have detailed data [9]. In some situations, assumptions are made that lead to uncertainties in the results [9, 10]. Hence, it is suggested to have a reliable and robust database for LCA studies. However, the collection of reliable data is not so easy [11].

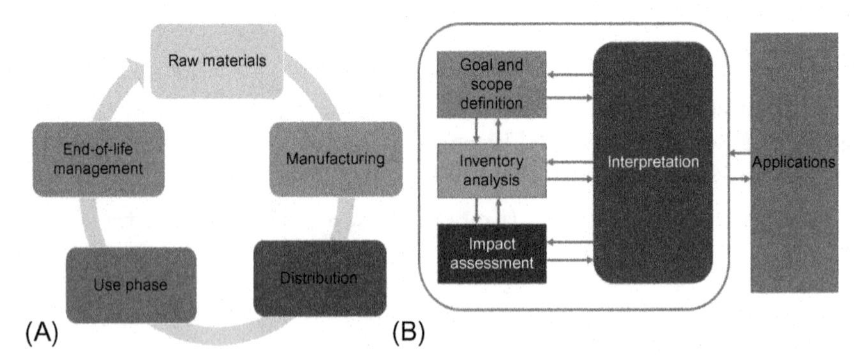

FIG. 12.2 (A) Life cycle stages of a typical product; and (B) the general framework of life cycle assessment.

TABLE 12.1 Description of stages in the LCA framework.

Stages in the LCA framework	Description	Tasks
Goal and scope definition	- This is the first stage in the LCA study. - Outlines the method selection, functional unit, assumptions, and targeted scope.	- Selection of an intended application. - Selection of methods. - Note of assumptions and limitations for an impact study. - The decision of the targeted audience. - The decision on whether it is a comparative study. - Selection of the functional unit.
Inventory analysis	- This is the second stage in the LCA study. - Process modeling considering all the material data in the entire life cycle of the product, starting from the raw material to the end-of-life management. - It mainly deals with the type of data required for the study, the collection procedures used, and validation.	- Identifying the system boundary. - A decision on cut-off criteria. - Primary and secondary data collection. - Modeling the process. - Evaluation of life cycle inventory results.
Impact assessment	- This is the third stage in the LCA study. - This step is mainly based on inputs and data used in the inventory analysis (i.e., second stage). - Outlines the various environmental impacts and the type of resources used in the study.	- Results classification. - Characterization of the results. - Studies on normalization and results grouping.

For example, with new products it is a challenge to collect data. Hence, we suggest the use of modern tools that have a reliable database and also the flexibility to design a new process with the available data. Below are the three most popular LCA tools used by environmentalists working in industries, academia, and research institutes.

SimaPro: This is a widely used LCA tool for industrial applications, and in most cases, it is considered to be one of the expert versions for useful LCA application. With this tool, decisions related to the product life cycle and design can be made effectively. Overall, these decisions will boost companies to meet the requirements of regulatory bodies. This tool was developed considering the scientific information related to almost every product and material. The information provided in this tool is transparent to an extent and mostly avoids the black-box process [12].

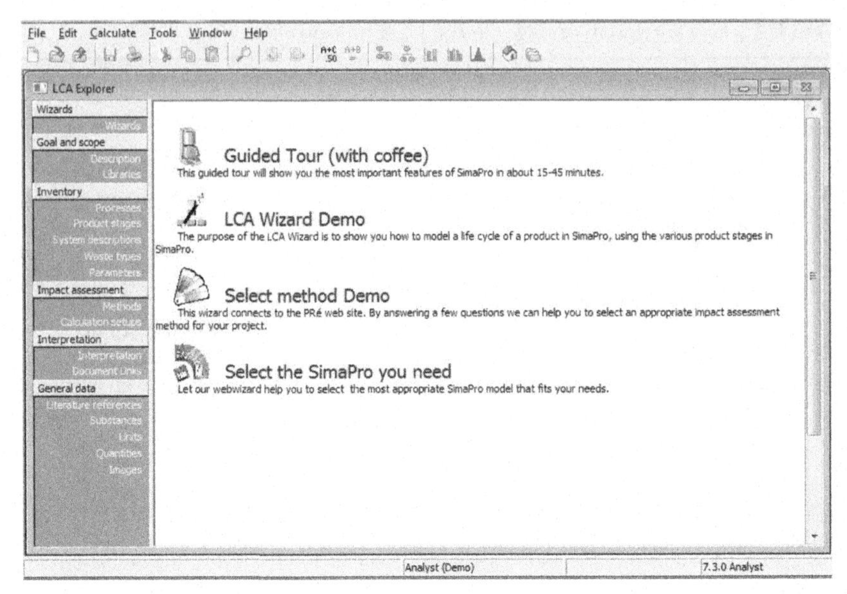

FIG. 12.3 Explorer window in SimaPro showing all the LCA framework stages [12].

In Fig. 12.3, the LCA explorer window highlighting the stages in LCA modeling is shown [12].

By using SimaPro, any user can make appropriate decisions by carrying out the analysis, based on the accuracy of obtained results. SimaPro offers the following applications and functions [12].

- Any complex life cycle process can be modeled more easily in SimaPro due to the systematic and transparent approach that is followed while modeling.
- The possibility of measuring environmental impacts as per the impact categories throughout the product life cycle.
- It offers a solution in the collection, analysis, and monitoring of any product data, along with its sustainability performance.
- In every supply chain, there exist numerous hotspots, and they might be critical depending on the product and its life cycle. Identifying them is crucial and needs expert modeling such as with SimaPro. SimaPro will make the process look more straightforward in an understandable way, which enables hotspot identification.

OpenLCA: This is an open-source LCA tool widely used in academic communities. Its performance and functionalities are almost similar to the commercially available LCA tools, with some exceptions. It allows users to have sustainability analysis reports of any product; a typical LCA explorer window of OpenLCA is shown in Fig. 12.4 [13].

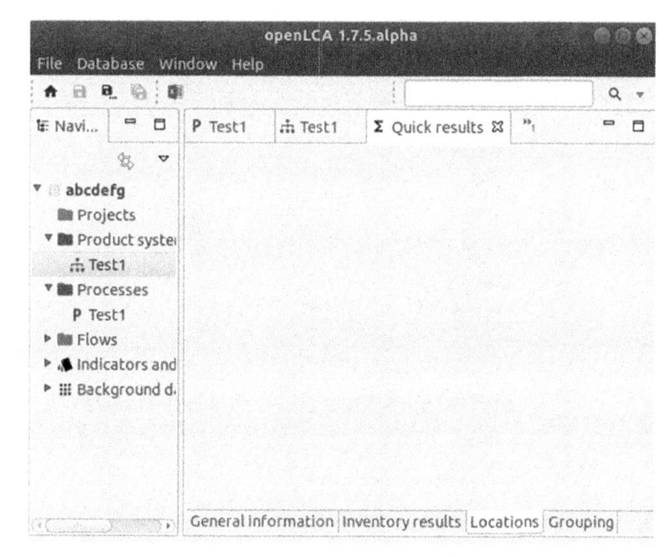

FIG. 12.4 LCA explorer window in the OpenLCA tool [13].

The various functionalities offered by OpenLCA are [13]:

- It provides a detailed LCA calculation report along with the results.
- It helps in identifying the main drivers and key parameters that influenced the LCA results throughout the product life cycle.
- It offers the service of visualizing the LCA results.
- Model sharing is more comfortable in OpenLCA due to its import and export capabilities.
- It also offers an analysis related to life cycle cost.
- Social assessment related to the product life cycle can also be carried out using OpenLCA.

GaBi: Like the other two LCA tools, GaBi also used for the sustainability assessment of any product. It is not an open-source tool, and it is mostly available on a subscription basis. GaBi offers various functionalities, and it mainly helps in database creation and management. Fig. 12.5 shows the database management option in the GaBi LCA tool version 4 [14].

Various business applications that are possible with the GaBi tool are briefly listed along with a description in Table 12.2.

12.2.3 Impact categories

For assessing environmental impacts, indicators are a must. In general, the LCA practitioners have a clear indicator, that is, tCO_2e/year, with respect to the unit for greenhouse gas (GHG) emissions. Similar to GHG emissions, there exist numerous other environmental impact categories, which are listed and described in Table 12.3.

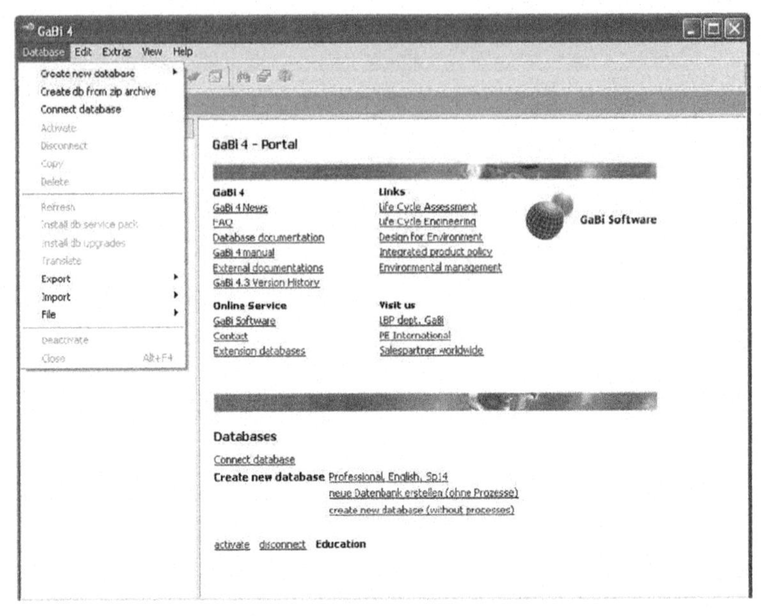

FIG. 12.5 Database management in the GaBi LCA tool [14].

TABLE 12.2 List of business applications possible with the use of the GaBi tool [14].

Name	Abbreviation	Descrption
Design for environment	DfE	Offer solutions to develop products using numerous methods by limiting the overall energy consumption as well as the material required. Besides, it also helps in meeting the regulatory body standards in terms of product design. Overall it promotes ecofriendly design and the efficient use of existing resources.
Ecodesign	ED	It helps in designing products with minimal carbon footprints as well as minimum resource consumption.
Efficient value chains	EVC	The value chain represents the involvement of various stakeholders in the product life cycle. GaBi provides detailed LCA results that enable the value chain by enhancing efficiency.
Life cycle costing	LCC	Reduction in cost is possible with the help of process optimization using the GaBi tool.
Sustainable product marketing	SPM	Based on the LCA results, SPM is possible by the use of sustainable labels. This is possible only when the product meets regulatory body declarations.
Life cycle working environment	LCWE	GaBi also offers solutions to the development of a product life cycle process based on social responsibilities.

TABLE 12.3 Indicators and their descriptions used in LCA [15–19].

Indicator		Description
Acidification	Soil	Nitrogen and sulfur oxides are considered the most harmful chemicals, whose release into the soil could potentially have intense impacts. This results in the acidification of soil.
	Water	Like soil, water bodies are prone to acidification due to the release of harmful chemicals such as nitrogen and sulfur oxides.
Aquatic ecotoxicity	Freshwater	Marine life in freshwater bodies is affected due to water contamination with toxic substances. Aquatic ecotoxicity is one of the indicators used to measure the overall impact of these poisonous substances on freshwater organisms.
	Marine	Similar to aquatic ecotoxicity in freshwater, there is a possibility for it in the marine ecosystem. The impact of toxic substances on marine aquatic life is measured using this indicator.
Depletion of resources	Elements	There are many naturally available resources, and they also seem to be depleting. Hence, measuring such trends is necessary.
	Fossil fuels	Fossil fuels are being depleted due to excessive use, and this measurement is essential.
Eutrophication		With the release of nitrogen and phosphor-related compounds into the aquatic ecosystem, nutritional levels are increasing. Eutrophication is one such indicator that measures the possible enrichment of nutritional elements.
Global warming		The release of various greenhouse gas emissions into the air is measured using a global warming potential indicator.
Human toxicity		The release of toxic elements into the environment has a very strong impact on human beings. An indicator that measures the effects of such chemicals on humans is called human toxicity.
Ozone depletion		Due to the release of emissions, the ozone layer is effected. This indicator is used to measure the impact of various emissions responsible for ozone depletion.
Ozone creation		Specific gas emissions have a negative effect on the ozone layer creation. This indicator is used to measure the impacts of such emissions.
Terrestrial ecotoxicity		Due to the release of toxic elements, there is a possibility of terrestrial ecotoxicity. It generally happens only with the release of such elements onto a land surface. In such cases, this indicator is used to measure the impact of toxic elements on land organisms.

Continued

TABLE 12.3 Indicators and their descriptions used in LCA [15–19]—cont'd

Indicator		Description
Pollution	Air	Chances of air pollution are very high due to the release of toxic elements. This indicator represents the quantification of the amount of air required to dilute the poisonous elements released.
	Water	The release of toxic chemicals into water bodies is common in current industrial systems. Water pollution is an indicator that measures the amount of water required to dilute poisonous chemicals.
	Soil	Soil pollution generally happens with the accumulation of various toxic elements. Usually, the mix of these elements will happen with the water. Hence, to measure soil pollution, this indicator is used. It involves the quantification of the amount of water required to dilute the released toxic chemicals into the soil.

12.2.4 Challenges in LCA and technical advances that support LCA

The conventional LCA method based on embodied energy and emission factors is subjective to uncertainty due to a lack of quality data. This leads to judgments on LCA outputs. In similar passions, LCA studies using simulation tools such as SimaPro, OpenLCA, and GaBi also have some uncertainty in the results. These studies mainly work based on the set boundaries for the environmental impact assessment. These boundaries are generally referred to as system boundaries, and the judgment on the boundary would vary based on the system, which leads to some uncertainty. While modeling the complex process, the LCA studies must account for detailed process data, but in practice, this is not valid. Even in the situation where accurate process data is considered for LCA, there exists uncertainty due to the data quality [9–11].

It is understood that data is the most challenging in LCA studies, and getting real-time quantitative data would solve the issue of uncertainty. In this regard, a real-time monitoring system that collects data should be used in LCA studies. Digital technologies such as the Internet of Things (IoT) can be used for collecting the data of any complex system [20]. The collected data using IoT systems can be stored in blockchain-based distributed ledgers [21]. This sort of monitoring system is called the blockchain-based Internet of Things (B-IoT). A B-IoT system integrated with data visualization will provide real-time data that can be used as a reliable source for the sustainability reporting of any product or process [22].

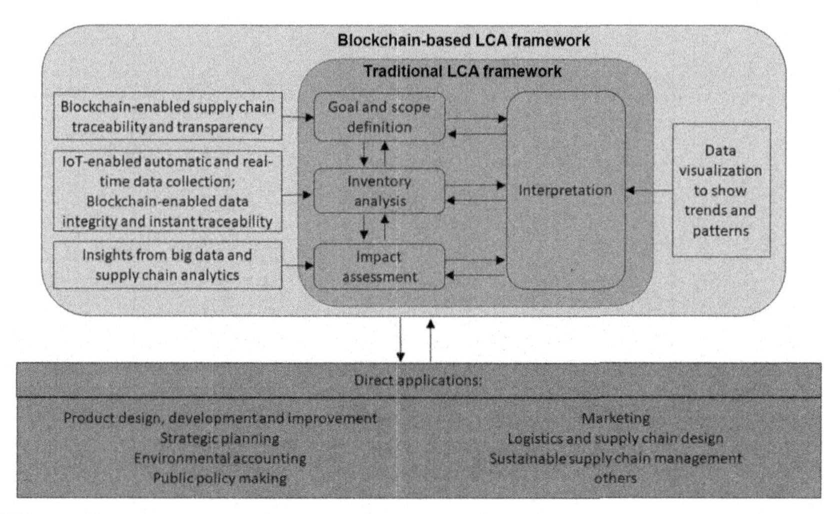

FIG. 12.6 Blockchain-based LCA [22].

In Fig. 12.6, the blockchain-based LCA framework is shown [22]. The framework is based on the general framework of LCA, which is shown in Fig. 12.2B. The goal and scope definition stage of LCA studies is further improved based on the two essential blockchain features, traceability and transparency. These two features enhance the overall product supply chain. The second stage of the LCA, is the inventory analysis where data is given more priority, at this stage based on the data needs the B-IoT system will operate. This system allows real-time data collection, which improves the input data quality as well as avoids uncertainty. In the third stage, impact assessment, the role of analytics would help in understanding the overall impact categories. Data visualization tools integrated with B-IoT systems can help in the interpretation of the data and the results.

12.3 Life cycle assessment of photovoltaics

Based on the methodology discussed in the previous section, life cycle emissions from the PV plant can be quantified. A few studies that involve such estimates are summarized in Tables 12.4 and 12.5 for silicon and thin-film PV systems, respectively. Also, a brief discussion on the studies shown in Tables 12.4 and 12.5 gives more understanding on how the overall environmental impact would vary depending upon the PV technology. It is a known fact that PV systems would perform differently for different locations. Apart from this, the type of technology chosen for installation will also have a significant influence on the overall

TABLE 12.4 Summary of LCA studies conducted on silicon PV systems [8].

Location	PV system type	Functional unit	Boundaries	Energy payback time	CO_2 mitigation or global warming potential	Reference
Multiple locations	Polycrystalline silicon	0.65 m^2 PV panel area	Manufacturing and operational phase	3.5–7 years	50–800 g/kWh	[23]
United States	Monocrystalline silicon	1 kWh	Manufacturing, including the balance of system and operational phase	3.8 years	10.2 g/kWh	[24]
Switzerland	Poly- and monocrystalline silicon	3 kWp	Manufacturing, including the balance of system and operational phase	3–6 years	136–100 g/kWh	[25]
Italy	Monocrystalline silicon	1 MWh	Manufacturing and operational phase	5.5 years	44.7 g/kWh	[26]
Few locations in South European Countries	Poly- and monocrystalline silicon	1 kWp	Manufacturing and operational phase	1.7–2.7 years	30–45 g/kWh	[27]

TABLE 12.5 Summary of LCA studies conducted on thin-film PV systems [7].

Location	Thin-film PV system type	Installed capacity	Energy payback time	CO_2 mitigation	Reference
Malaysia	CdTe	100 kW	0.94 years	0.76 g/kWh	[28]
Italy	CdTe	1 m^2	1.30 years	–	[29]
United States	a-Si	33 kW	3.2 years	34.3 g/kWh	[30]
Germany	CdTe	1 m^2	1.1 years	30 g/kWh	[31]
China	a-Si	100 MW	2.2 years	15.6 g/kWh	[32]

performance. In a similar way, the LCA results of a PV system in different locations seem to be varied.

From Table 12.4, it is understood that crystalline silicon modules have shown different energy payback times and global warming potential for different locations. On the other side, there is a clear observation of the variation of the results with respect to the boundary chosen. Like the crystalline silicon PV systems, the variation in LCA results of numerous thin-film PV modules is also observed. For example, a 1 m^2 area of CdTe PV module energy payback time in Italy and Germany is different.

12.4 Environmental impacts associated with PV systems

This section briefs the environmental impacts associated with PV systems in three broad categories of its life cycle: manufacturing, operational, and end-of-life management.

12.4.1 Manufacturing

The manufacturing stage of PV modules is the one crucial life cycle stage. It is considered one of the complex processes that involves the use of chemicals to process the material, depending upon the chosen technology. In current markets, the most commonly used PV module uses crystalline technology, which is silicon-based. The second most popular is the cadmium telluride (CdTe) module. In this section, various environmental impacts associated with crystalline silicon-based and cadmium-based PV modules are briefly given in Boxes 12.1 and 12.2, respectively [6, 33, 34].

BOX 12.1

Environmental impacts due to manufacturing of crystalline PV modules.

Crystalline silicon PV modules are formed by combining the PV cells in series and parallel configurations. These cells mainly consist of silica, which is a natural resource abundantly available on Earth. But only a minor share of this silica material is used in the electronic industries, especially in PV modules. The extraction of silicon from sand is a complex process that involves different stages. At first, silica is mined and then converted into crystals. But when we visualize the process of silica mining to a PV cell wafer, the most dangerous and harmful substance that is released into the environment is dust. The generated silica dust has severe effects on the workforce involved in cell fabrication. The main problem associated with silica dust is lung disease.

On the other side, in the manufacturing stage, a few harmful gases and other substances are also released. These include silane gas, silicon tetrachloride, and other wastes in the form of solids and liquids. These have a very strong impact on the environment as well as on the health conditions of the workforce.

BOX 12.2

Environmental impacts due to manufacturing of CdTe PV modules.

CdTe is on thin-film technologies, which have become the most popular in recent years. These modules have very high chemical stability. It is a cost-effective and very efficient technology. A solar PV module using this technology has thin layers that contain materials such as CdTe and CdS. Here, Cd is the most toxic substance. It has substantial environmental impacts and its release into the atmosphere causes health impacts. Cd emissions from CdTe are around 0.26 g/GWh. This is quite less when compared to fossil fuels such as natural gas, coal, and oil, whose Cd emissions are 0.3 g/GWh, 3.7 g/GWh, and 44.3 g/GWh, respectively [33].

12.4.2 Operational

The operational stage of the PV system is a prolonged run activity and typically ranges 20–25 years, depending upon the PV technology. The operational stage has considerably fewer environmental impacts when compared to the manufacturing and end-of-life stages [5, 6, 34]. The following are the environmental impacts:

- Land use is one of the significant problems in the operational stage. Small-scale PV installation generally does not require much land area, but large-scale PV installation requires a vast land area. At present, a typical utility-scale power plant requires around five acres of land for a 1 MW installation. In most cases, agricultural land is used for PV system installation, which in turn disturbs the energy-water-food nexus. The best area suitable for PV installation is one without agriculture [5, 34–36].
- Habitat issue is another impact. These issues arise due to the installation of PV plants in remote areas or in faraway locations that disturb nature [34, 35].
- In solar PV plants, the use of electrical and mechanical equipment is most common. For example, a transformer is the best electrical equipment used. Most times, there is a leakage of transformer oil, which has an environmental impact. Also, the regular maintenance of transformers results in vast amounts of transformer oil as waste.

12.4.3 End-of-life management

The lifetime of PV modules is usually 20–25 years. Depending upon the PV technology, the lifetime would vary. Once the lifetime is over, these modules turn out to be waste. But these waste PV modules contain many useful materials such as aluminum, silver, steel, and copper, and they have commercial value in the market. Based on the current cumulative installation capacities, the overall estimated PV panel waste would be huge. The cost associated with the materials present in the waste is also very high. The statistics related to PV panel waste are shown in Fig. 12.7 [37].

Based on the data shown in Fig. 12.7, it is estimated that cumulative PV capacity could increase up to 4500 GW by the end of 2050. The associated PV waste would be increased to a value of 70–80 million tons under the early loss scenario [37]. Dumping the generated PV waste in landfills would have many environmental issues. Hence, recycling and proper end-of-life management are the only options to avoid environmental impacts. If solar PV panels are not treated effectively, they will release

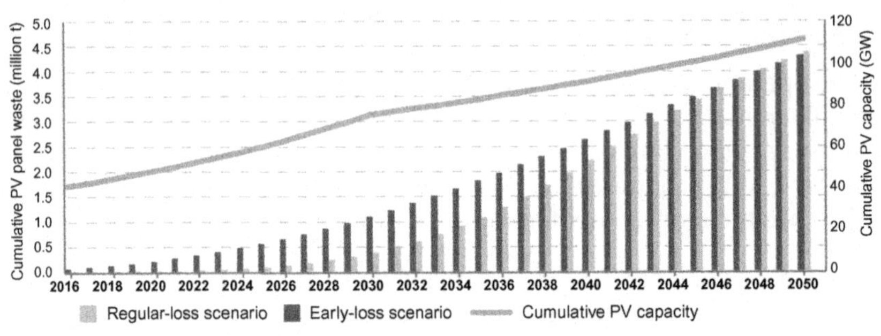

FIG. 12.7 Global estimated PV panel waste [37].

toxic materials into landfills, where they can percolate into groundwater and the air. However, there also exist a few environmental impacts due to PV waste recycling [38]. Considering the crystalline silicon solar cells, they generally contain a few materials that have a direct impact on human and animal health. In recycling, careful observations have to be made so that the treatment of harmful metals such as lead and chromium is effective [6, 33]. In recent years, thin-film PV technology has been widely used, and CdTe technology in particular has become very popular. The waste generated with CdTe PV modules contains cadmium-related toxic substances. These types of PV modules can be recycled based on the methods used for electronic waste such as cathode ray tubes and batteries. The best approach to recycle these PV modules is by using strong acids that help in the stripping of metals [6, 33, 39]. Two other thin-film modules that have become very popular in recent years are copper-indium selenide (CIS) and copper-indium gallium selenide (CIGS). These PV modules contain rare metals and recycling such PV modules would give valuable materials. The recovered materials from CIS and CIGS waste can be used in manufacturing television screens. In the end-of-life management of CIS and CIGS modules, elements such as selenium, gallium, and indium can be recovered [37, 40].

A case study given in ref. [41] reveals that the material recovery potential is high with the use of thermal and chemical treatment methods. The recovered materials from a 24 W PV module are shown in Fig. 12.8. The most dominant material in the PV module is glass, and its contribution is approximately 59.51%, s equal to 5474.92 g. Using the thermal and chemical treatment methods, around 5365.42 g of glass material is recovered. Aluminum is also another valuable material whose share in the PV module is 16.71%, which equals 1537.32 g and out of which, 1322.09 g is recovered. Similarly, 98% of the steel and 85% of the copper are recovered [41].

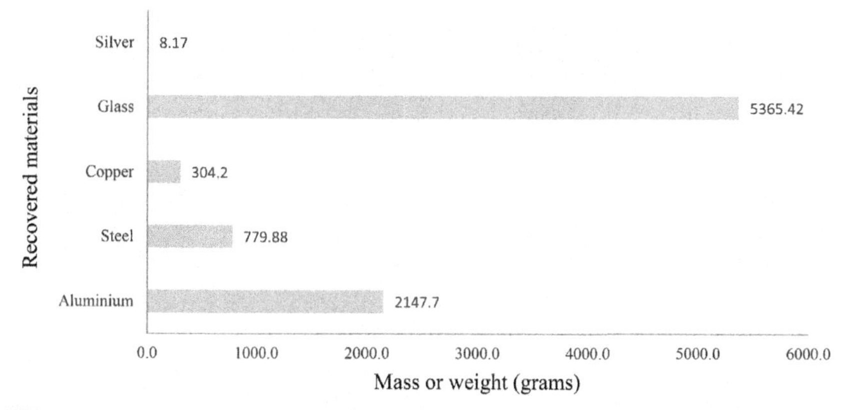

FIG. 12.8 Estimated material recovery potential from a 24W PV module [41].

12.4.4 Business models and their impact

PV technologies that are currently in the market are mostly under the linear business practice, which generally follows the take-make-use-waste principle shown in Fig. 12.9.

As per the cumulative PV installation that is estimated and shown in Fig. 12.7, it is understood that PV waste is also increasing. But with the current linear business model practice, effectiveness and value creation from PV waste are not possible. PV waste contains many useful materials, offering extraordinary opportunities for manufacturing industries with material supplements in a cost-effective manner. But with the current business model, huge waste is created, becoming a severe threat to the environment.

In this context, there is a strong need to change the PV business model. The change should include research and development activities toward waste management. Measures that are most suitable for the demanufacturing of PV modules should be considered. Also, activities that push the recycling, end-of-life material recovery, and reuse of decommissioned PV modules for various other functions should be explored. Such activities allow material circularization and ensure that PV waste is treated effectively [42]. In Fig. 12.10, a circular business model of a PV system

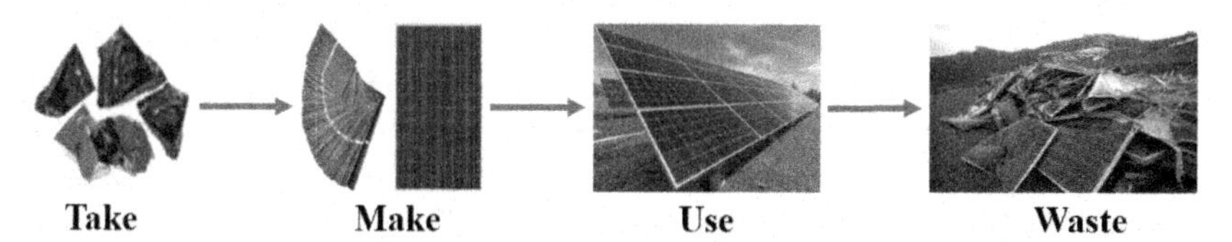

FIG. 12.9 The linear business model for PV modules and PV systems.

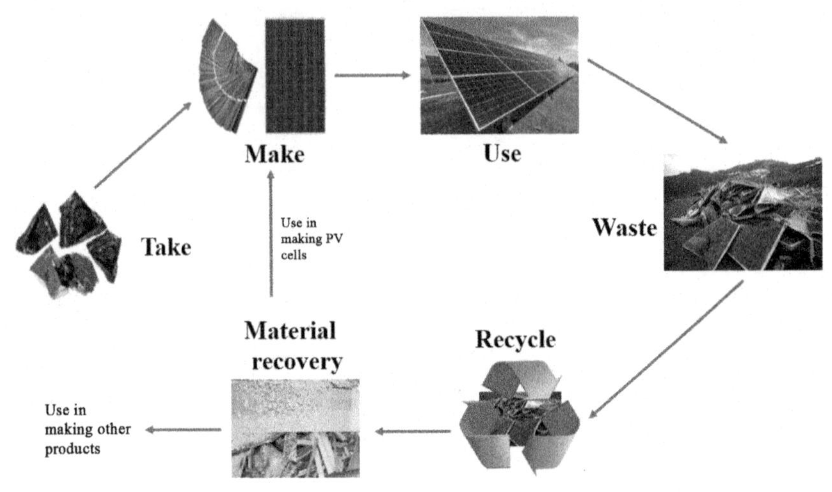

FIG. 12.10 The circular business model for PV systems.

is shown, which generally follows the linear business model principle coupled with end-of-life management principles.

The circular business model of the PV system allows us to tackle environmental impacts when compared to the linear business model. In a circular business model, the PV module is subjective to various end-of-life management methods for recovering valuable materials that can be used in the manufacturing of PV modules. In some cases, these recovered materials can also be used in the manufacturing of other products. Both the above-discussed business models have a few environmental impacts, which are clearly highlighted in Table 12.6.

12.5 Conclusion

There is considerable growth for PV systems in the future. This growth has been accelerated in recent years due to the continuous efforts taken on cost reduction and technical advancements seen in PV systems. On the other side, the environmental concerns of the PV system are really fearsome when their growth rate and cumulative installation capacities are considered. The total PV waste could reach up to a million tons by the end of 2019. Dumping PV waste in landfills and that long-term exposure to water, air, and land would cause severe health and environmental problems. Hence, the assessment of evaluating environmental impacts is necessary. Various methods used for assessing the environmental impacts associated with PV systems in both the conventional way as well with advanced simulation tools are given in this chapter. In addition, a

TABLE 12.6 Environmental effects due to business models followed in PV systems.

Business model	Environmental effects			
	Manufacturing stage	**Operational stage**	**End-of-life stage**	**Material recovery stage**
Linear business model	- Environmental impacts due to material processing. - Use of chemicals in PV cell fabrication. - Emissions due to the embodied energy. - Emissions related to module making.	- The leak of materials such as cadmium in oxide form. - Pollution related to land is also one of the considerable environmental impacts. - Effect on aquatic life due to floating and submerged photovoltaics.	Contamination of water, land, and air due to harmful emissions.	NA
Circular business model			No	Harmful acids and other chemicals are used.

NA—not applicable.

sophisticated framework that supports the LCA considering realistic data is presented. The environmental impacts of PV systems considering the life cycle stages of PV systems as per real-time conditions are given. Environmental impacts associated with PV systems are discussed in three broad categories: the manufacturing, operational, and end-of-life stages. This chapter has introduced two different business models: linear and circular. In addition, the environmental impacts due to the current linear business model in various life cycle stages are also presented. The circular business model implementation for PV systems and its role in mitigating environmental impacts were discussed.

References

[1] Wesoff E. IEA: global installed PV capacity leaps to 303 Gigawatts. Greentechmedia; 2017 [April 27].
[2] IEA PV Snapshot 2019 (PDF). International Energy Agency. Retrieved 2 June 2019.
[3] Solangi KH, Islam MR, Saidur R, Rahim NA, Fayaz H. A review on global solar energy policy. Renew Sust Energy Rev 2011;15(4):2149–63.
[4] Kumar NM, Singh AK, Reddy KVK. Fossil fuel to solar power: a sustainable technical design for street lighting in Fugar City, Nigeria. Procedia Comput Sci 2016;93:956–66.
[5] Turney D, Fthenakis V. Environmental impacts from the installation and operation of large-scale solar power plants. Renew Sust Energy Rev 2011;15(6):3261–70.
[6] de Wild-Scholten M, Alsema E. Towards cleaner solar PV: environmental and health impacts of crystalline silicon photovoltaics. Refocus 2004;5(5):46–9.

[7] Rajput P, Singh YK, Tiwari GN, Sastry OS, Dubey S, Pandey K. Life cycle assessment of the 3.2 kW cadmium telluride (CdTe) photovoltaic system in composite climate of India. Sol Energy 2018;159:415–22.
[8] Gerbinet S, Belboom S, Léonard A. Life cycle analysis (LCA) of photovoltaic panels: a review. Renew Sust Energy Rev 2014;38:747–53.
[9] Christensen TH, Bhander G, Lindvall H, Larsen AW, Fruergaard T, Damgaard A, Manfredi S, Boldrin A, Riber C, Hauschild M. Experience with the use of LCA-modelling (EASEWASTE) in waste management. Waste Manag Res 2007;25(3):257–62.
[10] Ciroth A, Fleischer G, Steinbach J. Uncertainty calculation in life cycle assessments. Int J Life Cycle Assess 2004;9(4):216.
[11] Notarnicola B, Sala S, Anton A, McLaren SJ, Saouter E, Sonesson U. The role of life cycle assessment in supporting sustainable agri-food systems: a review of the challenges. J Clean Prod 2017;140:399–409.
[12] SimaPro, URL. https://simapro.com/ [Accessed on 4th September 2019].
[13] OpenLCA URL. http://www.openlca.org/ [Accessed on 4th September 2019].
[14] GaBi Software URL: http://www.gabi-software.com/international/index/ [Accessed on 4th September 2019].
[15] Owens JW. LCA impact assessment categories. Int J Life Cycle Assess 1996;1(3):151–8.
[16] Evans A, Strezov V, Evans TJ. Assessment of sustainability indicators for renewable energy technologies. Renew Sust Energy Rev 2009;13(5):1082–8.
[17] Impact Categories (LCA)—Overview, URL. https://ecochain.com/knowledge/impact-categories-lca/ [Accessed 5th September 2019].
[18] Steinmann ZJ, Schippe AM, Hauck M, Huijbregts MA. How many environmental impact indicators are needed in the evaluation of product life cycles? Environ Sci Technol 2016;50(7):3913–9.
[19] Impact Assessment Methodology ReCiPe2016; 2017 March 13. URL. https://simapro.com/2017/updated-impact-assessment-methodology-recipe-2016/ [Accessed on 10th September 2019].
[20] Kumar NM, Mallick PK. The Internet of Things: insights into the building blocks component interactions and architecture layers. Procedia Comput Sci 2018;132:109–17.
[21] Kumar NM, Mallick PK. Blockchain technology for security issues and challenges in IoT. Procedia Comput Sci 2018;132:1815–23.
[22] Zhang A, Zhong RY, Farooque M, Kang K, Venkatesh VG. Blockchain-based life cycle assessment: an implementation framework and system architecture. Resour Conserv Recycl 2020;152:104512.
[23] Stoppato A. Life cycle assessment of photovoltaic electricity generation. Energy 2008;33(2):224–32.
[24] Perez MJ, Fthenakis V, Kim HC, Pereira AO. Façade–integrated photovoltaics: a life cycle and performance assessment case study. Prog Photovolt Res Appl 2012;20(8):975–90.
[25] Jungbluth N, Bauer C, Dones R, Frischknecht R. Life cycle assessment for emerging technologies: case studies for photovoltaic and wind power (11 pp). Int J Life Cycle Assess 2005;10(1):24–34.
[26] Desideri U, Zepparelli F, Morettini V, Garroni E. Comparative analysis of concentrating solar power and photovoltaic technologies: technical and environmental evaluations. Appl Energy 2013;102:765–84.
[27] Alsema E, de Wild MJ. Environmental impact of crystalline silicon photovoltaic module production. In: MRS Online Proceedings Library Archive 2005. p. 895.
[28] Kim H, Cha K, Fthenakis VM, Sinha P, Hur T. Life cycle assessment of cadmium telluride photovoltaic (CdTe PV) systems. Sol Energy 2014;103:78–88.
[29] Vellini M, Gambini M, Prattella V. Environmental impacts of PV technology throughout the life cycle: importance of the end-of-life management for Si-panels and CdTe-panels. Energy 2017;138:1099–111.

[30] Pacca S, Sivaraman D, Keoleian GA. Parameters affecting the life cycle performance of PV technologies and systems. Energy Policy 2007;35(6):3316–26.
[31] Held M, Ilg R. Update of environmental indicators and energy payback time of CdTe PV systems in Europe. Prog Photovolt Res Appl 2011;19(5):614–26.
[32] Ito M, Kato K, Komoto K, Kichimi T, Kurokawa K. A comparative study on cost and life-cycle analysis for 100 MW very large-scale PV (VLS-PV) systems in deserts using m-Si, a-Si, CdTe, and CIS modules. Prog Photovolt Res Appl 2008;16(1):17–30.
[33] Sundaram S, Benson D, Mallick TK. Potential environmental impacts from solar energy technologies. In: Sundaram S, Benson D, Mallick TK, editors. Solar photovoltaic technology production. Academic Press; 2016, p. 23–45, ISBN 9780128029534, [chapter 3]. https://doi.org/10.1016/B978-0-12-802953-4.00003-2.
[34] Kaygusuz K. Environmental impacts of the solar energy systems. Energy Sour Part A 2009;31(15):1376–86.
[35] Paul Breeze H. Solar integration and the environmental impact of solar power. In: Breeze P, editor. Solar power generation. Academic Press; 2016, p. 81–7, ISBN 9780128040041, [chapter 11]. https://doi.org/10.1016/B978-0-12-804004-1.00011-7.
[36] Kumar NM, Kanchikere J, Mallikarjun P. Floatovoltaics: towards improved energy efficiency, land and water management. Int J Civil Eng Technol 2018;9:1089–96.
[37] IRENA, IEA-PVPS. End-of-life management: solar photovoltaic panels. International Renewable Energy Agency and International Energy Agency Photovoltaic Power Systems; 2016.
[38] Lunardi M, Alvarez-Gaitan J, Bilbao J, Corkish R. Comparative life cycle assessment of end-of-life silicon solar photovoltaic modules. Appl Sci 2018;8(8):1396.
[39] Lunardi MM, Alvarez-Gaitan JP, Bilbao JI, Corkish R. A review of recycling processes for photovoltaic modules. In: Solar panels and photovoltaic materials. IntechOpen; 2018.
[40] Domínguez A, Geyer R. Photovoltaic waste assessment of major photovoltaic installations in the United States of America. Renew Energy 2019;133:1188–200.
[41] Gangwar P, Kumar NM, Singh AK, Jayakumar A, Mathew M. Solar photovoltaic tree and its end-of-life management using thermal and chemical treatments for material recovery. Case Stud Therm Eng 2019;100474.
[42] Sica D, Malandrino O, Supino S, Testa M, Lucchetti MC. Management of end-of-life photovoltaic panels as a step towards a circular economy. Renew Sust Energ Rev 2018;82:2934–45.

CHAPTER

13

Solar PV market and policies

Hyun Jin Julie Yu[a] *and Patrice Geoffron*[b]

[a]French Alternative Energies and Atomic Energy Commission (CEA Saclay), Institute for Techno-Economics of Energy Systems (I-tésé), Gif-sur-Yvette Cedex, France, [b]Paris-Dauphine University, PSL University, LEDa-CGEMP, UMR CNRS-IRD, Paris, France

13.1 Introduction

13.1.1 Dynamics of the photovoltaic sector and economic issues associated with its development

The Paris Agreement defined international climate objectives to keep the mean global temperature rise to well below 2 degrees above preindustrial levels and to limit the temperature rise even further to 1.5 degrees above preindustrial levels [1].

Photovoltaic (PV) energy has been identified as a solution for achieving these goals. The prices of PV modules have fallen rapidly over the past decade due to the globalization of the sector, which has greatly improved the competitiveness of this technology.

PV installations have grown significantly, from less than 1GW peak power (GWc) in 2000 to more than 0.5TW in 2018. PV prices have been divided by about a factor of 10 since 2005, falling to below 0.30€/Wp. Though European countries paved the way in the 2000s, it is now Asia (China, Japan) and the United States that are driving PV market growth. There is also great growth potential in emerging markets such as India and the Middle East.

The share of renewables reached about 25% of global electricity generation in 2017, of which solar PV accounted for 7% (see Fig. 13.1). However, the share of solar PV in the world's power supply is expected to increase rapidly in the coming decades. According to the IEA Sustainable Development Scenario, solar PV will account for 17% of the global power supply in 2040.

https://doi.org/10.1016/B978-0-12-819610-6.00013-2

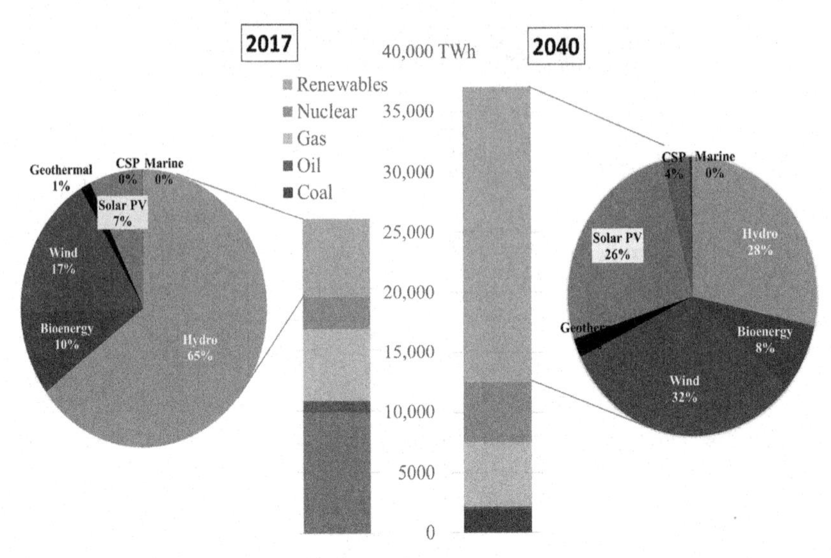

FIG. 13.1 Breakdown of renewables in 2017 and in 2040 [2].

However, the development of PV combined with the industrial dynamics of the sector have raised new economic issues. The PV market has been suffering from overcapacity since 2008 and serious trade disputes among major PV countries are still under way. In addition, there are increasingly more questions about systemic effects (deoptimization of the mix, impact on electricity markets) led by the large-scale integration of PV into existing energy systems.

In this chapter, we thus discuss the major economic issues affecting the solar PV market and the role of policies for its sustainable growth in the future.

13.2 Current global solar PV markets

13.2.1 Demand-side: PV installations

The global PV supply has demonstrated exponential growth in the last 20 years. The world's cumulative PV installed capacity has increased sharply from a mere 1 GWp in 2000 to more than 0.5 TWp of solar power capacity in 2018. The winning bids of recent tenders around the world are now commonly less than €50/MWh (including Europe), and the lowest PV prices without support have fallen below 3 c$/kWh in 2016 (i.e., Abu-Dhabi).

The solar industry has undoubtedly undergone a paradigm shift during its growth from a niche market to a primary electricity source in the energy system (Fig. 13.2).

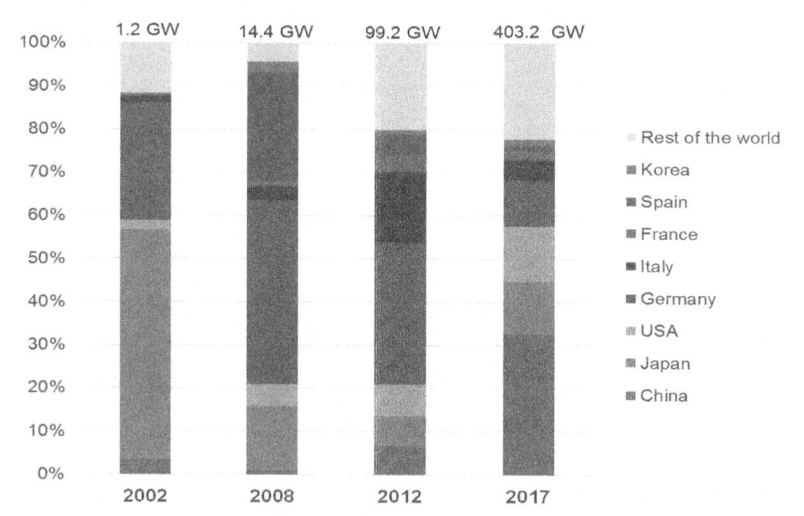

FIG. 13.2 Demand-side: cumulative installed PV capacity in the world [3].

Japan was the PV market leader in the early 2000s, accounting for more than 50% of the world's cumulative installations and more than 40% of global annual growth. However, Europe took the leading position in the global PV market from the mid-2000s, with Germany in pole position; it accounted for around 70% of the world's newly installed capacity in 2005. In addition, there were installation peaks in Spain (2008) and Italy (2010). In 2013, Europe represented almost 60% of the global cumulative PV capacity with 81 GW.

However, Europe has since been losing market share after the economic recession in 2008 and the entry of cheap Chinese products into the global PV market. The PV paradigm change started in 2013 as new growth began in non-European countries (China, Japan, United States, India). More than 60% of all new installations in 2013 were from China, Japan, and the United States. Asian countries, with China, Japan and India at the forefront, are currently developing the PV market faster than any other European market.

China has been the largest PV installer in the global PV market. China alone represented about 35% of the cumulative capacity of PV installations in 2018 (Fig. 13.3) while the sum of PV installations in Europe accounted for only 23% in 2018. In addition, the top three countries (China, Japan, and the United States) accounted for almost 50% of the world's cumulative PV installed capacity in 2018. Other regions such as Africa, the Middle East, Southeast Asia, and Latin America began making an effort in PV market development. PV has great potential in these regions with respect to the increasing demand for energy and the energy poverty problem.

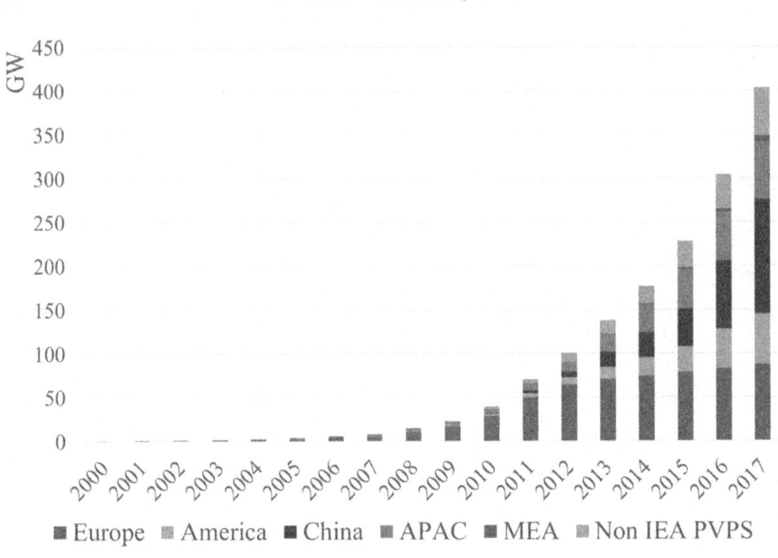

FIG. 13.3 Cumulative installed PV capacity in the world [3, 4].

13.2.2 Supply side: PV industry (PV cells and modules)

China has the largest manufacturing industry capacity for PV cells and modules in the world. In 2017, China accounted for almost 70% of total world production (72GW, Fig. 13.4); this was almost equivalent to the world's annual PV installation in 2016. However, China was a latecomer to this market, having only started in the mid-2000s. The Chinese share has since rapidly increased, occupying almost 60% of the world's total

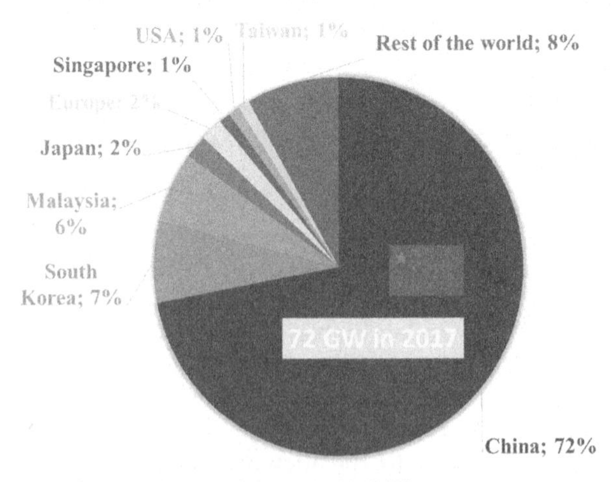

FIG. 13.4 Share of PV module production in 2017 [5].

production in 2012. Other Asian countries such as Taiwan, South Korea, and Malaysia have also increased their production. Before China, Japan and Germany played important roles in PV manufacturing from the early 2000s.

In the current PV market, wafer-based crystalline silicon PV production remains dominant (>90%), even though other technologies have appropriate advantages, such as building integration. However, the technology that enters a market first can take advantage of the inertia created in the market by past investment. It can be favored due to accumulated experience, economies of scale, established infrastructure, culture, and organization of stakeholders (technology lock-in).

The production of a thin-film module only accounts for a small portion of the global solar cell markets (Fig. 13.5). For example, around 5 GW of thin-film modules (CdTe, CIGS) were produced in 2016 [5]. Malaysia, Japan, Germany, Italy, and the United States are the major producing countries of thin-film technologies. The world's largest thin-film PV producer is First Solar, which is based in the United States. It produced around 3 GW of CdTe thin-film PV modules in 2016 via its production lines in the United States and Malaysia. In Japan, Solar Frontier produced around 1 GW of thin-film PV modules in 2016. There are few incentives to make long-term investments in other technologies due to the high risks and barriers (commercialization, systemic, cultural, and institutional). Moreover, some risks (toxicity and raw material supply) related to these technologies hinder market development. To overcome this technology lock-in, policies play an essential role in supporting the development of these technologies.

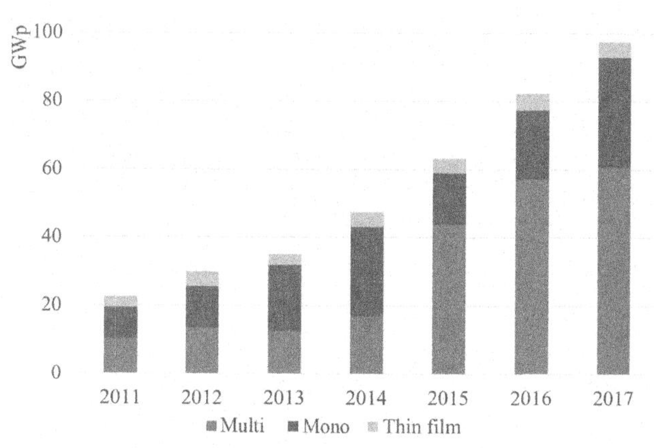

FIG. 13.5 PV production by technology [6].

13.3 Solar PV policy mechanisms: a multicriteria decision analysis

In this section, we present an approach for analyzing solar PV policy success based on a multicriteria decision mechanism.

13.3.1 Overview of solar PV policy mechanisms

An approach based on a schematic map can help us to visualize better how policy inputs and resources turn to specific outputs with a long-term impact to achieve targeted policy objectives. The following traditional schematic model (Fig. 13.6) can be used to understand the impact of PV policy support on society in many countries.

For a complete evaluation, it is important to consider the PV policy system as a whole (systemic analysis). First, we need to know why we want to develop solar PV energy in our society. Most policy support for renewable energy generally aims at providing a sustainable energy system with environmental, social, and economic benefits. This not only involves advancing the cost-competitiveness of renewable technologies or increasing the share of renewable energy sources in the mix, but also creating economic benefits (job creation) or improving the quality of life, for example, health. Policy priorities will vary from one country to another, depending on the situation.

The policy objectives to support renewable energies such as solar PV generally focus on: (1) energy security (energy supply diversification), (2) climate change mitigation (energy transition, GHG emission reduction), (3) improved access to energy (energy equity), and (4) socioeconomic

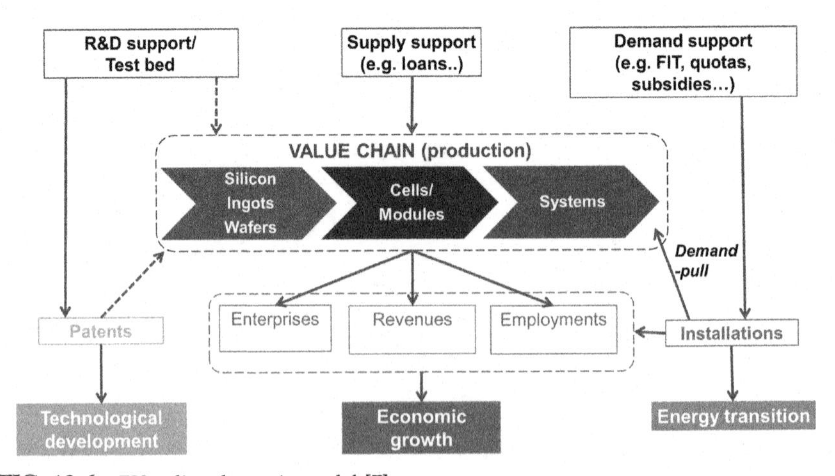

FIG. 13.6 PV policy dynamic model [7].

TABLE 13.1 Measurable policy results.

Supply-side	Demand-side
• R&D sector: publications, patents, efficiency, material reduction, etc. • Industry: number of firms, production capacity, reduction of modules and system prices	• PV installation: installation capacity, usage • Social acceptance, training capacity, investments, administration process

development (jobs, economic growth) [8–10]. According to defined policy objectives, policy inputs are decided together with the allocation of resources. Solar PV policy inputs can be classified into supply-side support (R&D and production) and demand-side aspects (diffusion of solar PV energy and PV integration) [11]. PV policy instruments can generally be categorized into three groups: fiscal incentives, public financing, and regulations[a] [9, p. 197]. The policy results can then be determined as outputs using measurable variables (generated results such as products or services in terms of R&D activities, industry production, and PV installations, Table 13.1).

In addition, the policy results can also be determined based on the direct or indirect impact (short-term and long-term), including technological, economic, and energy aspects.

A comprehensive analysis of **key contextual factors**[b] should be conducted to fully understand PV policy mechanisms in a given country. There are various aspects to be considered, for example, technology progress, environmental and economic situations, geopolitics, energy prices, electricity markets, industry competition, consumer behavior, natural resources, and human resources. They generally have a significant impact on the overall system and are not usually under control. The most commonly used standards to determine the success of policy instruments are efficiency and effectiveness. The efficiency of policies can be evaluated by comparing the results with the financial resources committed. In the longer term, policy effectiveness compares outcomes to initial policy objectives to assess which targeted policy objectives are reached. In addition, the PV policy mechanism evolves with a feedback loop.

[a]**Fiscal incentives**: reduced contributions to the public treasury through tax deductions (such as income tax or other taxes), rebates, grants. **Public financing**: public support such as loans, equity, or financial reliability such as guarantees. **Regulations**: rules to guide or control.

[b]There are various factors affecting the mechanisms; for example, human resource factors such as the price of labor or education, the quality of the electrical grid, the scarcity of the domestic energy supply, the manufacturing capabilities of fossil fuels, and the social opinion of energy sources and energy price changes.

Besides, **stakeholder analysis** is another important step for systemic analysis. There are several ways of doing this. A straightforward method is to use an "interest-influence" matrix to identify stakeholder positions for solar PV development. Defined stakeholders can be organized into a 2×2 interest-influence matrix[c]. This method classifies stakeholders into four groups: promoters, defenders, apathetics, and latents. This analytical framework helps us to identify possible risks that should be managed to promote solar PV energy [12].

Latent groups, which have no particular interest in solar PV development but a strong influence on the energy market, should be carefully examined because they represent a significant potential threat. When PV policy results are expected to conflict with their interests, they can raise strong opposition or even block development. Therefore, it is necessary to properly understand their needs and integrate them into political decisions. This group often includes traditional energy industries (grid operators, generators) and price-sensitive electricity consumers. The current PV market is dynamic and fast-changing, and stakeholder positions can thus evolve over time. There may be new entrants or existing players moving from one group to another (for example, stakeholders from the latent group can become defenders if they are searching for new PV business opportunities). In this regard, the analysis should be updated regularly to follow the market dynamics.

13.3.2 Political support for solar PV installations (feed-in tariffs and other alternative instruments)

In this section, we describe the key mechanisms of financial support for the installation of solar photovoltaic systems [13]. Many countries have implemented policy support to support the energy transition (for example, feed-in tariffs), and this has played an essential role in developing the PV sector. Feed-in tariffs (FITs) indicate fixed electricity prices that are paid to renewable electricity generators for each unit of energy produced and injected into the electricity grid. It guarantees grid access with priority focusing on renewable electricity. This public policy instrument was designed to stimulate investment in renewables, and the payment is based on fixed prices during a specified number of years; this is often related to the economic lifetime of renewable projects.

This system developed mainly in Europe from the 1990s, in parallel with subsidies. An important turning point came in the early 2000s when the system set long-term purchase agreements (usually 20 years) based on attractive tariffs covering the real cost of renewable electricity. This

[c]Author's analysis based on World Bank's definition.

mechanism attracted investors because it reduced risks by ensuring a return on investment. Germany was the first country to make a significant investment in PV energy based on this mechanism, and this case has since been adopted as a model by many other countries. The establishment of a correct purchase price and an evolution mechanism is a crucial element of success.

However, the FIT scheme is a very tariff-sensitive policy instrument and policymakers do not control installation volumes [13, 14]. The pricing system must reflect the dynamics of the PV industry and cost reduction trends. It is easy to control the tariff system in a closed market, and it is more complicated in an open innovative market because of the uncertainty of PV prices influenced by rapidly changing market conditions, competition, and customer demand for new products and services. In this regard, sensitivity to high prices makes the system costly and risky for policymakers. There is a significant information asymmetry in the FIT system: policymakers only have access to production data without precise information on the purchase price of PV systems. It is, therefore, difficult to predict the profitability of PV based on this mechanism, and this involves certain risks (for example, overcompensation, undercompensation, installation peaks, irregular installations, etc.). It can thus result in windfall effects, bubbles, and nonperennial market growth; these situations have been observed historically in several countries. Accordingly, some countries have tried to associate this mechanism with quantity-driven methods, that is, elaborate pricing with volume monitoring, FIT adjustment according to targets of installations, etc.

In addition, the variability of PV electricity costs by location calls the effectiveness of a single national tariff into question (not an appropriate economic incentive in a place where it would be most useful). We can see that these risks increase when utility-scale PV plants are included in a single national tariff mechanism [13].

In addition, when the FIT system is financed by electricity bills (EEG in Germany, CSPE in France), an uncontrolled increase in PV installations raises electricity tariffs, which can raise energy poverty problems (increase of energy-poor households) or weaken corporate competitiveness [14]. In this regard, Yu et al. concluded that this instrument does not guarantee sustainable and stable PV growth, in particular, faced with uncertainty about the future PV market evolution.

Feed-in-Premium (FIP) is a more market-oriented solution that better reflects market price dynamics. PV power is sold based on the electricity spot market price, and generators receive a premium on top of the market price. Because the government only pays the premium, policy costs will be largely reduced compared with the FIT system.

Another effective way to limit these risks led by FITs is to use **tenders** based on the price mechanism of purchasing electricity (**power purchase**

agreement, PPA) produced by renewable energy sources. The company that suggests the lowest price wins the contract. We can control the level of installation volume by tender (a quantity-driven mechanism). Even though the state does not control the price, this mechanism makes it possible to avoid windfall effects.

13.4 PV globalization and market dynamics

13.4.1 Chinese entry into the global PV market

Since the mid-2000s, the increase in PV demand in line with policy supports in Europe has attracted Chinese players into the PV manufacturing market. Chinese production soared in a short time, and it became the largest manufacturer in the solar PV market in 2007. Since then, module prices have reduced with Chinese mass production, for example, from $4.73/Wp in 2007 to $0.67/Wp in 2013 [15]. The current module prices are less than $0.30/Wp.

In this regard, it is interesting to review historical changes in Chinese PV policies. The Chinese PV market development followed different strategies from other major PV countries such as Germany and Japan (a balance between industry perspective and PV installation). China entered the global PV market relatively late. China launched solar PV R&D in the late 1950s and entered the application stage in the 1970s; however, it was not until the mid-1980s that the industrialization of PV materials started. In contrast with Germany, Chinese PV policy was export-oriented. It first focused on easy-to-follow technologies establishing production lines of labor-intensive downstream manufacturing (modules and cells) rather than conducting serious R&D for technology development. However, despite its leading position in PV manufacturing, China was dependent on imported refined silicon and equipment for its massive production due to technological barriers [16]. After China became the major PV manufacturer in the world, it started focusing more on its capital-intensive upstream industry such as silicon production through R&D to advance related-technologies that had been lagging since 2009, with the goal of catching up with the major producing countries [17–20].

The Chinese government, at both central and local levels, supported PV manufacturing investment through various forms of subsidies: innovation funds, regional investment support policies (2009) issued by some Chinese city governments, free or low-cost loans, tax rebates, research grants, cheap land, energy subsidies, easy credit, and technological, infrastructure and personnel support [21]. China's low labor costs and low energy prices facilitated the industry's expansion by reducing production costs [22]. Moreover, faced with intense global competition after 2009,

China's easy access to credit and permissive standards gave local manufacturers the possibility of taking advantage of scale effects to build gigawatt-scale plants [23]. In this manner, Chinese-subsidized credit supported PV producers in their capacity expansion regardless of their productivity levels, even if some loans included a high risk of default. However, China's expansion of its production capacity was heavily dependent on overseas markets (for example, Europe) without establishing a domestic market; for example, China exported around 97% of module production in 2009 [24] (Fig. 13.7).

Chinese PV development first encouraged the industry before it was decided to expand domestic installations to overcome the industry slowdown [3, 19]. Chinese solar PV power generation started in the 1960s but the dramatic progress is a recent event over the last 15 years (the Chinese installed capacity was only 140 MWp in 2008) [25]. Since the late 2000s, China's on-grid solar PV installations have rapidly increased based on the strength of the incentive programs to grid-connected rooftop and BIPV systems; for example, central government subsidy programmes such as the "Rooftop Subsidy Program" (2009), the "Golden Sun Demonstration Program" (2009), and the "Solar PV Concession Program" (2009) [17, 26]. In 2011, the national FIT scheme started to support domestic PV market growth. China contributed significantly to the world's PV capacity; the country's new growth in global installation represented about 45% with almost 45 GW in 2018.

Even though China rapidly expanded its installations, thereby becoming a major driver of market growth, its PV contribution to the electricity generation mix is still small: 3.3% in 2018.

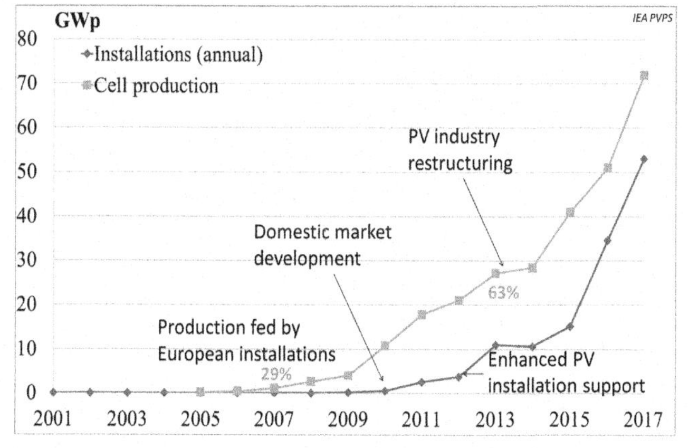

FIG. 13.7 Annual installations versus cell production in China [3, 17, 18].

13.4.2 International trade effects after the mass entry of Chinese products

Chinese manufacturers attempted to take a share of the market from their competitors based on an export-oriented strategy, and finally surpassed their German and Japanese competitors (since 2007), occupying a dominant market share in the global PV market.

China's dominance was beneficial in terms of economies of scale through the increase in market size with the mass production of solar cells and modules, which again lowered the global price of cells and modules. This price level is much lower than the European government (Germany) expected for the policy design (for example, FIT scheme), leading to some unexpected consequences with the open trade system.

Before the mass entry of Chinese products, the European market benefitted from open trade while pacing with the growing domestic installations. However, this mechanism began to be threatened by the mass entry of Chinese production.

First, the German production quantity was largely reduced, beaten by the competitive price offered by Chinese manufacturers. Second, with the mass inflow of cheap products into the German market, German installations rose much faster compared with the expected quantity under the FIT system. This became a financial burden for Germany, therefore inducing an unpredicted increase in FIT costs. Local installers started to use cheaper components and this distorted the local FIT mechanism at the end (speculation). Accordingly, some local key players closed down.

Furthermore, China also encountered industrial problems. China depends heavily on the overseas market to absorb its mass production because the country extended its PV industry without domestic market growth, with the majority of its mass production being exported to the overseas markets. Moreover, China was also dependent on imported silicon for its mass production. In this context, the chain-reaction bankruptcies observed in the last decade can be understood when the European market growth shrank due to its economic downturn. In addition, the Chinese government's decision to expand the domestic market can be seen as a natural result to resolve the national economic problem (Golden Sun program in 2009, FIT scheme in 2011) [27].

With large-scale Chinese inputs based on the GW-scale production capacity, the global PV market experienced mass PV industrialization and an unprecedented decrease in the price of solar PV cells and modules. Nevertheless, the European PV industry was suffering from fierce competition from Chinese players, and the global PV market encountered both excessive supply capacity in the global market and a PV industry crisis. Furthermore, most EU countries have a large trade deficit in solar components and equipment due to competition from China and other non-EU countries. Germany is no exception. China immediately absorbed a considerable portion of the German market share (job losses, industry decrease) faced with this fierce price competition.

Faced with the global economic crisis, Chinese PV manufacturers continued to produce large quantities of solar PV products, thereby aggravating the global PV industry situation with a global supply-demand imbalance. The global PV industry reacted to the oversupply with even fiercer international competition [28]. Chinese PV manufacturers also encountered a difficult period in the globalized market due to a lack of outlets for its production and the PV industry went through a restructuring process. Many PV firms in the world have since gone bankrupt [29]. The largest Chinese solar company, Suntech, filed for bankruptcy in March 2013, even though they had received billions of dollars in direct loans from the Chinese government [29]. Furthermore, China recently faced obstacles for imports of PV products going through trade disputes with the United States and the EU. This issue has become a major economic concern for our global society. China started to relocate its production lines (for example, Taiwan) but needed to explore new avenues for market growth. Under these conditions, the interactions of Chinese supply-side policy decisions (easy access to capital, subsidies) and European demand-side policy (FITs) have led to unexpected consequences for the globalized PV sector, such as overcapacity, an industrial crisis, and trade disputes. The globalization impact on German solar PV policies thus reflects the importance of a systemic approach that considers diverse aspects of PV markets when designing a country's PV support scheme [19].

13.4.3 PV growth potential in new regions

We have seen that the solar PV sector has experienced strong market growth supported by favorable political reactions in the energy transition context. Despite these positive conditions, however, the global PV market went through a chaotic period due to overproduction, the industry crisis, and long-lasting trade disputes [19]. In addition, long-lasting trade disputes between countries (for example, China versus the United States, China versus the EU) narrowed the scope of the PV market for the relevant countries. Nationwide PV installations are usually insufficient to feed the GW-scale supply volumes. In 2018, the largest PV firm's production was almost equivalent to the total demand for PV installations in Europe. Faced with globalization, the nationwide PV innovation system, which aims at creating a virtuous circle [30] between R&D, market growth, and price reduction, has been somewhat broken [19]. Therefore, we need to look for new opportunities that provide the PV industry with new outlets for the current oversupply of PV products.

Energy access issues in Africa and PV growth opportunities

Even though several countries in sub-Saharan Africa (SSA) have made progress in expanding electricity access in recent years, it has the lowest energy access rates in the world. Only about half the population (roughly 600 million people) manages to have access to electricity while about 890 million people still cook with traditional fuels in these regions (the fuels and technologies that households use for cooking represent a serious health problem, that is, inhaling carbon monoxide and particulates from traditional biomass cooking stoves). The limited access to a reliable power supply and the endemic shortage that grid-connected communities often encounter impose significant restrictions on their economic activities and improvement of the quality of life. Such a situation is paradoxical for a continent particularly dense and diverse in potential primary energy resources while a huge renewable potential (solar, wind, hydro, geothermal) remains untapped across all Africa. Mechanically, the progress toward electrification by renewables in these countries will have the greatest impact on global outcomes.

However, the promise of rapid GDP growth based on the "carbon-intensive" model in many countries led to local and global environmental concerns. The Paris Climate Agreement is not compatible with the adoption in Africa of the same energy development path as in China or India [31]. Therefore, through the establishment of a new paradigm based on sustainable energy supply and consumption, these regions can develop an innovative low-carbon energy system that enables them to take another economic development route in a more sustainable way [19, 32]. For example, sub-Saharan countries could directly foster innovative strategies mimicking, in the energy field, the adaptation of mobile technologies combining low-carbon technologies: renewable sources (that is, solar PV), energy efficiency improvements, storage, microgrids, etc. [31].

More than 1 billion people in the world still have no access to electricity. We can aspire to further deploy solar PV systems in less developed and developing countries that are faced with energy poverty problems. According to the World Bank, energy access problems are concentrated in Africa and Southeast Asia [33]. Interestingly, however, there are also significant solar energy resources in these regions and solar power is the most competitive in these regions (Fig. 13.8). However, these regions are more likely to be reluctant to invest in PV installations due to their difficult financial situations. This explains why these countries may prefer to continue supplying diesel-based power to residents despite the high costs. In this regard, international efforts will be necessary if we intend to roll out this electrification program. Such actions should involve not only governmental levels but also the private sector and civil contributions.

This approach expands the scope of the global PV market within the international context to solve the current PV industry's anxiety [32]. Furthermore, new regions could also benefit from a sustainable energy supply system for their socioeconomic development. In particular, this solution provides an interesting option to address the problem of world energy poverty. It would increase the world's electrification rate and eventually have a positive impact on global economic growth. By broadening the scope of the potential PV market to cover the entire international arena

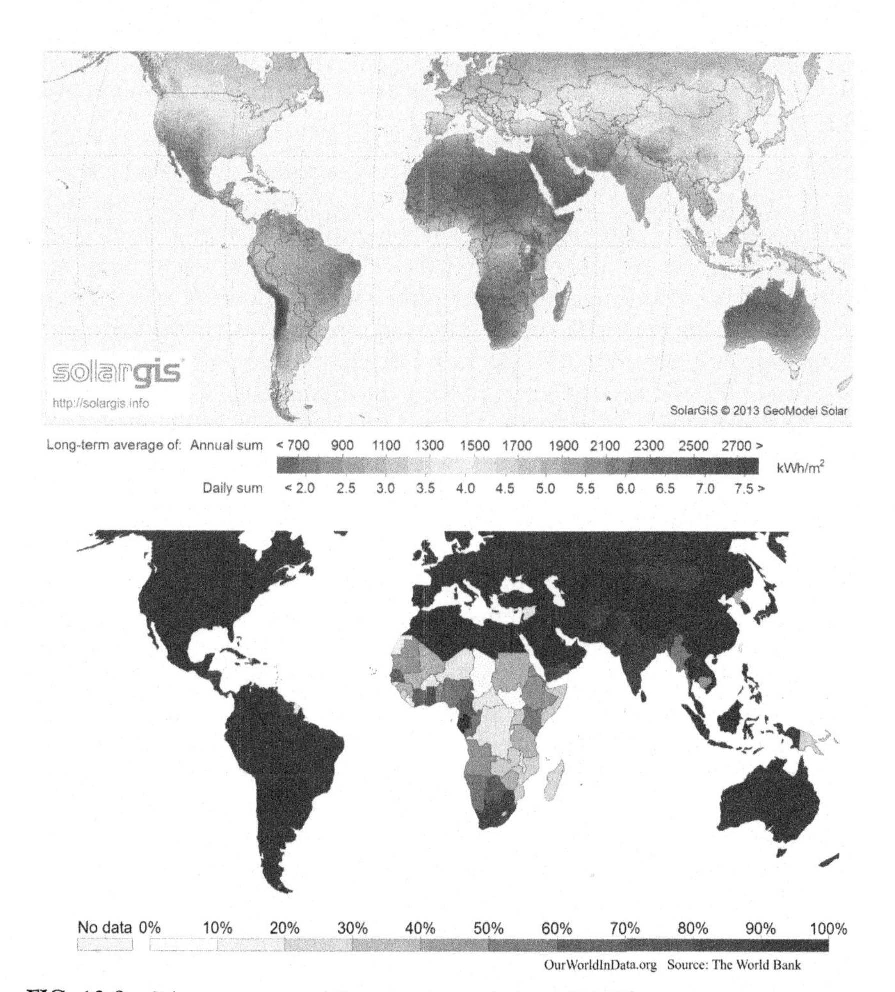

FIG. 13.8 Solar resources and the energy poverty issue [34, 35].

within an open economy, the investment to increase the foreign demand of PV installations would be partially returned to the domestic industry growth of participating countries. In addition, PV costs would be reduced thanks to the enlarged market size and experience. It is important to note that the enhanced competitiveness of PV power would eventually contribute to future national-based installations in all relevant countries with reduced PV costs. Therefore, the energy transition can be implemented within an international context. All stakeholders would benefit from the approach that encompasses new regions with improved energy access regardless of the political objective (industry or energy transition).

13.5 Systemic effects associated with large PV integration

13.5.1 Systemic effects of PV integration into electricity systems

PV electricity is not dispatchable and is not able to meet the electricity demand during all seasons of the year. The intermittency of variable PV power and the unique characteristics of the electricity supply-demand mechanism lead to systemic effects. The integration of PV into the existing grid system requires more effort to deal with its intermittency compared with dispatchable technologies. These efforts include not only an engineering perspective to ensure the operation of all physical systems but also economic aspects concerning the systemic value of PV integration. Therefore, the value of PV power in a society needs to be discussed in a more comprehensive manner by taking into account the systemic effects involved.

The systemic effects of variable PV integration can be classified according to three levels. The first level concerns the impact on **technical aspects such as infrastructure, the grid, and the electricity production mix** to maintain the operation of electricity systems. The second level of systemic effects concerns the **indirect financial impacts related to the regulatory mechanisms of electricity systems,** for example, the electricity tariff system and electricity price formation. The last level involves **different types of externalities with respect to PV integration into society**. Various positive or negative aspects, which influence the national system and social welfare, should be considered, that is, environment, technology, economy, jobs, and strategic position. The higher the level of systemic effects, the broader the scope of analysis is expected because of diverse correlations with other contextual, social, or systemic variables.

Assessments of the integration efforts associated with PV penetration and their dynamic impact on the electricity systems has been provided in various studies [36–41].

For the first level of systemic effects, the OECD/NEA [38] largely divided the systemic costs (**grid-level costs**) of PV integration into two parts: (1) additional investments to extend and upgrade the existing grid, and (2) the costs for increased short-term balancing and to maintain the long-term adequacy of the electricity supply to integrate variable energies. **Short-term balancing** concerns the second-by-second balancing of electricity supply and demand (for example, real-time adjustment, day-before forecasting). It is closely related to the accuracy of the weather forecast and the predictability of supply and demand. Improved forecast and prediction would decrease uncertainty on production planning and enhance the management of production capacities for a day. More importantly, the level of flexible capacity in the electricity mix and the size of the interconnected electricity system both influence the balancing task in terms of instantaneous adjustments to match changes in demand. Therefore, countries that have a large share of flexible technology capacities (for example, hydropower) in their energy mix do not need to balance costs as much.

Intermittent PV systems require **long-term dispatchable back-up capacity** to meet the electricity demand at all times [38, 42]. Nondispatchable energies like PV contribute very little to generating system adequacy in many European countries (the capacity credit of PV power in these countries is very low). The long-term back-up costs include investment as well as operation and maintenance costs to provide additional adequacy capacities (increase in demand) or to keep existing capacities available (constant demand). These costs are necessary to maintain a certain level of system reliability when variable energies are integrated into the electricity mix. When the capacity credit of PV power is low, the back-up costs account for the majority of the grid-level costs.

13.5.2 Financial impact on the electricity mix

The large-scale integration of PV power into the energy mix has a significant impact on the existing electricity mix. It raises a number of issues such as changes in the market price formation, and deoptimization of the electricity mix.

PV integration into the existing electricity mix reduces the operating hours (capacity factor) of dispatchable conventional plants and eventually influences their profitability. This issue is critical because they are compulsory to maintain the security of electricity systems. The negative financial impact on existing dispatchable capacities related to the **current electricity price formation** can be discussed. In Europe, the current management of the electric power system usually ranks the capacities in ascending order of marginal costs of production (merit order). This ranking is organized on the basis of the day-ahead declaration of available capacities. The electricity price is determined by the highest marginal costs of production units to satisfy demand. The price is imposed on all other producers. Base-load capacities have low variable costs and are ranked first (for example, run-of-the-river hydroelectricity, nuclear). Peaking capacities have high variable costs and are ranked last (for example, oil, gas).

PV electricity with zero marginal variable costs is ranked first in the merit order before base-load capacities and the merit order shifts to the right. However, the electricity demand is very inelastic; the price variability does not have much impact on consumption. Therefore, the electricity price is reduced with the same demand curve (Fig. 13.9) [43, 44]. It raises an issue with the payment of the initial investment (*losses of infra marginal rent*). In addition, in terms of the temporarily reduced demand, it is sometimes technically too difficult to shut down capacity for only a short time. This occurs when PV production is at its maximum (that is, summer daytime). In these extreme cases, the market price can be negative. In addition, the value of PV electricity varies according to time of production and location because of the unique feature of the electricity system. Wholesale electricity prices can be reduced significantly when all PV power plants in a country generate electricity at the same time on a clear, sunny day.

It also concerns the reduced use of peak capacities, which in turn reduces the revenues of conventional power plants. With the deterioration in the peak coefficient,[d] the extreme peaking capacities become an issue because they cannot cover their fixed costs [46] (*missing money*). This would thus exacerbate the problem in terms of future investment choices; investors are reluctant to build conventional plants because of the uncertainty in recovering the capital invested. This threatens the energy supply security. In this regard, the current electricity mechanism needs to be reconsidered to better integrate large-scale renewable energies.

Another issue should be discussed with regard to **electricity tariffs**. Electricity retail tariffs are often composed of electricity generation costs,

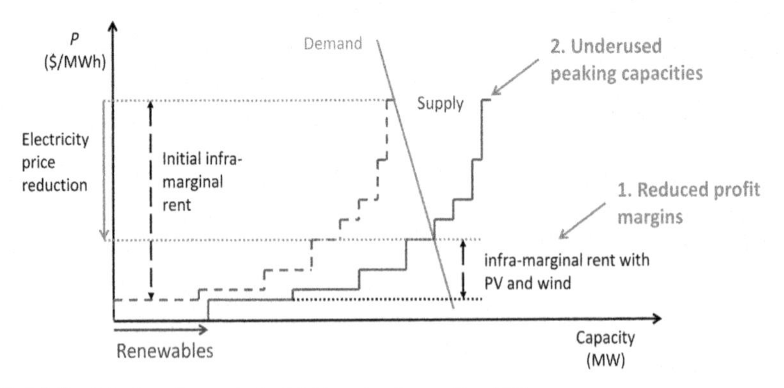

FIG. 13.9 Merit order and electricity price formation [13].

[d]The peak coefficient is the ratio between the average hourly production during the year and the peak production [45].

grid management costs, and taxes. The large-scale integration of PV power into the system can lead to revenue loss for stakeholders when fewer consumers purchase electricity from the grid [12]. However, the maximum grid capacity must be kept to maintain the security of the power supply when the capacity credit of PV is low, and grid operators will have more activities to manage the integration of variable PV energies. In this regard, it is essential to secure the budget for grid financing when designing large-scale PV integration policies.

13.5.3 Systemic benefits of PV self-consumption and limits

PV self-consumption offers an interesting way of reducing systemic effects with regard to high-scale PV integration. PV self-consumption can be defined as the local consumption of PV-generated electricity, which reduces the distance between electricity generation and consumption through the onsite consumption of power. This idea has been around since the 60s and 70s for power supply in nongrid-connected areas where grid extension was expensive; however, it has remained a niche market with few installations.

Faced with a rapid reduction in PV system prices, end users would naturally adapt the mode of self-consumption of PV electricity when it helps them reduce their electricity bills compared with the conventional way of purchasing electricity from the grid. PV system owners who produce electricity for their own needs and whose systems are connected to the main distribution network are called prosumers (consumers/producers). A combination of increasing power grid costs, decreasing PV system costs, and reduced feed-in tariffs (for example, Germany) provides enough economic incentives to encourage PV self-consumption.[e]

PV self-consumption offers diverse systemic benefits compared with PV integration based on a full grid injection of PV power. First, PV self-consumption facilitates limit the generated solar PV electricity injected into the power grid. This makes it possible to limit the cost of public support because it avoids long-term purchasing contracts based on fixed tariffs (for example, FITs). The ratio of self-consumption, which defines the rate between onsite consumption and the total production of the system installed on the site, is a very important factor in terms of deciding the economics of self-consumed models of PV power. When there is no compensation scheme to purchase the surplus PV output, the PV installation on which it is possible to best correlate the pattern of onsite energy use with PV system output is more economically advantageous. The full self-consumption is more suitable for sectors such as supermarkets or

[e]The concept of grid parity is valid for 100% PV self-consumption.

office buildings and it is more difficult to achieve a significant level of self-consumption in the residential sector (a weak correlation between solar power output and consumption). Its correlation can be increased through methods such as demand-side management or coupling with storage technology, but PV system costs will increase.

Next, despite higher system costs, distributed PV self-consumption can help reduce PV systemic effects compared with utility-scale PV systems. As well as using existing land, the systemic benefits of developing PV self-consumption compared with utility-scale PV power plants include various aspects with regard to reducing PV integration costs. They usually concern power losses during transmission and distribution (T&D), additional grid-related investment, balancing via geographic spreading of PV installations, optimal sizing of PV systems, diverse coupling with storage solutions (batteries), etc. The balance between the PV power cost and the systemic effect should be finely analyzed in their context before implementing a PV deployment strategy.

However, PV self-consumption does not directly solve the question of the seasonal reserve capacity needed to supply electricity during annual demand peaks. The contribution of PV to an adequate reserve capacity differs from one country to another. For example, in France, the annual demand peaks usually occur in the winter evenings while Greece has them at midday during the summer. Therefore, this issue should be also considered in accordance with national PV deployment plans [47].

In addition, PV self-consumption needs to solve another problem, the loss of network funding. Self-consumption deployment can be done in parallel by reorganizing the financing of the network (increase in the fixed part, time-based tariffs) or, possibly, an assessment of solutions to reduce network demand.

The power sector is transforming from a traditional centralized system to locally producing and consuming systems with more proactive customers. The coupled dynamics of the PV market and Li-ion batteries will improve the economics of residential PV consumption in the near future [47]. A rupture could influence the national power system if the transition of PV self-consumption in the residential sector occurs massively or suddenly. Such change will influence the interests of electricity market stakeholders and can be problematic for the national energy system. Therefore, policymakers would have to focus on an optimal mix of PV power to achieve a careful balance with the other energy technologies and grid issues.

13.6 PV future deployment scenarios and perspectives

The IEA Sustainable Development Scenario suggests that even greater efforts are required to shift to a low-carbon energy system based on the larger integration of renewable energies. The IEA has proposed a new scenario of sustainable development that aims at providing universal access to energy by 2030 to meet climate change objectives while improving human health. According to the IEA Sustainable Development Scenario, 17% of global electricity will be supplied by solar PV power by 2040. Table 13.2 illustrates the IEA solar PV diffusion objectives with respect to the PV installed capacity and PV electricity generation by 2025, 2035, and 2040 [2].

The IEA expects that the global cumulative PV installed capacity will reach 1–1.5 TW by 2025 and 2–4 TW by 2040. Solar Power Europe's PV market outlook anticipates that we can pass the milestone of 1 TW by 2022. The Shell sky scenario expects 1 TW of solar power by 2022 and 5 TW by 2030.

There is great potential for growth in new regions such as the Middle East, Southeast Asia, India, and Africa (Terrawatt initiative and the International Solar Alliance (ISA) of COP21). Supported by political support toward sustainable development, populations without access to electricity in Asian countries are diminishing. However, despite this recent progress, more efforts in sub-Saharan African countries should be made to provide a universal sustainable energy supply.

Module prices are expected to continue to decline over the next decade as the global market grows, and there is room for continuous institutional innovations and technological advances. These prices are expected to fall to less than $0.25/Wp when 1 TW capacity of PV electricity is reached in a few years and to around $0.10/Wp by 2035 [48].

Because the PV system price is a key variable of the initial investment when calculating PV electricity costs, this decline in module prices will lead to low PV power generation costs in the future. Over the last decade, the PV system price drop mainly correlated with the module price reduction. Therefore, until recently, the reduced module prices were the most focused driver to enhance the economic competitiveness of PV electricity. However, it seems difficult to expect the future PV system price to be reduced by means of module price drops alone, as we have seen with historical data. Other factors become more important, such as soft costs. It is also possible to improve the economic competitiveness of PV systems by focusing on nonmodule sectors driven by policy actions (for example, standardization, training, etc.) to prepare for the large-scale penetration of decentralized PV systems.

In the future system, solar PV energy will play a central role as one of the cheapest energy sources in the future decarbonized economy (decarbonization, decentralization, and digitization). A radical change will occur in energy systems, and synergies between different low-carbon technologies (batteries, smart grids, or the Internet of Things) can occur at a systemic level via coupling possibilities and associated services. In addition, subsidy-free demand for the PV self-consumption sector will naturally increase in many countries in the near future. End users will

TABLE 13.2 IEA solar PV diffusion objectives [2].

	PV installed capacity (GW)					PV electricity generation (TWh)				
	2017	**2025**	**2030**	**2035**	**2040**	**2017**	**2025**	**2030**	**2035**	**2040**
Current Policy Scenario (CPS)	403	1008	1290	1596	1951	531	1334	1782	2291	2956
New Policy Scenario (NPS)		1109	1589	2033	2540		1463	2197	2935	3839
Sustainable Development Scenario (SDS)		1472	2346	3300	4240		1940	3268	4806	6409

become significantly more demanding in the future energy system (increasing demand for green energy and autonomy). Therefore, it is increasingly important to analyze the needs and behaviors of prosumers and small energy companies, both at an individual level and a collective one. Some of these behaviors become influential behind decision-making processes and energy policy design (for example, nudging, choice architecture, time preferences). In addition, customer-oriented business and new value creation around PV energy must be properly aligned with infrastructure needs (hardware, software/artificial intelligence) and welfare optimization. In this regard, we also need to understand how human decision-making and behaviors (for example, cultural change, social movement) are formed to properly design the future energy system based on cooperation with diverse expertise (natural sciences, engineering, public policy, and economic and social sciences).

13.7 Conclusion

The global energy transition is under way and the energy sector now faces challenges in predicting the evolution of energy markets. PV energy is transforming from one of the electricity generation sources to the basis for sustainable energy systems on the international energy transition pathway. Subsidy-free market demand for solar PV energy in many sectors will emerge in many countries in the near future, and energy policy decisions will continue to have a significant impact on the development of the solar sector. However, future policy mechanisms for PV energy will differ from the traditional mechanisms; we need to focus more on a systemic approach to find a systemic balance between stakeholders in the energy system. They must understand the international dynamics of PV markets, and market insight should be included in policy dialogues to prepare the national energy transition plans.

PV will play a significant role in the international energy transition by addressing several global issues such as energy poverty and climate change. Global collaborative actions that widen the energy security frontier based on abundant PV resources are highly recommended, not only for environmental sustainability but also for global economic benefits. In addition, new cross-disciplinary business models can emerge based on systemic innovations (technological, economic, and institutional); for example, a mix of decentralized energy, green mobility, integration in buildings/territories, hydrogen, and storage. At the same time, the large-scale development of PV can raise new economic and social issues in our society, and we need new frameworks to address these risks and challenges. In this regard, the fundamental questions on the role of PV energy in our future society need to be considered based on a multidirectional approach that combines the complete energy system and each stakeholder's position.

References

[1] UNFCCC. Adoption of the Paris agreement—FCCC/CP/2015/L.9/Rev.1; 2015.
[2] IEA. World energy outlook 2018; 2018.
[3] IEA PVPS, "Trends in photovoltaic applications," 2002 to 2017.
[4] Solar Power Europe. Global market outlook for solar power 2019-2023; 2019.
[5] IEA PVPS. Trends in photovoltaic applications; 2017.
[6] Fraunhofer, "Photovoltaics report," 2011 to 2019.
[7] Yu HJJ, Popiolek N. A comparative study on the consequence and impact of public policies in favor of solar photovoltaic (PV) development. In: IEPEC International Energy Program Evaluation Conference, Berlin; 2014.
[8] Macintosh A, Wilkinson D. Searching for public benefits in solar subsidies. Energy Policy 2011;39(6):3199–209.
[9] IPCC. IPCC Special Report on Renewable Energy Sources and Climate Change Mitigation. Cambridge: Cambridge University Press; 2011.
[10] Byrne J, Kurdgelashvili L. The role of policy in PV industry growth: past, present and future. In: Handbook of Photovoltaic Science and Engineering. 2nd ed. John Wiley & Sons, Ltd; 2011. p. 39–81.
[11] Finon D. L'inadéquation du mode de subvention du photovoltaïque à sa maturité technologique. CIRED & Gis LARSEN Working Paper, 2008.
[12] Yu HJJ, Popiolek N. Opportunities and challenges of photovoltaic (PV) growth with Self-consumption Model. In: *38th International Association for Energy Economics (IAEE) International Conference*, Antalya (Turkey); 2015.
[13] Yu HJJ. [PhD Thesis]. Public policies for the development of solar photovoltaic energy and the impacts on dynamics of technology systems and markets. Paris-Dauphine University; 2016.
[14] Yu HJJ. Le financement du développement de l'énergie photovoltaïque en France et les questions liées aux coûts de l'intermittence. In: *La lettre de l'Itésé 31*, Eté; 2017.
[15] IEA PVPS. Trends in photovoltaic applications; 2013.
[16] de La Tour A, Glachant M, Ménière Y. Innovation and international technology transfer: the case of the Chinese photovoltaic industry. Energy Policy 2011;39(2):761–70.
[17] IEA PVPS China. National survey report of PV power applications in China; 2011.
[18] IEA PVPS China. National survey report of PV power applications in China; 2012.
[19] Yu HJJ, Popiolek N, Geoffron P. Solar photovoltaic energy policy and globalization: a multiperspective approach with case studies of Germany, Japan, and China. Prog Photovolt Res Appl 2016;24(4):409–587.
[20] Campillo J, Foster S. Global solar photovoltaic industry analysis with focus on the Chinese market. Malardalen University: The Department of Public Technology; 2008.
[21] Gang C. China's Solar PV manufacturing and subsidies from the perspective of state capitalism. Copenhagen J Asian Stud 2015;33(1):90–106.
[22] Grau T, Huo M, Neuhoff K. Survey of photovoltaic industry and policy in Germany and China. Energy Policy 2012;51:20–37.
[23] Goodrich A, James T, Woodhouse M. Solar PV manufacturing cost analysis: US competitiveness in a global industry; 2011.
[24] IEA PVPS. Annual report 2010; 2010.

[25] Zhao ZY, Zhang SY, Hubbard B, Yao X. The emergence of the solar photovoltaic power industry in China. Renew Sustain Energy Rev 2013;21:229–36.
[26] Zhang S, Zhao X, Andrews-Speed P, He Y. The development trajectories of wind power and solar PV power in China: a comparison and policy recommendations. Renew Sustain Energy Rev 2013;26:322–31.
[27] Zhang S, He Y. Analysis on the development and policy of solar PV power in China. Renew Sustain Energy Rev 2013;21:393–401.
[28] European Commission. Memo: EU imposes provisional anti-dumping duties on Chinese solar panels. Brussels, 2013.
[29] Bloomberg new energy finance. Solar insight—research note: PV production in 2013: an all-Asian affair; 2014.
[30] Watanabe C, Wakabayashi K, Miyazawa T. Industrial dynamism and the creation of a "virtuous cycle" between R&D, market growth and price reduction: the case of photovoltaic power generation (PV) development in Japan. Technovation 2000;20(6):299–312.
[31] Geoffron P. Energy access in Africa through 'low carbon' technologies as a way to meet the sustainable development goals—analysis and polocy brief. Paris (France): Dauphine Université Paris - LEDa; 2019.
[32] Yu HJJ. Virtuous cycle of solar photovoltaic development in new regions. Renew Sustain Energy Rev 2017;78:1357–66.
[33] The World Bank. Access to energy is at the heart of development; 2018 April 18 [Online]. Available: https://www.worldbank.org/en/news/feature/2018/04/18/access-energy-sustainable-development-goal-7. [Accessed 15 November 2019].
[34] The World Bank. Access to electricity (% of population); 2016 [Online]. Available: http://data.worldbank.org/indicator/EG.ELC.ACCS.ZS/countries?display=default. [Accessed 18 May 2016].
[35] Solargis. World Map of Global Horizontal Irradiation; 2013 [Online]. Available: http://solargis.info/doc/free-solar-radiation-maps-GHI.
[36] Borenstein S. The private and public economics of renewable electricity generation. J Econ Perspect 2012;26(1):67–92.
[37] Joskow PL. Comparing the costs of intermittent and dispatchable electricity generating technologies. Am Econ Rev 2011;238–41.
[38] Keppler JH, Cometto M. Nuclear energy and renewables: system effects in low-carbon electricity systems. OECD, Paris, France: Nuclear Energy Angency; 2012.
[39] Hirth L. The economics of wind and solar variability: How the variability of wind and solar power affects their marginal value, optimal deployment, and integration costs [Thesis dissertation]. Fakultät VI, Berlin: Technische Universität Berlin; 2014.
[40] Hirth L, Ueckerdt F, Edenhofer O. Integration costs revisited—an economic framework for wind and solar variability. Renew Energy 2015;74:925–39.
[41] Ueckerdt F, Hirth L, Luderer G, Edenhofer O. System LCOE: what are the costs of variable renewables? Energy 2013;63:61–75.
[42] Pudjianto D, Djapic P, Dragovic J, Strbac G. Grid integration cost of photovoltaic power generation. London: Energy Futures Lab, Imperial College; 2013.
[43] Haas R, Lettner G, Auer H, Duic N. The looming revolution: how photovoltaics will change electricity markets in Europe fundamentally. Energy 2013;57:38–43.
[44] Commissariat Général à la Stratégie et la Prospective (CGSP). The crisis of the European electricity system: diagnosis and possible ways forward; 2014.
[45] Heinen S, Elzinga D, Kim SK, Ikeda Y. Impact of smart grid technologies on peak load to 2050. IEA; 2011.
[46] Hogan WW. On an "Energy only" electricity market design for resource adequacy; 2005.
[47] Yu HJJ. A prospective economic assessment of residential PV self-consumption with batteries and its systemic effects: the French case in 2030. Energy Policy 2018;113:673–87.
[48] Green M. Current overview of PV technologies and visions for the future. Brussel; 2018.

Index

Note: Page numbers followed by *f* indicate figures, *t* indicate tables, and *b* indicate boxes.

E

F

G

H

M

N

O

P

Q

R

S

Printed in the United States
By Bookmasters